T0135389

# Studies in Systems, Decision and Control

Volume 149

**Series editor**

Janusz Kacprzyk, Polish Academy of Sciences, Warsaw, Poland
e-mail: kacprzyk@ibspan.waw.pl

The series "Studies in Systems, Decision and Control" (SSDC) covers both new developments and advances, as well as the state of the art, in the various areas of broadly perceived systems, decision making and control- quickly, up to date and with a high quality. The intent is to cover the theory, applications, and perspectives on the state of the art and future developments relevant to systems, decision making, control, complex processes and related areas, as embedded in the fields of engineering, computer science, physics, economics, social and life sciences, as well as the paradigms and methodologies behind them. The series contains monographs, textbooks, lecture notes and edited volumes in systems, decision making and control spanning the areas of Cyber-Physical Systems, Autonomous Systems, Sensor Networks, Control Systems, Energy Systems, Automotive Systems, Biological Systems, Vehicular Networking and Connected Vehicles, Aerospace Systems, Automation, Manufacturing, Smart Grids, Nonlinear Systems, Power Systems, Robotics, Social Systems, Economic Systems and other. Of particular value to both the contributors and the readership are the short publication timeframe and the world-wide distribution and exposure which enable both a wide and rapid dissemination of research output.

More information about this series at http://www.springer.com/series/13304

Cengiz Kahraman · Gülgün Kayakutlu
Editors

# Energy Management—Collective and Computational Intelligence with Theory and Applications

Springer

*Editors*
Cengiz Kahraman
Department of Industrial Engineering
Istanbul Technical University
Istanbul
Turkey

Gülgün Kayakutlu
Energy Institute
Istanbul Technical University
Istanbul
Turkey

ISSN 2198-4182 ISSN 2198-4190 (electronic)
Studies in Systems, Decision and Control
ISBN 978-3-030-09299-3 ISBN 978-3-319-75690-5 (eBook)
https://doi.org/10.1007/978-3-319-75690-5

© Springer International Publishing AG, part of Springer Nature 2018
Softcover re-print of the Hardcover 1st edition 2018
This work is subject to copyright. All rights are reserved by the Publisher, whether the whole or part of the material is concerned, specifically the rights of translation, reprinting, reuse of illustrations, recitation, broadcasting, reproduction on microfilms or in any other physical way, and transmission or information storage and retrieval, electronic adaptation, computer software, or by similar or dissimilar methodology now known or hereafter developed.
The use of general descriptive names, registered names, trademarks, service marks, etc. in this publication does not imply, even in the absence of a specific statement, that such names are exempt from the relevant protective laws and regulations and therefore free for general use.
The publisher, the authors and the editors are safe to assume that the advice and information in this book are believed to be true and accurate at the date of publication. Neither the publisher nor the authors or the editors give a warranty, express or implied, with respect to the material contained herein or for any errors or omissions that may have been made. The publisher remains neutral with regard to jurisdictional claims in published maps and institutional affiliations.

Printed on acid-free paper

This Springer imprint is published by the registered company Springer International Publishing AG part of Springer Nature
The registered company address is: Gewerbestrasse 11, 6330 Cham, Switzerland

*To our beloved parents*

# Preface

It was a chilly spring day when a total blackout is experienced by a population of seventy million in Turkey. Some big cities were without power for 8 hours and more. No production could be made in twelve industrial zones for a whole day just because of some technical problems in transmission. Complexity of managing the national grid showed up. The distribution has to be improved, micro-grids are to be constructed, energy security has to be reviewed, more renewable energy investments are to be allowed, energy load has to be rearranged, energy trading is to be regulated based on better predictions and more. Hence, every single citizen understood that there are too many angles of taking decisions in the energy investment, management, and operations.

This book is focused on different levels of complexity in energy applications responded by a variety of computational methods. The book is organized to prepare for decisions by forecasting, giving decisions on different levels, and measuring the success of those decisions as well as getting ready for the fore coming changes represented in eight parts. Economic, strategic, and operational analyses are three different angles of managerial approach to the energy applications. Forecasting supports the managerial decisions by feeding in all the possible alternatives and scenarios. Performance analysis validates if the managerial decisions are taken alongside the market realities. A special chapter is reserved for the collective intelligence applications with two chapters. Future trends give some light for the unknowns to arrive.

The first part, Introduction, is constructed by defining the complexity in energy systems and fuzzy approaches to complexity. In Chap. 1 of this part, Kayakutlu gives the general concepts and classifications of complexity before describing why the energy systems are complex. She classifies the impacts of energy complexity. The second article of Kahraman et al. gives a detailed literature review on fuzzy sets to demonstrate the necessity and support of fuzzy logic in responding the complex system problems. The fuzzy applications are studied for a large variety of energy sources.

Forecasting, the second part, includes three chapters showing that any forecasting exercises on technology, planning, or scheduling can be of great help in economic decision as well as energy efficiency efforts. In Chap. 3, Lavoie et al. work on super-efficient dryers, important technology where there is a room for energy savings. The likelihood of using the technology is combined with the Bass diffusion model for 15 years to show the efficiency improvements and the need for incentives. Oztaysi et al. give a review of basic fuzzy forecasting techniques where statistical analysis like time series analysis and regression are included as well as a machine learning technique, ANFIS. An example for the energy expenditure forecasting using Fuzzy Time Series is given. Yunusov et al. work on smart schedule of storage on low voltage feeders. They use the autoregressive forecasting with linear approach and with trends removed approach considering the seasonality. They compare the results with random forest regression and support vector machine approach.

The third part, Economic Analysis, starts with the photovoltaic metering investments that Cristofaridis and Ioannis have studied. Electricity generation with photovoltaics (PV) can show different patterns, and k-means clustering is used to cluster those patterns. Then, PV generation for each pattern is analyzed for the prosumers using both internal rate of return and net present value analysis. Both the costs and the revenues are based on total energy billing. Öztürk et al. study the lifelong economic analysis of a micro-grid installation with hybrid renewable energies. The model is constructed by using mixed integer nonlinear programming before the net present value is performed. Case study for this chapter is realized for an industrial zone in Turkey. Kahraman et al. define a variety of fuzzy net present worth calculation to analyze the investments. An application on wind turbine life cycle analysis is performed using the intuitionistic fuzzy approach. Carbon tax is an important issue for any kind of energy generation in Europe. Gouveia and Climaco have structured two models for carbon emission taxes and worked on the models applying data envelopment analysis for each country. Solutions for the models are given using piecewise linear functions. The robustness for carbon taxes in European countries is compared.

Strategic analysis for the energy systems will be the long-term decisions that are critical for an investor in the energy field or crucial for the energy economy of a country. Hence, it will be the subject of the fourth part. We have four chapters in this part. In the first one, Coban and Onar analyze the solar energy prices in scenarios created as a result of hesitant fuzzy cognitive maps. Büyüközkan et al. analyze the investment strategies for wind, solar, biomass, hydro-, and geothermal energies using a multi-criteria decision-making approach. This study is a decision support for the energy investor for all the possible renewable energy resources using hesitant fuzzy algorithm. In Chap. 13, the third strategic analysis is realized by Fragkogios and Sahadiris on the crude oil refineries. The chain of scheduling for refineries is critical for the crude oil business; therefore, the review of mathematical, heuristic, and hybrid approaches will support strategic decisions on the subject. The last paper of the part is another multi-criteria decision-making approach using

VIKOR and TODIM to evaluate investment plans in wind, solar, hydro-, and landfilled gas alternative resources.

Energy system processes are critical in the daily decisions. Energy load plans and schedules, energy demand smoothing to avoid the expense of the peak hours and/or electrical charging or storage planning are daily practices and will be covered in the fifth part. Valentin et al. give a real-time planning model for the process control systems. They compared an exact optimization approach modeled using branch and cut and the genetic algorithm approaches. Application is shown with the case studies. Oztaysi et al. give an overview of all intuitionistic fuzzy techniques used for performance analysis in energy management. As a case study, six wind energy alternatives are evaluated based on eight attributes using axiomatic design. The third article of the performance analysis part is the study of Nguyen et al., where electrical car charging at residential side is analyzed. A predictive framework is presented for the charge plans, and the case studies for the real-time application are given. The last article of Bektaş et al. is an example of demand site management. A two-step approach of mixed integer model and a Bayesian game are designed in order to realize peak shaving for a homogeneous management of the multi-site users.

Performance analysis part starts with Chap. 18 written by Asan et al. on Energy Service Companies (ESCO) Market of Turkey. The effectiveness of the market is shown by using a new systematic approach of analyzing the key barriers and drivers by examining the direct and indirect causal relations. The second article of the part is the study of Oztaysi et al., where multi-attribute axiomatic design is given on wind energy alternatives and the performance of different techniques according to those design parameters is measured by using triangular intuitionistic fuzzy sets. In the last study of this part, Karabulut and Büyüközkan apply a group decision making using fuzzy set theory and VIKOR techniques to measure the sustainability performance of different energy resources.

The seventh part of the book includes two articles on collective intelligence applications in the energy field. There are two articles presented. The first one gives a general overview of all the collective intelligence techniques and applications through different dimensions of energy applications, whereas the second gives the fuzzy implementation of the same algorithms for forecasting, economic analysis, strategic and operational analysis, and performance approaches in the energy field.

The eighth and last part of the book is reserved for the future trends where two chapters are included. In the study of Yanık and Kılıç, the impact of block chains is presented and the possibilities for a power and utility block chains are discussed. The interrelationships of different factors in a distributed generation are evaluated. Chapter 24 is designed to value the innovation-based energy future. In this study, Mercier-Laurent and Kayakutlu propose an intelligence-based computational model for considering the time, synergy, and systematic impacts of innovation factors on the energy systems.

This book gives a broad review of computational techniques used in energy management. We strongly believe that approaches presented in this book will open a new dimension for both the decision makers and the academicians.

Istanbul, Turkey Cengiz Kahraman
March 2015 Gülgün Kayakutlu

# Contents

# About the Editors

**Prof. Cengiz Kahraman** is a Full Professor at Istanbul Technical University. His research areas are engineering economics, quality control and management, statistical decision making, multi-criteria decision making, and fuzzy decision making. He published about 200 journal papers and about 150 conference papers. He became the guest editor of many international journals and the editor of many international books from Springer. He is the member of editorial boards of 20 international journals.

**Gülgün Kayakutlu** is currently teaching Operations Research and Intelligent Optimization courses in the Industrial Engineering Department of ITU. She worked for IEA, founded Sybase Turkey, and was one of the 100 experts of Entovation International. She conducts and supervises joint university-company Ph.D. on energy optimization.

# Part I
# Introduction

# Chapter 1
# Complexity in Energy Systems

**Gülgün Kayakutlu**

**Abstract** This chapter is designed to define the complexity concepts and reviews the use of these concepts in the energy field. Our aim is to give a summary for the motivation of this book and overview the issues and the approaches to analyze and understand those issues. Energy applications have the wide arena for complexity and therefore there is a huge variety of collaborative and computational approaches. This chapter will only review the methods considered in this book, but there are a lot more that would add value to the energy industry.

## 1.1 Introduction

In the beginning of this century Mark Buchanan said "Physics is not Physics any more" (Buchanan and Aldana-Gonzalez 2003). He was trying to introduce the concept of "Nexus" as an opponent for complexity. Lots of environmental or energy economists have taken it as Buchanan defined it and created Nexus in different segments of energy applications despite the fact that complexity and solution approaches were not yet sufficiently studied.

Majority of the business systems are recently considered to include complex systems. As technology improved, there has been a computational analysis on the complexity and therefore, intelligent algorithms are developed to find solutions. Global economy and the growing information network have complicated the analysis. Energy systems, energy efficiency, energy planning and management, decision making in the energy field are all found complex. Following the use of multiple resources (fossil and clean energy together) and provision of distributed solutions with self-energy production (cogeneration, tri-generation, local energy and micro-grids), energy applications have become the most challenging applications of the network economy.

---

G. Kayakutlu (✉)
İstanbul Technical University Energy Institute, Maslak, Istanbul, Turkey
e-mail: kayakutlu@itu.edu.tr

© Springer International Publishing AG, part of Springer Nature 2018

C. Kahraman and G. Kayakutlu (eds.), *Energy Management—Collective and Computational Intelligence with Theory and Applications*, Studies in Systems, Decision and Control 149, https://doi.org/10.1007/978-3-319-75690-5_1

This chapter aims to clarify the concepts and solutions for complexity with a focus on energy applications. It is so organized that the concepts will be defined in the next chapter and the application of these concepts in the energy field will follow. The third chapter will be overviewing the search for analyzing and understanding complexity. The last chapter will give a future view for the trends on the subject.

## 1.2 Complexity Concepts

The word complexity makes researchers think of neural and/or intelligent systems immediately. Whereas, one of the earliest definitions made by O'Sullivan relates the complexity to the original systems theory definition of Bertalanffy in 1960 and extends it for the structural and dynamic properties of a system (O'Sullivan 2009). Several researchers gave the interaction of brain, society and the business world interactions as a good example of complexity. Since 2010 the complex system represents the network of things where nodes are the components and the branches are the interactions. Mainly researchers are focused on emergence, resilience, transitions, predictability and control.

Overview of the literature shows that the following features of complex events are generally accepted.

**Collective Behavior**: Networks of individual components with no central control, hard to predict, changing patterns. Natural resource management is a good example that has never been static. Though they are disrupted by the humanity in the past, they show a transition in a dynamic network evolution. This system shows integrated focus on the various linkages between the dynamic system and static structural configurations. Berkes emphasizes the natural resources before emphasizing the need for including the geographical states, cultures and ecosystems in ecological management (Berkes and Berkes 2009). In any network there are subsystems or even agents used but they work with mutual interactions which cause a collaborative action and hence cause adaptive optimization processes improving the collective performance (Rammel et al. 2007).

**Signaling and Information Processing**: Systems produce and use information and signals from both internal and external environments. Simultaneous non-linear interactions are observed among the sub systems which are too heavy for the simple information systems (Kwapień and Drożdż 2012). Using the similarities of biological complexity and the grids cause new methods to handle heterogeneous sources of distributed information (Strizh et al. 2007). Thus, even the classical signalling methods are to be improved to cope with the parametric state based approaches of the dynamic systems (Frank Pai and Palazotto 2008).

**Adaptiveness**: Systems change their behaviour to improve the chances of survival or success through learning or evolutionary processes composed of many interacting parts, giving rise to emergent patterns. The behaviour is said to be emergent because some complex properties do not show up at local level of each component (Rammel et al. 2007). Social dynamics like constructing group opinion have been studied in adaptive networks; in social games many empirical or theoretical models are proposed (Sayama et al. 2013). There is no single rule for any component of the system but rather dynamically changing actions for each event at different state. The immune system is a good example of a complex adaptive system (Holland 2010). The anticipation of sudden blackouts is expected to rise the power prices, yet, it does in USA but not in China. Continuous learning through the defects is a need where direct differential solutions are no more sufficient. Transient states need to be analysed to find solutions for complex adaptive systems where both supervised and unsupervised learnings are beneficial (Marchiori et al. 2011).

**Uncertainty**: Deterministic systems are easy to predict. Dynamism however causes unexpected event states not even known by the agent or component of the system. It is known that the uncertainty is not only caused by lack of information or knowledge but unexpected happenings. Changes in the stock market indices based on a horrible joke made a Global leader is the best example of uncertainty. It is well seen that the social and natural systems can not be taken as Newtonian machines. "they are self-organizing systems whose properties emerge from the non-linear interactions among the agents" (Driebe and McDaniel 2005). Any problem in the network economy needs political, technical, economical social interactions and conflicts to be solved (Rapaport and Ireland 2012). As the volume of unknowns increases, the prediction for the next status becomes more difficult (Koutsourelakis 2008).

**Randomness**: Randomness is included for each individual component, since each one works with its current power at the state of evaluation with arrangements of co-operation (Kramarz and Kramarz 2011). Multi-level, multi agent systems aggregate the results in such a way that the gaps occur more frequently. Bayesian approach to cover the gaps with probabilities might not always be possible but can be created by using random number generations (Chawla et al. 2015). It is also observed that large scale co-operations are built by involving "simulation like" interactions at the local levels (Hadzibeganovic et al. 2015). Usually, randomness is observed in agent based approaches either as input data, agent-to-agent transitions and or random variations in the results achieved.

**Collaboration**: Any business or economic system is in need for interactions and feedbacks among the social, ecological and knowledge networks (Sayama et al. 2013). Andersson et al's research on societal complexity defines the concept graphically (Andersson et al. 2014) as in Fig. 1.1. We do not let the systems to go wicked since we have systematic approaches to achieve the results. However, human related systems are approached by the majority of researchers with a systemic approach where there is a mess and rules cannot be defined easily. Since the success of collaborative actions are more difficult to prove new methods like network based game theory are being developed (Pacheco et al. 2014).

Globalism was handled only by trading and sales worldwide. A new era approaches with consideration of collaboration to construct ecological systems

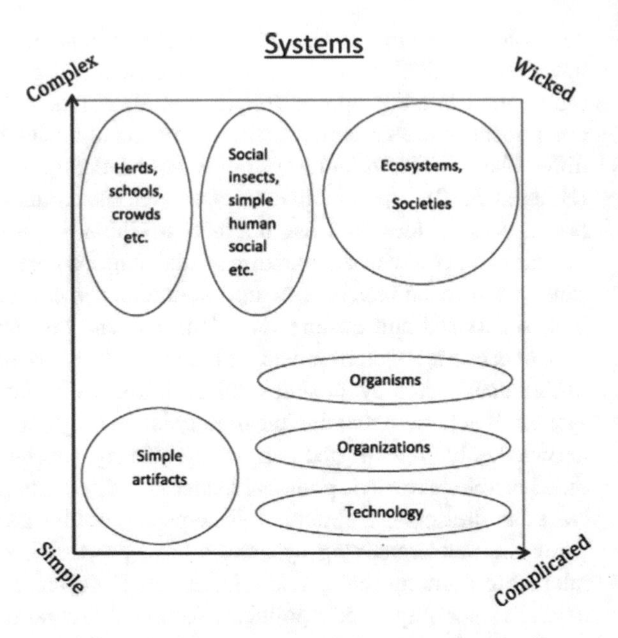


**Fig. 1.1** System definition in reference to more complication (Andersson et al. 2014)

representing the interactive behaviour of the human life and the nature. As Lezotre points for the pharmaceutical industry, geographical-political-economical integration is not anymore sufficient (Lezotre 2014).

## 1.3 Complexity in the Energy Markets

Energy systems are no more run with a single objective of minimizing the costs while balancing the supply and demand. Although the original energy models were linear, new updates are made to include adaptability to the ever-changing dynamic markets. The fact that energy systems are physically integrated through multi-energy technologies, that energy markets force the connectivity of different role players, the balances can be provided with a combined view of demand and supply makes those systems highly complex. Good suggests to incorporate the energy systems with the trading systems to avoid misleading insights (Good et al. 2017). Socio-technical networks with new transition arenas have to consider sustainability, low-carbon life, green and wise economy (Kayakutlu and Mercier-Laurent 2017). New energy types, hybrid use of energy resources, distributed energy deliveries, socio-ecological approaches increase the difficulties to respond to uncertainties in capacities, demand and prices. We can analyse the complexities in energy systems through three levels Resources, Distribution and Interactions.

### 1.3.1 Energy Resources

In 2013 USA has declared to be focused on internal resources not only for the immense shale gas reserves but also hydraulic fracking and related technologies. In their foreign policies they declared to approach Brasil for immense bio-energy resources and Africa which is to discover the natural resources yet (Zakheim 2014). As the variety of energy sources increases, the obligation for hybrid use of renewable and fossil energies grow. Renewable energies are given more importance with global warming abatement concerns but the uncertainties cause the need for the use of natural gas, fuel or coal sites to back the unexpected losses. Though innovative techniques are found to control and manage hybrid use of energy resources, the difficulty of improving skill, cultural reactions, research and development investments and technology integration cause complexities (Cainelli et al. 2015). After 2015 COP21 agreement in Paris, it has been extremely difficult to use distributed energy resources to achieve the goals of keeping warming under 2 °C, in parallel with providing sustainable environment, having competitive advantages in economy and guarantee the energy security. Power storage technologies have been facilitator to respond to these difficulties. Even using the vehicles as an alternative for energy sustainability have become the support for the management (Haddadian et al. 2016).

### 1.3.2 Energy Distribution

Uncertainty in demand and efficiency concerns caused the efforts and research for the demand side management and smart grids. The use of demand site management to provide flexibility in pricing by changing the peak loads has also given the countries the advantage in carbon emissions. It is shown by the researchers that the rebound effect can be avoided by smoothing the energy load based on demand site management, which causes reductions in power consumption prices (Bergaentzlé et al. 2014).

In natural gas operations, it is shown that the use of hubs reduces the prices and causes the energy saving (Kazakos et al. 2016).

District energies with multiple resource use and emission considerations combined allow the support for both energy efficiency and emission mitigation efforts (Fazlollahi et al. 2015). Starting with the first district heating in Denmark, the smaller grid instead of huge national grids. Development of the renewable energy technologies allowed micro grid design for islands or isolated districts. Interrelations among the national and private grids, moreover, regulations to allow block purchase in intra-day market would also have positive impacts in energy saving. However, demand site management and dynamic pricing need smart measuring, intelligent control systems, advanced load management and smart performance monitoring (Siano 2014). Distributed energy use with micro-grids or self-energy generation forced to integrate the supply and the demand of power, heat and cooling to minimize the load and costs (Bahl et al. 2017).

### *1.3.3 Socio-ecological approaches*

The current energy market models are redefined to maximize the total surplus of wealth considering the climate policies and change of the nature, energy securities in the region as well as global economic balance. The new optimization models are focused on handling the time and space problem (dynamism), balancing the use of resources to avoid uncertainty increased by hybrid use of fossil and renewable sources and integrated social risks (Pfenninger et al. 2014). As a matter of fact, ecological dynamics- observed as a totality of human interactions, human-nature interactions and the feedback from all interactions- adds variability to the energy systems (Parrott 2011). Each stakeholder in the energy world is expected to have a vision of sustainability that will guarantee the development potential of ecology, culture and economy in the long run. The complexity increases exponentially when integrative decision subsystems based on integration of nature, economy and culture (Zhao and Wen 2012). Currently the conflicts of socio-ecological approaches and the energy economics are causing complexities growing exponentially (Weber and Cabras 2018).

## 1.4 Response to Complexity: Computational and Collective Intelligence

Lyapunov defined the mathematics of stability using partial differentials of a function in the early years of twentieth century (Gass and Harris 2001). Since then, the scientists tried to find the balance for the growing complexity. Energy system models have been upgraded to be one of the following (Pfenninger et al. 2014):

1. Models providing alternatives for evolving the system using combinatorial optimization;
2. Models providing forecasts by handling uncertainties with probabilities used in simulations;
3. Scenarios based on econometric or statistical methods;
4. Normative or narrative scenarios using intelligent approaches.

Bale et al. (2015) suggests that a more generally accepted approach is to create a computational model of the system. Computational model can use data mining or knowledge based approaches, uses fuzzy and neural or hybridized neuro-fuzzy methods, evolutionary algorithms. Furthermore, it can be both theoretical and application driven. Uncertainty and nonlinearity in complex problems are handled generally in multi-input single output model, but fuzzy logic and neural networks allow the improvement to multi input multi output designs. Optimizing the solution of these models is called computational optimization. Computational optimization is defined to use single or multiple mathematical techniques to choose the best available alternative given in a domain based on certain criteria and a set of constraints (Gamarra and Guerrero 2015). Either discrete or continuous NP-Hard models are solved using learning and adaptation, which makes those methods computational intelligence. A recent study of intelligence techniques used on renewable energies shows that the problems analysed are scheduling, environmental management, operations control and management, flow control, allocation or size optimization, strategic, tactical or operational decision making (Jha et al. 2017). In other words, every single application in the energy field is enclosed with NP-hard problems.

Computational intelligence has an issue of taking long computational time; adoptive and reinforced learning algorithm are creating solutions for that issue (Ruano et al. 2014). Ruano also emphasizes the collaborative use of knowledge would allow the computational techniques to be more effective. Majority of the computational intelligence work include:

(a) Probabilistic Reasoning
(b) Neural Networks
(c) Machine Learning Techniques
(d) Fuzzy Logic
(e) Bayesian Networks
(f) Evolutionary Algorithms
(g) Particle Swarm Optimization
(h) Intelligent Agents
(i) Case Based Reasoning
(j) Hybridized techniques of the above.

Das and Gosh (2017) gives the comparison of most of the above on the time series analysis. Though it is very difficult to distinguish the computational and collective intelligence algorithms, the given methods can be extended to collective intelligence with a variety of colony and swarm studies like Ant Colony, Honey Bees, Fish Swarm, Sheep Swarm, Insect Colonies and so on.

As it is observed in Fig. 1.2, the research on both computational and collective intelligence continue to increase based on data from SCOPUS database. Though artificial intelligence research has started in 1950s, we see the application of computational and collective methods only after 1976. Data shows the biggest number of research in 2016 that is only because 2017 is not completed. According to this graph the peak has yet to be reached since, the interest continues to grow.

A review on hybrid renewable energy use planning and configuration shows that both for operation planning and operation control shows that computational is heavily used in local grids (Siddaiah and Saini 2016). When it comes to distribution

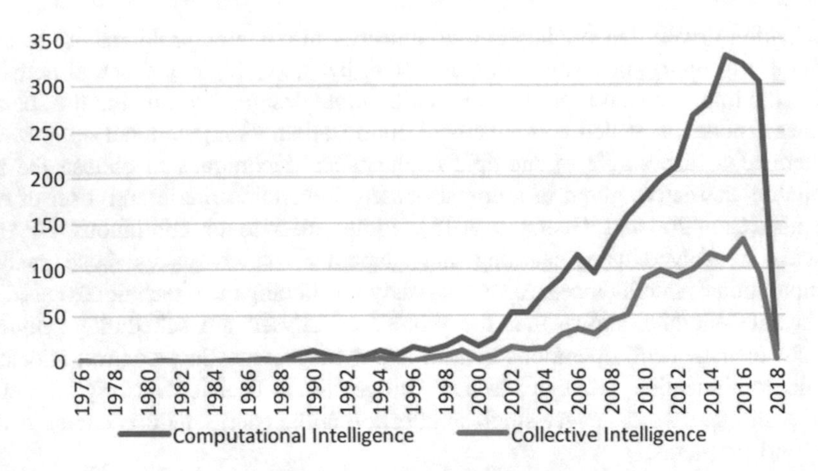


**Fig. 1.2** Research on computational and collective intelligence. *Source* SCOPUS database

multi agent approaches are used to control multiple carrier lines. In management of smart grids, demand site management in the network can be applied by implementation of the an intelligent agents to provide advantages (Nunna et al. 2016). Kazakos et al. has shown that balancing the operational costs, to reduce the scheduling errors, to reduce the emissions intelligent agent system can be fruitful (Kazakos et al. 2016).

Social concerns in energy management have started with residential use and the smart buildings. Behavioral analysis refines the efficient energy load in residences (Popoola 2018). But when energy production, distribution and use are considered decision support has to support the socio-ecological systems. Parrott claims that when socio-ecological models are concerned, either human agents should be included in the model to pause an action at any moment he likes or environmental change should be realized based on the predicted trends or result of human activities (Parrott 2011). Public perception surveys also provide good data to be analyzed by using the computational intelligence. Predicting the future of socio-ecological systems cannot be realized in precision without the adaptive systems. Forecasts of those systems can be realized in scenarios considering the future states instead of well-defined environment used for statistical analysis. Besides, the models designed for complex systems can only be sustainable if the adaptive capacity of the system can be protected (Gaziulusoy and Brezet 2015).

## 1.5 Conclusion

Complex systems cause NP problems and energy applications are sustained in a complex environment. Computational Intelligence models aim to respond specific research questions in a complex environment. Some previously defined method like particle swarm optimization is used to analyze the collected evidences and select some alternatives. As a new approach, adaptiveness and learning is included so that some previously unknown results can be achieved.

Energy innovations with the renewable techniques and distributed energy resources, smart use of energy and environmental concerns arc all analyzed in a complex environment. Thus, the computational intelligence brings a lot to decisions, planning, policy designing and operation of the energy systems.

This chapter has been an introduction to the book where computational intelligence applications in the energy field are gathered. The concepts are defined as to construct a frame for all the analysis made. Furthermore, the summary of energy needs for computational intelligence is given to clarify the motivations for bringing this book together. Further studies will also be recommended in the conclusive chapter, the Future of Energy.

## References

Andersson, C., Törnberg, A., & Törnberg, P. (2014). Societal systems—Complex or worse? *Futures., 63,* 145–157.

Bahl, B., Lampe, M., Voll, P., & Bardow, A. (2017). Optimization-based identification and quantification of demand-side management potential for distributed energy supply systems. *Energy, 135,* 889–899.

Bale, C. S. E., Varga, L., & Foxon, T. J. (2015). Energy and complexity: New ways forward. *Applied Energy [Internet], 138,* 150–9. Available from: http://www.sciencedirect.com/science/article/pii/S0306261914011076. [cited 21 Jan 2015]

Bergaentzlé, C., Clastres, C., & Khalfallah, H. (2014). Demand-side management and European environmental and energy goals: An optimal complementary approach. *Energy Policy, 67,* 858–869.

Berkes, F., & Berkes, M. K. (2009). Ecological complexity, fuzzy logic, and holism in indigenous knowledge. *Futures, 41*(1), 6–12.

Buchanan, M., & Aldana-Gonzalez, M. (2003). *Nexus: Small worlds and the groundbreaking science of networks [Internet]. Physics Today* (Vol. 56, 240 p). Available from: http://scitation.aip.org/content/aip/magazine/physicstoday/article/56/3/10.1063/1.1570777.

Cainelli, G., De Marchi, V., & Grandinetti, R. (2015). Does the development of environmental innovation require different resources? Evidence from Spanish manufacturing firms. *Journal of Cleaner Production, 94,* 211–220.

Chawla, S., Malec, D., & Sivan, B. (2015). The power of randomness in Bayesian optimal mechanism design. *Games and Economic Behavior, 91,* 297–317.

Das, M., & Gosh, S. K. (2017). Data-driven approaches for meteorological time series prediction: A comparative study of the state-of-the-art computational intelligence techniques. *Pattern Recognit Letters* 1–10.

Driebe, D., & McDaniel, R. (2005). *Uncertainty and surprise in complex systems [Internet],* 19–30 p. Available from: http://www.springerlink.com/content/fh184l6783385723.

Fazlollahi, S., Becker, G., Ashouri, A., & Maréchal, F. (2015). Multi-objective, multi-period optimization of district energy systems: IV—A case study. *Energy [Internet].* Available from: http://www.sciencedirect.com/science/article/pii/S0360544215002856 [cited 19 Apr 2015].

Frank Pai, P., & Palazotto, A. N. (2008). HHT-based nonlinear signal processing method for parametric and non-parametric identification of dynamical systems. *International Journal of Mechanical Sciences, 50*(12), 1619–1635.

Gamarra, C., & Guerrero, J. M. (2015). Computational optimization techniques applied to microgrids planning: A review. *Renewable and Sustainable Energy Reviews., 48,* 413–424.

Gass, S. I, & Harris, C. M. (Eds.). (2001). *Encyclopedia of operations research & management science*. Kluwer Aca. Los Angeles, 745 p.

Gaziulusoy, A. I., & Brezet, H. (2015). Design for system innovations and transitions: A conceptual framework integrating insights from sustainability science and theories of system innovations and transitions. *Journal of Cleaner Production, 108,* 1–11.

Good, N., Martínez Ceseña, E. A., & Mancarella, P. (2017). Ten questions concerning smart districts. *Building and Environment, 118,* 362–376.

Haddadian, G., Khalili, N., Khodayar, M., & Shahidehpour, M. (2016). Optimal coordination of variable renewable resources and electric vehicles as distributed storage for energy sustainability. *Sustain Energy, Grids Networks, 6,* 14–24.

Hadzibeganovic, T., Stauffer, D., & Han, X. P. (2015). Randomness in the evolution of cooperation. *Behavioural Processes, 113,* 86–93.

Holland, J. H. (2010). Complex adaptive systems. *Daedalus, 121*(1), 17–30.

Jha, S. K., Bilalovic, J., Jha, A., Patel, N., & Zhang, H. (2017). Renewable energy: Present research and future scope of artificial intelligence. *Renewable and Sustainable Energy Reviews, 77,* 297–317.

Kayakutlu, G., & Mercier-Laurent, E. (2017). 5—Future of Energy. In *intelligence in energy [Internet]*. pp. 153–198. Available from: http://www.sciencedirect.com/science/article/pii/B9781785480393500055.

Kazakos, S. S., Papadopoulos, P., Grau Unda, I., Gorman, T., Belaidi, A., & Zigan, S. (2016). Multiple energy carrier optimisation with intelligent agents. *Applied Energy, 167,* 323–335.

Koutsourelakis, P. S. (2008). Design of complex systems in the presence of large uncertainties: A statistical approach. *Computer Methods in Applied Mechanics and Engineering, 197*(49–50), 4092–4103.

Kramarz, M., & Kramarz, W. (2011). Simulation modelling of complex distribution systems. *Procedia—Social and Behavioral Sciences [Internet], 20,* 283–291. Available from: http://www.sciencedirect.com/science/article/pii/S1877042811014145.

Kwapień, J., & Drożdż, S. (2012). Physical approach to complex systems. *Physics Reports [Internet], 515*(3–4), 115–226. Available from: http://linkinghub.elsevier.com/retrieve/pii/S0370157312000166.

Lezotre, P.-L. (2014). Part II—Value and influencing factors of the cooperation, convergence, and harmonization in the pharmaceutical sector. In *International cooperation, convergence and harmonization of pharmaceutical regulations* (pp. 171–219).

Marchiori, S. C., da Silveira, Maria do Carmo, G., Lotufo, A. D. P., Minussi, C. R., & Lopes, M. L. M. (2011). Neural network based on adaptive resonance theory with continuous training for multi-configuration transient stability analysis of electric power systems. *Applied Soft Computing [Internet], 11*(1), 706–715. Available from: http://www.sciencedirect.com/science/article/pii/S1568494609002890.

Nunna, H. S. V. S. K., Saklani, A. M., Sesetti, A., Battula, S., Doolla, S., & Srinivasan, D. (2016). Multi-agent based demand response management system for combined operation of smart microgrids. *Sustain Energy, Grids Networks, 6,* 25–34.

O'Sullivan, D. (2009). Complexity theory, nonlinear dynamic spatial systems. In *International encyclopedia of human geography [Internet]*. pp. 239–244. Available from: http://www.sciencedirect.com/science/article/pii/B9780080449104004144/pdfft?md5=8d3212b4ddbbfd6924cd04f313f17024&pid=3-s2.0-B9780080449104004144-main.pdf%5Cn.

Pacheco, J. M., Vasconcelos, V. V., & Santos, F. C. (2014). Climate governance as a complex adaptive system: Reply to comments on "climate change governance, cooperation and self-organization". *Physics of Life Reviews, 11,* 595–597.

Parrott, L. (2011). Hybrid modelling of complex ecological systems for decision support: Recent successes and future perspectives. *Ecological Informatics, 6,* 44–49.

Pfenninger, S., Hawkes, A., & Keirstead, J. (2014). Energy systems modeling for twenty-first century energy challenges. *Renewable and Sustainable Energy Reviews, 33,* 74–86.

Popoola, O. M. (2018). Computational intelligence modelling based on variables interlinked with behavioral tendencies for energy usage profile—A necessity. *Renewable and Sustainable Energy Reviews, 82,* 60–72.

Rammel, C., Stagl, S., & Wilfing, H. (2007). Managing complex adaptive systems—A co-evolutionary perspective on natural resource management. *Ecological Economics, 63*(1), 9–21.

Rapaport, B., & Ireland, V. (2012). Understanding the dynamics of system-of-systems in complex regional conflicts. *Procedia Computer Science, 12,* 43–48.

Ruano, A. E., Ge, S. S., Guerra, T. M., Lewis, F. L., Principe, J. C., & Colnarič, M. (2014). Computational intelligence in control. In *IFAC Proceedings Volumes* (IFAC-PapersOnline), pp. 8867–8878.

Sayama, H., Pestov, I., Schmidt, J., Bush, B. J., Wong, C., Yamanoi, J., et al. (2013). Modeling complex systems with adaptive networks. *Computers & Mathematics with Applications, 65* (10), 1645–1664.

Siano, P. (2014). Demand response and smart grids—A survey. *Renewable and Sustainable Energy Reviews, 30,* 461–478.

Siddaiah, R., & Saini, R. P. (2016). A review on planning, configurations, modeling and optimization techniques of hybrid renewable energy systems for off grid applications. *Renewable and Sustainable Energy Reviews, 58,* 376–396.

Strizh, I., Joutchkov, A., Tverdokhlebov, N., & Golitsyn, S. (2007). Systems biology and grid technologies: Challenges for understanding complex cell signaling networks. *Future Generation Computer Systems, 23*(3), 428–434.

Weber, G., & Cabras, I. (2018). The transition of Germany's energy production, green economy, low-carbon economy, socio-environmental conflicts, and equitable society. *Journal of Cleaner Production, 167,* 1222–1231.

Zakheim, D. S. (2014). Facing the challenges of the 21st century. *Orbis, 58*(1), 8–14.

Zhao, Q. J., & Wen, Z. M. (2012). Integrative networks of the complex social-ecological systems. *Procedia Environmental Sciences [Internet], 13,* 1383–94. Available from: http://linkinghub.elsevier.com/retrieve/pii/S1878029612001326.

# Chapter 2
# Fuzzy Sets Applications in Complex Energy Systems: A Literature Review

**Cengiz Kahraman, Başar Oztaysi, Sezi Çevik Onar and Sultan Ceren Öner**

**Abstract** With the emergence of new energy-related technologies and new energy sources, energy planning has become even more vital and complex. Decision making and optimization are very important for complex energy systems. Efficient decision making requires the involvement of various stakeholders which makes the decision problem even more difficult. Fuzzy sets provide tools for mathematically representing vagueness and imprecision in the data or the linguistic stakeholder evaluations. In this chapter an extended literature on fuzzy sets application of complex energy systems. The main issues emphasized in the literature review can be summarized as prediction and modelling the energy configuration conditions, interactions among the various critical design parameters, and solving power systems challenges under uncertainty. The fuzzy application on complex energy systems is presented for different energy types, such as bioenergy, wave energy, photovoltaic systems, hydrogen energy, nuclear energy, wind and thermal energy.

## 2.1 Introduction

Complex systems are composed of many components, which interact with each other. Thus, modelling the behavior of such systems are intrinsically difficult to model due to the dependencies, relationships, or interactions between their components. Energy systems are complex in nature since it involves a continuous and integrated process, which aims transfer of the energy from the source to final customer's location (Ligtvoet and Chappin 2012). Mostashari (2011) emphasize that for effective decision making in complex energy systems (CESs), stakeholder involvement is very critical since CESs are embedded within a complex social setting with uncertain and often emergent long-term social, economic, and environmental impacts.

C. Kahraman (✉) · B. Oztaysi · S. Çevik Onar · S. C. Öner
Istanbul Technical University, Maçka, Istanbul, Turkey
e-mail: kahramanc@itu.edu.tr

© Springer International Publishing AG, part of Springer Nature 2018

C. Kahraman and G. Kayakutlu (eds.), *Energy Management—Collective and Computational Intelligence with Theory and Applications*, Studies in Systems, Decision and Control 149, https://doi.org/10.1007/978-3-319-75690-5_2

For effective decision making, especially in complex systems, considering stakeholders points of view and evaluating the alternative solutions based on these viewpoints is very important. However, the points of view may conflict with each other, and in some cases, the criterion may contain subjective and linguistic evaluations. In traditional formulation, human judgements are formulated in mathematical models using crisp numbers. However, in practice, decision-makers may have difficulties in assigning numerical values to the evaluations, or there may be imprecision or vagueness in the data used for the solution of the model. Fuzzy sets theory was specifically designed to handle such cases by mathematically representing uncertainty and vagueness. Besides, fuzzy sets provide formalized tools to generate decision models with imprecise data. Kahraman et al. (2003) state that in decision models when imprecise parameters are treated as imprecise values instead of prices ones, the decision process tend to provide more powerful and credible results. On the other hand, knowledge can be expressed more naturally by using fuzzy sets so that decision problems may be simplified (Kahraman and Kaya 2010).

The objective of this chapter is to provide an extended literature review of fuzzy sets applications on complex energy systems. To this end, a literature survey with three main focus is conducted. The first focus is predicting and modelling the energy configuration conditions, the second focus is the interactions among the various critical design parameters, and the final focus is solving power systems challenges with large penetrations of technologies under uncertain parameters. The literature survey is presented for different energy types, namely, bioenergy, wave energy, photovoltaic systems, hydrogen energy, nuclear energy, wind and thermal energy.

The organization of the chapter is as follows: In Sect. 2.2, fuzzy set theory is briefly introduced, Sect. 2.3 summarizes complex energy systems. Literature review on fuzzy application of complex energy systems is given in Sect. 2.4, and finally, the conclusion is given in the last section.

## 2.2 Fuzzy Sets Theory

### *2.2.1 Ordinary Fuzzy Sets and Their Extensions*

Zadeh (1965) introduced fuzzy sets in 1965 and since then more than eight extensions of fuzzy sets have been developed. An ordinary fuzzy set $\tilde{A}$ in X where X is a collection of objects denoted generically by x can be defined as follows:

$$\tilde{A} = \left\{ \left( x, \mu_{\tilde{A}}(x) | x \in X \right) \right\} \tag{2.1}$$

where $\mu_{\tilde{A}}(x)$ is the membership function which maps X to the membership space. Its range is the subset of nonnegative real numbers whose supremum is finite. Zadeh (1965) introduced fuzzy sets as a class of objects with a continuum of grades of membership.

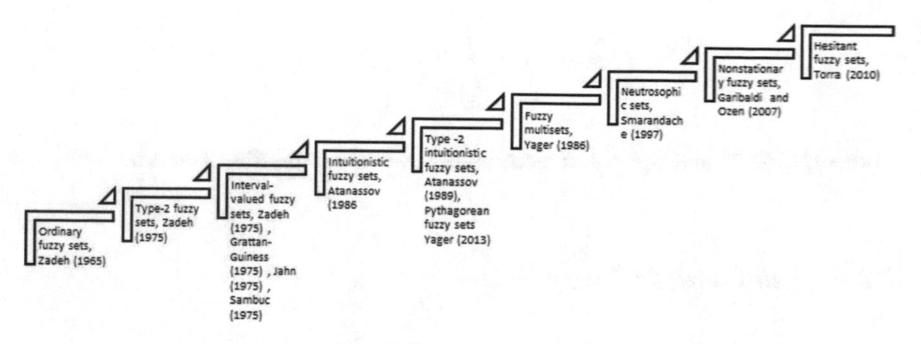


**Fig. 2.1** The history of fuzzy sets

Ordinary fuzzy sets are extensively used in production research areas whereas the extensions of fuzzy sets are recently started to be used in production problems. Therefore, the number of the works on production research using these new extensions is relatively low. Figure 2.1 shows the extensions of the fuzzy sets and their history.

### 2.2.2 *Interval-Valued Fuzzy Sets*

An interval-valued fuzzy set (IVFS) is a special case of type 2 fuzzy set. An IVFS is defined by a mapping F from the universe U to the set of closed intervals in [0, 1]. Let $\mathrm{F(u)} = [\mathrm{F_*(u)}, \mathrm{F^*(u)}]$. The union, intersection, and complementation of IVFSs are obtained by canonically extending fuzzy set-theoretic operations to intervals.

### 2.2.3 *Type-n Fuzzy Sets*

Zadeh (1975) introduced type-2 fuzzy sets in 1978. A type-2 fuzzy set lets us incorporate uncertainty about the membership function into the fuzzy set theory. A type-2 fuzzy set $\stackrel{\approx}{A}$ in the universe of discourse X can be represented by a type-2 membership function $\mu_{\stackrel{\approx}{A}}$ shown as follows:

$$\stackrel{\approx}{A} = \left\{ \left( (x,u), \mu_{\stackrel{\approx}{A}}(x,u) \right) \middle| \forall x \in X, \forall u \in J_x \subseteq [0,1],\ 0 \le \mu_{\stackrel{\approx}{A}}(x,u) \le 1 \right\} \tag{2.2}$$

where $J_x$ denotes an interval [0, 1]. The type-2 fuzzy set $\stackrel{\approx}{A}$ also, can be represented as follows:

$$\tilde{\tilde{A}} = \int_{x \in X} \int_{u \in J_x} \mu_{\tilde{\tilde{A}}}(x,u)/(x,u) \quad (2.3)$$

where $J_x \subseteq [0,1]$ and the $\int\int$ denote union over all admissible x and u.

### 2.2.4 Intuitionistic Fuzzy Sets

Atanassov's (1986) intuitionistic fuzzy sets (IFSs) include the membership value as well as the non-membership value for describing any x in X such that the sum of membership and non-membership is at most equal to 1.

Let $X \neq \emptyset$ be a given set. An intuitionistic fuzzy set in X is an object A given by

$$\tilde{A} = \{\langle x, \mu_{\tilde{A}}(x), v_{\tilde{A}}(x)\rangle; x \in X\}, \quad (2.4)$$

where $\mu_{\tilde{A}} : X \to [0,1]$ and $v_{\tilde{A}} : X \to [0,1]$ satisfy the condition

$$0 \leq \mu_{\tilde{A}}(x) + v_{\tilde{A}}(x) \leq 1 \quad (2.5)$$

for every $x \in X$.

### 2.2.5 Fuzzy Multisets

Let X be a nonempty set. A fuzzy multiset $\tilde{A}$ drawn from X is characterized by a function, "count membership" of $\tilde{A}$ denoted by $CM_A$ such that $CM_A : X \to Q$ where Q is the set of all crisp multisets drawn from the unit interval [0, 1]. Then for any $x \in X$, the value $CM_A(x)$ is a crisp multiset drawn from [0, 1]. For each $x \in X$, the membership sequence is defined as the decreasingly ordered sequence of elements in $CM_A(x)$. It is denoted by $\left(\mu^1_{\tilde{A}}(x), \mu^2_{\tilde{A}}(x), \ldots, \mu^n_{\tilde{A}}(x)\right)$, where $\mu^1_{\tilde{A}}(x) \geq \mu^2_{\tilde{A}}(x) \geq \cdots \geq \mu^n_{\tilde{A}}(x)$.

### 2.2.6 Nonstationary Fuzzy Sets

Let $\tilde{A}$ denote a fuzzy set of a universe of discourse X characterized by a membership function $\mu_{\tilde{A}}$. Let T be a set of time points $t_i$ (possibly infinite) and f: $T \to \Re$ denote a perturbation function. A non-stationary fuzzy set $\tilde{A}$ of the universe of discourse X is characterized by a non-stationary membership function $\mu_{\tilde{A}}$: T ×

$X \to [0, 1]$ which associates each element (t, x) of $T \times X$ with a time-specific variation of $\mu_{\tilde{A}}(x)$. The non-stationary fuzzy set Ã is denoted by:

$$\tilde{A} = \int_{t \in T} \int_{x \in X} \mu_{\tilde{A}}(t, x)/x/t \tag{2.6}$$

### 2.2.7 *Hesitant Fuzzy Sets*

Hesitant fuzzy sets (HFSs), initially developed by Torra (2010), are the extensions of normal fuzzy sets which handle the situations where a set of values are possible for the membership of a single element.

Torra (2010) defines hesitant fuzzy sets (HFSs) as follow: Let *X* be a solid set, an HFS on *X* is in terms of a function that when applied to *X* returns a subset of [0, 1]. The mathematical expression for HFS is as follows:

$$E = \{\langle x, h_E(x) \rangle | x \in X\} \tag{2.7}$$

where $h_E(x)$ is a set of some values in [0, 1], denoting the possible membership degrees of the element $x \in X$ to the set *E*.

### 2.2.8 *Neutrosophic Theory*

Smarandache (1999) introduced the level of indeterminacy/neutrality (i) as an independent component and defined the neutrosophic set on three elements

$$(t, i, f) = (\mathit{truth}, \mathit{indeterminacy}, \mathit{falsehood}) \tag{2.8}$$

The words "neutrosophy" and "neutrosophic" were coined/invented by F. Smarandache in his 1999 book. Etymologically, "neutro-sophy" (noun) [French neutre < Latin neuter, neutral, and Greek sophia, skill/wisdom] means knowledge of neutral thought while "neutrosophic" (adjective), means having the nature of, or having the characteristic of Neutrosophy.

### 2.2.9 *Pythagorean Fuzzy Sets-Type 2 Intuitionistic Fuzzy Sets*

Type-2 Intuitionistic Fuzzy Sets originally introduced by Atanassov (1989) are renamed and re-introduced by Yager (2013) as the Pythagorean fuzzy sets. These sets similar to the intuitionistic fuzzy sets consider both membership and non-membership values. Yager (2013) claims that under some conditions the sum of membership and non-membership values can exceed one. Unlike Neutrosophic fuzzy sets, the relation between membership and non-membership values should satisfy the following condition:

$$\mu 2 + v2 \leq 1 \quad (2.9)$$

## 2.3 Complex Energy Systems

There are various definitions of complex systems in the literature. In general, a system is complex if it has a group of interrelated component and subsystems, for which the degree and nature of the relationships is imperfectly known, with varying directionality, magnitude and time-scales of interactions. Three types of complexity in systems are defined in the literature: behavioral complexity, internal complexity and evaluative complexity.

Behavioral complexity arises when the emergent behavior of a system is difficult to predict and may be difficult to understand even after the fact. Internal or structural complexity is a measure for the interconnectedness in the structure of a complex system, where small changes made to part of the system can result in major changes in the system output and even result in system-wide failure. Evaluative complexity is due to the existence of stakeholders in a complex system, and is an indication of the different normative beliefs that influence views on the system. Thus, even in the absence of the two former types of complexity, and even if one were able to model the outputs and the performance of the system, it would still be difficult to reach an agreement on what "good" system performance signifies. This type of complexity is one of the primary motivators for engaging stakeholders in systems modeling and policy design, and is an essential part of this book (Sussman 2003).

These complexities all exist in the energy systems. The problems arising from behavioral complexity are often faced because it is very difficult to guarantee the continuity of sources such as wind and water. Evaluative complexity of energy systems is another problem decision-makers often face. For the evaluation of an energy investment, linguistic terms can be used in the decision matrix, rather than exact numerical values. This requires a fuzzy approach to be employed in the evaluation process.

## 2.4 Literature Review: Complex Energy Systems and Fuzzy Sets Applications

Complex energy systems are continuous and integrated process that initiated with attaining the required energy supply and ends up with the transfer of the energy to final customer's location regarding one or more energy component such as solar, thermal, waste, biomass and wind power (Ligtvoet and Chappin 2012). Hence, complex energy systems include not only energy supply, energy production and energy distribution but also the relations and contentions between these elements and also analyzing storage and usage of the energy to evaluate energy supply and demand (Oluwamayowa et al. 2017).

In recent years, due to the insufficient energy supply, production, distribution, storage, and use of energy has become a critical point for sustainability and affected significant changes. Demand and production balance are being amended ineradicably by the developments in "smart grids," renewable energy generation processes, hybrid vehicles and energy storage technologies. Explaining the relations between such systems causes numerous challenges, especially configuring mathematical expressions considering energy parameters and nature of the energy supply structure ranging from multiple spatial and temporal variables to future operating conditions. To address these challenges, various mathematical models including stochastic models and dynamic system behavior modelling are adapted to develop, analyze, and integrate predictive models of system behavior. On the other hand, some of these system parameters can be indefinite or contradictory that can complicate system modelling. Additionally, due to the emerging and dynamic relations between energy system elements, models should be flexible to sudden changes and cope with uncertainties and provide continuous monitoring. From the point of this view, fuzzy based models that formulize energy systems could deal with the difficulties of instant variations and continuous monitoring of the entire system (Toffolo and Lazzaretto 2008). In addition to the static structure, interactions and variations could not be shown in the mathematical modelling. To cope with imprecise information appeared in complex energy implementation processes, authors and practitioners mainly emphasized on defining indefinite variables which provide identification of the problem for continuous monitoring. With this respect, fuzzy based models obtain membership functions between 0 and 1 enable the formalization of complex energy implementation processes reflecting the instant variations of complex system behavior.

As mentioned before, various renewable energy resources such as wind, biomass and solar power are becoming increasingly substantial sources of energy on the electricity based power systems. The consistent growth of renewable energy systems and complex energy systems necessitate a change in attitude for implementing energy systems technologies. This section addresses three important research questions in the area of modelling complex energy systems: (i) how do we predict and model the energy configuration conditions, and evaluate the uncertainty in the diversified resource and energy production at a specific site? (ii) how do we focus on the complex interactions among the various critical design parameters and natural elements in planning high performing complex energy systems? (iii) how do we efficiently solve power systems challenges with large penetrations of complex energy efficiency technologies considering uncertain parameters?

### 2.4.1 *Bioenergy and Fuzzy Sets*

Biomass is one of the renewable energy sources that can involve oil, starch, sugar and cellulosic biomass feedstocks and microalgae, macroalgae and cyanobacteria which has an appropriate way for storage, utilization and substitution quickly without natural resources depletion (Khishtandar et al. 2017). The main necessity for biomass usage is heating and transportation fuel consumption. For instance, biomass for the transportation sector necessitates the usage and storage of ethanol, biodiesel and advanced hydrocarbon biofuels (Yue et al. 2014).

A whole of the developing countries is dealing with the increasing amount of fuel consumption for the satisfaction of energy needs and import fuel from other countries with ultrahigh prices which can obstruct their economic growth and reduces the gathering of the opportunities for new energy implementation projects. Thus, developing countries can provide energy procurement by biofuel production for sustainable energy planning (Demirbas and Demirbas 2007).

As seen from the previous studies in literature, bioenergy production technologies evaluation, sustainability of bioenergy systems, bioenergy supply chain design, biomass location selection, biomass source planning and optimizing biofuels production are mainly modelled by a fuzzy set theory based approaches such as fuzzy multi-objective linear programming models, fuzzy information axiom and fuzzy MCDM methods. For example, Cebi et al. (2016) proposed fuzzy information axiom based method for location selection of biomass power plant. Ubando et al. (2016) suggested fuzzy mixed integer non-linear programming model select prospective support tenants for planning bioenergy based industrial symbiosis. Diversified from these studies, Ziolkowska (2014) used PROMETHEE approach combined with fuzzy LP model to represent available resources for optimizing biofuels production. Other applications are represented in Table 2.1.

### 2.4.2 *Wave Energy and Fuzzy Sets*

Wave energy is captured directly from surface waves or from pressure fluctuations below the surface that contains tremendous energy potential. For extracting necessary power, wave technologies have been designed to be installed in the near-shore, offshore, and far offshore locations. While wave energy technologies are intended to be installed at or near the water's surface, there can be major differences in their technical concept and design. Thus, for predicting the behavior of wave

**Table 2.1** Fuzzy set applications in bioenergy

| Main subject | Author(s) | Application area | Type of fuzzy sets/numbers | Contribution |
|---|---|---|---|---|
| Bioenergy production technologies evaluation | Khishtandar et al. (2017) | Multi-criteria decision making | Hesitant fuzzy sets | Hesitant fuzzy sets based outranking method by considering multi-actor and multi-criteria |
| Assessment of the sustainability of bioenergy systems | Buchholz et al. (2009) | Multi-criteria analysis | Triangular fuzzy numbers | Fuzzy set theory based NAIADE and comparison of this method by other group decision making approaches (SuperDecisions, DecideIT, Decision Lab) |
| Bioenergy supply chain design | Yılmaz Balaman and Selim (2014) | Optimization MILP | Triangular fuzzy numbers based fuzzy goal programming | Fuzzy multiobjective linear programming model |
| Planning bioenergy-based industrial symbiosis | Ng et al. (2014) | Optimization | Triangular fuzzy numbers based disjunctive fuzzy optimization approach | Fuzzy optimization approach for economic assessment of bioenergy based systems |
| Bioenergy project viability decisions | Wright et al. (2013) | Energy economics | Triangular fuzzy numbers | Fuzzy levelised energy cost (F-LEC) methodology to incorporate the cost of financing a project from debt and equity sources |
| Biomass location selection | Cebi et al. (2016) | Multiple criteria decision making | Trapezoidal fuzzy numbers | Fuzzy information axiom based method for location selection of biomass power plant |
| Suitability of bioenergy to mitigate greenhouse gases | Muench (2015) | Qualitative comparative analysis (QCA) | Trapezoidal fuzzy numbers | Life cycle assessments (LCAs) of biomass systems for electricity Generation with fuzzy conditions |
| Potential biomass source planning | Kuhmaier et al. (2014) | Spatial multicriteria decision analysis (SMCDA) | Linear membership functions | Combination of spatial analysis, fuzzy system, and AHP to define the best localization of wood terminal in Austria |

(continued)

**Table 2.1** (continued)

| Main subject | Author(s) | Application area | Type of fuzzy sets/numbers | Contribution |
|---|---|---|---|---|
| Planning bioenergy-based industrial symbiosis | Ubando et al. (2016) | Optimization | Fuzzy trapezoidal membership function | Fuzzy mixed integer non-linear programming model to select prospective support tenants |
| Assessment of biomass energy sources and technologies | Cutz et al. (2016) | MCDM | Triangular fuzzy numbers | Fuzzy transforming matrix based on fuzzy linear 0–1 programming |
| Optimizing biofuels production | Ziolkowska (2014) | MCDM | Triangular fuzzy numbers | PROMETHEE approach combined with fuzzy LP model to represent available resources |
| Biogas plant location selection | Franco et al. (2015) | MCDM | Interval-valued fuzzy numbers | Fuzzy weighted overlap dominance (FWOD) and AHP combined procedure for aggregating GIS-based data |

energy converter, the design of wave energy converter and the prediction of ocean wave energy parameters, fuzzy set theory based models can be experimented. In this context, especially soft computing approaches such as Adaptive Network-based Fuzzy Inference System (ANFIS) models and Takagi–Sugeno (TS) Fuzzy Inference System for rule extraction and prediction of system performance. For instance, Özger and Şen (2007) adapted Takagi–Sugeno (TS) fuzzy modelling principles to predict the changes in wave characteristics. Abed-Elmdoust and Kerachian (2012) implemented rough set theory based ANFIS model for Wave height prediction. Different from these studies, Stefanakos (2016) practiced fuzzy time series forecasting combined with ANFIS model for predicting wind and wave parameters. Other applications are shown in Table 2.2.

### 2.4.3 *Photovoltaic Systems and Fuzzy Sets*

Renewable energy adaptation in daily life has caused a fast improvement of the usage in natural sources. One of these sources is solar energy and power generation from solar energy using Photo Voltaic (PV) systems noticed a fast development in PV systems which is used for high power generation consists of several combinations of series and parallel PV modules. It consists of three basic parts: PV panel, power converter and tracking controller (Rajesh and Mabel 2015). The PV system

**Table 2.2** Fuzzy sets applications in wave energy

| Main subject | Author(s) | Application area | Type of fuzzy sets/numbers | Contribution |
|---|---|---|---|---|
| Predict the behavior of wave energy converter | Amarkarthik and Sivakumar (2016) | Soft computing | Triangular fuzzy numbers | Adaptive network-based fuzzy inference system (ANFIS) model for predicting device behavior |
| Design of wave energy converter | Ahn et al. (2012) | Soft computing | Triangular membership functions | FIS model to adjust the pump displacement |
| | Truong and Ahn (2014) | Soft computing | Triangular membership functions | Adaptive fuzzy PID controller (AFPID) based on grey FIS method |
| Prediction of ocean wave energy and parameters | Özger and Şen (2007) | Soft computing | Triangular membership functions | Takagi–Sugeno (TS) fuzzy modelling principles to predict the changes in wave characteristics |
| | Kazeminezhad et al. (2005) | Soft computing | Triangular membership functions | Adaptive network-based fuzzy inference system (ANFIS) model for predicting wave parameters |
| | Özger (2010) | Soft computing | Generalized bell function | Wavelet fuzzy logic algorithm for wave height forecasting |
| | Özger (2011) | Soft computing | Triangular membership functions | Takagi–Sugeno (TS) type fuzzy inference system was employed to predict wave energy amount from meteorological variables |
| | Sylaios et al. (2009) | Soft computing | Generalized bell function | Takagi–Sugeno-rule-based fuzzy inference system (FIS) was developed aiming at forecasting wave parameters |
| | Abed-Elmdoust and Kerachian (2012) | Soft computing | Triangular membership functions | Rough set theory based ANFIS model for wave height prediction |
| | Akpınar et al. (2014) | Soft computing | Triangular membership functions | ANFIS model for forecasting wave parameters and comparison of other parametric methods |
| | Stefanakos (2016) | Soft computing | Interval-valued fuzzy numbers | Fuzzy time series forecasting combined with ANFIS model for predicting wind and wave parameters |

characteristics depend on the environmental condition, insolation, temperature, capacity of the system, etc. The variation of these external factors highly affects the output power characteristics of PV generators. Thus, fuzzy logic based approaches are successfully adapted to the following areas: (1) Fault detection, (2) Solar Parameter identification (3) Maximum power point tracking (4) Efficiency maximization. To solve the drawbacks of conventional modelling methods, Artificial Intelligence (AI) techniques and evolutionary algorithms are employed with fuzzy based approaches such as Mamdani, ANFIS and Takagi-Sugeno fuzzy inference systems. For instance, Singh and Agrawal (2015) conducted Genetic Algorithm–Fuzzy System (GA–FS) approach to identify the optimized parameters of the glazed photovoltaic thermal (PVT) system. As an example of system modelling, Rahrah et al. (2015) developed fuzzy Logic Controller (FLC) with the neuro-Fuzzy algorithm (NF) for the comparison of gathering maximum output power operating point of the photovoltaic generator. Other examples are given in Table 2.3.

### 2.4.4 Hydrogen Energy and Fuzzy Sets

Hydrogen energy represents the future of renewable energy and can be used in a fuel cell to produce electricity, as an emissions-free alternative. Natural hydrogen is always associated with other elements in compound form such as water, coal and petroleum. There are significant challenges to be overcome in order to make hydrogen viable, in production, storage and power generation. These circumstances prompt researchers to deal with design and control of hydrogen-based systems, sustainability of hydrogen fuel cell implementation and risk and safety assessment of hydrogen extraction systems (Chang 2017). As a result of reflecting uncertain parameters in system design and measuring control performance by minimizing human errors, fuzzy set theory based studies have been adopted widely. To illustrate hydrogen economy assessment, Lee et al. (2011) conducted integrated fuzzy AHP/DEA approach for long-term strategic energy technology roadmap of hydrogen energy implementation. Another example can be given from sustainability assessment of hydrogen energy application as seen in Afgan and Carvalho (2004)'s fuzzy sustainability index rating compared with other complex energy systems study. Additionally, econometric analysis of the R&D performance, budget allocation problems and evaluation and control of hydrogen storage systems are also evaluated as fuzzy MCDM problem. Other applications can be realized from Table 2.4.

### 2.4.5 Nuclear Energy and Fuzzy Sets

Due in large part to the high manufacturing industry growth rates and increased number of energy consumption, developing countries need growing energy markets

**Table 2.3** Fuzzy sets applications in photovoltaic systems

| Main subject | Author(s) | Application area | Type of fuzzy sets/numbers | Contribution |
|---|---|---|---|---|
| Fault detection | Serdio Fernández et al. (2016) | Prediction/soft computing | Triangular membership functions | Mamdani fuzzy inference system for determining the degree of clipping of the PV solar plant power production |
| | Bonsignore et al. (2014) | Soft computing | Triangular membership functions | ANFIS PV model simulator for photovoltaic module modelling |
| Parameter identification | Singh and Agrawal (2015) | Prediction | Triangular membership functions | Genetic algorithm–fuzzy system (GA–FS) approach to identify the optimized parameters of the glazed photovoltaic thermal (PVT) system |
| Photovoltaic pumping system modelling | Rahrah et al. (2015) | Optimization | Trapezoidal membership function | Fuzzy logic controller (FLC) and neuro-fuzzy algorithm (NF) comparison for gathering maximum output power operating point of the photovoltaic generator |
| Maximum power point tracking | Chekired et al. (2014) | Soft computing | Triangular membership functions | Neural networks (NN), fuzzy logic (FL), genetic algorithm (GA) and hybrid systems (e.g. neuro-fuzzy and ANFIS comparison to improve the efficiency of PV systems under variable weather conditions |
| | Palaniswamy and Srinivasan (2016) | Soft computing/ power optimization | Triangular membership functions | Takagi-Sugeno fuzzy approach to maximum power point tracking |
| | Das et al. (2017) | Soft computing and optimization | Triangular membership functions | Comparison of different MPPT techniques including fuzzy logic based controllers |
| | Chiu and Ouyang(2011) | Soft computing | Triangular membership functions | Takagi-Sugeno fuzzy approach to model output tracking control problem |

(continued)

**Table 2.3** (continued)

| Main subject | Author(s) | Application area | Type of fuzzy sets/numbers | Contribution |
|---|---|---|---|---|
| | Kottas et al. (2006) | Soft computing | Triangular membership functions | Fuzzy cognitive networks to measure energy conversion efficiency |
| | Khaehintung et al. (2010) | Soft computing | Triangular membership functions | Particle swarm optimization based fuzzy logic controller for current-mode boost converter (CMBC) |
| | Subiyanto et al. (2012) | Soft computing | Triangular membership functions | Hopfield neural network (HNN) based fuzzy logic controller to overcome the limits of fuzzification |
| Concentrating photovoltaic technology | Carmona et al. (2013) | Soft computing | Triangular membership functions | Multiobjective evolutionary algorithm approach based on GA and fuzzy inference system |
| Efficiency maximization | Singh and Agrawal (2016) | Soft computing and optimization | Triangular membership functions | Fuzzified genetic algorithm for overall exergy efficiency in hybrid dual channel semitransparent photovoltaic-thermal module |

**Table 2.4** Fuzzy sets applications in hydrogen energy

| Main subject | Author(s) | Application area | Type of fuzzy sets/ numbers | Contribution |
|---|---|---|---|---|
| Control performance and design | Chang (2017) | Optimization | Trapezoidal membership function | Fuzzy gain scheduling adaptation to digital signal processing algorithm for Rapid-convergent sliding mode proportional-integral (PI) technology applications |
| | Coteli et al. (2017) | Optimization | Type 2 fuzzy numbers | Type-2 fuzzy neural system controller for PWM rectifiers |
| Hydrogen economy assessment | Lee et al. (2011) | Multi-criteria decision making | Interval valued membership function | Integrated fuzzy AHP/DEA approach for long-term strategic energy technology roadmap |
| | Lee et al. (2011) | Multi-criteria decision making | Triangular membership functions | Fuzzy analytic hierarchy process (AHP) for R&D budget allocation |
| Econometric analysis of the R&D performance | Lee et al. (2010) | Multi-criteria decision making | Triangular membership functions | Fuzzy AHP/DEA integrated model approach for relative efficiency of the R&D performance |
| Sustainability assessment of hydrogen energy | Afgan and Carvalho (2004) | Multi-criteria decision making | Linear membership functions | Sustainability index rating compared by other complex energy systems |
| | Afgan et al. (2007) | Multi-criteria decision making | Linear membership functions | Scenario analysis of sustainability assessment |
| Analysis of operator human errors | Castiglia and Giardina (2013) | Risk analysis | Trapezoidal fuzzy numbers | Fuzzy HEART (human error assessment and reduction technique) to evaluate the probability of erroneous actions |
| Assessment of hydrogen fuel cell applications | Chang et al. (2012) | Multi-criteria decision making | Triangular membership functions | Fuzzy ranking method for the selection of criteria and preferred hydrogen fuel cell product |
| Evaluation and control of hydrogen storage systems | Gim and Kim (2014) | Multi-criteria decision making | Triangular membership functions | Hydrogen storage systems for automobiles are evaluated using the fuzzy analytic hierarchy process (AHP) |
| | Safari et al. (2013) | Optimization | Triangular membership functions | Particle swarm optimization based fuzzy logic controller for autonomous green power energy system |

(continued)

**Table 2.4** (continued)

| Main subject | Author(s) | Application area | Type of fuzzy sets/ numbers | Contribution |
|---|---|---|---|---|
| Sustainability of hydrogen supply chain | Ren et al. (2013) | Multi-criteria decision making/ optimization | Triangular membership functions | Extension theory and AHP are combined to prioritize and classify the sustainability of hydrogen supply chains |
| Hybrid energy management | Tabanjat et al. (2017) | Soft computing | Triangular membership functions | Neural networks (NN) based fuzzy logic control (FLC) is applied to the hydrogen power systems for minimizing the energy production cost |

in the world. To this end, nuclear energy projects may be considered particularly due to the following advantages of nuclear energy. (1) It does not lead to carbon emissions, (2) its fuel can be obtained easily, economically, and be stored, (3) as long as appropriate security measures are taken and implemented, the risks to humans or nature are low.

The key issues that must be addressed as the main part of nuclear energy applications are (1) safety and reliability analysis of nuclear energy systems (2) nuclear energy reactor and critical parameter estimation (3) nuclear power plant selection (4) load and demand regulation. These problems cause critical effects to the environment, public and also stakeholders of nuclear energy investments. For the analysis of these problems, fuzzy set theory based models can reflect uncertainties appeared in the system operations or decision-making process. For instance, nuclear power plant selection can be assessed by fuzzy MCDM methods when considering uncertain geological and seismological issues, cost and risk factors (Erol et al. 2014). Other applications are given in Table 2.5.

## 2.4.6 Wind-Thermal Energy and Fuzzy Sets

In recent years, wind energy is becoming a significant component of the power generation throughout the world. Its probabilistic nature impacts the system operation due to its uncertain system parameters. For instance, wind generation is mainly dependent on wind speed. Thus fluctuations of wind generation should be considered in advance (Reddy and Abhyankar 2013). To overcome this problem, power systems need some resources to compensate the wind power generation forecasting uncertainty. One of these resources is thermal energy when the system faces a shortage of production caused by lack of wind. Thus, studies related to thermal wind energy have been increased especially in demand response, wind energy generation prediction, market clearing and emission management (Falsafi et al. 2014). These subjects are modelled using stochastic programming,

**Table 2.5** Fuzzy sets applications in nuclear energy

| Main subject | Author(s) | Application area | Type of fuzzy sets/numbers | Contribution |
|---|---|---|---|---|
| Safety analysis of nuclear energy | Woo (2014) | Soft computing and optimization | Triangular membership functions | Modified fault tree analysis for clear hydrogen resource pursuit |
| Parameter estimation | Moon and Kang (1999) | MCDM | Triangular membership functions | Aggregation method for extracting appropriate results |
| Nuclear power plant selection | Erdogan and Kaya (2016) | MCDM | Interval type-2 fuzzy numbers | Interval type-2 fuzzy analytical hierarchy process (AHP) for weights of criteria and interval type-2 fuzzy TOPSIS for ranking alternatives |
| | Erol et al. (2014) | MCDM | Trapezoidal fuzzy numbers | Fuzzy entropy method to identify the weights of the relevant criteria and Fuzzy Compromise Programming to rank alternatives |
| Safety assessment | Purba (2014) | Optimization | Triangular membership functions | The fuzzy-based reliability approach generates the basic event failure probabilities of fault trees |
| | Woo and Lee (2010) | Optimization | Triangular and circular membership functions | Probabilistic model and non-linear fuzzy set algorithm is compared for failure analysis |
| Safety culture assessment | Dos Santos Grecco et al. (2014) | MCDM | Triangular membership functions | Fuzzy Delphi technique based on similarity aggregation method for the assessment of safety performance indicators |
| Process improvement | Guimarães and Lapa (2004) | Soft Computing | Triangular membership functions | FIS for the assessment of nuclear power plant (NPP) transients |
| Nuclear supplier selection | Wu et al. (2016) | MCDM | Triangular membership functions | Fuzzy VIKOR based potential supplier selection model |
| Load regulation | Luan et al. (2011) | Soft Computing | Nonlinear fuzzy membership functions | Takagi-Sugeno (T-S) fuzzy control system for load-following operation |

**Table 2.6** Fuzzy sets applications in wind thermal energy

| Main subject | Author(s) | Application area | Type of fuzzy sets/numbers | Contribution |
|---|---|---|---|---|
| Emission management | Reddy and Abhyankar (2013) | Optimization/MCDM | Linear membership functions | Pareto based best compromise solution selection based on fuzzy min max approach |
| | Azizipanah-Abarghooee et al. (2012) | Optimization/MCDM | Linear membership function | fuzzy-based clustering technique is utilized to reduce the size of the repository |
| Wind-thermal generation scheduling | Falsafi et al. (2014) | Optimization/MCDM | Triangular membership functions | Fuzzy entropy method for selecting the best solution |

probabilistic multi objective optimization techniques and evolutionary algorithms. To cope with uncertain parameters and stochastic variables, fuzzy set theory based methodologies have been utilized in recent years as seen from Table 2.6.

From Table 2.6, fuzzy set theory based methods are mainly implemented in solution selection procedure which is appeared after one of the optimization techniques adapted to the problem. These studies are limited as an application of fuzzy set theory based applications for wind thermal energy modelling. For instance, Azizipanah-Abarghooee et al. (2012) conducted fuzzy clustering for grouping similar solutions after multi objective stochastic search algorithm is applied for probabilistic wind-thermal economic emission dispatch problem. Reddy and Abhyankar (2013) used fuzzy min-max approach for best compromise solution selection when multi-objective strength Pareto evolutionary algorithm is performed for market clearing model. The main motivation of these studies is uncertainty modelling using an optimization method and after that, selection of the best alternative considering uncertain parameters by utilizing fuzzy set theory based applications.

## 2.5 Conclusion

A complex system is typically adaptive or evolutionary and influenced by social and political, as well as physical, processes. Energy production systems involve high level of complexity with their physical, social, and political dimensions. Energy systems are complex systems since they have interrelated, heterogeneous elements. They exhibit complex social and technological dynamics. Hence, their behaviors cannot be predicted by understanding each of the components separately.

It is hard to model the complex energy systems by using classical logic-based approaches. Fuzzy logic based-approaches have been employed for the evaluations of complex energy systems such as nuclear energy, hydrogen energy, bioenergy, wave energy, wind-thermal energy systems. Fuzzy control, fuzzy inference systems, and multi-criteria decision making are the most used applications of fuzzy sets in complex energy systems.

For further research, we suggest classification of complex energy systems with respect to the extension types of fuzzy sets. For instance, the energy studies related to intuitionistic fuzzy sets, hesitant fuzzy sets, etc.

## References

Abed-Elmdoust, A., & Kerachian, R. (2012). Wave height prediction using the rough set theory. *Ocean Engineering, 54,* 244–250.

Afgan, N. H., & Carvalho, M. G. (2004). Sustainability assessment of hydrogen energy systems. *International Journal of Hydrogen Energy, 29*(13), 1327–1342.

Afgan, N. H., Veziroglu, A., & Carvalho, M. G. (2007). Multi-criteria evaluation of hydrogen system options. *International Journal of Hydrogen Energy, 32*(15), 3183–3193.
Ahn, K. K., Truong, D. Q., Tien, H. H., & Yoon, J. I. (2012). An innovative design of wave energy converter. *Renewable Energy, 42,* 186–194.
Akpınar, A., Özger, M., Kömürcü, M. İ. (2014). Prediction of wave parameters by using fuzzy inference system and the parametric models along the south coasts of the Black Sea. *Journal of Marine Science and Technology, 19,* 1–14.
Amarkarthik, A., & Sivakumar, K. (2016). Investigation on modeling of non-buoyant body typed point absorbing wave energy converter using adaptive network-based fuzzy inference system. *International Journal of Marine Energy, 13,* 157–168.
Atanassov, K.T. (1986). Intuitionistic fuzzy sets. *Fuzzy Sets and Systems, 20*(1), 87–96.
Atanassov, K.T. (1989). On intuitionistic fuzzy sets and their applications. In *Actual Problems of Sciences, Bulgarian Academy of Sciences* (Vol. 1, pp. 1–53) (in Bulgarian).
Azizipanah-Abarghooee, R., Niknam, T., Roosta, A., Malekpour, A. R., & Zare, M. (2012). Probabilistic multi objective wind-thermal economic emission dispatch based on point estimated method. *Energy, 37*(1), 322–335.
Bonsignore, L., Davarifar, M., Rabhi, A., Tina, G. M., & Elhajjaji, A. (2014). Neuro-fuzzy fault detection method for photovoltaic systems. *Energy Procedia, 62,* 431–441.
Buchholz, T., Rametsteiner, E., Volk, T. A., & Luzadis, V. A. (2009). Multi criteria analysis for bioenergy systems assessments. *Energy Policy, 37,* 484–495.
Carmona, C. J., González, P., García-Domingo, B., del Jesus, M. J., & Aguilera, J. (2013). MEFES: An evolutionary proposal for the detection of exceptions in subgroup discovery. An application to concentrating photovoltaic technology. *Knowledge-Based Systems, 54,* 73–85.
Castiglia, F., & Giardina, M. (2013). Analysis of operator human errors in hydrogen refuelling stations: Comparison between human rate assessment techniques. *International Journal of Hydrogen Energy, 38*(2), 1166–1176.
Cebi, S., Ilbahar, E., & Atasoy, A. (2016). A fuzzy information axiom based method to determine the optimal location for a biomass power plant: A case study in Aegean Region of Turkey. *Energy, 116,* 894–907.
Chang, E. (2017). Rapid-convergent sliding mode proportional-integral technology with fuzzy gain scheduling for hydrogen energy applications. *International Journal of Hydrogen Energy, 42*(29), 18216–18222.
Chang, P. L., Hsu, C. W., & Lin, C. Y. (2012). Assessment of hydrogen fuel cell applications using fuzzy multiple-criteria decision making method. *Applied Energy, 100,* 93–99.
Chekired, F., Mellit, A., Kalogirou, S., & Larbes, C. (2014). Intelligent maximum power point trackers for photovoltaic applications using FPGA chip: A comparative study. *Solar Energy, 101,* 83–99.
Chiu, C. S., & Ouyang, R. Y. L. (2011). Maximum power tracking control of uncertain photovoltaic systems: A unified T-S fuzzy model-based approach. *IEEE Transactions on Control Systems Technology, 19,* 1516–1526.
Coteli, R., Acikgoz, H., Ucar, F., & Dandil, B. (2017). Design and implementation of type-2 fuzzy neural system controller for PWM rectifiers. *International Journal of Hydrogen Energy, 42* (32), 20759–20771.
Cutz, L., Haro, P., Santana, D., & Johnsson, F. (2016). Assessment of biomass energy sources and technologies: The case of Central America. *Renewable and Sustainable Energy Reviews, 58,* 1411–1431.
Das, S. K., Verma, D., Nema, S., & Nema, R. K. (2017). Shading mitigation techniques: State-of-the-art in photovoltaic applications. *Renewable and Sustainable Energy Reviews, 78,* 369–390.
Demirbas, A. H., & Demirbas, I. (2007). Importance of rural bioenergy for developing countries. *Energy Conversion and Management, 48*(8), 2386–2398.
Dos Santos Grecco, C. H., Vidal, M. C. R., Cosenza, C. A. N., Dos Santos, I. J. A., & De Carvalho, P. V. R. (2014). Safety culture assessment: A fuzzy model for improving safety performance in a radioactive installation. *Progress in Nuclear Energy, 70,* 71–83.

Erdoğan, M., & Kaya, İ. (2016). A combined fuzzy approach to determine the best region for a nuclear power plant in Turkey. *Applied Soft Computing, 39,* 84–93.
Erol, İ., Sencer, S., Özmen, A., & Searcy, C. (2014). Fuzzy MCDM framework for locating a nuclear power plant in Turkey. *Energy Policy, 67,* 186–197.
Falsafi, H., Zakariazadeh, A., & Jadid, S. (2014). The role of demand response in single and multi-objective wind-thermal generation scheduling: A stochastic programming. *Energy, 64,* 853–867.
Franco, C., Bojesen, M., Hougaard, J. L., & Nielsen, K. (2015). A fuzzy approach to a multiple criteria and geographical information system for decision support on suitable locations for biogas plants. *Applied Energy, 140,* 304–315.
Gim, B., & Kim, J. W. (2014). Multi-criteria evaluation of hydrogen storage systems for automobiles in Korea using the fuzzy analytic hierarchy process. *International Journal of Hydrogen Energy, 39*(15), 7852–7858.
Guimarães, A. C. F., & Lapa, C. M. F. (2004). Nuclear transient phase ranking table using fuzzy inference system. *Annals of Nuclear Energy, 31*(15), 1803–1812.
Kahraman, C., & Kaya, İ. (2010). Fuzzy acceptance sampling plans. In C. Kahraman & M. Yavuz (Eds.), *Production engineering and management under fuzziness* (pp. 457–481). Berlin: Springer.
Kahraman, C., Ruan, D., & Dogan, I. (2003). Fuzzy group decision-making for facility location selection. *Information Sciences, 157,* 135–153.
Kazeminezhad, M. H., Etemad-Shahidi, A., & Mousavi, S. J. (2005). Application of fuzzy inference system in the prediction of wave parameters. *Ocean Engineering, 32*(14–15), 1709–1725.
Khaehintung, N., Kunakorn, A., & Sirisuk, P. (2010). A novel fuzzy logic control technique tuned by particle swarm optimization for maximum power point tracking for a photovoltaic system using a current mode boost converter with bifurcation control. *International Journal of Control, Automation and Systems, 8,* 289–300.
Khishtandar, S., Zandieh, M., & Dorri, B. (2017). A multi-criteria decision-making framework for sustainability assessment of bioenergy production technologies with hesitant fuzzy linguistic term sets: The case of Iran. *Renewable and Sustainable Energy Reviews, 77,* 1130–1145.
Kottas, T. L., Boutalis, Y. S., & Karlis, A. D. (2006). New maximum power point tracker for PV arrays using fuzzy controller in close cooperation with fuzzy cognitive networks. *IEEE Transactions on Energy Conversion, 21*, 793–803.
Kuhmaier, M., Kanzian, C., & Stampfer, K. (2014). Identification of potential energy wood terminal locations using a spatial multi criteria decision analysis. *Biomass and Bioenergy, 66,* 337–347.
Lee, S. K., Mogi, G., Lee, S. K., Hui, K. S., & Kim, J. W. (2010). Econometric analysis of the R&D performance in the national hydrogen energy technology development for measuring relative efficiency: The fuzzy AHP/DEA integrated model approach. *International Journal of Hydrogen Energy, 35*(6), 2236–2246.
Lee, S. K., Mogi, G., Lee, S. K., & Kim, J. W. (2011a). Prioritizing the weights of hydrogen energy technologies in the sector of the hydrogen economy by using a fuzzy AHP approach. *International Journal of Hydrogen Energy, 36*(2), 1897–1902.
Lee, S. K., Mogi, G., Li, Z., Hui, K. S., Lee, S. K., Hui, K. N., et al. (2011b). Measuring the relative efficiency of hydrogen energy technologies for implementing the hydrogen economy: An integrated fuzzy AHP/DEA approach. *International Journal of Hydrogen Energy, 36*(20), 12655–12663.
Ligtvoet, A., & Chappin, E. J. L. (2012). Experience-based exploration of complex energy systems. *Journal of Futures Studies, 17*(1), 57–70.
Luan, X., Young, A. G., Han, W. S., & Zhai, Y. (2011). Load-following control of nuclear reactors based on Takagi-Sugeno fuzzy model. *IFAC Proceedings Volumes, 44*(1), 8253–8258.
Moon, J. H., & Kang, C. S. (1999). Use of fuzzy set theory in the aggregation of expert judgments. *Annals of Nuclear Energy, 26*(6), 461–469.

Mostashari, A. (2011). *Collaborative modeling and decision-making for complex energy systems*. World Scientific.

Muench, S. (2015). Greenhouse gas mitigation potential of electricity from biomass. *Journal of Cleaner Production, 103,* 483–490.

Ng, R. T. L., Ng, D. K. S., Tan, R. R., & El-Halwagi, M. M. (2014). Disjunctive fuzzy optimisation for planning and synthesis of bioenergy-based industrial symbiosis system. *Journal of Environmental Chemical Engineering, 2*(2014), 652–664.

Oluwamayowa, O. A., Shearing, P. R., & Fraga, E. S. (2017). On the design of complex energy systems: Accounting for renewables variability in systems sizing. *Computers & Chemical Engineering, 103,* 103–115.

Özger, M. (2010). Significant wave height forecasting using wavelet fuzzy logic approach. *Ocean Engineering, 37*(16), 1443–1451.

Özger, M. (2011). Prediction of ocean wave energy from meteorological variables by fuzzy logic modeling. *Expert Systems with Applications, 38*(5), 6269–6274.

Özger, M., & Şen, Z. (2007). Prediction of wave parameters by using fuzzy logic approach. *Ocean Engineering, 34*(3), 460–469.

Palaniswamy, A. M., & Srinivasan, K. (2016). Takagi-Sugeno fuzzy approach for power optimization in standalone photovoltaic systems. *Solar Energy, 139,* 213–220.

Purba, J. H. (2014). A fuzzy-based reliability approach to evaluate basic events of fault tree analysis for nuclear power plant probabilistic safety assessment. *Annals of Nuclear Energy, 70,* 21–29.

Rahrah, K., Rekioua, D., Rekioua, T., & Bacha, S. (2015). Photovoltaic pumping system in Bejaia climate with battery storage. *International Journal of Hydrogen Energy, 40*(39), 13665–13675.

Rajesh, R., & Mabel, M. C. (2015). A comprehensive review of photovoltaic systems. *Renewable and Sustainable Energy Reviews, 51,* 231–248.

Reddy, S. S., Bijwe, P. R., & Abhyankar, A. R. (2013). Multi-objective market clearing of electrical energy, spinning reserves and emission for wind-thermal power system. *International Journal of Electrical Power & Energy Systems, 53,* 782–794.

Ren, J., Manzardo, A., Toniolo, S., & Scipioni, A. (2013). Sustainability of hydrogen supply chain. Part II: Prioritizing and classifying the sustainability of hydrogen supply chains based on the combination of extension theory and AHP. *International Journal of Hydrogen Energy, 38* (32), 13845–13855.

Safari, S., Ardehali, M. M., & Sirizi, M. J. (2013). Particle swarm optimization based fuzzy logic controller for autonomous green power energy system with hydrogen storage. *Energy Conversion and Management, 65,* 41–49.

Serdio Fernández, F., Muñoz-García, M. A., & Saminger-Platz, S. (2016). Detecting clipping in photovoltaic solar plants using fuzzy systems on the feature space. *Solar Energy, 132,* 345–356.

Singh, S., & Agrawal, S. (2015). Parameter identification of the glazed photovoltaic thermal system using genetic algorithm-fuzzy system (GA–FS) approach and its comparative study. *Energy Conversion and Management, 105,* 763–771.

Singh, S., & Agrawal, S. (2016). Efficiency maximization and performance evaluation of hybrid dual channel semi-transparent photovoltaic thermal module using fuzzyfied genetic algorithm. *Energy Conversion and Management, 122,* 449–461.

Smarandache, F. (1999). A unifying field in logics: neutrosophic logic, philosophy, 1–141.

Stefanakos, C. (2016). Fuzzy time series forecasting of nonstationary wind and wave data. *Ocean Engineering, 121,* 1–12.

Subiyanto, S., Mohamed, A., Hannan, M. A. (2012). Intelligent maximum power point tracking for PV system using Hopfield neural network optimized fuzzy logic controller. *Energy and Buildings, 51,* 29–38.

Sussman, J. (2003). Collected views on complexity in systems. In *Engineering Systems Division Working Paper Series ESD-WP-2003–01.06-ESD Internal Symposium, Massachusetts Institute of Technology*.

Sylaios, G., Bouchette, F., Tsihrintzis Vassilios, A., & Denamiel, C. (2009). A fuzzy inference system for wind-wave modeling. *Ocean Engineering, 36*(17), 1358–1365.

Tabanjat, A., Becherif, M., Hissel, D., & Ramadan, H. S. (2017). Energy management hypothesis for hybrid power system of H2/WT/PV/GMT via AI techniques. *International Journal of Hydrogen Energy*. https://doi.org/10.1016/j.ijhydene.2017.06.085. Available online July 6, 2017.

Toffolo, A., & Lazzaretto, A. (2008). Energy system diagnosis by a fuzzy expert system with genetically evolved rules. *International Journal of Thermodynamics, 11*(3), 115–121.

Torra, V. (2010). Hesitant fuzzy sets. *International Journal of Intelligent Systems*, *25*(6), 529–539.

Truong, D. Q., & Ahn, K. K. (2014). Development of a novel point absorber in heave for wave energy conversion. *Renewable Energy, 65*, 183–191.

Ubando, A. T., Culaba, A. B., Aviso, K. B., Tan, R. R., Cuello, J. L., Ng, D. K. S., et al. (2016). Fuzzy mixed integer non-linear programming model for the design of an algae-based eco-industrial park with prospective selection of support tenants under product price variability. *Journal of Cleaner Production, 136,* 183–196.

Woo, T. H. (2014). Modified fuzzy algorithm based safety analysis of nuclear energy for sustainable hydrogen production in climate change prevention. *International Journal of Electrical Power & Energy Systems, 61,* 192–196.

Woo, T. H., & Lee, U. C. (2010). The statistical analysis of the passive system reliability in the Nuclear Power Plants (NPPs). *Progress in Nuclear Energy, 52*(5), 456–461.

Wright, D. G., Dey, P. K., & Brammer, J. G. (2013). A fuzzy levelised energy cost method for renewable energy technology assessment. *Energy Policy, 62,* 315–323.

Wu, Y., Chen, K., Zeng, B., Xu, H., & Yang, Y. (2016). Supplier selection in nuclear power industry with extended VIKOR method under linguistic information. *Applied Soft Computing, 48,* 444–457.

Yager, R. R. (2013). Pythagorean fuzzy subsets. In *Proceedings of the Joint IFSA Congress and NAFIPS Meeting, Edmonton, Canada* (pp. 57–61).

Yılmaz Balaman, S., & Selim, H. (2014). A fuzzy multi objective linear programming model for design and management of anaerobic digestion based bioenergy supply chains. *Energy, 74,* 928–940.

Yue, D., You, F., & Snyder, S. W. (2014). Biomass-to-bioenergy and biofuel supply chain optimization: Overview, key issues and challenges. *Computers & Chemical Engineering, 66,* 36–56.

Zadeh, L.A. (1965). Fuzzy sets. *Information and Control*, *8*, 338–356.

Zadeh, L.A. (1975). The concept of a linguistic variable and its application to approximate reasoning-I. *Information Sciences*, *8*, 199–249.

Ziolkowska, J. R. (2014). Optimizing biofuels production in an uncertain decision environment: Conventional vs. advanced technologies. *Applied Energy, 114*, 366–376.

# Part II
# Forecasting

# Chapter 3
# Forecasting Super-Efficient Dryers Adoption in the Pacific Northwest

**Joao Lavoie, Husam Barham, Apeksha Gupta, Tania Lilja, Tin Nguyen, Jisun Kim and Tugrul U. Daim**

**Abstract** Energy efficiency (EE) is an important source of electricity in the USA, by ways of saving electricity and curbing demand growth through the use of more efficient products, and the Pacific NW is a leading region in the EE efforts in America. Some of these efforts include studies and policies aiming to introduce energy efficient home appliances into the market and boosting its adoption, and organizations such as the Northwest Energy Efficiency Alliance (NEEA) are focused on the development of those studies and policies. As a way to assist and inform NEEA, the present chapter uses the Bass Model as a methodology to predict the adoption of Super-Efficient Clothes Dryers (SED) in the Pacific NW. A literature review is conducted to better understand the role of NEEA and clothes dryers in the EE realm, the model inputs and assumptions are explained and its results are discussed. Conclusions, for both NEEA and for the general EE community are drawn, and future work opportunities are identified.

---

J. Lavoie · H. Barham · A. Gupta · T. Lilja · T. Nguyen · J. Kim · T. U. Daim (✉)
Portland State University, Portland, USA
e-mail: tugrul.u.daim@pdx.edu

J. Lavoie
e-mail: jlavoie@pdx.edu

H. Barham
e-mail: hbarham@pdx.edu

A. Gupta
e-mail: apgupta@pdx.edu

T. Lilja
e-mail: tania.lilja@pdx.edu

T. Nguyen
e-mail: tinn@pdx.edu

J. Kim
e-mail: jisunk@pdx.edu

© Springer International Publishing AG, part of Springer Nature 2018
C. Kahraman and G. Kayakutlu (eds.), *Energy Management—Collective and Computational Intelligence with Theory and Applications*, Studies in Systems, Decision and Control 149, https://doi.org/10.1007/978-3-319-75690-5_3

## 3.1 Introduction

Energy efficiency is now estimated to be the third source of electricity in the USA (in the form of virtual power capacity equal to the savings resulted from efficiency) (Alliance to Save Energy 2013).

The Pacific Northwest is among the leading regions in the US when it comes to energy efficiency (Alliance to Save Energy 2013; Walton 2015), in fact, it is estimated that energy efficiency will meet all the new demands in the Pacific Northwest region through 2035 (Walton 2015).

The Northwest Energy Efficiency Alliance (NEEA), is an institution that represents a joint effort from hundreds of public utilities and related entities in the Pacific Northwest of USA, and aims to boost energy efficiency initiatives, ultimately to bring more energy efficient options to the market—in the form of appliances and related products and advanced processes (NEEA 2016a).

One product NEEA has been evaluating recently is super-efficient dryers, a product that promises energy efficiency as much as 40% of the current energy being consumed by a typical dryer in the region (Lee et al. 2015). Before NEEA can promote and support this product, they need to understand its market potentials, so two separated studies were conducted to forecast this product future, however; the studies indicated different results (Lee et al. 2015).

In this project, the objectives are to forecast the adoption and market share of super-efficient dryers (SED's) and to provide additional information that could help organizations like NEEA to promote and disseminate SED's, creating market transformations towards the adoption of energy efficient products. More specifically, our goal is to help NEEA by providing them with yet another counter-factual model of the super efficient dryers adoption (focusing on a specific market segment), and also to enable NEEA to use this model as a tool to convince manufacturers to increase their marketing efforts towards a larger adoption. The next section brings a literature review on technology forecasting, energy efficiency and other important topics for this study, followed by methodology, analysis and results, conclusions, limitations and future research.

## 3.2 Literature Review

### *3.2.1 Technology Forecasting*

As it happens with most of the major technological changes and advancements, war was the event that propelled people to start thinking about technology forecasting. First came the acknowledgement that technology was powerful enough to change not only our quality of life but also to fundamentally change the whole relationship between humans, then the necessity of having assessment methods in place to deal with those technologies. Right after that, the necessity of forecasting technological development was recognized. The first efforts towards technology forecasting began back in the 1950s and 1960s (starting at the end of the second World War and intensifying during the Cold War), but the works started to be more structured and powerful after 1972, with the launch of the Office of Technology Assessment (OTA)—a U.S. Congress organization created to deal with all aspects related to technology.

According to Bettis and Hitt (1995), technology development has created a new competitive landscape, in which organizations have to act in a much more fast-paced fashion, and also have to commit astronomic amount of capital without being sure if the markets will develop properly. Nevertheless, in order to understand the technological advancements and try to predict them, it is essential to understand and regard past, as pointed out in Barley (1998).

Although several benefits derive from technology forecasting works, those are not simple tasks, and achieving a highly accurate result is not easy as well. According to Bowonder et al. (1999), major issues related to forecasting in general are the way people usually discount the future, the natural biases and traps humans can be led to and the different ways or perspectives that experts can regard a situation through.

The methods used in technological forecasting works are plentiful. Linstone (1999), mentions the most recurrent technology forecasting methods at that time, citing trend extrapolation, growth curves, extrapolation, leading indicators, causal models, technological substitution, technology measurement, scenarios and Delphi. Porter (1999), points out the most prominent methods at the end of the 20th century: creativity methods, monitoring, trend analysis, modeling, expert opinion and scenarios. In 2005, Coccia (2005) has presented an in-depth study of Technometrics, a set of techniques used in and derives from a variety of disciplines to measure and understand technological changes overtime.

Several examples of research pieces on technology forecasting can be found in the literature, including a well-known and recognized journal that is dedicated to the subject—Technology Forecasting and Social Change. In Guice (1999), for instance, the author tries to understand the trends of emerging technologies, through the eyes of a particular organization—DARPA, within the U.S. Department of Defense (DoD). In a completely different arena, Harold Linstone brilliantly uses its three perspectives—Technical, Organizational and Personal—to analyze the terrorist threats of the 21st century, in the light of technological advancements (Linstone 2003). Devezas et al. (2005) analyze and try to forecast the growth of the internet through the lens of K-Waves. In a thorough research piece, Martino (2003) goes over several different techniques that can be applied to technology forecasting, including Delphi, scenarios, probabilistic forecasts, growth curves and others (Martino 2003). Coates et al. (2001) also provide a very interesting overview of the history of technology forecasting and its emerging trends and methods (Coates et al. 2001). The technology Futures Analysis Methods Group, led by Professor Alan Porter, has published a comprehensive analysis of technology forecasting methods along with a framework describing a process on how to apply the methods and have better results on a forecasting endeavor (Porter et al. 2004). Table 3.1 lists some of

**Table 3.1** Technology forecasting applications and methods

| Authors/year | Application | Method |
|---|---|---|
| Martino (1993) | Jet fighters | Scoring method/planar tradeoff surfaces |
| Richard et al. (2003) | Energy consumption, efficiency and carbon efficiency | Scenarios |
| Zhu and Porter (2002) | Nanotechnology, internet | Empirical technology forecasting methods |
| Mann (2003) | Hydraulic system bearings, material system design | Theory of inventive problem solving—TRIZ |
| Yoon and Park (2007) | Thin film transistor-liquid crystal display | Morphology analysis and conjoint analysis |
| Daim et al. (2006) | Fuel cell, food safety, optical storage | Bibliometrics and Patent analysis |
| Anderson et al. (2002) | Microprocessors | Data envelopment analysis |
| Watts et al. (1997) | Ceramic engine parts | Bibliometrics and patent analysis |
| Kayal (1999) | Semiconductor | Technology cycle time indicator |
| Inman et al. (2006) | Jet fighters | Technology forecasting data envelopment analysis—TFDEA |
| Sager (2001) | Biotechnology | Scenarios |
| Phaal (2004) | Printing and automotive sector | Technology roadmapping |
| Rowe and Wright (1999) | – | Delphi |
| Winebrake and Creswick (2003) | Hydrogen fueling systems | Analytic hierarchy process (AHP) and Scenarios |
| Ilonen et al. (2006) | Diffusion of innovations | Bass Model and Self-organizing Map (SOM) |
| Martino (1993) | Jet fighters | Scoring method/planar tradeoff surfaces |
| Richard et al. (2003) | Energy consumption, efficiency and carbon efficiency | Scenarios |
| Zhu and Porter (2002) | Nanotechnology, internet | Empirical technology forecasting methods |
| Mann (2003) | Hydraulic system bearings, material system design. | Theory of inventive problem solving—TRIZ |
| Yoon and Park (2007) | Thin film transistor-liquid crystal display | Morphology analysis and conjoint analysis |
| Daim et al. (2006) | Fuel cell, food safety, optical storage | Bibliometrics and Patent analysis |
| Anderson et al. (2002) | Microprocessors | Data envelopment analysis |

the areas technology forecasting has been applied to, along with the techniques/methodologies used.

### 3.2.2 Energy Efficiency

The last forty years in America have seen significant gains in efforts towards energy conservation. The US's energy efficiency (and vulnerability) were brought to the nation's attention after the 1973 and 1979 oil shocks, prompting local, state, regional and national policies to develop. Nonprofits, for-profit companies, and government regulators have begun to work towards a greener future that will lower energy spending and increase energy availability. Thankfully, even though economic output in the US has tripled since the 70s, energy consumption has only increased 50% (Alliance to Save Energy 2013). The reason for this is estimated by economists to be predominantly a result of the adoption of energy efficient products and services. The American Council for an Energy Efficient Economy (ACEEE) reports that 60–75% of the increased energy consumption has been absorbed and neutralized thanks to clean energy efforts (Walton 2015).

Even more progress has been made more recently, since 2010. Over 20 federal programs have been enacted towards increased energy efficiency (NEEA 2016a). The Pacific NW is a leader in American energy efficiency power resources (Lee et al. 2015). Only hydroelectricity ranks higher in the region (Alliance to Save Energy 2016). The NW Power and Conservation Council has enacted policies and regulations to bring the region to a staggering 90% carbon free electricity by 2030 (Northwest Power and Conservation Council 2016).

### 3.2.3 Super Efficient Dryers

Super efficient dryers, or heat pump dryers, are described as being more environmentally friendly and cheaper to run, as it uses less than 50% of the energy used by conventional dryers. Not only that but also, super efficient dryers use less peak power consumption than conventional dryers. According to government's energy star, heat pump dryers from Europe are around five times more energy-efficient than conventional dryers from America—in terms of peak electricity consumption. Furthermore, heat pump dryers have low air temperature inside the drum compared to conventional dryers and don't produce any noise (BEKO 2016; Rice et al. 2015).

According to Beko Company, a European SED manufacturer:

> [H]eat pump tumble dryers use hot air to absorb moisture from your clothes in order to get them dry after a wash. After this, air passes through the drum, it goes through the evaporator which removes the moisture, which is collected as condensation and stored in into a tank. The remaining air is re-heated and sent back to the drum to start the cycle again and continue drying your clothes. (BEKO 2016)

Also, while typical dryers are vented (the dryer evaporates water from the wet clothing and then vents the moist air to the outside); the heat pump dryers use a different technology and are called unvented (a reservoir is used to collect water removed from the clothes) (Denkenberger et al. 2013). SED dryers have been available and popular in Europe for years, however, they just became available in the US market recently, largely due to the efforts of the Super Efficient Dryer Initiative (Rice et al. 2015; SEDI 2016b).

#### 3.2.3.1 Super Efficient Dryer Initiative (SEDI)

SEDI was introduced in 2010 by the New Jersey Clean Energy Program (SEDI 2016b), with the goal of improving dryer energy efficiency by developing new technologies capable of generating high-energy savings for consumers in North America. An ultimate goal of SEDI was introduced advanced clothes dryers to North American market through the key stakeholder's engagement that SED was better choice for North America to adopt it. In order to bring Super Efficient Dryers to North American market, SEDI needs to address these three questions below (Badger et al. 2012):

- Are there market and technical drivers strong enough to justify forming such an initiative?
- Are there stakeholders open to the idea and ready to back it up by actively participating—not only in America but also across other countries?
- Would the initiative be granted with enough resources in order to accomplish its objectives and change the market into accepting this new type of dryer?

A market research conducted by SEDI indicated that 85% of U.S households have clothes dryers, furthermore; the research revealed that dryers make up 6% of residential electricity consumption, adding a total cost of $9 billion on consumers every year (SEDI 2016b). Another study by SEDI indicates clothes dryers energy consumption in U.S market and also it project the sale growth rate between 2010 and 2030 to be 1.8% (SEDI 2016a).

SEDI formed partnerships with the industry and got sponsorship from environment and energy efficiency organizations, to find and promote alternative efficient dryers to the North American markets. In 2013, SEDI has an operating budget of $266,000 for energy efficient program across United States and Canada (Granda 2013). SEDI have six sponsors by leading efficiency programs such as SEDI (2016b), Super Efficient Dryer Initiative (2013)

- Northeast: New Jersey Clean Energy Program; National Grid; Efficient Vermont Mid-Atlantic; Long Island Power Authority
- Pacific Northwest and Canada: BC Hydro; Northwest Energy Efficiency Alliance.

The above-mentioned sponsors have a budget of $3.2 billion in Energy Efficient programs, serve around 64 million people and have approximately $600 million in budget for residential programs.

Under SEDI, a team studied current heat pump technologies, and found that they are 50–60% more efficient than traditional dryers currently available in North America (Coates et al. 2001). According to their market research analysis, there are over 25 different models of heat pump dryers that are being used in the European markets (Badger et al. 2012). On the other hand, market penetration for Super-Efficient Dryer Initiative in North America is very low due to several reasons, among them; the price of SED dryers; a SED dryer normally costs around $1600 while conventional dryer price range is from $200 to $800, in addition, Pricing is one of the big consumer preference when it's to choose dryers. There are top three purchase criteria when consumers decide to purchase super efficient dryers such reliability, function and price. Pricing plays a major role in consumers' preference. According to Consumers Report Buying Guide, top twenty-five recommend such as

- Electric Dryers: average cost approximately $1089 (It's can be ranging from $600 to $1600) (Evergreen Economics 2016)
- Gas Dryers: average slightly more at $1185 (It's can be ranging from $600 to $1700) (Evergreen Economics 2016)
  - Gas models costs around $100 more than electric counterparts (Evergreen Economics 2016)
  - Some gas utilities offers consumer rebate ($50–100) installing gas dryers (Energystar.gov. 2011)
  - $50 more than electric dryers (Energystar.gov. 2011)
  - Typical unit cost around $950–$2050 for small to medium capacities (Energystar.gov 2011).
- High-end hybrid and heat pump(super-efficient) dryers (LGEcoHybridTM and Whirlpool HybridCareTM) costs around $1600 unless discount or rebate come along with incentives (Evergreen Economics 2016).

Furthermore, heat pump tumble dryers that have been used in Europe have small external dimension and drum compared to conventional dryers that are being used in North America.

In order to persuade consumers in North America to purchase Super-Efficient Dryers, the SEDI team conducted a cost effectiveness analysis to see the benefit of energy saving in dryers. Table 3.2 illustrates the cost effectiveness analysis:

Based on Table 3.2, consumers can save at least $54 annually from energy efficiency of the SED dryers more than traditional dryers annual saving cost around $76, making it a good incentive for consumers, but not enough to justify the price difference from traditional dryers. So, there is still a need for strong government support, funding, and regulations to achieve high market penetration.

**Table 3.2** Cost-effectiveness of advanced clothes dryer (Badger et al. 2012)

| Electric clothes dryer | Conventional dryer | Efficient dryer |
|---|---|---|
| Savings per year (kWh/year) | 462 | 332 |
| Savings per year ($/year) | 76 | 54 |
| Savings on lifetime (kWh) | 5541 | 3987 |
| Savings on lifetime ($) | 909 | 654 |
| Price premium ($) | 405 | 253 |
| Payback on price premium (year) | 5.3 | 4.6 |
| Present value of net benefits ($) | 297 | 235 |
| Benefit-to-cost ratio | 1.86 | 2.09 |

SEDI suggested several approaches to achieve market transformation (Foster et al. 2014):

- Energy savings potential identification
- Make good use of information in order to enhance the decision-making process
- Have sound enough testing processes
- Identify criteria that translates what the market expects
- Engaging with stakeholders in the industry looking for support
- Favoring consumers' interests and decrease risk for industry players at the same time
- Create and keep resilient and agile programs
- Being dynamic so as to bring programs to fruition.

Anytime new technology emerges, there will be market barriers. According to SEDI, there are five market barriers that Super efficient dryers have to overcome (Badger et al. 2012):

- No product available
- Unproven product performance
- Unknown Energy Savings
- Likely Higher Product Pricing
- Low Consumer Awareness.

So, to achieve market transformation, SEDI conducted several activities between 2012 and 2013 (Badger et al. 2012):

- Leverage on the Success Europe has achieved
- Build relationships with industry players in America
- Support Energy Star and Emerging Tech Award
- Help industry players in getting approvals from regulators
- Perform tests—both in labs and in the field
- Enhance testing processes of Dept. of Energy

- Identify and delineate incremental costs
- Assist programs into creating incentives
- Help Efficiency Programs Market New Dryers
- Support Efficiency Programs in addressing multifamily and retail markets.

## 3.2.4 *Other Emerging Types of Dryers*

The literature review revealed other types of emerging dryer technologies, among those, it seems that microwave dryers and solar clothes dryers are the most promising. Following is a brief about each of them:

### 3.2.4.1 Microwave Dryers

Most dryers use the concept of passing warm air on clothes to dry them, however, microwave dryers use microwave energy to evaporate moisture directly, saving up to 25% of the energy consumed by traditional dryers (Okey et al. 1994), and although the concept is not new (Yoon 1988), the technology still have shortcomings, especially, the impact on metal objects within the clothes, and hence didn't materialize into commercial use yet (Levy 1991).

### 3.2.4.2 Solar Clothes Dryer

This type of dryers turns solar energy into hot water, and then uses that hot water to run the dryer and dry clothes (Kitzmiller 1985; Off-grid.net 2011). The idea behind this technology is also not new (Kitzmiller 1985), but it started to see commercial use recently, and it is still very expensive (Sullivan et al. 2013).

## 3.2.5 *NEEA*

As aforementioned, NEEA is a joint effort representing hundreds of utility companies in the Pacific Northwest, and its initiatives benefit around 13 million utility customers. They aim to increase the adoption of energy efficient products, services and practices for gas and electric energy efficiency. Since 1996, NEEA has cost-effectively delivered over 1275 aMW of energy efficiency through market transformation, which can power around 900,000 houses (NEEA 2016b).

Northwest has around 6 million dryers which is among the highest sources of energy consumption among appliances. Super efficient dryers can lead to average 180 MW savings per region. NEEA's role is to engage with industry players in order to push product development towards energy-efficient solutions, while reducing the risk that manufacturers would face by doing so. Also, NEEA supports those players after the products are developed, by helping them in introducing the products into the market. Another role of NEEA is to work on the legal and standards side, aiming to push policies and testing towards the promotion of energy-efficient solutions (NEEA 2016c).

Representatives of SEDI Super Efficient Dryers Initiative) are working with appliance industry and other energy efficient advocates like NEEA to explore and promote the more energy efficient dryers. To explore the market, NEEA has initiated the following study which focuses on the below issues (Evergreen Economics 2016):

1. Should we focus specifically on promoting only super-efficient dryers or rather on ENERGY STAR-labeled dryers?
2. Are consumers willing to pay for more efficient dryers?
3. Which non-energy benefits and features of super-efficient dryers are the end users interested in?
4. The target market for super efficient dryers should be entire market or a niche market, like multifamily buildings with venting constraints?
5. Do we need to focus on just clothes dryers, or is there a need to address clothes washer as well?
6. What is the value proposition for supply chain actors in regional initiative?

NEEA did a characterization study about super efficient dryers and found out that cost is coming out to be a major barrier. They estimate that the production of heat pump dryers costs around $500–$550 more than conventional dryers with comparable features. While the, ENERGY STAR dryers are costing $25–$50 more. However, the estimate is that if heat pump models were produced at higher volumes, the total production costs would drop from $500 to $200 to $300 at the retailer level which would be similar to the increment cost of heat pump dryers in Europe which is about $300 more than conventional dryers (Evergreen Economics 2016). This resulted in a need to introduce an incentive plan to increase the production volume and reduce the manufacturing cost.

NEEA has collaborated with many organizations across the country for various rebates/incentive plans in order to promote energy efficiency. The incentive plans are divided in 2 categories as per the receiver of benefits

#### 3.2.5.1 Midstream and Upstream Incentives

NEEA focuses on Midstream incentives to corporate retail partners or manufacturers rather than end users. This is more effective when consumers do not have a lot of information about products and they tend to ask more questions from the sales people. So, this approach can be more valuable for products that have a huge variety of features/options to choose from (such as TVs or larger appliances

(Dunsky Energy Consulting 2016; NEEA 2015). They focused on upstream incentives for Low Wattage Replacement Lamps that shifts the stocking and marketing practices (NEEA 2015).

#### 3.2.5.2 Downstream Incentives

This type of incentive is focused on end users. NEEA collaborates with many organizations for rebate and other incentives at the user level as well. For example, they collaborated with Office of Energy Efficiency & Renewable Energy for incentive plans for IDAHO to promote energy efficient products. Idaho budgeted $30 million to promote energy efficiency and load management programs through various utilities and regional programs (Energy.gov 2016). Avista Energy company offers rebates on a wide array of equipment efficiency including equipment, commercial lighting and variable frequency drive retrofits, such as site-specific incentives and power management for PC networks (Energy.gov 2016). Also, Idaho power sponsors the Energy Efficiency for Businesses, which offers various types of rebates, for companies implementing energy-efficiency enhancements in their lighting, HVAC and other equipment (Energy.gov 2016).

## 3.3 Methodology

As stated earlier, the objectives of this paper are to forecast the adoption and market share of super-efficient dryers (SED's) and to provide additional information that could help organizations such as NEEA to promote and disseminate SED's, creating market transformations towards the adoption of energy efficient products. The technology forecasting method chosen to tackle this issue was the Bass model, further explained in this section.

The Bass Model (Bass 1969) is an attempt to understand and forecast the adoption and diffusion of technologies and products in the market, translating this phenomenon into a mathematical formula that generates a distribution similar to that of an S-shaped curve. As noted in Wright et al. (1997), This method became popular to model diffusion of products and technologies, and has been deployed—with good results—in several different cases.

As defined by its own creator, Frank Bass, the model would be an empirical generalization, which is a repeating pattern able to be represented by mathematical models (Bass et al. 1995). To put it in simple words, the adoption and diffusion of technologies usually follow a similar pattern (that of an S-shaped curve), and therefore it is possible to generalize and forecast the adoption and diffusion of technologies by ways of creating such a model. In another research work of Frank Bass, the author provides a comprehensive list of examples of other empirical generalization methods used in marketing and business, for several different topics such as R&D, customer satisfaction, brand awareness, etc. (Bass and Wind 1995).

The Bass model assumes two distinct types of consumers (or adopters): the innovators and the imitators. Innovators are those who adopt the new technology without any influence from others whatsoever. These consumers are usually in smaller numbers (especially for high-tech markets) and do not compose the mainstream market. They are technology enthusiasts, more prone to take risks and want to check and experiment new products and technologies by themselves—sooner rather than later. Conversely, imitators are more conservative and skeptical. These consumers are less prone to take risks and would rather wait until they are more certain about the performance and reliability of the new technology or product. Imitators usually compose the mainstream market and are strongly influenced by innovators—innovators opinions and "reviews" would, to a great extent, dictate how imitators will behave towards the new technology or product. Also, imitators will also be affected by other imitators who have already adopted the technology.

The model has essentially five components, among which, three are the model's parameters (p, q and m):

- p = coefficient of innovation
- q = coefficient of imitation
- m = total market
- N(t) = cumulative amount of consumers who adopted the technology up to period t
- N(t − 1) = cumulative amount of consumers who adopted the technology up to period t − 1
- S(t) = amount of new consumers that adopted the technology during period t; S (t) is also written as N(t) − N(t − 1).

The parameter 'm' is the total market penetration of the new technology over time. It sets the maximum 'height' of the adoption curve—the total amount of customers not to be exceeded during the adoption process. This parameter must be cautiously set, once it could lead to mistaken decisions to take into account a too large or too narrow amount of people who are willing to adopt the technology. The parameter 'p' and 'q' set the shape of the curve—how fast or slow it would take to ramp up, reach the inflexion point and reach the maximum point ('m'). As summarized in (Ofek 2009), 'm' sets the adoption scale, while 'p' and 'q' set the adoption pace. The coefficient of innovation is the rate at which innovators will adopt the technology. It does not change and depicts the willingness of technology enthusiasts to purchase the product. The coefficient of imitation is the rate at which imitators will adopt the technology. The more time goes by, the more imitators adopt the technology (for more social interactions and influences will occur between innovators and imitators and between imitators that have already adopted the technology and imitators that have not yet adopted it).

The following Eq. (3.1) depicts how likely a new customer is to adopt the new technology or product in period t:

Likelihood of technology adoption

$$p+\left(\frac{q}{m}\right)N(t-1) \tag{3.1}$$

The Bass model is represented when one multiplies the likelihood of a new consumer to adopt the technology (Eq. 3.1) by the total number of consumers that still have not adopted that technology. The total number of consumers still to adopt the technology on period t is found by subtracting the consumers who have so far adopted the technology (N(t − 1)) from the total consumers (m)—as shown in the following equation (Eq. 3.2):

Consumers still to adopt the technology on period *t*

$$m-N(t-1) \tag{3.2}$$

Therefore, the Bass model is given by the following equation:

The Bass Model

$$S(t)=\left[p+\left(\frac{q}{m}\right)N(t-1)\right][m-N(t-1)] \tag{3.3}$$

Super Efficient Dryers are relatively new appliances, its penetration in the US market is very timid and there is virtually no historical sales data on which a researcher would base its forecast on. Given these characteristics, it is reasonable to believe that the Bass model is a proper method to be used as a forecasting tool for SED's. According to (Ofek 2009), one assumption of the model is that consumers will only adopt the technology or product once—therefore the model is more successfully applied to products that are durable, and not to technologies or products that are purchased often.

## 3.4 Analysis

As stated in the methodology section, we have chosen the Bass Model because it is recognized as a good fit to forecast the adoption of a new product, for which there are no historical data available. Notwithstanding the methodology being meaningful, it can be tricky to create a meaningful model—in order to do that, the parameters have to make sense and the reasoning behind the determination of the parameters also has to make sense. As the conventional Bass Model was chosen to be used, the parameters to be estimated were 'p', 'q' and 'm' (please refer to the methodology section). In the following paragraphs we proceed to discuss the determination of the model's parameters and to present and discuss the results.

Super Efficient Dryers are a very new product in the U.S., with virtually no penetration in the American market. Therefore, there are few research works conducted on this product and no coefficients of innovation and imitation available to be used, leaving the researchers with the task of estimating these parameters to the best of their abilities. One very common and effective strategy of estimating these coefficients is by choosing analogous products (for which there are ‘p’ and ‘q’ available) and creating a system to weight those coefficients, ultimately arriving at the estimation of the coefficients for the desired product. As a requisite of this strategy, the analogous products have to be somehow similar to the reference product—for instance, in terms of market; function; technical specifications; price; regulation; competition; supply-chain; etc. Having chosen the analogous products and gathered data about them, the researcher then proceeds to consult with experts, who will rate the similarity of the analogous products with the reference product. After the rating, data is computed and the final ‘p’ and ‘q’ are determined.

For this paper, five analogous products were chosen, as following:

- Conventional clothes dryers (Bass et al. 1994)
  - p = 0.0134
  - q = 0.3317
- Clothes washers (Van den Bulte 2000)
  - p = 0.016
  - q = 0.49
- Solar panels (Agarwal et al. 2015)
  - p = 0.00005
  - q = 0.0807
- Major appliances (Kohli et al. 1999)
  - p = 0.0059
  - q = 0.245
- Solar water heaters (Yamaguchi et al. 2013)
  - p = 0.029
  - q = 0.265

The conventional clothes dryers were chosen because both market and function dimensions are identical to those of super efficient dryers. The clothes washers were chosen for those are complement products to dryers, sharing the same market base. Although solar panels do not share any commonality when it comes to market, function or technical attributes, it was chosen because these products have the same ‘energy efficiency’ component. Solar panels are similar to super efficient dryers because it is an attempt of substituting an incumbent solution (electricity provided by the power grid) by a much more energy efficient solution, and it also faces the same entry barriers as solar panels (higher price, strong incumbent and entrenched solutions). Major appliances are chosen for similar market and technical

**Table 3.3** Experts' ratings for analogous products

| | Clothes dryer | Clothes washer | Solar panels | Major appliances | Solar water heater |
|---|---|---|---|---|---|
| Expert 1 | 8 | 7 | 9 | 7 | 3 |
| Expert 2 | 7 | 7 | 10 | 6 | 8 |
| Expert 3 | 10 | 8 | 8 | 5 | 5 |
| Expert 4 | 8 | G | 9 | 6 | 4 |
| Expert 5 | 10 | 8 | 9 | 6 | 5 |
| Avg | 8.6 | 7.2 | 9 | 6 | 5 |

characteristics. Finally, solar water heaters are also chosen due to its energy efficiency component.

Having the parameters for the analogous products, a panel of five experts was assembled in order to weight the analogous products and come up with the final parameters for super efficient dryers. Each expert was asked to rate each analogous products, assigning to them values between 1 and 10; 1 being not related at all and 10 being extremely related. The rating was done following the dimensions explained earlier in this section, with special focus to function characteristics and energy efficiency characteristics. Table 3.3 shows the rating for each expert and the average for each analogous product.

The above averages, in turn, were multiplied by the respective 'p' and 'q', generating the weighted parameters for each analogous products, as follows:

- Conventional clothes dryers
  - Weighted p = 0.11524
  - Weighted q = 2.85262
- Clothes washers
  - Weighted p = 0.1152
  - Weighted q = 0.3528
- Solar panels
  - Weighted p = 0.00045
  - Weighted q = 0.7263
- Major appliances
  - Weighted p = 0.0354
  - Weighted q = 1.47
- Solar water heaters
  - Weighted p = 0.145
  - Weighted q = 1.325.

The above values, then, were summed up and divided by the summation of the averages of the analogous products, generating the final 'p' and 'q' for the super efficient dryers:

- Super Efficient Dryer
  - p = 0.011
  - q = 0.188.

The next parameter to be estimated is 'm' (total population). According to President and BCG (2012), the annual shipment of clothes dryers in the US is around 5.6 million units. For the purpose of this research, this number is considered to be the total market for clothes dryers in the US, although some other metrics and reasoning could be considered. Having in mind that one of the purposes of this research is to provide arguments to convince manufacturers to attack the market more fiercely with super efficient dryers, it does not make much sense to consider the total market as the 'm' parameter. That is because the price difference between a conventional dryer and a super efficient one is still very large. Moreover, the energy savings (on an individual basis) and the technical specifications do not seem to justify that price difference at this point. That leads us to conclude that, as of now, super efficient dryers do not compete with low-priced conventional dryers, rather they should be competing in a different market segment within the total market. Of course, once manufacturers invest more in the development of these products and also once the sales numbers grow, efficiencies in production and economies of scale will kick in, dragging the prices down and eventually enabling super efficient dryers to compete for the total market.

The market segment that would be a good fit for super efficient dryers, at this point, would be the clothes dryers 'upper market'. That would be the market segment where more expensive dryers are sold. Notwithstanding the fact that, in this segment, SED's would compete with high end and more sophisticated products, the idea is that customers that compose this segment would not be spooked by SED's prices and would be more open to make the purchase—customers composing this segment have a higher income and therefore are willing to spend more money on a clothes dryer. According to U.S. household income distribution (2016), 26.4% of American households have an income equal or higher than $100,000. Again, for the purposes of this research, this section of American households is considered to compose the 'upper market'. Multiplying this percentage by the total market for clothes dryers, the total market for SED's is determined:

- Super Efficient Dryers market in the US
  - m = 1.48 million units per year.

Having determined the parameters, it is possible to create the adoption curve. The first curve to be presented uses a time span of 25 years (Fig. 3.1), along with its associated data (Table 3.4).

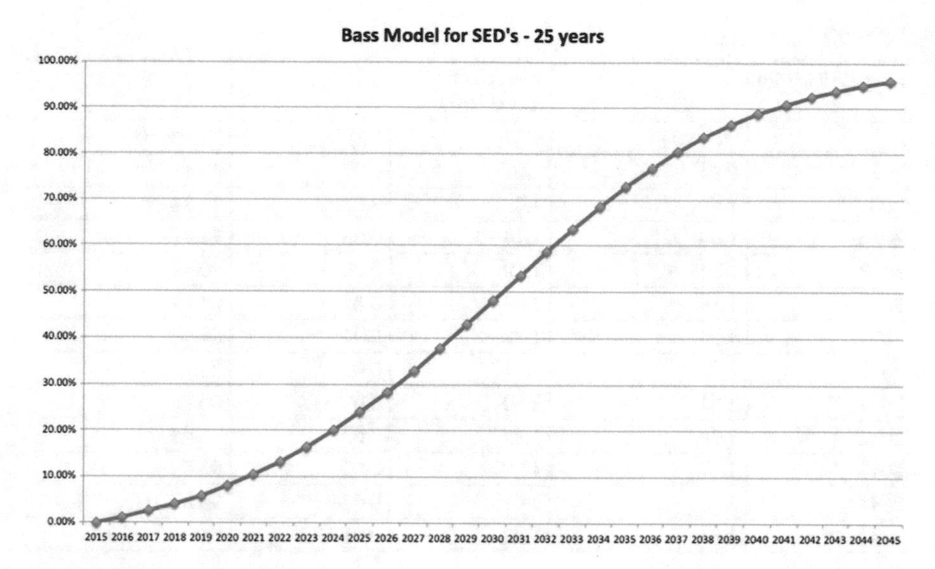


**Fig. 3.1** Bass model for 25 years

**Table 3.4** Bass model data

| m = 1.48 Million | | P = 0.011<br>q = 0.188 | | |
|---|---|---|---|---|
| t | N(t − 1) | N(t) | S(t) | N(t)/m (%) |
| 2015 | – | 0 | – | 0.00 |
| 2016 | 0.00 | 0.02 | 0.02 | 1.15 |
| 2017 | 0.02 | 0.04 | 0.02 | 2.50 |
| 2018 | 0.04 | 0.06 | 0.02 | 4.08 |
| 2019 | 0.06 | 0.09 | 0.03 | 5.91 |
| 2020 | 0.09 | 0.12 | 0.03 | 8.04 |
| 2021 | 0.12 | 0.16 | 0.04 | 10.48 |
| 2022 | 0.16 | 0.20 | 0.04 | 13.28 |
| 2023 | 0.20 | 0.24 | 0.05 | 16.44 |
| 2024 | 0.24 | 0.30 | 0.05 | 19.98 |
| 2025 | 0.30 | 0.35 | 0.06 | 23.90 |
| 2026 | 0.35 | 0.42 | 0.06 | 28.19 |
| 2027 | 0.42 | 0.49 | 0.07 | 32.82 |
| 2028 | 0.49 | 0.56 | 0.07 | 37.73 |
| 2029 | 0.56 | 0.63 | 0.08 | 42.86 |
| 2030 | 0.63 | 0.71 | 0.08 | 48.12 |
| 2031 | 0.71 | 0.79 | o.o8 | 53.41 |
| 2032 | 0.79 | 0.87 | 0.08 | 58.62 |
| 2033 | 0.87 | 0.94 | 0.07 | 63.65 |

(continued)

**Table 3.4** (continued)

| m = 1.48 Million | | P = 0.011<br>q = 0.188 | | |
|---|---|---|---|---|
| t | N(t − 1) | N(t) | S(t) | N(t)/m (%) |
| 2034 | 0.94 | 1.01 | 0.07 | 68.42 |
| 2035 | 1.01 | 1.08 | 0.07 | 72.84 |
| 2036 | 1.08 | 1.14 | 0.06 | 76.87 |
| 2037 | 1.14 | 1.19 | 0.05 | 80.48 |
| 2038 | 1.19 | 1.24 | 0.05 | 83.65 |
| 2039 | 1.24 | 1.28 | 0.04 | 86.41 |
| 2040 | 1.28 | 1.31 | 0.03 | 88.77 |
| 2041 | 1.31 | 1.34 | 0.03 | 90.77 |
| 2042 | 1.34 | 1.37 | 0.02 | 92.45 |
| 2043 | 1.37 | 1.39 | 0.02 | 93.85 |
| 2044 | 1.39 | 1.41 | 0.02 | 95.01 |
| 2045 | 1.41 | 1.42 | 0.01 | 95.96 |

The curve above starts in 2015—the year NEEA launched SEDI—and ends in 2045. The curve reaches its inflexion point between 2032 and 2033, when the adoption is around 63%. At the end of this forecasting curve (in 2045), the adoption reaches 1.42 million units sold per year (95.96% of the market).

A forecasting curve such as the first one is very valuable in order to understand how the market will behave on its own—without any influence from external parties. Institutions that try to create market changes (such as NEEA) use curves like this one—this curve could inform NEEA on what would happen had they not tried to influence the market. Moreover, taking this curve together with other studies, NEEA can develop policy strategies. Nevertheless, a very important point should be made regarding this first curve. Although it is possible to believe that SED's will still be still actively and strongly sold in 2045, it is likely that, by then, new technologies under development now (some of those technologies are already under development as explained in the literature review part of this report) will be available in the market, competing and gaining market-share from incumbent technologies (such as SED's) For that reason, in order to engage in conversations with market players, it is more appropriate to consider a shorter period of time for this adoption forecast.

When the same data is plotted considering a time span of 15 years, this is the curve (Fig. 3.2).

The second curve also starts in 2015, but ends in 2030. The inflection point, as already mentioned, is around the year of 2033, and the adoption peak for a single period is between 2029 and 2032. At the end of the curve, the adoption percentage is 48.12%, representing 710,000 units sold in the year of 2030.

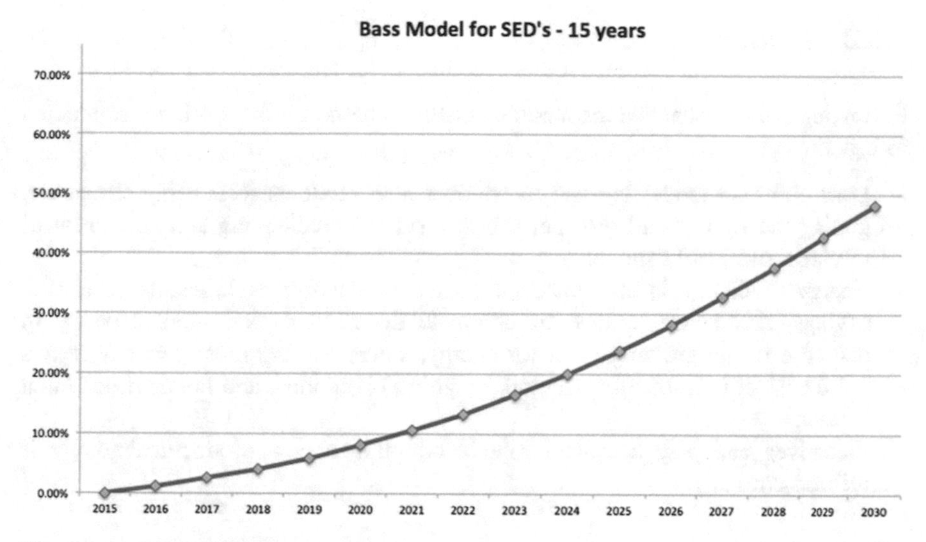


**Fig. 3.2** Bass model for 15 years

## 3.5 Conclusions and Recommendations

### *3.5.1 For NEEA*

Following are our conclusions/recommendations towards NEEA:

- The adoption curves presented in this study can serve as another source of information for NEEA to try to understand the potential market for super efficient dryers. Although they have already conducted their own research to arrive to the counterfactual adoption, this piece of research can complement that work and help NEEA in the formulation of policy strategies and characteristics, towards the advancement of energy efficiency in the Pacific Northwest (in this case, through the broader adoption of more energy efficient products).
- Specifically with regards to the second curve (15 years), it could serve as a tool to engage clothes dryer's manufacturers in conversation, show them the market potential and convince them to invest in those products, to offer them and to attack the market more fiercely. By segmenting the market, the argument in favor of offering SED's is much more compelling, since manufacturers will see the opportunity of focusing their marketing efforts, ultimately gaining almost 50% of that market segment. Conversely, if manufacturers focus on the entire market, other considerations and forces come into play (most importantly the price gap between conventional dryers and SED's), and market players will likely feel like it is going to demand too much marketing and R&D effort, especially if, in the end, the market-share will be around 5%.

### *3.5.2 General*

Following are our conclusions/recommendations based on the work we conducted in this project:

- Bass diffusion model is ideal to conduct new products forecasting. However, getting the right p and q to get robust results is challenging and need a lot of research and justifications.
- Energy efficiency is an important source of energy; as it results in energy savings, and hence, reduce or eliminate the need to add more capacity in response to the growing need for energy, since; Adding more capacity means more cost in term of infrastructure, ongoing operations, and the environmental impact.
- Incentives can play a crucial role in adopting a new product/technology if utilized correctly.

## 3.6 Future Work

Following are suggested areas for future work based on the findings of this project:

- Influencing the market: Our project's findings indicates that there is a market potential of SED in the PNW region. NEEA should go ahead and conduct a follow-up study on what type of incentive packages should be offered to accelerate the market adoption. Such study could utilize tools like sensitivity analysis (Saltelli et al. 2000) or conjoint analysis (Green and Srinivasan 1978) to decide on the proper incentives and what outcome they could achieve. With the help of consumer analysis to find out more about the process of buying a dryer and how consumers making this decision and how being part of the 'upper-income' category affect the customer' preferences.
- In this project, and in other similar projects, the degree to which the analogous products are relevant, affect the accuracy of Bass diffusion model results; as the p and q parameters have direct impact on the quality of the forecast, hence; we suggest building an HDM model based on an experts panel to make more accurate decisions when choosing analogous products for the Bass diffusion model.
- There are different versions of Bass diffusion model, like the generalized model, which consider other factors including prices and advertisement (Bass et al. 1994), and a future study could try to run this modified version and compare the results for more accurate forecast.

## References


Agarwal, A., Cai, D., Shah, S., Chandy, M., Sherick, R. (2015). A model for residential adoption of photovoltaic systems. In *2015 IEEE Power & Energy Society General Meeting*, pp. 1–5.

Alliance to Save Energy. (2013, January). *The history of energy efficiency*. Available: https://www.ase.org/sites/ase.org/files/resources/Media%20browser/ee_commission_history_report_2-1-13.pdf. Accessed: October 07, 2016.

Alliance to Save Energy. (2016). Energy legislation in the 110th congress, 2009–2010. Available: http://www.ase.org/resources/energy-legislation. [Accessed: November 07, 2016].

Anderson, T., Färe, R., Grosskopf, S., Inman, L., & Song, X. (2002). Further examination of Moore's law with data envelopment analysis. *Technological Forecasting and Social Change, 69*(5), 465–477.

Badger, C., Foster, R., Wold C., Granda C. (2012). Super efficient dryer initiative. Available: https://www.energystar.gov/sites/default/files/asset/document/SEDI.pdf. [Accessed: December 08, 2016].

Badger, C., Foster, R., Granda, C., & Wold, C. (2012) Bringing North American clothes dryers into the 21st century: A case study in moving markets—0193-000286.pdf. Available: http://aceee.org/files/proceedings/2012/data/papers/0193-000286.pdf. [Accessed: December 08, 2016].

Barley, S. R. (1998). What can we learn from the history of technology? *Journal of Engineering and Technology Management, 15,* 237–255.

Bass, F. M. (1969). A new product growth for model consumer durables. *Management Science, 15* (5), 215–227.

Bass, F. M. (1995). Empirical generalizations and marketing science: A personal view. *Marketing Science, 14*(3), G6–G19.

Bass, F. M., & Wind, J. (1995). Introduction to the special issue: Empirical generalizations in marketing. *Marketing Science, 14*(3), G1–G5.

Bass, F. M., Krishnan, T. V., & Jain, D. C. (1994). Why the Bass model fits without decision variables. *Marketing Science, 13*(3), 203–223.

BEKO. (2016). Benefits of a heat pump tumble dryer infographic | Beko UK. Available: http://www.beko.co.uk/lifestyle/benefits-of-a-tumble-dryer-heat-pump. [Accessed: November 03, 2016].

Bettis, R. A., & Hitt, M. A. (1995). The new competitive landscape. *Strategic Management Journal, 16,* 7–19.

Bowonder, B., Miyake, T., & Muralidharan, B. (1999). Predicting the future: Lessons from evolutionary theory. *Technological Forecasting and Social Change, 62*(1–2), 51–62.

Coates, V., Farooque, M., Klavans, R., Lapid, K., Linstone, H. A., Pistorius, C., et al. (2001). On the future of technological forecasting. *Technological Forecasting and Social Change, 67*(1), 1–17.

Coccia, M. (2005). Technometrics: Origins, historical evolution and new directions. *Technological Forecasting and Social Change, 72*(8), 944–979.

Daim, T. U., Rueda, G., Martin, H., & Gerdsri, P. (2006). Forecasting emerging technologies: Use of bibliometrics and patent analysis. *Technological Forecasting and Social Change, 73*(8), 981–1012.

Denkenberger, D., Calwell, C., Beck, N., Trimboli, B., Driscoll, D. (2013) Analysis of potential energy savings from heat pump clothes dryers in North America. Available: http://www.ecosresearch.com/wp-content/uploads/2015/12/2013_Analysis-of-Potential-Energy-Savings-from-Heat-Pump-Clothes-Dryers-in-North-America.pdf. [Accessed: November 08, 2016].

Devezas, T. C., Linstone, H. A., & Santos, H. J. S. (2005). The growth dynamics of the internet and the long wave theory. *Technological Forecasting and Social Change, 72*(8), 913–935.

Dunsky Energy Consulting. (2016). *Upstream, downstream, or midstream incentives? It depends*. Available: http://www.dunsky.com/up-down-or-midstream-incentives/. [Accessed: December 08, 2016].

Energy.gov. (2016). *Energy incentive programs, Idaho | department of energy*. Available: http://energy.gov/eere/femp/energy-incentive-programs-idaho. [Accessed: December 08, 2016].

Energystar.gov. (2011, November). *ENERGY STAR scoping report—Residential clothes dryers*. Available: https://www.energystar.gov/sites/default/files/asset/document/ENERGY_STAR_Scoping_Report_Residential_Clothes_Dryers.pdf. [Accessed: October 08, 2016].

Evergreen Economics. (2016). *NEEA dryer market characterization final report*. Available: https://neea.org/docs/default-source/reports/characterization-of-the-super-efficient-dryer-market.pdf?sfvrsn=6. [Accessed: November 15, 2016].

Foster, R., Badger, C., Banwell, P., & Granda, C. (2014). *Emerging technologies: A case study of the super efficient dryers initiative*. Available: http://aceee.org/files/proceedings/2014/data/papers/9-137.pdf. [Accessed: December 08, 2016].

Granda, C. (2013). Market transformation for clothes dryers: Lessons learned from the european experience. Available: http://www.topten.eu/uploads/File/CGranda418_13%2012032013.pdf. [Accessed: December 08, 2016].

Green, P. E., & Srinivasan, V. (1978). Conjoint analysis in consumer research: issues and outlook. *Journal of Consumer Research, 5*(2), 103.

Guice, J. (1999). Designing the future: the culture of new trends in science and technology. *Research Policy, 28*(1), 81–98.

Ilonen, J., Kamarainen, J. K., Puumalainen, K., Sundqvist, S., & Kälviäinen, H. (2006). Toward automatic forecasts for diffusion of innovations. *Technological Forecasting and Social Change, 73*(2), 182–198.

Inman, O. L., Anderson, T. R., & Harmon, R. R. (2006). Predicting U.S. jet fighter aircraft introductions from 1944 to 1982: A dogfight between regression and TFDEA. *Technological Forecasting and Social Change, 73*(9), 1178–1187.

Kayal, A. (1999). Measuring the pace of technological progress. *Technological Forecasting and Social Change, 60*(3), 237–245.

Kitzmiller, G. R. (1985) *Solar clothes dryer*, US4514914 A, May 07, 1985.

Kohli, R., Lehmann, D. R., & Pae, J. (1999). Extent and impact of incubation time in new product diffusion. *Journal of Product Innovation Management, 16*(2), 134–144.

Lee, A., Cofer, S., & McCormack, R. (2015, July 21). Super efficient clothes dryer market baseline. Available: https://neea.org/docs/default-source/reports/neea-super-efficient-clothes-dryer-market-baseline.pdf?sfvrsn=8. [Accessed: December 07, 2016].

Levy, C. J. (1991, September 15). Tech notes; using microwaves to dry clothes. *The New York Times*.

Lilien, G. L., Rangaswamy, A., & Van den Bulte, C. (2000). 12. Diffusion models: Managerial applications and software. *New-product diffusion models, 11*.

Linstone, H. A. (1999). Tfsc: 1969–1999. *Technological Forecasting and Social Change, 62*(1–2), 1–8.

Linstone, H. A. (2003). The 21st century: Everyman as faust technology, terrorism, and the multiple perspective approach. *Technological Forecasting and Social Change, 70*, 283–296.

Mann, D. L. (2003). Better technology forecasting using systematic innovation methods. *Technological Forecasting and Social Change, 70*(8), 779–795.

Martino, J. P. (1993). A comparison of two composite measures of technology. *Technological Forecasting and Social Change, 44*(2), 147–159.

Martino, J. P. (2003). A review of selected recent advances in technological forecasting. *Technological Forecasting and Social Change, 70*(8), 719–733.

NEEA. (2015). *NEEA 2015–2019 business plan supporting document initiative descriptions*. Available: https://neea.org/docs/default-source/default-document-library/neea-initiative-descriptions-2015-19.pdf?sfvrsn=12. [Accessed: December 08, 2016].

NEEA. (2016a). *About northwest energy efficiency alliance (NEEA)*. Available: http://neea.org/about-neea. Accessed: October 22, 2016.

NEEA. (2016b). *About Northwest Energy Efficiency Alliance (NEEA)*. Available: http://neea.org/about-neea. [Accessed: November 08, 2016].

NEEA. (2016c). *Super-efficient dryers work*. Available: http://neea.org/initiatives/residential/super-efficient-dryers. [Accessed: December 08, 2016].

Northwest Power and Conservation Council. (2016). NW councel—Energy efficiency. Available: http://www.nwcouncil.org/energy/energy-efficiency/home/. [Accessed: November 08, 2016].

Ofek, E. (2009). *Forecasting the adoption of a new product*. USA: Harvard Business School Publishing.

Off-grid.net. (2011, July 15). *"Solar powered clothes dryer," Living off the grid: Free yourself.* Available: http://www.off-grid.net/solar-powered-clothes-dryer/. [Accessed: December 07, 2016].

Okey, M. C., Foslien, W., & Kesselring, J. P. (1994). *Fuzzy logic controls for the EPRI microwave clothes dryer*. pp. 1348–1353.

Phaal, R. (2004). Technology roadmapping—A planning framework for evolution and revolution. *Technological Forecasting and Social Change, 71,* 5–26.

Porter, A. L. (1999). Tech forecasting: An empirical perspective. *Technological Forecasting and Social Change, 62*(1), 19–28.

Porter, A. L., Ashton, W. B., Clar, G., Coates, J. F., Cuhls, K., Cunningham, S. W., et al. (2004). Technology futures analysis: Toward integration of the field and new methods. *Technological Forecasting and Social Change, 71*(3), 287–303.

President, S. V., & BCG. (2016). *Automatic washer/dryer shipments United States 2012–2016 | Statistic*. Statista. Available: https://www.statista.com/statistics/271593/automatic-washer-and-dryer-shipments-in-the-us-since-2009-by-month/. [Accessed: December 06, 2016].

Rice, J. (2015, January 29). Heat-pump clothes dryers, GreenBuildingAdvisor.com. Available: http://www.greenbuildingadvisor.com/blogs/dept/guest-blogs/heat-pump-clothes-dryers. [Accessed: October 20, 2016].

Richard, S., Anders, H., & Peter, S. (2003). Analysis of US energy scenarios: Meta-scenarios, pathways, and policy implications. *Technological Forecasting and Social Change, 70*(4), 297–315.

Rowe, G., & Wright, G. (1999). The Delphi technique as a forecasting tool: issues and analysis. *International Journal of Forecasting, 15*(4), 353–375.

Sager, B. (2001). Scenarios on the future of biotechnology. *Technological Forecasting and Social Change, 68*(2), 109–129.

Saltelli, A., Chan, K., & Scott, E. M. (2000). *Sensitivity analysis*. Available: http://tocs.ulb.tu-darmstadt.de/134259246.pdf. [Accessed: November 03, 2016].

SEDI. (2016a). Potential energy savings from heat pump dryers in North America—SEDI_Fact_Sheet. Available: https://www.energystar.gov/ia/partners/pt_awards/SEDI_Fact_Sheet_H.pdf. [Accessed: December 08, 2016].

SEDI. (2016b). SEDI Transforms North American Market for Super-Efficient Dryers. Available: http://clasp.ngo/en/Resources/Resources/Headlines/2014/SEDI-Transforms-North-American-Market-for-Super-Efficient-Dryers. [Accessed: November 04, 2016].

Sullivan, D. (2013, October 04). The world's first solar-powered laundry dryer, Houselogic. Available: https://www.houselogic.com/by-room/bathroom-laundry/miele-introduces-first-solar-dryer/. [Accessed: December 07 2016].

Super Efficient Dryer Initiative. (2013). *2013 sedi clothes dryers summit 1 opening slides*.

U.S. Household Income Distribution. (2016). *Statista*. Available: https://www.statista.com/statistics/203183/percentage-distribution-of-household-income-in-the-us/. [Accessed: December 06, 2016].

Walton, R. (2015, December 18). *Report: Energy efficiency can meet Pacific Northwest demand through 2035*," Utility Dive. Available: http://www.utilitydive.com/news/report-energy-efficiency-can-meet-pacific-northwest-demand-through-2035/411041/. Accessed: November 04, 2016.

Watts, R. J., & Porter, A. L. (1997). Innovation forecasting. In *Innovation in Technology Management. The Key to Global Leadership. PICMET '97* (Vol. 47, pp. 25–47).

Winebrake, J. J., & Creswick, B. P. (2003). The future of hydrogen fueling systems for transportation: An application of perspective-based scenario analysis using the analytic hierarchy process. *Technological Forecasting and Social Change, 70*(4), 359–384.

Wright, M., Upritchard, C., & Lewis T. (1997). *A validation of the Bass new product diffusion model in New Zealand*. Mark. Bull.-Dep. Mark. MASSEY University (Vol. 8, pp. 15–29).

Yamaguchi, Y., Akai, K., Shen, J., Fujimura, N., Shimoda, Y., & Saijo, T. (2013). Prediction of photovoltaic and solar water heater diffusion and evaluation of promotion policies on the basis of consumers' choices. *Applied Energy, 102,* 1148–1159.

Yoon, B., & Park, Y. (2007). Development of new technology forecasting algorithm: Hybrid approach for morphology analysis and conjoint analysis of patent information. *IEEE Transactions on Engineering Management, 54*(3), 588–599.

Yoon, C.-H. (1988). *Microwave clothes dryer*, US4765066 A, August 23, 1988.

Zhu, D., & Porter, A. L. A. L. (2002). Automated extraction and visualization of information for technological intelligence and forecasting. *Technological Forecasting and Social Change, 69*(5), 495–506.

# Chapter 4
# Fuzzy Forecasting Methods for Energy Planning

**Basar Oztaysi, Sezi Çevik Onar, Eda Bolturk and Cengiz Kahraman**

**Abstract** For energy planning, forecasting the energy demand for a specific time interval and supply of a specific source is very crucial. In the energy sector, forecasting may be long term, midterm or short term. While traditional forecasting techniques provide results for crisp data, for data with imprecision or vagueness fuzzy based approaches can be used. In this chapter, fuzzy forecasting methods such as, fuzzy time series (FTS), fuzzy regression, adaptive network-based fuzzy inference system (ANFIS) and fuzzy inference systems (FIS) as explained. Later, an extended literature review of fuzzy forecasting in energy planning is provided. Finally, a numerical application is given to give a better understanding of fuzzy forecasting approaches.

## 4.1 Introduction

Energy is one of the scarcest sources in the world. Energy planning helps you get better control over your energy resources. Thus, facing with an energy scarcity or excessive energy consumption costs in the future can be prevented. Forecasting the energy consumption costs or the required energy levels for a firm is very helpful for the planning of future.

Forecasting methods can be divided into two main categories. Qualitative forecasting methods and quantitative forecasting methods. Qualitative forecasting methods are based on judgments, intuition, or personal experiences and subjective in nature. They are not based on hard mathematical computations. Qualitative forecasting methods are based on mathematical models, and objective in nature. Classical forecasting methods use crisp data and generally do not care the possible

B. Oztaysi · S. Çevik Onar · C. Kahraman (✉)
Istanbul Technical University, Macka, Istanbul, Turkey
e-mail: kahraman@itu.edu.tr

E. Bolturk
Istanbul Takas ve Saklama Bankası A.S, Istanbul, Turkey

© Springer International Publishing AG, part of Springer Nature 2018

C. Kahraman and G. Kayakutlu (eds.), *Energy Management—Collective and Computational Intelligence with Theory and Applications*, Studies in Systems, Decision and Control 149, https://doi.org/10.1007/978-3-319-75690-5_4

changes in the data, which the future forecasts are based on. In case of incomplete data or uncertain data, we need some extensions in the classical approaches.

Incomplete and/or vague forecasting data require fuzzy forecasting methods to be used. The advantage of the use of fuzzy logic is in processing imprecision, uncertainty, vagueness, semi-truth, or approximated and nonlinear data. Forecasting data generally involve these kinds of characteristics. Ordinary fuzzy forecasting, intuitionistic fuzzy forecasting, hesitant fuzzy forecasting, and type-2 fuzzy forecasting techniques have been developed and applied to some forecasting problems.

Carvalho and Costa (2017) propose a fuzzy forecasting methodology of time series for electrical energy prices. They use triangular fuzzy membership functions and apply the extended autocorrelation function. Kumar and Gangwar (2015) use fuzzy sets induced by intuitionistic fuzzy sets to develop a fuzzy time series forecasting model to incorporate degree of hesitation (nondeterminacy).

Bisht and Kumar (2016) propose a fuzzy time series forecasting method based on hesitant fuzzy sets. The proposed method addresses the problem of establishing a common membership grade for the situation when multiple fuzzification methods are available to fuzzify time series data. Hassan et al. (2016) present a novel design of interval type-2 fuzzy logic systems by using the theory of extreme learning machine for electricity load demand forecasting.

The rest of the chapter is organized as follows. Section 4.2 summarizes the literature on fuzzy forecasting. Section 4.3 presents the fuzzy forecasting methods. Section 4.4 gives a numerical application on fuzzy forecasting. Finally, Sect. 4.5 concludes this chapter.

## 4.2 Literature Review

Forecasting is one of the most precious activities in planning because of define company's strategies. Planning activities in the energy sector are very important because the cost of investment is very critical. There are different techniques in energy planning in electricity (Piras et al. 1995), wind power (Lou et al. 2008), power system planning (Holmukhe et al. 2010), forecasting energy and diesel consumption (Neto et al. 2011), electricity consumption (Bolturk et al. 2012) short term load forecasting (Liu et al. 2010; Jain and Jain 2013), electricity demand estimation (Zahedi et al. 2013), long term load forecasting (Akdemir and Cetinkaya 2012). While traditional methods use crisp data to make predictions, Zadeh (1965) introduced fuzzy sets in order to integrate expert evaluations into the problem and to deal with imprecision or vagueness intrinsic to the decision problem (Kahraman et al. 2016). In this section, a literature review on fuzzy forecasting methods is provided.

Piras et al. (1995) examined heterogeneous neural network architecture in order to forecast electrical load forecasting in energy planning and use artificial neural networks (ANN) to get results. Mori and Kobayashi (1996) recommended an optimal fuzzy inference method for short-term load forecasting (STLF) which mentions a structure of the simplified fuzzy inference. The structure is modeled to clasp nonlinear manner in short-term loads and its aim is to minimize errors.

Electric power system load forecasting has a significant function in the energy management that has a big effect in operation, controlling and planning of electric power system. The load estimation used in electrical systems planning has to think future loads and their geographical positions in order to allowing the creator for situating the electrical equipment. The load forecasting affects different features aspects like peak load demand period, transformer sizing, conductor sizing, capacitor placement, and etc. Cartina et al. (2000) used nonlinear fuzzy regression approach in distribution networks for forecast peak load for STLF is an important subject in the operative planning activities of companies devoted to the allocation and trade of energy. Tranchita and Torres (2004) proposed an original method which consists of LAMDA-fuzzy-clustering techniques, regression trees, classification and regression trees algorithm and fuzzy inference for the peak power, daily energy and load curve forecast. Another STLF paper is studied by Hayati and Karami (2005). They explore the use of computational intelligence methods and use the three important architectures of the neural network and a hybrid neuro-fuzzy network named Evolving Fuzzy Neural Network to model STLF systems. Zhao et al. (2006) use the ANN and fuzzy theory for STLF. An algorithm—a pace search for optimum to renew the weight value of fuzzy neurons applied STLF method to some area's power system. Lou et al. (2008) use similarity theory and fuzzy clustering method in order classifying various periods and select a result to substitute to the output of wind power. Holmukhe et al. (2010) use fuzzy logic systems for power system planning for STLF and aimed to define improved method for load forecasting.

Liu et al. (2010) use time-varying slide FTS method that reduces the load forecasting error in STFL. The proposed technique fits a study framework of FTS to exercise trend estimator and uses estimator to obtain forecasting values at forecasting step.

Neto et al. (2011) develop a determination support system for forecasting the cost of electricity production using non-stationary data by integrating the methodology of FTS in order to see uncertainness intrinsic in study of diesel fuel consumption. Li and Choudhury (2011) present a technique to embody a fuzzy and probabilistic load model in transmission energy loss evaluation to overcome uncertainness of load forecast.

Long-term forecasting is a leading issue in energy planning especially in the size of energy plant and location. Akdemir and Cetinkaya (2012) use adaptive neural fuzzy inference system using real energy data in long-term load forecasting. Bolturk et al. (2012) examine electricity consumption to predict possible electricity consumption in energy planning for a company and they use FTS and compare total values of three periods.

Zahedi et al. (2013) estimate electricity demand with ANFIS to get more reliable and accurate planning. Electricity demand is studied with ANFIS and the study includes some parameters such as occupation, gross domestic product, people, dwelling count and two meteorological parameters. Jain and Jain (2013) develop a fuzzy model and similarity-based STLF using swarm intelligence because of uncertainties in planning and operation of the electric power system is the complex, nonlinear, and non-stationary system. Another paper is presented by Chen et al. (2013) to show a solar radiation estimate model established on fuzzy and neural networks to get well results.

Bain and Baracli (2014) investigate the practice of the ANFIS to predict the energy demand planning. The results showed that hybrid ANFIS technique based upon fuzzy logic and ANNs perform efficiently in forecast accuracy. Li et al. (2015) use STLF with the grid method and FTS estimating technique to better forecast truth in energy planning.

Atsalakis et al. (2015) forecast energy export to plan potential energy demand to the importance of accuracy. The forecasting system is established on two ANFIS use to forecast the optimum energy export forecast parameters. In order to obtain the upper and lower bounds of wind power, Zhang et al. (2016) use an ANFIS to get wind power interval forecasts and based on the system and singular spectrum analysis and the paper develops a hybrid uncertainty forecasting model, Interval Forecast- ANFIS -Singular Spectrum Analysis-Firefly Algorithm. Another wind prediction study present by Okumus and Dinler (2016) with ANFIS and ANN for one h ahead to predict wind speed. Matthew and Satyanarayana (2016) use fuzzy logic along with the results for checking the accuracy of the proposed work in load forecasting. Because there is a requirement of power planning for the proper utilization of electrical energy in electrical power system.

Monthly forecast of electricity demand in the housing industry is studied by Son and Kim (2017) on a precise model and the proposed method is consists of support vector regression (SVR) and fuzzy-rough feature selection with particle swarm optimization (PSO) algorithms. Zhang et al. (2017) present a new variable-interval reference signal optimization approach and a fuzzy control-based charging/ discharging scheme to wind power system. Arcos-Aviles et al. (2017) propose a strategy based on a low complexity Fuzzy Logic Control for grid power profile smoothing of a residential grid-connected microgrid in the design of an energy management in order to show the presented work minimizing fluctuations and power peaks while keeping energy stored in battery between secure limits results, a simulation comparison highlighted. Chahkoutahi and Khashei (2017) propose a direct optimum parallel hybrid model is consists of multilayer perceptrons neural network, ANFIS and Seasonal ARIMA to electricity load forecasting in electricity load forecasting and aim of this study is put into practice upper hands of ANFIS and Seasonal ARIMA in modelling composite and equivocal systems in energy planning.

Fuzzy forecasting methods in energy planning have been extensively used and the analysis of them in Figs. 4.1, 4.2, 4.3 and 4.4; Table 4.1.

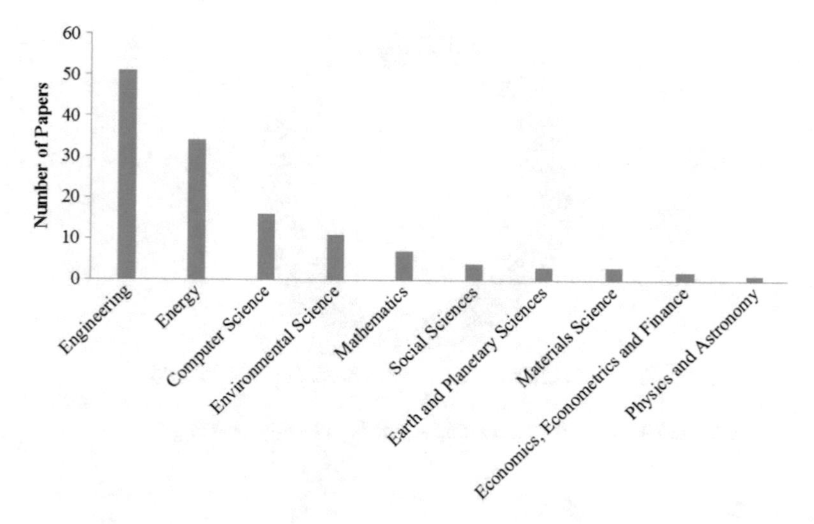


**Fig. 4.1** Distribution of fuzzy forecasting methods in energy planning by subject area

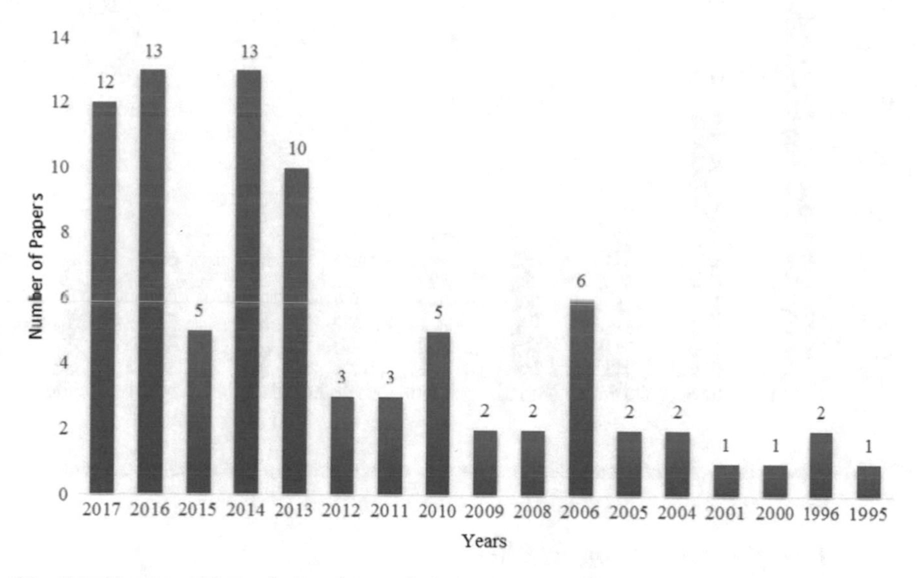


**Fig. 4.2** Number of fuzzy forecasting methods in energy planning papers by years

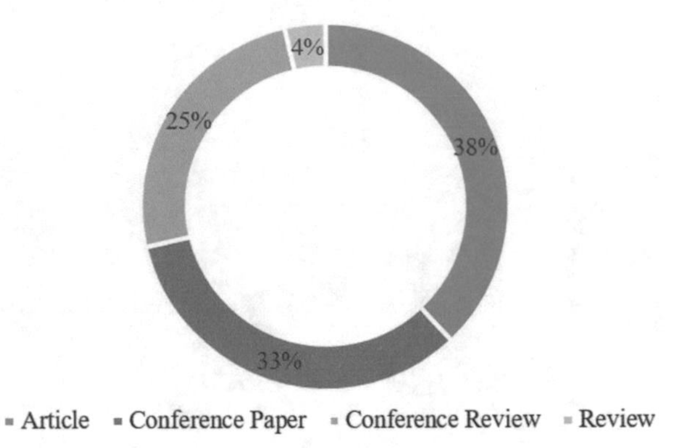


Fig. 4.3 Document type of fuzzy forecasting methods in energy planning

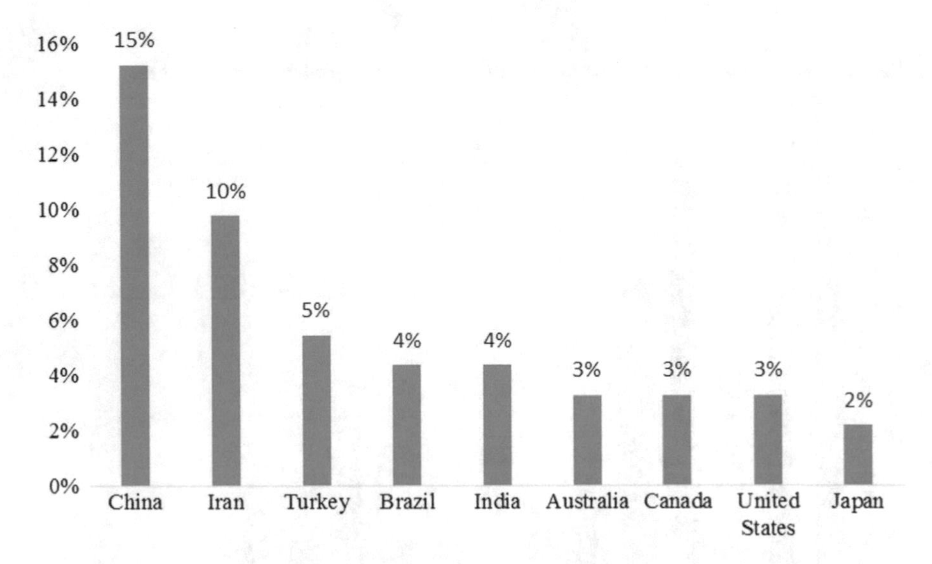


Fig. 4.4 Percentage of fuzzy forecasting methods in energy planning papers based on countries

## 4.3 Fuzzy Forecasting Methods

### *4.3.1 Fuzzy Time Series*

A sequence of data is called time series if they are listed in time order. A time series refers to a sequence taken at successive equally spaced points in time. In statistics, time series models assume which forecast for the next time interval can be made

**Table 4.1** Summarized literature review

| Author (Year) | Subject | Method |
|---|---|---|
| Piras et al. (1995) | Electrical load forecasting | ANNs and weighted fuzzy average |
| Mori and Kobayashi (1996) | STFL | Simplified fuzzy inference |
| Cartina et al. (2000) | Peak load estimation | Nonlinear fuzzy regression approach |
| Tranchita and Torres (2004) | STFL | CART classification, fuzzy inference, LAMDA-fuzzy-clustering, regression trees |
| Hayati and Karami (2005) | STLF | Hybrid neuro-fuzzy network |
| Zhao et al. (2006) | STLF | ANN and fuzzy theory |
| Lou et al. (2008) | Wind power | Fuzzy clustering method and similarity theory |
| Holmukhe et al. (2010) | STLF | Fuzzy logic |
| Liu et al. (2010) | STLF | FTS |
| Neto et al. (2011) | Electricity production | FTS and optimization techniques |
| Li and Choudhury (2011) | Load forecast | Load model combined fuzzy and probabilistic |
| Bolturk et al. (2012) | The consumption of electricity | FTS |
| Oztaysi and Ucal Sari (2012) | The forecasting of electricity demand | Seasonal FTS |
| Akdemir and Cetinkaya (2012) | Long-term load forecasting | ANFIS |
| Zahedi et al. (2013) | Electricity demand | Adaptive neuro-fuzzy network |
| Jain and Jain (2013) | STLF | Particle swarm optimization |
| Chen et al. (2013) | Solar radiation forecast | Fuzzy and neural networks |
| Bain and Baracli (2014) | Energy demand | ANFIS |
| Oztaysi and Bolturk (2014) | Literature review on fuzzy forecasting | FTS, fuzzy regression, fuzzy inference systems, ANFIS |
| Li et al. (2015) | STLF | FTS forecasting method |
| Atsalakis et al. (2015) | Energy export forecast | ANFIS |
| Zhang et al. (2016) | Get daily average wind power interval forecasts | ANFIS |

(continued)

**Table 4.1** (continued)

| Author (Year) | Subject | Method |
|---|---|---|
| Okumus and Dinler (2016) | One h ahead wind speed forecasts | ANFIS and ANN |
| Matthew and Satyanarayana (2016) | STFL | Fuzzy logic |
| Son and Kim (2017) | Electricity demand | SVR and PSO algorithm |
| Zhang et al. (2017) | Battery energy storage system | Variable-interval reference signal optimization approach |
| Arcos-Aviles et al. (2017) | Battery | Fuzzy logic |
| Chahkoutahi and Khashei (2017) | Load forecast | ANFIS and seasonal ARIMA |

using the past set of values observed at the same time interval (Kahraman et al. 2010). Traditional time series model has been extended to fuzzy sets. Song and Chissom (1993) provide one of the initial FTS and propose a technique for linguistic data using fuzzy relation equations.

Let $Y(u)(u = ..0, 1, 2, 3..)$ be a subset of R1, the universe of discourse. Fuzzy sets $f_i(tu)(i = 1, 2, 3, \ldots)$ are identified on universe of discourse. A FTS, F (u), on Y (u) is defined as a collecting of $f_i(u), f_2(u), \ldots f_n(u)$. If F (u) is affected only by F (u − 1), it is represented by $F(u-1) \to F(u)$. In this case we can say that there is a fuzzy relationship between $F(u)$ and $F(u-1)$, and it can be shown as in Eq. 4.1. The relation R shown in Eq. 4.1, is called the fuzzy relation between $F(u)$ and $F(u-1)$.

$$F(u) = F(u-1) \circ \mathrm{R}(\mathrm{u}, \mathrm{u}-1) \tag{4.1}$$

Song and Chissom (1993) define two types of FTSs. If relation R (u, u − 1) is individual of u, so F (u) is called a time-changing FTS, in other cases, the relation is a time-variant FTS. Assume a FTS, F(u), is stired up by F(u − 1), F(u − 2),…, and F(u − n). Equation 4.2 epitomise this fuzzy relationship (FLR) is defined the nth order FTS forecasting mockup.

$$F(u-n), \ldots, F(u-2), F(u-1) \to= F(u). \tag{4.2}$$

Chen (1996) outlines the steps of FTS forecasting chniques. Process starts with partitioning the universe of discourse into exact intervals. Then historical data is fuzzified. The third step is to build the fuzzy relationship between historical fuzzy data. Finally, the forecasts are calculated using this relation.

### *4.3.2 Fuzzy Regression*

Regression analysis is a statistical technique which is studied on widely. The technique focuses on exploring and modeling the relationship between an output factor and input factors. Traditional statistical linear fixation model is as given in 3.

$$y(x) = \beta_0 + \beta_1 x_{i1} + \ldots \beta_k x_{ik} + \varepsilon_i, i = 1, 2, \ldots, m \tag{4.3}$$

In this Eq. 4.3, y(x) is the output variable, $x_{ij}$ are the input variables. In the equation, $\beta$ j represents the coefficients of the formula and $\varepsilon_i$ shows the random error term. In the original technique, all of the parameters, coefficients and variables are crisp numbers.

The classical technique is widely adopted both by the academia and the professionals. However, Shapiro (2004) reports some shortcomings of the classical model for example an insufficient couple of observing, or missing data. Fuzzy regression models are proposed to overcome these issues. The literature provides various fuzzy regression models (Georg 1994; Sakawa and Hitoshi 1992; Tanaka et al. 1989; Wang and Tsaur 2000). In this chapter, we briefly explain Buckley's fuzzy regression model (Buckley 2004) study. In this model, the forecast is done based on confidence intervals. Buckley defines the fuzzy regression model is as in the following:

$$y(x) = a + b(x_i - \overline{x}) + \varepsilon_i \tag{4.4}$$

In this equation $\overline{x}$ shows the mean value of $x_i$. Confidence intervals of a, b and $\sigma$ are obtained using crisp numbers according to the technique $(1 - \beta)$ 100%. For this purpose the crisp estimators of the coefficients $\left(\widehat{a}, \widehat{b}\right)$ should be found. $\widehat{a} = \overline{y}$ and $\widehat{b} = \frac{B1}{B2}$ are the values of the estimators where

$$B1 = \sum_{i=1}^{n} y_i(x_i - \overline{x}) \tag{4.5}$$

$$B2 = \sum_{i=1}^{n} (x_i - \overline{x})^2 \tag{4.6}$$

$$\sigma^2 = \left(\frac{1}{n}\right) \sum_{i=1}^{n} \left[y_i - \widehat{a} - \widehat{b}(x_i - \overline{x})\right]^2 \tag{4.7}$$

Using the above mentioned Equations, $(1 - \beta)100\%$ confidence interval for a and b are obtained using 4.8 and 4.9:

$$\widehat{a} - t_{\frac{\beta}{2}} \sqrt{\frac{\widehat{\sigma}^2}{(n-2)}}, \widehat{a} + t_{\frac{\beta}{2}} \sqrt{\frac{\widehat{\sigma}^2}{(n-2)}} \tag{4.8}$$

$$\widehat{b} - t_{\frac{\beta}{2}} \sqrt{\frac{n\widehat{\sigma}^2}{(n-2)\sum_{i=1}^{n} (x_i - \widehat{x})^2}}, \widehat{b} + t_{\frac{\beta}{2}} \sqrt{\frac{n\widehat{\sigma}^2}{(n-2)\sum_{i=1}^{n} (x_i - \widehat{x})^2}} \tag{4.9}$$

The fuzzy regression equation is can be written as;

$$\widetilde{y}(x) = \widetilde{a} + \widetilde{b}(x_i - \overline{x}) \tag{4.10}$$

$\widetilde{y}(x)$, $\widetilde{a}$ and $\widetilde{b}$, are fuzzy, and $x$ and $\overline{x}$ are crisp in (4.10). For prediction, new fuzzy values for dependent variable can be calculated by new $x$ values. Using the interval arithmetic and ($\alpha$)-cut operationthe predictions can be obtain using 4.11.

$$\widetilde{Y}[x](\alpha) = \begin{cases} y(x)_1(\alpha) = \alpha_1(\alpha) + (x - \widetilde{x})b_1(\alpha) & \textit{if } (x - \overline{x}) > 0 \\ y(x)_2(\alpha) = \alpha_2(\alpha) + (x - \widetilde{x})b_2(\alpha) & \\ y(x)_1(\alpha) = \alpha_1(\alpha) + (x - \widetilde{x})b_1(\alpha) & \textit{if } (x - \overline{x}) > 0 \\ y(x)_2(\alpha) = \alpha_2(\alpha) + (x - \widetilde{x})b_2(\alpha) & \end{cases} \tag{4.11}$$

### *4.3.3 Fuzzy Inference Systems*

Fuzzy inference systems (FIS) utilize expert evaluations expressed by rules and a reasoning mechanism for forecasting. Another usage of this label are "Fuzzy rule-based systems" or "fuzzy expert systems" (Jang et al. 1997). A set of rules, a database and an argumentative machine are three main components in FIS. The if-then rules used for reasoning is stored in the rule base. The membership functions used in these rules are stored in the database. The output of the system is obtained by the reasoning mechanism which uses rules and the given input values.

The expert evaluations about different conditions are transferred to the system by using fuzzy rules. The rules are defined by using if and then clauses. For example, "If the service is good then the tip is high" is an if-then rule representing the number of tips in a restaurant. Here, "service" and "tip" are linguistic variables, good and high are linguistic values.

Fuzzification, fuzzy rules, fuzzy inference and defuzzification are four steps of a typical FIS (Oztaysi et al. 2013). The first step involves fuzzification in which all crisp input data is transformed to fuzzy values. In the second step, the rules are obtained from the experts using linguistic terms and fuzzy operator. As the number linguistic variables and associated linguistic variable raise the amount of the rules rise exponentially. The third step is inference procedure which provides a conclusion based on rules above. The literature provides a various model for inference including Mamdani's model (Mamdani and Assilian 1975), Sugeno's model (Sugeno and Kang 1988) and Tsukamoto Fuzzy Model (Tsukamoto 1979). As the inference procedure obtains the results, the next step is defuzzification which transforms the fuzzy output to crisp values.

### *4.3.4 ANFIS*

ANFIS uses expert evaluations, a dataset, and a learning mechanism to provide a relationship between inputs and outputs (Jang 1993). The system utilizes Sugeno inference model and artificial neural networks (Yun et al. 2008). ANFIS learns the membership function parameters of linguistic variables utilizing input/output data set.

Jang (1993) defines the system with five layer feed forward neural network as follows:

1. 1st layer is composed of adaptive nodes which have a node function such as, $O_{1,i} = \mu_{A_i}(x)$ (for i = 1, 2). x shows input to node I, and lingual term is $A_i$. $O_{1,i}$ refers to the membership degree of a fuzzy set A.
2. 2nd layer is composed of fixed nodes which produce an output showing the firing strength of a rule.

$$O_{2,i} = w_i = \mu_{A_i}(x)\mu_{B_i}(y), \quad i = 1, 2. \tag{4.12}$$

3. 3th layer is composed of linked nodes labelled N. The ratio of the ith rule's firing strength is calculated in order to sum of all rules' firing strengths as shown in Eq S. This layer produces normalized firing strengths as an output.

$$O_{3,i} = \overline{w}_i = \frac{w_i}{w_1 + w_2}, \quad i = 1, 2. \tag{4.13}$$

4. 4th layer is composed of adaptive nodes with node functions given in 4.14. In this equation $\overline{w}_i$ shows a regularized firing strength from 3th layer and $p_i$, $q_i$, and $r_i$ show the parameter set for this node

$$O_{4,i} = \overline{w}_i f_i = \overline{w}_i(p_i x + q_i y + r_i) \tag{4.14}$$

5. Sum of all arriving signals is calculated by a single node as an overall output value in 5th layer.

$$overall\ output = O_{4,i} = \sum_i \overline{w}_i f_i = \frac{\sum_i w_i f_i}{\sum_i w_i} \tag{4.15}$$

### 4.3.5 *Hwang, Chen, Lee's Fuzzy Time Series Method*

A FTS technique proposed by Hwang et al. (1998) is for handling forecasting problems. To find the future demands via Hwang, Chen, Lee's FTS technique (Hwang et al. 1998) and the demand is known by yearly, the following steps are applied:

1. The variations are found in two consecutive years. For example the demand of year q is d and the demand of year p is e, then the variation is e–d.
2. Minimum increase $D_{min}$ and maximum increase $D_{max}$ are found.
3. Define the universe of discourse U, U = [$D_{min}$ − D1, $D_{max}$ + D2], (D1 and D2 are appropriate positive numbers.)
4. Partition the universe of discourse U into several even length intervals.
5. Deciding linguistic terms delineated by fuzzy sets. The linguistic terms for the intervals can be described as follows: Big decrease, decrease and, etc.
6. Fuzzifying values of historical data.
7. Determining an appropriate window basis w, and output is calculated from the operation matrix $O^w(T)$. Criterion matrix C(T), that T is year for which we want to forecast the data.
8. Fuzzy forecasted variations are defuzzified.
9. Final forecasted data is calculating by forecasted data plus the number of last year's actual data.

## 4.4 A Numerical Application

In this section we apply FTS Using Hwang, Chen, Lee's Method (Hwang et al. 1998) to energy forecasting problem. The actual spending values, variations and linguistic terms are given in Table 4.2.

Maximum variation = 6.504.394, and minimum variation is −3.268.290. Universe of discourse [−3.500.000, 7.000.000] The linguistic terms for the intervals: {Big decrease, decrease, no change, increase, big increase, very big increase, too big increase}.

When we try to forecast Month 22 with a window size 5, the calculations are as follows:

$$O^5(22) = \begin{bmatrix} 0 & 0 & 0 & 0.5 & 1 & 0.5 & 0 \\ 0 & 0 & 0 & 0.5 & 1 & 0.5 & 0 \\ 0 & 0.5 & 1 & 0.5 & 0 & 0 & 0 \\ 1 & 0.5 & 0 & 0 & 0 & 0 & 0 \\ 0 & 0 & 0 & 0.5 & 1 & 0.5 & 0 \end{bmatrix} \tag{4.16}$$

$$C(22) = [0 \quad 0,5 \quad 1 \quad 0.5 \quad 0 \quad 0 \quad 0] \tag{4.17}$$

**Table 4.2** Actual spending and variations of historical data

| Months | Actual spending | Variation | Linguistic terms |
|---|---|---|---|
| Month 1 | 16.247.156 | | |
| Month 2 | 14.253.194 | −1.993.962 | A2 |
| Month 3 | 20.757.588 | 6.504.394 | A7 |
| Month 4 | 20.192.911 | −564.677 | A2 |
| Month 5 | 19.218.643 | −974.268 | A2 |
| Month 6 | 18.242.756 | −975.887 | A2 |
| Month 7 | 21.051.875 | 2.809.119 | A5 |
| Month 8 | 18.017.455 | −3.034.420 | A5 |
| Month 9 | 19.851.723 | 1.834.268 | A4 |
| Month 10 | 18.690.696 | −1.161.027 | A2 |
| Month 11 | 18.259.420 | −431.276 | A2 |
| Month 12 | 21.060.161 | 2.800.741 | A5 |
| Month 13 | 20.228.593 | −831.568 | A2 |
| Month 14 | 22.324.932 | 2.096.339 | A4 |
| Month 15 | 28.007.488 | 5.682.556 | A7 |
| Month 16 | 30.911.740 | 2.904.252 | A5 |
| Month 17 | 34.024.478 | 3.112.738 | A5 |
| Month 18 | 34.038.857 | 14.379 | A3 |
| Month 19 | 30.770.567 | −3.268.290 | A1 |
| Month 20 | 33.148.083 | 2.377.516 | A5 |
| Month 21 | 32.873.582 | −274.501 | A3 |
| Month 22 | 33.862.352 | 988.770 | A3 |
| Month 23 | 30.933.481 | −2.928.871 | A1 |
| Month 24 | 34.080.836 | 3.147.355 | A5 |

**Table 4.3** Membership of months

| Months | Actual | Forecast |
|---|---|---|
| Month 22 | A3 | A3 |
| Month 23 | A1 | A3 |
| Month 24 | A5 | A5 |

By applying the steps

$$F(22) = 0 \quad 0.25 \quad 1 \quad 0.25 \quad 0 \quad 0 \quad 0 \tag{4.18}$$

Since the maximum membership belongs to A3 (No change) the forecast for the next period is A3.

When we apply the method to the remaining three periods, we get the values as in Table 4.3.

## 4.5 Conclusion

Fuzzy forecasting methods present excellent tools for forecasting the future when incomplete, vague, and imprecise data exist in the considered problem. Fuzzy time series, fuzzy regression, fuzzy inference systems, and ANFIS are the most used fuzzy forecasting techniques in the literature. Forecasts for energy costs and energy production levels of the future are excessively important for both the energy producers and the energy consumers. Fuzzy forecasting provides the limits of possibilities in case of incomplete and vague data and a wider and deeper perspective of the uncertain future.

For further research, the recent extensions of fuzzy sets can be employed in the forecasting methods. Intuitionistic fuzzy forecasting techniques, hesitant fuzzy forecasting techniques, type-2 fuzzy forecasting techniques, and Pythagorean fuzzy forecasting techniques are possible research areas for future work directions.

**Acknowledgements** As a specialist in Information Technologies Department, Eda Bolturk thanks İstanbul Takas ve Saklama Bankası A.S. for getting support for this study.

## References

Akdemir, B., & Cetinkaya, N. (2012). Long-term load forecasting based on adaptive neural fuzzy inference system using real energy data. *Energy Procedia, 14,* 794–799.

Arcos-Aviles, D., Pascual, J., Guinjoan, F., Marroyo, L., Sanchis, P., & Marietta, M. P. (2017). Low complexity energy management strategy for grid profile smoothing of a residential grid-connected microgrid using generation and demand forecasting. *Applied Energy, 205,* 69–84.

Atsalakis, G., Frantzis, D., & Zopounidis, C. (2015). Energy's exports forecasting by a neuro-fuzzy controller. *Energy Systems, 6*(2), 249–267.

Bain, A., & Baracli, H. (2014). Modeling potential future energy demand for Turkey in 2034 by using an integrated fuzzy methodology. *Journal of Testing and Evaluation, 42*(6), 1466–1478.

Bisht, K., & Kumar, S. (2016). Fuzzy time series forecasting method based on hesitant fuzzy sets. *Expert Systems with Applications, 64,* 557–568.

Bolturk, E., Oztaysi, B., & Sari, I. U. (2012). Electricity consumption forecasting using fuzzy time series. In: *2012 IEEE 13th International Symposium on Computational Intelligence and Informatics (CINTI)*, pp. 245–249.

Buckley, J. J. (2004). *Fuzzy statistics*. Heidelberg: Springer.

Cartina, G., Alexandrescu, V., Grigoras, G., & Moshe, M. (2000). Peak load estimation in distribution networks by fuzzy regression approach, In: *Proceedings of the Mediterranean Electrotechnical Conference—MELECON* (Vol. 3, pp. 907–910).

Carvalho, J. G., & Costa, C. T. (2017). Identification method for fuzzy forecasting models of time series. *Applied Soft Computing, 50,* 166–182.

Chahkoutahi, F., & Khashei, M. (2017). A seasonal direct optimal hybrid model of computational intelligence and soft computing techniques for electricity load forecasting. *Energy, 140,* 988–1004.

Chen, S. X., Gooi, H. B., & Wang, M. Q. (2013). Solar radiation forecast based on fuzzy logic and neural networks. *Renewable Energy, 60,* 195–201.

Chen, S. M. (1996). Forecasting enrollments based on fuzzy time series. *Fuzzy Sets and Systems, 81*(3), 311–319.

Georg, P. (1994). Fuzzy linear regression with fuzzy intervals. *Fuzzy Sets and Systems, 63*(1), 45–55.

Hayati, M., & Karami, B. (2005). Application of computational intelligence in short-term load forecasting. *WSEAS Transactions on Circuits and Systems, 4*(11), 1594–1599.

Holmukhe, R. M., Dhumale, S., Chaudhari, P. S., & Kulkarni, P. P. (2010). Short term load forecasting with fuzzy logic systems for power system planning and reliability-a review. *AIP Conference Proceedings, 1298*(1), 445–458.

Hwang, J.-R., Chen, S.-M., & Lee, C.-H. (1998). Handling forecasting problems using fuzzy time series. *Fuzzy Sets and Systems, 100,* 217–228.

Jain, A., & Jain, M. B. (2013). Fuzzy modeling and similarity based short term load forecasting using swarm intelligence—a step towards smart grid. *Advances in Intelligent Systems and Computing, 202,* 15–27.

Jang, J. S. R. (1993). ANFIS: Adaptive-network-based fuzzy inference system. *Transactions on Systems, Man, and Cybernetics, 23*(3), 665–685.

Jang, J. S. R., Sun, C. T., & Mizutani, E. (1997). *Neuro-fuzzy and soft computing: A computational approach to learning and machine intelligence*. New Jersey: Prentice Hall.

Kahraman, C., Oztaysi, B., & Cevik, Onar S. (2016). A comprehensive literature review of 50 years of fuzzy set theory. *International Journal of Computational Intelligence Systems, 9,* 3–24.

Kahraman, C., Yavuz, M., & Kaya, I. (2010). Fuzzy and grey forecasting techniques and their applications in production systems. In C. Kahraman & M. Yavuz (Eds.), *Production engineering and management under fuzziness* (pp. 1–24). Heidelberg: Springer.

Khosravi, S. A., Jaafar, J., & Khanesar, M. A. (2016). A systematic design of interval type-2 fuzzy logic system using extreme learning machine for electricity load demand forecasting. *International Journal of Electrical Power & Energy Systems, 82,* 1–10.

Kumar, S., & Gangwar, S. S. (2015). A fuzzy time series forecasting method induced by intuitionistic fuzzy sets. *International Journal of Modeling, Simulation, and Scientific Computing, 6*(4).

Li, H., Zhao, Y., Zhang, Z., & Hu, X. (2015). Short-term load forecasting based on the grid method and the time series fuzzy load forecasting method. In: *International Conference on Renewable Power Generation (RPG 2015)* (pp. 1–6).

Li, W., & Choudhury, P. (2011). Including a combined fuzzy and probabilistic load model in transmission energy loss evaluation: Experience at BC hydro, In: *IEEE Power and Energy Society General Meeting* (pp. 1–8).

Liu X., Bai E., Fang J., Luo L. (2010). Time-variant slide fuzzy time-series method for short-term load forecasting. In: *Proceedings—2010 IEEE International Conference on Intelligent Computing and Intelligent Systems, ICIS 2010* (pp. 65–68).

Lou, S., Li, Z., & Wu, Y. (2008). Clustering analysis of the wind power output based on similarity theory. In: *3rd International Conference on Deregulation and Restructuring and Power Technologies, DRPT 2008* (pp. 2815–2819).

Mamdani, E. H., & Assilian, S. (1975). An experiment in linguistic synthesis with a fuzzy logic controller. *International Journal of Man-Machine Studies, 7*(1), 1–13.

Matthew, S., & Satyanarayana, S. (2016). An overview of short term load forecasting in electrical power system using fuzzy controller. In: *2016 5th International Conference on Reliability, Infocom Technologies and Optimization, ICRITO 2016: Trends and Future Directions* (pp. 296–300).

Mori, H., & Kobayashi, H. (1996). Optimal fuzzy inference for short-term load forecasting. *IEEE Transactions on Power Systems, 11*(1), 390–396.

Neto, J. C. D. L., da Costa Junior, C. T., Bitar, S. D. B., & Junior, W. B. (2011). Forecasting of energy and diesel consumption and the cost of energy production in isolated electrical systems in the Amazon using a fuzzification process in time series models. *Energy Policy, 39*(9), 4947–4955.

Okumus, I., & Dinler, A. (2016). Current status of wind energy forecasting and a hybrid method for hourly predictions. *Energy Conversion and Management, 123,* 362–371.

Oztaysi, B., Behret, H., Kabak, O., Sari, I. U., & Kahraman, C. (2013). Fuzzy inference systems for disaster response. In J. Montero, B. Vitoriano, & D. Ruan (Eds.), *Decision aid models for disaster management and emergencies*. San Diego: Atlantis Press.

Oztaysi, B., & Bolturk, E. (2014). Fuzzy methods for demand forecasting in supply chain management. *Supply chain management under fuzziness, 312,* 243–268.

Oztaysi, B., & Sari, I. U. (2012). Forecasting energy demand using fuzzy seasonal time series. *Computational Intelligence Systems in Industrial Engineering, 6,* 251–269.

Piras, A., Germond, A., Buchenel, B., Imhof, K., & Jaccard, Y. (1995). Heterogeneous artificial neural network for short term electrical load forecasting. *IEEE Transactions on Power Systems, 11*(1), 397–402.

Sakawa, M., & Hitoshi, Y. (1992). Multiobjective fuzzy linear regression analysis for fuzzy input-output data. *Fuzzy Sets and Systems, 47*(2), 173–181.

Son, H., & Kim, C. (2017). Short-term forecasting of electricity demand for the residential sector using weather and social variables. *Resources, Conservation and Recycling, 123,* 200–207.

Song, Q., & Chissom, B. S. (1993). Forecasting enrollments with fuzzy time series. *Fuzzy Sets and Systems, 54*(1), 1–9.

Sugeno, M., & Kang, G. T. (1988). Structure identification of fuzzy model. *Fuzzy Sets and Systems, 28*(1), 15–33.

Tanaka, H., Isao, H., & Junzo, W. (1989). Possibilistic linear regression analysis for fuzzy data. *European Journal of Operational Research, 40*(3), 389–396.

Tranchita, C., & Torres, Á. (2004). Soft computing techniques for short term load forecasting. In: *2004 IEEE PES Power Systems Conference and Exposition* (pp. 497–502).

Tsukamoto, Y. (1979). An approach to fuzzy reasoning method. In M. M. Gupta & R. R. Yager (Eds.), *Advances in fuzzy set theory and applications*. Amsterdam: North-Holland.

Wang, H.-F., & Tsaur, R.-C. (2000). Resolution of fuzzy regression model. *European Journal of Operational Research, 126*(3), 637–650.

Yun, Z., Quan, Z., Caixin, S., Shaolan, L., Yuming, L., & Yang, S. (2008). RBF neural network and ANFIS-based short-term load forecasting approach in real-time price environment. *IEEE Transactions on Systems, Man, and Cybernetics, 23*(3), 853–858.

Zadeh, L. A. (1965). Fuzzy sets. *Information and Control, 8,* 338–358.

Zahedi, G., Azizi, S., Bahadori, A., Elkamel, A., & Wan Alwi, S. R. (2013). Electricity demand estimation using an adaptive neuro-fuzzy network: A case study from the Ontario province—Canada. *Energy, 49*(1), 323–328.

Zhang, F., Meng, K., Xu, Z., Dong, Z., Zhang, L., Wan, C., et al. (2017). Battery ESS planning for wind smoothing via variable-interval reference modulation and self-adaptive SOC control strategy. *IEEE Transactions on Sustainable Energy, 8*(2), 695–707.

Zhang, Z., Song, Y., Liu, F., & Liu, J. (2016). Daily average wind power interval forecasts based on an optimal adaptive-network-based fuzzy inference system and singular spectrum analysis. *Sustainability (Switzerland), 8*(2), 1–30.

Zhao, Y., Tang, Y., & Zhang, Y. (2006). Short-term load forecasting based on artificial neural network and fuzzy theory. *Gaodianya Jishu/High Voltage Engineering, 32*(5), 107–110.

# Chapter 5
# Smart Storage Scheduling and Forecasting for Peak Reduction on Low-Voltage Feeders

**Timur Yunusov, Georgios Giasemidis and Stephen Haben**

**Abstract** The transition to a low carbon economy will likely bring new challenges to the distribution networks, which could face increased demands due to low-carbon technologies and new behavioural trends. A traditional solution to increased demand is network reinforcement through asset replacement, but this could be costly and disruptive. Smart algorithms combined with modern technologies can lead to inexpensive alternatives. In particular, battery storage devices with smart control algorithms can assist in load peak reduction. The control algorithms aim to schedule the battery to charge at times of low demand and discharge, feeding the network, at times of high load. This study analyses two scheduling algorithms, model predictive control (MPC) and fixed day-ahead scheduler (FDS), comparing against a set-point control (SPC) benchmark. The forecasts presented here cover a wide range of techniques, from traditional linear regression forecasts to machine learning methods. The results demonstrate that the forecasting and control methods need to be selected for each feeder taking into account the demand characteristics, whilst MPC tends to outperform the FDS on feeders with higher daily demand. This chapter contributes in two main directions: (i) several forecasting methods are considered and compared and (ii) new energy storage control algorithm, MPC with half-hourly updated (rolling) forecasts designed for low voltage network application, is introduced, analysed and compared.

---

T. Yunusov (✉)
TSBE Centre, School of Build Environment, University of Reading,
Reading RG6 6AF, UK
e-mail: t.yunusov@reading.ac.uk

G. Giasemidis
CountingLab LTD and CMoHB, University of Reading,
Reading RG6 6AX, UK

S. Haben
Mathematical Institute, University of Oxford, Oxford OX2 6GG, UK

© Springer International Publishing AG, part of Springer Nature 2018

C. Kahraman and G. Kayakutlu (eds.), *Energy Management—Collective and Computational Intelligence with Theory and Applications*, Studies in Systems, Decision and Control 149, https://doi.org/10.1007/978-3-319-75690-5_5

## 5.1 Introduction

In a low carbon economy, storage devices are going to be an essential component of any future energy network. With the electrification of heating and transport and the move from fossil fuels to more renewable, distributed sources of generation, there are increasing requirements for new and novel network solutions (Evans 2016).

Battery energy storage systems (BESS) are one potential solution, and can be deployed to control the networks, ameliorate the disruption caused by low carbon technologies and increase flexibilities. Although BESS have traditionally been an expensive solution, the rapid reduction in cost in recent years is beginning to make them competitive with traditional reinforcement such as replacing existing assets (Schmidt et al. 2014). BESS can be deployed for a wide variety of network solutions including, demand smoothening, voltage control, and phase balancing (Nair and Garimella 2010). In this chapter, the primary focus will be on the application of peak demand reduction for the purposes of increasing the network headroom and maintaining the thermal capacity of the network (Joshi and Pindoriya 2015).

To be optimally utilised, the storage devices must accurately anticipate the future state of the system and be ready to respond to the stochastic nature of the power flows. In particular, this means that the optimal control of the BESS must incorporate accurate forecasts of the network in addition to the battery constraints (Megel et al. 2015; Rowe et al. 2014). At the low voltage (LV) level, this is much more challenging than at the higher voltage levels because of the increased volatility and irregularity of demand (Yunusov et al. 2017).

In this chapter, the focus is on utilising forecast techniques within battery control methodologies with the aim to reduce the peak demand on low voltage substation feeders. First, a number of forecasting methods are developed and analysed and then incorporated into the control algorithms which are then applied to a simulated BESS and applied to monitoring data from real LV feeders. Several control methodologies are considered, namely set-point control, a fixed day-ahead scheduler and two variants on the model predictive control.

## 5.2 Forecasting Methods

### 5.2.1 Data

This chapter uses the demand data from the low voltage feeders, monitored as part of the New Thames Valley Vision (Scottish and Southern Energy Power Distribution: New Thames Valley Vision 2014) project, in the Bracknell area, UK. The feeders typically supply electricity for about 10 to over 150 generally residential customers with a half hourly demand of no more than around 40 kWh.

**Table 5.1** Summary features for the feeders considered for the analysis in this chapter

| Feeder | Mean demand | STD | Max demand |
|---|---|---|---|
| S1 | 7.57 | 3.32 | 36.20 |
| S2 | 5.42 | 2.89 | 30.00 |
| S3 | 11.15 | 5.86 | 42.14 |
| M1 | 11.11 | 6.05 | 64.65 |
| M2 | 16.43 | 11.14 | 63.60 |
| M3 | 21.58 | 9.40 | 111.50 |
| L1 | 30.97 | 14.43 | 204.00 |
| L2 | 37.13 | 18.04 | 205.15 |
| L3 | 24.26 | 9.74 | 110.00 |

This includes average half hourly demand, standard deviation of demand, the maximum recorded half-hourly demand (measured over a year). All values are in kWh

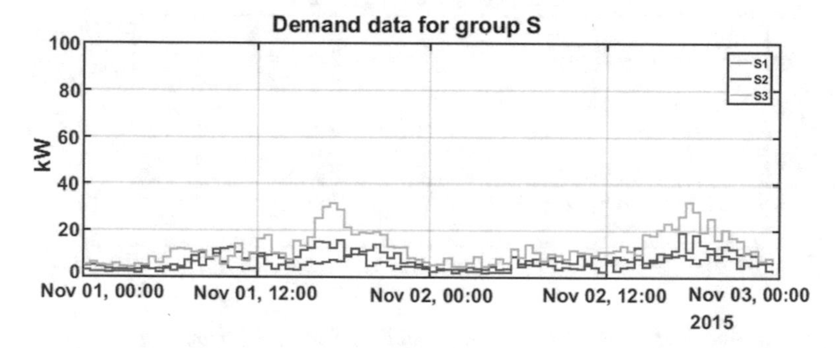


**Fig. 5.1** Example of demand for feeders in group S (Sunday and Monday)

For the methods we demonstrate in this chapter a number of feeders of varying mean demand and volatility (through the standard deviation of the load) are selected. Some of the attributes of the chosen feeders are presented in Table 5.1. The chosen feeders are roughly labelled according to their size with S1, S2, S3 representing the smaller feeders, M1, M2, M3 the medium sized feeders and L1, L2, L3 the larger feeders. Examples of the demands for each of these feeders is shown in Figs. 5.1, 5.2 and 5.3. Notice that feeder L2 is dominated by commercial customers and hence there is a very low demand on the Sunday (1st November 2015).

The monitored data for the selected feeders consists of half hourly energy demand (kWh) for the period from 10th March 2014 to 15th November 2015. A two-week period, from 1st to 14th of November 2015, is used as a test period for assessing the storage control algorithms with the remaining data used for parameter selection (via cross-validation) and training of the forecast models.

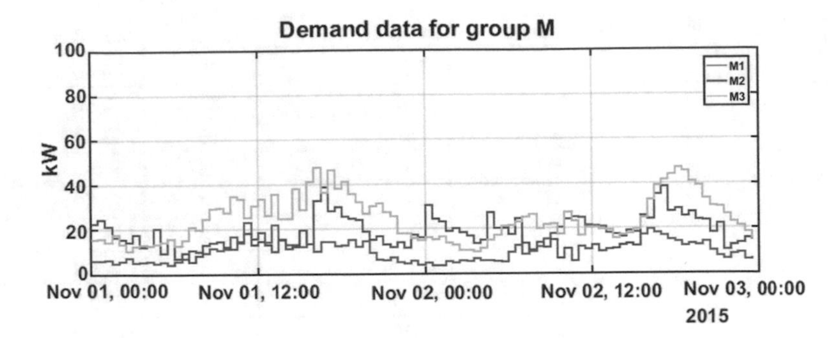


**Fig. 5.2** Example of demand for feeders in group M (Sunday and Monday)

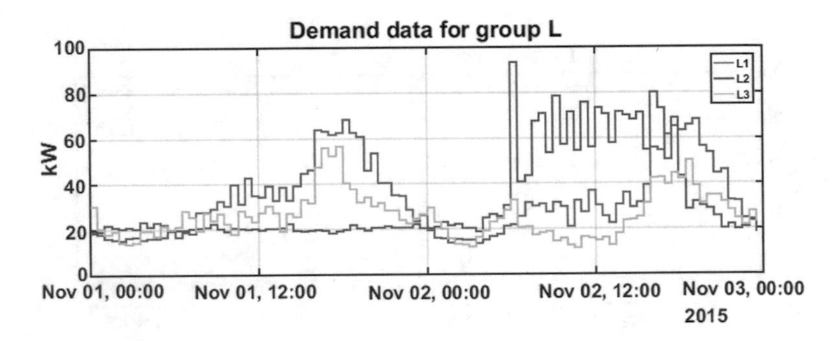


**Fig. 5.3** Example of demand for feeders in group L (Sunday and Monday)

## 5.2.2 Methods

Several forecast methods for the scheduling algorithms are considered, which we describe in following sections. First, useful notation and terminology is introduced.

Let $d(t)$ be the monitored demand at time-step $t$ and $d_f(t)$ is the predicted value at the same time-step (in our case half hourly). Given a time-series of demand observations, $\mathbf{D} = \{d(1), \ldots, d(N)\}$, the forecast value at horizon $h$ from the last available observation at time step $t = N$ is given by $d_f(N+h)$.

Some forecasts are referred to as fixed or daily forecasts, because they are updated at the beginning of each day in the test period, i.e. using all observations until the last half-hour of the previous day, for horizons $h = 1, \ldots, 96$, i.e. up to two days ahead. In contrast, a rolling forecast is updated half-hourly, i.e. using all observations up to the most recent half-hour, and provides a forecast for horizons $h = 1, \ldots, 96$.

#### 5.2.2.1 A Simple Seasonal Method

The method is based on an update of the simple seasonal model presented in (Haben and Giasemidis 2016). Suppose a given time series $d(t)$, for the feeder load at time-step $t = 1, \ldots, N = D \cdot H$, where $D$ is the number of days in thc training set and $H$ is resolution of the data, i.e. the number of time-steps in the day. In the case considered here the data is half-hourly and so $H = 48$. First a mean model, $\mu(t)$, is constructed of the form

$$\mu(t) = \sum_{k=1}^{H} \mathcal{D}_k(t)\left(a_k + b_k\eta(t) + \sum_{p=1}^{P} c_k^p \sin\left(\frac{2\pi p\eta(t)}{365}\right) + d_k^p \cos\left(\frac{2\pi p\eta(t)}{365}\right)\right) + \sum_{l=1}^{7H} f_l \mathcal{W}_l(t). \tag{5.1}$$

Here $\eta(k) = \lfloor \frac{t}{H} \rfloor + 1$, is the day of the trial and there are two dummy variables identifying the period of the day, $\mathcal{D}_k(t)$, and the period of the week, $\mathcal{W}_l(t)$, defined by

$$\mathcal{D}_j(t) = \begin{cases} 1, & t \bmod H = j, \\ 0, & \text{otherwise}, \end{cases}$$

and

$$\mathcal{W}_j(t) = \begin{cases} 1, & t \bmod 7H = j, \\ 0, & \text{otherwise}, \end{cases}$$

respectively. Thus, essentially there are 336 models rcprcscnting each period of the week. The $a_k$ terms represent the average demand for that half hourly period (which is augmented based on the day of the week by $f_l$), a linear trend term $b_k$, and annual seasonality terms defined by $c_k$ and $d_k$. The coefficients are found by a least squares minimisation on the historical information. The model is labelled *ST*. A variation of this model is also considered by removing the trend term (i.e. $b_k = 0, \forall k$). We label this method *SnT* to indicate no linear trend is utilized.

These fixed forecasting methods are quite effective but are limited for intra-day forecasting since they are only updated daily (Yunusov et al. 2017). A standard method for improving these forecasts and also creating rolling forecasts is to update the forecast with autoregressive terms. Once the mean equations are found we create a residual timeseries defined by

$$r(t) = \sum_{m=1}^{M_{max}} \phi_m r(t-k) + \varepsilon(t), \tag{5.2}$$

where $r(t) = d(t) - \mu(t)$ and $\varepsilon(t)$ is the error term. The auto-regressive forecasts are easily found via the Yule-Walker equations. The optimal order $M_{max}$ is found by minimising the Akaike information criterion (AIC) for a maximum order of $m = 15H$ (i.e. we consider an optimal order of up to 15 days.). A rolling forecast of the residual time series is produced by applying the regression to a translating window of the historical data. The autoregressive terms define two further models which are labelled as *STAR* and *SnTAR* depending on if trend is included, or not respectively in the mean Eq. (5.1).

#### 5.2.2.2 Random Forest Regression

Random Forest is a popular machine learning method for classification and regression, based on an ensemble of decision trees (Breiman 2001; Breiman et al. 1984). The ensemble consists of a number of estimators, the number of decision trees in the forest, each of which uses a bootstrap sample of the observations with a subset of features. Each decision tree is considered a weak predictor, but the collection of the trees gives rise to an accurate value (regression) or class (classification). The number of trees in the ensemble is an important parameter and must be tuned to its optimal value via cross-validation. Random Forest Regression (*RFR*) has been considered for time-series prediction and particularly for short-term load forecasting in (Dudek 2014; Gajowniczek and Zbkowski 2017; Lahouar and Slama 2015).

Most of the machine learning algorithms can be expressed in terms of the real-valued observations $Y = \{Y_1, \ldots, Y_N\}$ and their features $X = \{\mathbf{X}_1, \ldots, \mathbf{X}_N\}$, where $\mathbf{X}_i = (x_1, \ldots, x_m) \in \mathbb{R}^m$ is the feature vector for the observation $i$. For the case study presented here, $Y = \{d(1), \ldots, d(N)\}$. The features of each observation are split into three main categories. To forecast time $t$ at horizon $h$, i.e. last monitored value is at time $t - h$, the features of the model are:

- Lag. The $H = 48$ (a day) past values are considered giving the lag features $\mathbf{X}_t^{(1)}(h) = d(t-h), d(t-h-1), \ldots, d(t-h-47)$.
- Past weeks. The load at the same time of the week for the past four weeks. The features are $\mathbf{X}_t^{(2)} = (d(t-n_w), d(t-2n_w), d(t-3n_w), d(t-4n_w))$, where $n_w$ is the number of observations in a week-period. For load data at half-hourly resolution, $n_w = 336$.
- The time of the day. This is a scalar feature, $1 \le X_t^{(3)} \le H$.

The final feature vector for an observation at time $t$ at horizon $h$ is

$$\mathbf{X}_t(h) = \left( \mathbf{X}_t^{(1)}(h), \quad \mathbf{X}_t^{(2)}, \quad X_t^{(3)} \right).$$

To select the optimal number of trees in the random forest, a validation period of one week prior to the test period is used. Ensembles with varying number of trees from 5 to 100 in increments of 5 are considered. Figure 5.4 shows the error

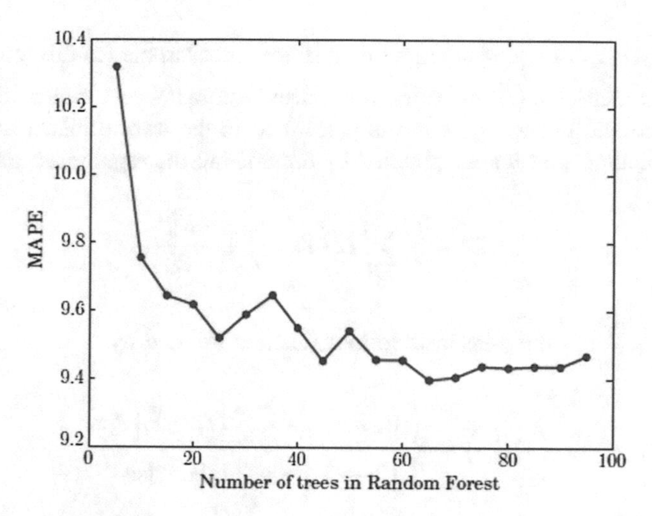


**Fig. 5.4** MAPE error score as a function of the number of trees in the ensemble for one-half-hour ahead forecast of feeder L2

(MAPE) as a function of the number of trees for feeder L2. Similar results hold for the other feeders too. Clearly the MAPE error steeply decreases as the number of trees increases beyond a few tens of trees and then fluctuates around a constant value for a large number of trees. In conclusion, considering all feeders, a value of 30 trees in the ensemble of a Random Forest is optimal between forecasting accuracy and performance, as larger numbers of trees make the algorithm computationally expensive.

Using the optimal number of trees, the final Random Forest forecast is trained using one year prior to the test period, 1st November 2014 to 31st October 2015. A total of 96 Random Forest Regression models are trained, one for each horizon in the test period.

#### 5.2.2.3 Support Vector Regression

Support Vector Regression (*SVR*) is a popular machine learning method used for time-series prediction (Hong 2009; Hu et al. 2013; Sapankevych and Sankar 2009; TIAN et al. 2004; Trkay and Demren 2011). As defined in the previous section, given a set of observations and features $\{(Y_i, \mathbf{X}_i)\}, i = 1, \ldots, N$, the SVR algorithm aims to fit a linear (5.3), or non-linear (5.4) regression function $f$ (Drucker et al. 1997; Vapnik 1995),

$$\widehat{Y} = f(\mathbf{X}) = \mathbf{w} \cdot \mathbf{X} + b, \tag{5.3}$$

$$\widehat{Y} = f(\mathbf{X}) = \mathbf{w} \cdot \phi(\mathbf{X}) + b, \tag{5.4}$$

with $\widehat{Y} = \left\{ \widehat{Y}_1, \ldots, \widehat{Y}_N \right\}$. If the input data is non-linear in the feature space, the use of a kernel function, $\phi(\cdot)$, maps the non-linear features $\mathbf{X}$ to a higher dimensional feature space and linear regression is performed in the transformed feature space. The coefficients $\mathbf{w}$ and $b$ are estimated by minimising the regularised risk function

$$R(C) = \frac{C}{N} \sum_{i=1}^{N} L_{\varepsilon}\left(Y_i, \widehat{Y}_i\right) + \frac{\|\mathbf{w}\|^2}{2}, \tag{5.5}$$

where $L_{\varepsilon}\left(Y_i, \widehat{Y}_i\right)$ is the ε-insensitive loss function defined by

$$L_{\varepsilon}\left(Y_i, \widehat{Y}_i\right) = \begin{cases} 0, & \left|Y_i - \widehat{Y}_i\right| \leq \varepsilon, \\ \left|Y_i - \widehat{Y}_i\right| - \varepsilon, & \text{otherwise.} \end{cases} \tag{5.6}$$

Support Vector Regression has two free parameters, $C$ and $\varepsilon$, that require tuning. $C$ is called as the regularisation constant and controls the flatness (or complexity) of the model, a trade-off between empirical error and model flatness, and $\varepsilon$ determines the amount of error allowed by the model. In addition, the choice of kernel $\phi(\cdot)$ is also important for the final model. The parameter selection is a challenging task and several advanced methods, based on evolutionary algorithms, have been developed to try and solve this problem, see Hong (2009), Hu et al. (2013). These methods go beyond the scope of this study and hence the parameters are instead found via a grid search. To simplify the task the error allowance term is set to $\varepsilon = 0.1$, and only three kernels are considered for the regression, a linear, a radial basis function (RBF) and a polynomial. The regularisation constant $C$ is restricted to vary from 0.1 to 100. The *RBF* kernel has an extra free parameter, $\gamma$, which controls the width of the kernel, and varies from 0.01 to 100. Finally, the degree of the polynomial kernel requires tuning too, changing from 2 to 5.

A validation of the results is performed using the week prior to the test-period. We find that the linear kernel outperforms the RBF and polynomial kernels for all values of the $C$ parameter. With the linear kernel, large values of $C > 20$ seem to reduce the model accuracy as shown in Fig. 5.5 for feeder L1. Similar conclusions hold for all feeders. As a result, the regularisation constant is fixed at $C = 1$ for all feeders.

Since the Support Vector Regression forecast is computationally more intensive than the Random Forest Regression, a shorter training period is used corresponding to eight weeks prior to the test period, i.e. 5th November 2015 to 31st October 2015. Similar to the Random Forest Regression forecast, 96 SVR models are trained, one for each horizon in the test period. This method will be denoted as the SVR method.

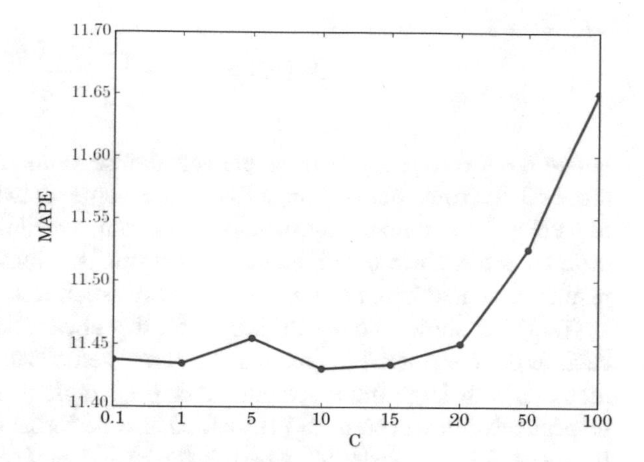


**Fig. 5.5** MAPE error score against the regularisation constant $C$ for one-half-hour ahead forecast of feeder L1

#### 5.2.2.4 Benchmark Methods

To understand the accuracy and main drivers of the forecast methods, informative benchmarks are required. In this section, standard but accurate forecast methods are described which will be used to compare to the more complicated methods described in the previous sections.

The first simple model is defined as

$$d(t) = \frac{1}{p}\sum_{k=1}^{p} d(t - kn_w), \tag{5.7}$$

where $n_w = 336$ is the number of time steps in a weekly period. In other words, a weekly average estimate is made for each week. Testing for $p = 1, ..., 8$ shows that using $p = 5$ weeks of data to construct the average produces the most accurate. The model is denoted *7SAV*. This model is mainly motivated by the fact that only recent past is important for the actual behaviour.

The other benchmark we consider is the persistence model, which is a special case of the average model above but only using the last week as the current week

$$d(t) = d(t - n_w). \tag{5.8}$$

This special case is denoted *LW*.

### 5.2.3 Analysis of Forecasts

In this section the accuracy of the forecasts is analysed. A standard error measure, the mean absolute percentage error (MAPE), is used given by

$$MAPE(\mathbf{a}, \mathbf{f}) = \frac{1}{n} \sum_{k=1}^{n} \frac{|a_k - f_k|}{|a_k|}, \tag{5.9}$$

where $\mathbf{a} = (a_1, \ldots, a_n)^T \in \mathbb{R}^n$ is the actual/observation and $\mathbf{f} = (f_1, \ldots, f_n)^T \in \mathbb{R}^n$ is the estimate/forecast. Although there are some drawbacks to this method, the MAPE gives a simple way of comparing and combining the errors of feeders of different sizes, since the differences are normalised by the size of the demand. The results presented here are for the entire two-week test period.

Table 5.2 shows the MAPE scores for day ahead forecasts for each method and each feeder considered in this trial. The best methods tend to be STAR and SnTAR methods with large improvements over the simple average versions ST and SnT respectively. However, the SVR method also performs well, being the best forecast for feeder S2. For feeder S3, which is the basic linear seasonal model without trend (SnT), adding the autoregressive effects (SnTAR) reduces the accuracy of the forecast. This suggests there is not a strong correlation with recent time periods for this. Another surprising result is the reasonably good performance of the simple average forecast, 7SAV. In addition, it is clear that some feeders are relatively easy to forecast compare to others. For example, feeders L1 and L2 all have very good forecasts, often with less than 12% errors. There does not seem to be a clear relationship between size of feeder and the accuracy of the forecast, although it has been shown that often larger feeders are on average easier to forecast (Yunusov et al. 2017). We expect the performance of the day ahead storage plan to perform best with the methods with the best forecast accuracy. On average over all the feeders the SnTAR method has the best day ahead forecast accuracy with a MAPE of 16.31%, just ahead of STAR with 16.40%. The average errors across all feeders are shown in Table 5.3.

**Table 5.2** MAPE for day ahead forecasts

| Feeder | Methods | | | | | | | |
|---|---|---|---|---|---|---|---|---|
| | LW | 7SAV | ST | STAR | SnT | SnTAR | RFR | SVR |
| S1 | 20.20 | 15.54 | 15.29 | **14.98** | 16.10 | 15.24 | 17.41 | 16.53 |
| S2 | 34.72 | 26.81 | 29.15 | 28.28 | 28.75 | 27.97 | 38.86 | **26.75** |
| S3 | 24.49 | 18.97 | 17.49 | 17.57 | **17.44** | 17.86 | 25.20 | 21.06 |
| M1 | 21.11 | 14.78 | 15.94 | **13.69** | 16.39 | 13.90 | 16.73 | 14.98 |
| M2 | 27.92 | 25.69 | 29.02 | 24.33 | 29.19 | **23.92** | 33.03 | 38.43 |
| M3 | 17.77 | 13.58 | 13.98 | 12.76 | 13.45 | **12.43** | 16.32 | 13.87 |
| L1 | 15.04 | 11.30 | 10.71 | **10.13** | 11.38 | 10.44 | 14.81 | 12.18 |
| L2 | 30.86 | **10.76** | 13.75 | 11.22 | 14.02 | 11.27 | 12.83 | 11.33 |
| L3 | 18.63 | 15.37 | 17.48 | 14.62 | 15.96 | **13.71** | 22.79 | 19.33 |

The best score for each feeder is highlighted in bold

Table 5.3 The average day-ahead MAPE for all 9 feeders

| Feeder | Methods | | | | | | | |
|---|---|---|---|---|---|---|---|---|
| | LW | 7SAV | ST | STAR | SnT | SnTAR | RFR | SVR |
| MAPE (%) | 23.42 | 16.98 | 18.09 | 16.40 | 18.08 | 16.31 | 21.01 | 19.38 |

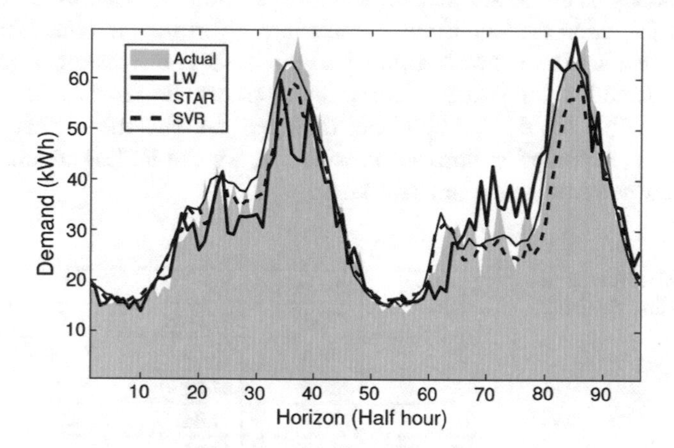


Fig. 5.6 Example of 2 day ahead forecasts for selected methods including benchmark LW for feeder L1. Actual demand is shaded

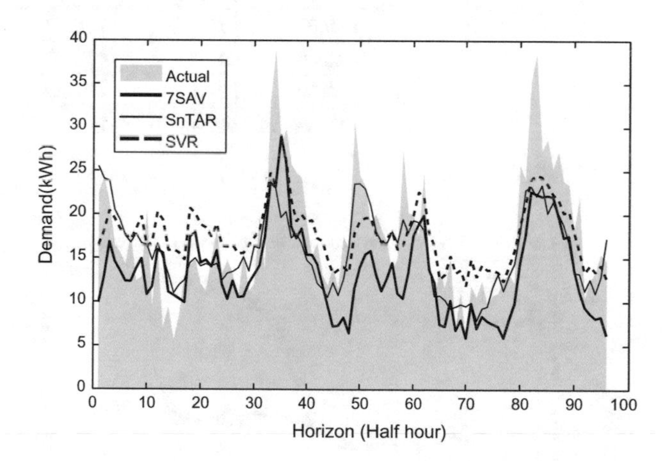


Fig. 5.7 Example of 2 day ahead forecasts for selected methods including benchmark 7SAV for feeder M2. Actual demand is shaded

Examples of two accurate day-ahead forecasts are shown in Fig. 5.6 for the first two days of the test set for feeder L1. Most methods scored MAPEs of less than 15% on average for this feeder. Even the naive LW forecasts performs satisfactorily. In contrast M2 is less accurately estimated, with errors greater than 20% MAPE for all methods. An example of the day ahead forecasts for M2 are shown in Fig. 5.7.

The forecast errors for each rolling forecast method are shown in Table 5.4 over all horizons (up to 96 half hours ahead) and starting from all time periods in the 14-day period. Similarly, to the day ahead forecasts the STAR and SnTAR methods are the best performing. This suggests that these methods should give the best peak reduction when implemented with the rolling MPC control.

The accuracy of the STAR method with forecast horizon is shown in more detail in Fig. 5.8 for three feeders. Similar results are given for the other feeders. The typical features are the reduced accuracy as the horizon increases before typically stabilising around a fixed value. As the figure shows the most rapid drop off in accuracy is within the first 10 half hour time periods. This means that within the rolling control, considering horizons beyond 5 h, we can be less certain about the demand then within the first few half hours.

**Table 5.4** MAPE over all horizons (rolling forecasts only)

| Feeder | Methods | | | |
|---|---|---|---|---|
| | STAR | SnTAR | RFR | SVR |
| S1 | **15.10** | 15.45 | 20.52 | 21.09 |
| S2 | 28.23 | **27.81** | 37.25 | 37.20 |
| S3 | **17.29** | 17.44 | 27.22 | 29.07 |
| M1 | **14.08** | 14.22 | 18.78 | 18.63 |
| M2 | 25.04 | **24.85** | 36.34 | 34.87 |
| M3 | 13.27 | **12.85** | 18.28 | 17.12 |
| L1 | **10.03** | 10.39 | 15.77 | 16.06 |
| L2 | **11.49** | 11.84 | 13.07 | 11.83 |
| L3 | 15.59 | **14.64** | 21.88 | 19.16 |

Best forecasts for each feeder are in bold

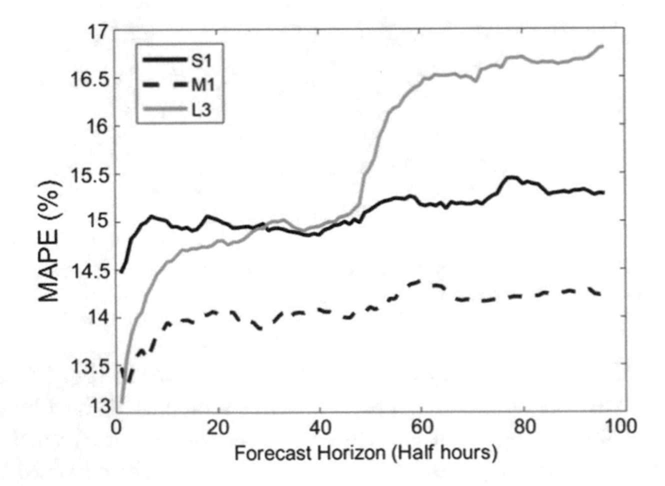


**Fig. 5.8** MAPE with Horizon for the STAR forecast method for 3 feeders

### 5.2.4 *Discussion*

From the analysis and comparison performed in the previous section, we observe that the seasonal method with auto-regressive terms often outperforms the machine learning methods, the SVR and RFR. The latter methods can be further improved in many ways.

First, both machine learning methods considered in this chapter have a few more parameters that require tuning. The RFR has several parameters that control the growth of the trees in the forest, such as the maximum depth of the trees, splitting criteria at the nodes, etc. The SVR requires tuning of the $\varepsilon$ parameter that controls the amount of error. A value too small and the model overfits the training observations, a value too large and the model misses essential patterns and correlations. Parameter selection for such methods is a non-trivial task and advanced evolutionary algorithms are often applied to the problem. However, this is beyond the scope of this chapter.

Second, the addition of further features, such as past mean daily demand, the day of the week, might also improve these models.

Finally, these models are quite accurate for one time-step ahead prediction. Instead of having one model per horizon, a single model can be trained on one time-step ahead observations and the forecast values can be used for the rolling forecast and horizons greater than one.

## 5.3 Application of Forecasts in Energy Storage Control

The control methodology for an energy storage system is governed by the energy storage technology, rating and capacity, location within the network levels and objective of the energy storage device. In this section, three control methods are applied to a simulated Battery Energy Storage System (BESS) sized to deal with the peak demand using demand data for a set of Low Voltage (LV) feeders. The peak reduction functionality for LV networks gives a relief to the thermal constraints in the LV feeder and the secondary transformer. Furthermore, choosing an appropriate location of the BESS on the LV feeder could also improve the voltage profile and reduce losses (Yunusov et al. 2016).

Three energy storage control algorithms are considered: Fixed Day-ahead Scheduler (FDS), Model Predictive Control (MPC) with persistent error model and MPC with rolling forecast model. As a benchmark method, Set-Point Control (SPC) is also considered. Each control methods is assessed on each individual day separately. The output of each control method is a charge-discharge schedule matching the temporal resolution of the forecasts (i.e. 30 min).

To ensure fair comparison of the control methods, the BESS is sized to deal with 20% of the actual peak demand on each day of the study period for each of the feeders (i.e. assuming perfect foresight for the purpose of determining the required

**Table 5.5** Power rating and energy storage capacity of BESS allocated to each feeder, given as a range for maximum and minimum values across the 14 days of the test period

| Feeder | kW | kWh |
|---|---|---|
| S1 | 3.98–5.9 | 2.55–16.17 |
| S2 | 5.78–8.32 | 3.2–27.9 |
| S3 | 10.2–12.84 | 11.08–39.27 |
| M1 | 8.26–11.64 | 11.64–85.32 |
| M2 | 11.04–15.54 | 7.25–71.61 |
| M3 | 16.04–19.56 | 9.56–82.3 |
| L1 | 21.7–32.26 | 76.02–108.72 |
| L2 | 9.34–37.36 | 74.72–362.12 |
| L3 | 16.66–22.58 | 25.26–80.64 |

BESS configuration). The range of BESS configuration values across all days for each feeder is given in Table 5.5. Each day in the study period is treated individually to assess the ability of the algorithms to deal with the peak of a day. BESS is assumed to be at 50% state-of-charge at the midnight.

### 5.3.1 Set-Point Control

SPC is a control method that dictates the energy storage when to charge and discharge in reaction to the measured signal crossing a particular threshold—the setpoint. In the case presented in this section, the controller is monitoring the power flow at the head of the feeder and maintains the power flow at a level specified by the set-point. If the power flow is above the set-point the controller instructs the BESS to inject power into the network to maintain the power flow as close to the set-point as possible. Same applies if the power flow is below the set-point consequently charging the BESS to its full capacity.

Operation of the SPC in subject to:

$$C_{min} \leq c(t) \leq C_{max}, \tag{5.10}$$

$$P_{min} \leq p(t) \leq P_{max}, \tag{5.11}$$

$$c(t+1) = c(t) + (p(t)\mu - \lambda)\tau, \tag{5.12}$$

$$\mu = \begin{cases} \mu, & \text{if } p(t) \geq 0, \\ \frac{1}{\mu}, & \text{if } p(t) < 0, \end{cases} \tag{5.13}$$

where $C_{min}$ and $C_{max}$ are the minimum and maximum capacity, respectively, in kWh, $P_{min}$ and $P_{max}$ are minimum and maximum power rating in kW for charge and discharge, respectively, $\mu$ is the efficiency (96% in each direction), $\lambda$ is the continuous losses within BESS (assumed to be 100 W), $\tau$ is the control period duration in hours (in this study 0.5), $p(t)$ is the BESS power scheduled for time step $t$ and

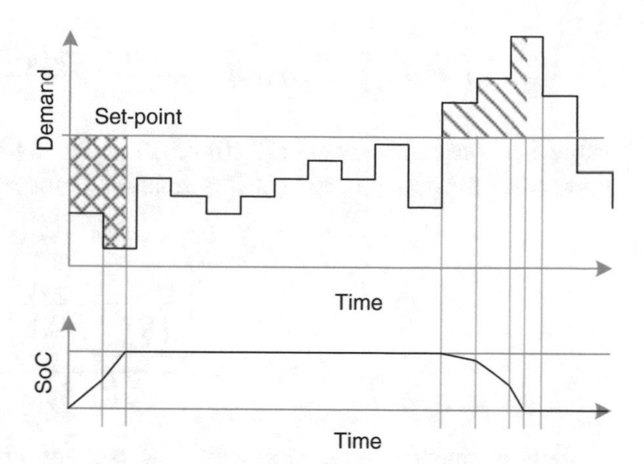


**Fig. 5.9** Example of SPC failing to meet the requirements for peak reduction at the given set-point threshold

$c(t)$ is the energy stored in the BESS at the end of the time period $t$, during which $p(t)$ was applied.

Effectively, SPC for network support applications has a limited performance as it cannot anticipate the future states (Hida et al. 2010). Furthermore, the performance is driven by carefully chosen set-point corresponding to the rating and capacity of the BESS. As the result, there is a danger of imposing additional peaks from charging for multiple BESS and not having sufficient capacity to deal with the peak at the given threshold (Fig. 5.9). The simplicity of the SPC also means that it is not possible to impose limits on the depth of discharge and the number of charge-discharge cycles. Unnecessary cycling of BESS could lead to shorter life span and, hence, lower cost efficiency.

For this section the set-point value for SPC is chosen per day using actual demand values per feeder. Using actual demand means that the SPC has a perfect foresight and will provide 20% peak reduction as long as there is a sufficient energy storage and peaks are not in close proximity.

## 5.3.2 *Fixed Day-Ahead Schedule*

Fixed day-head scheduling offers the combination of simplicity and day ahead planning the operation of the BESS for the future states. Assuming 100% accuracy of the forecasts, fixed-day ahead schedule is a good measure of forecast accuracy on the timing of the peaks.

The aim of the schedule optimisation is to find a schedule, $p$, for the forecasted demand, $d_f$, over a period of $N$ half-hours (in this case 48), such that the cost function in (5.14) is minimised:

$$F(p, d_f) = \xi_p(p, d_f) + \frac{\alpha \xi_{cd}(p)}{w} + \frac{\beta \xi_{sc}(c)}{w} + \frac{\gamma \xi_{ts}(c)}{w}, \tag{5.14}$$

subject to constraints given in (5.10)–(5.13). $\xi_p(p, d_f)$ is the Peak-to-Average cost component for peak reduction, self-normalised to the initial conditions, defined as:

$$\xi_p(p, d_f) = \frac{\left( \frac{\max_{t=1}^{N}(d(t) + p(t))}{\left( \frac{\sum_{t=1}^{N} d(t) + p(t)}{N} \right)} \right)^2}{\xi_p(p_i, d_f)}. \tag{5.15}$$

Cost component, $\xi_{cd}$, represents the cost of charge dynamics, aimed at smoothing the charging of energy storage and is defined as:

$$\xi_{cd} = \frac{\max\left( \left| \frac{\delta I_{[0, p_{max}]}(p)}{\delta t} \right| \right)}{p_{max}}. \tag{5.16}$$

The storage cycling cost component, $\xi_{sc}$, aims to allow at most only one full charge and discharge cycle per day and is defined as:

$$\xi_{sc} = \frac{1}{2} \left( \sum_{t=1}^{N} \left| \frac{\delta I_{(0, p_{max}]}(c)}{\delta t} \right| + \sum_{t=1}^{N} \left| \frac{\delta I_{[-p_{max}, 0)}(c)}{\delta t} \right| \right). \tag{5.17}$$

At the end of the schedule, the energy should reach 50% State-of-Charge (SoC), which is achieved with the target SoC cost component, $\xi_{ts}$, defined as:

$$\xi_{ts}(c) = \frac{(c(N) - 0.5 C_{max})^2}{\xi_{ts}(c_i)}. \tag{5.18}$$

Scaling factor, $w$, ensures that sum of cost components $\xi_{cd}$, $\xi_{sc}$ and $\xi_{ts}$ is at the same scale as $\xi_p$:

$$w = \alpha + \beta + \gamma, \quad \text{where } \alpha, \beta, \gamma \in [0, 1]. \tag{5.19}$$

It is unavoidable to have errors in the forecast and it is possible that the day-ahead fixed schedule could increase the actual peak by instructing the BESS to charge during the peak. To reduce the risk of increasing the peaks, it would be beneficial to update the forecast using the latest observations and calculate the new schedule with every update.

### 5.3.3 Model Predictive Control

MPC, also known as receding horizon control, computes an optimal schedule for the duration of the horizon by updating the demand model with recent observations and deploying only the first control value from the current horizon (García et al. 1989). Once the control for the first time step is deployed and observation for the current time step is received, the controller moves to the next horizon to compute a new optimal schedule on the updated demand model. Updates on the demand model could improve the accuracy of the forecast and hence increase the likelihood of successful peak reduction. Receding horizon control also has an advantage of reducing peaks in each horizon, allowing peak reductions both at the beginning of the 24-period and at the end.

Typically, MPC includes a model to predict the state of the system for the duration of the horizon. In the case of energy storage control, it is not always possible since the demand forecasts could be provided by an external system and the forecasts cannot be updated using the original model. In this section we consider two methods for updating the demand model: (i) persistent error update on a fixed day-ahead forecast and (ii) rolling demand forecast using demand modelling. The first method assumes strong correlation in errors between the forecast and updates the forecast within the control horizon based on the forecast error in the previous time step (Fig. 5.11) (Yunusov et al. 2017). System diagram for the MPC with persistent error model (MPC-PE) is depicted in Fig. 5.10.

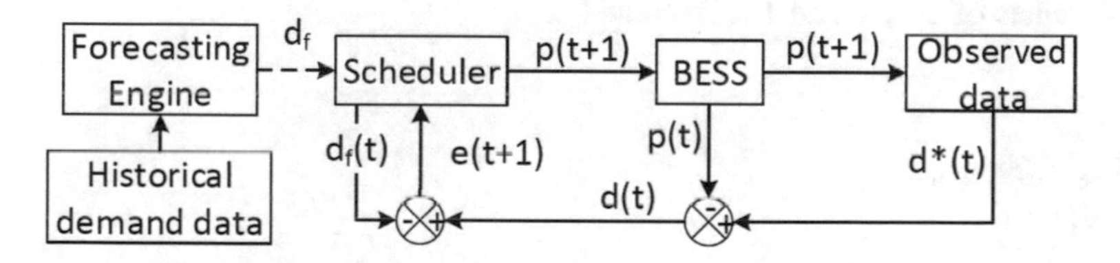


**Fig. 5.10** Schematic of the MPC system with persistent error model

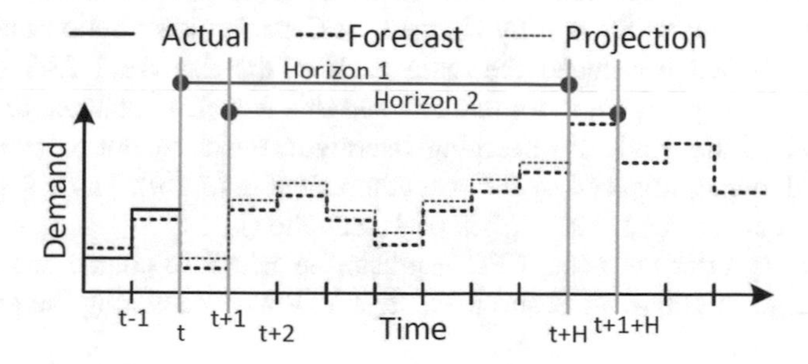


**Fig. 5.11** Demand forecast update using persistent error model

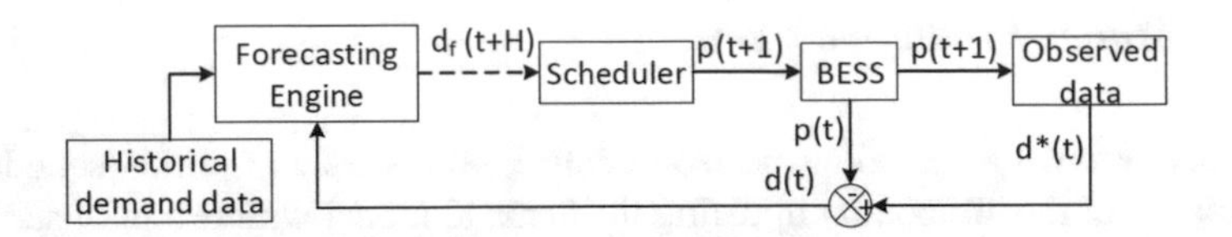


**Fig. 5.12** Schematic of the MPC system with rolling forecast model

The second method uses access to the forecast models to update the forecast profile based on the latest observations. At each time step, the new forecast is generated and used for scheduling the operation of the BESS for the duration of control horizon. System diagram for the MPC with rolling forecasts is depicted in Fig. 5.12.

For both MPC methods, MPC with persistent error model (MPC-PE) and MPC with rolling model, BESS scheduling is achieved through minimisation of the cost function defined in Eq. (5.14) subject to constraints given in (5.10)–(5.13). Schedule is optimised only for the duration of the control window of 27 half-hours. Such length of the control window captures the evening peak from early morning hours, allowing the BESS to be charged during the periods on low demand.

As a result, the storage cycling component given in (5.17) has to be scaled down to a proportion of the control window to the half-hours in the day. To ensure that the cost associated with the target state-of-charge is reflected appropriately, the cost component defined in (5.18) is only contributing to the total cost function when the end of the day is included in the control window.

The cost function weighting, $\alpha$, $\beta$ and $\gamma$ are fixed for both variants of MPC and have values of 0.5, 1 and 1, respectively.

### 5.3.4 Results

Figure 5.13 demonstrates the operation of the control methods assuming perfect foresight of the demand the on feeder S1.

SPC immediately starts to charge the BESS as the demand is below the set-point of 18.4 kW. As the BESS is fully charged, SPC performs no action until the first peak at 5:30 when it reduced the main peak of the day from 23 kW down to 18.4 kW—the set-point value for this day and this feeder. In contrast to SPC, both FDS and MPC are gradually charging overnight, ensuring not to create sudden increase in demand (imposed by the cost component in (5.16)). The first peak of the day is also reduced by 4.6 kW (20% peak reduction).

Immediately after the peak, SPC instructs the BESS to charge and brings the state-of-charge to 100%. FDS and MPC, again slowly recovering the energy used

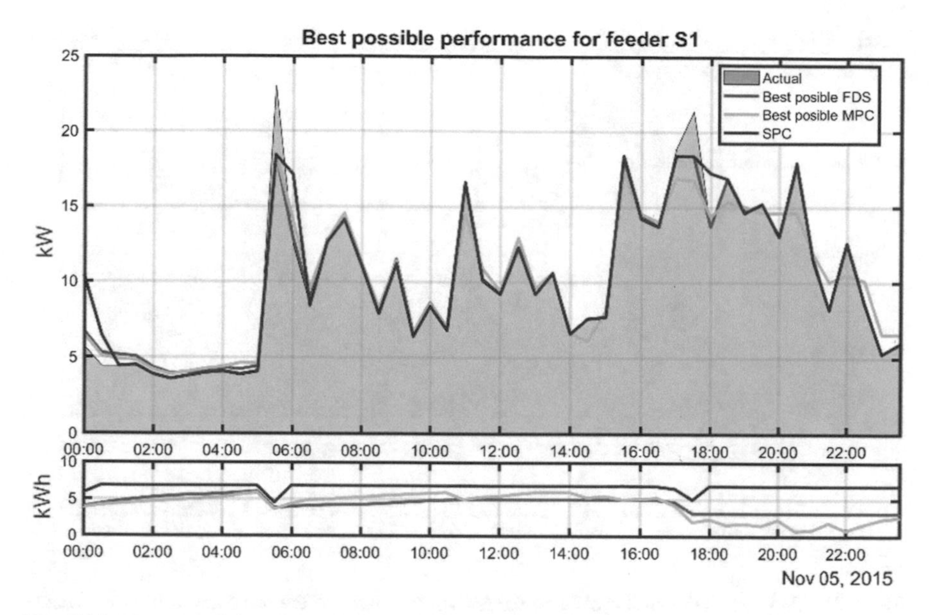


**Fig. 5.13** Top: actual demand profile on feeder S1 compared to SPC, best possible peak reduction based on FDS and MPC. Bottom: corresponding state-of-charge profile

for the first peak. At 11:00, MPC shows its first difference to FDS. FDS is only capable of reducing the highest peak of the day, whereas MPC has reduced the peak at 11 AM by an additional 2 kW due to the receding horizon operating outside the area of the main peak. FDS and SPC have reduced the evening peak at 17:30 from 21.4 kW down to 18.4 kW (14%). For the FDS, the evening peak cannot be reduced below the morning peak as the cost function is concerned with the highest peak of the day. SPC will only reduce the peak to the level of the set-point. MPC, however, achieved a 4.6 kW reduction (limited by the rating of the BESS) on the peak (21.5%) since this peak is considered separately from the morning peak. MPC also continued to smooth out the demand profile by reducing peaks at 20:30 and 22:00 by 3.1 and 2 kW respectively, whereas the SPC and FDS took no action. Finally, both FDS and MPC reached near 50% state of charge at the end of the day and with SPC BESS remained at 100%.

Figure 5.14 gives an example of several control methods applied to SnTAR forecast compared against the SPC and best possible of FDS and MPC. The performance of the both fixed day-ahead (SnTAR-f) and rolling (SnTAR-r) forecast is evident since the general shape of the profile is captured and the timing of the peak is off by a half-hour. The shape of resultant profiles from FDS, MPC-PE and MPC is consistent with the shape of the forecasts, meaning their performance on peak reduction is similar, 12.8, 15.43 and 16.4% respectively.

Average peak reduction performance for each feeder for FDS on all day-ahead forecasts is given in Table 5.6. Under the chosen perfect conditions, SPC provides the highest peak reduction for all feeders. Such performance is expected since the

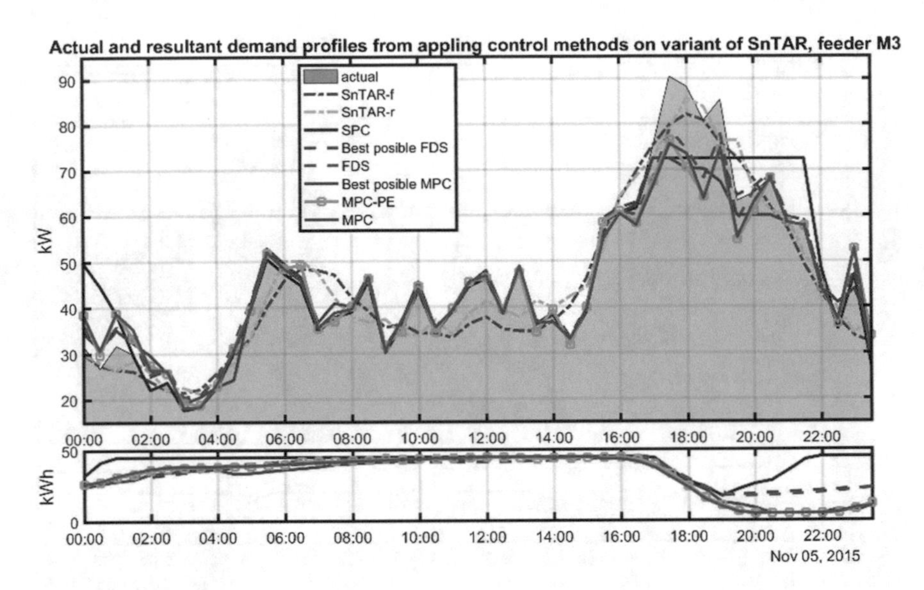


**Fig. 5.14** Top: actual demand profile on feeder M3 compared to the fixed and rolling variants of SnTAR, resultant profile from SPC, best possible peak reduction based on FDS, MPC-PE on day-ahead variant of SnTAR and MPC on rolling variant of SnTAR. Bottom: corresponding state-of-charge profiles

SPC and BESS are configured to achieve 20% reductions on the highest peaks. Peak reduction below 20% is caused by close proximity of smaller peaks (albeit still above the set-point values) to the main peak. Limitation in capacity does not allow the SPC to reduce the main peak to the expected values. Best possible FDS on average for each feeder is below the expected peak-reduction values (see Sect. 5.3.5 for discussion on this discrepancy).

Overall, feeders with lower demand (S1 to M2) have achieved relatively poor peak reduction for under the FDS method. It is likely due to the low number of customers on those feeders and, hence, the higher variability in demand, which is more difficult to forecast accurately. In practice, though, those feeders typically would not be experiencing stress with such level of demand and unlikely to be considered for deployment of BESS as reinforcement. Feeder L2 also demonstrates very poor performance in peak reduction for all control methods. Predominantly supplying commercial loads, feeder L2 has a distinctive demand profile which includes high variability during the day and individual short-term spikes of high demand.

On average FDS have performed the best with the fixed day-ahead variant of SnTAR, closely followed by fixed day-ahead variant of STAR. This is consistent with the accuracy of the forecasts utilised as shown in Table 5.3 with STAR and SnTAR, on average, giving the best estimates.

**Table 5.6** The overall peak reduction (%) with FDS per feeder for each day-ahead forecast (including day-ahead variant of the rolling forecasts)

| Feeder | SPC | Best | LW | 7SAV | ST | SnT | STAR | SnTAR | RFR | SVR |
|---|---|---|---|---|---|---|---|---|---|---|
| S1 | 19.85 | 19.36 | 0.75 | 0.83 | **2** | 1.65 | 2 | 1.82 | 1.86 | 1.56 |
| S2 | 19.9 | 18.83 | 0.44 | 2.41 | **5.09** | 5 | 4.84 | 4.63 | 2.63 | 3.34 |
| S3 | 20 | 19.91 | 3.03 | 5.39 | 7.86 | **8.2** | 7.92 | 7.9 | 4.65 | 7.34 |
| M1 | 20 | 18.75 | 2.42 | 2.03 | **4.28** | 3.78 | 3.76 | 3.71 | 4.15 | 2.36 |
| M2 | 19.97 | 19.41 | 2.35 | **4.56** | 1.8 | 1.65 | 3.95 | 4.21 | 1.37 | 4.24 |
| M3 | 20 | 19.76 | 7.57 | 10.14 | 10.56 | **10.59** | 10.43 | 10.58 | 6.8 | 10.5 |
| L1 | 20 | 19.12 | 7.91 | 7.74 | 12.23 | **12.42** | 11.16 | 11.83 | 11.37 | 9.7 |
| L2 | 20 | 13.68 | 1.94 | 1.12 | 1.12 | 1.73 | 2.75 | 2.09 | **3.22** | 0.25 |
| L3 | 20 | 19.49 | 6.02 | 7.89 | 8.36 | 9.12 | 9.1 | **9.81** | 6.04 | 8.97 |
| Ave. | 19.97 | 18.7 | 3.52 | 4.65 | 5.92 | 6.02 | 6.21 | **6.29** | 4.68 | 5.36 |

Best results for each feeder are highlighted in bold

The corresponding average peak reduction values for each feeder using MPC method (both persistent error mode on fixed day-ahead forecasts and standard MPC with rolling forecasts) are given in Table 5.7.

Similarly to FDS, feeders S1 to M2 and L2 achieved lower peak reduction compared to other feeders. Overall, on average MPC have performed better than FDS, except for L2. Comparing the best peak reductions for each feeder typically utilising the MPC reduces the peak by an extra 18%, and for M1 and M2 there is in fact an extra 42% reduction in the peak size. Interestingly, MPC using the persistent error model out performs, on average, the MPC with rolling forecasts in reducing the peaks on feeder L1. Further also note that despite some forecasts being fixed, e.g. SnT and ST, the MPC-PE (with persistent error) still improves the peak reduction. Hence for this method the improvement is mainly due to the extra information becoming available to the controller and potentially also the increased emphasis on the peak due to the error correction.

We would perhaps expect the rolling forecast to perform even better than the day ahead fixed control but as shown in Fig. 5.8 the forecast accuracy for the rolling forecast drops off rapidly beyond 10 half hours and hence is less effective for the 27 half hour horizon we deploy here.

On average across all feeders, the MPC with the rolling SnTAR performs the best, closely followed by MPC-PE with the day-ahead variant of STAR.

### *5.3.5 Discussion*

This study has applied a number of assumptions that would not be possible in the real-world deployment of the BESS. The BESS configuration and set-point values selection was based on the perfect foresight of demand. In practice, of course, the

**Table 5.7** The overall peak reduction with MPC per feeder for all forecasts

| Feeder | Best | MPC-PE | | | | | | | | MPC | | | |
|---|---|---|---|---|---|---|---|---|---|---|---|---|---|
| | | LW | 7SAV | ST | SnT | STAR | SnTAR | RFR | SVR | STAR | SnTAR | RFR | SVR |
| S1 | 18.47 | −0.42 | 1.12 | 2.02 | **3.17** | 2.89 | 2.8 | 1.89 | −0.72 | 2.19 | 2.46 | 1.8 | 1.62 |
| S2 | 17.67 | 2.01 | 2.99 | 5.84 | 5.62 | **6.81** | 5.9 | 1.92 | −5.18 | 6.01 | 6.44 | 2.38 | 4.01 |
| S3 | 19.61 | 5.35 | 7.96 | 9.58 | 9.57 | 8.92 | 9.7 | 6.94 | 8.39 | 9.64 | **10.18** | 8.58 | 8.7 |
| M1 | 16.89 | 2.69 | 3.91 | 5.47 | 4.73 | 5.37 | 4.7 | 4.89 | 4.42 | **6.26** | 5.34 | 4.79 | 4.5 |
| M2 | 18.46 | 3.43 | **6.68** | 3.53 | 3.05 | 5.47 | 5.29 | 3.12 | 6.25 | 6.12 | 6.09 | 2.12 | 6.37 |
| M3 | 19.43 | 11.72 | 12.89 | 12.69 | 12.99 | 12.71 | 12.12 | 8.4 | 13.29 | 13.2 | **13.49** | 11.4 | 12.13 |
| L1 | 19.07 | 10.81 | 8.82 | 14.03 | 13.58 | 14.52 | 14.12 | **14.72** | 12.26 | 13.32 | 13.68 | 11.79 | 11.81 |
| L2 | 7.21 | −0.29 | 1.09 | 1.49 | 1.62 | 2.01 | 1.56 | 2.47 | −0.27 | 2.03 | 0.56 | **2.6** | 0.69 |
| L3 | 19.29 | 6.77 | 8.92 | 10.84 | 11.14 | 10.98 | 10.74 | 7.84 | 10.19 | 10.3 | **11.44** | 6.44 | 9.57 |
| Ave. | 17.35 | 4.68 | 6.04 | 7.28 | 7.27 | 7.7 | 7.04 | 5.8 | 5.4 | 7.69 | **7.74** | 5.77 | 6.73 |

For the fixed day-ahead forecasts persistent error model was used

Best results for each feeder are highlighted in bold

perfect foresight of demand would not be available. Instead, suitable forecasts of future demand would guide the selection of the necessary BESS configuration for each feeder in consideration, taking into account the network safety margins and the assessment of low probability high risk events. Similarly, to achieve the presented performance of SPC it would be necessary to select a significantly oversized BESS, ensuring that all possible peaks on the feeder are equipped for. Such arrangement is likely to be economically inefficient. However, SPC does have an appropriate application for feeders with low demand and high variability that require reinforcement. It could be more cost effective to deploy a larger BESS with SPC to ensure that there is enough capacity to deal with a wide range of peaks.

Looking at Table 5.6, the reader would expect the best possible performance of FDS (assuming perfect foresight) to be on par with the SPC. However, unlike FDS and MPC, SPC does not take care of the BESS health and does not consider the context of the network—two essential aspects of application and deployment of BESS on real networks. SPC performs better than FDS under a number of conditions because of the complexity of the cost function given in Eq. (5.14) and the schedule optimisation method. The cost function components are normalised to the initial conditions and constructed from two types: peak-reduction (given in Eq. (5.15)) and non-peak reduction. The non-peak reduction components of the cost functions (charge dynamics in Eq. (5.16), reaching 50% target state-of-charge as defined in Eq. (5.18) and aiming for no more than one full charge-discharge cycle a day, Eq. (5.17)) are weighted to have the same range as the peak reduction component. This could result in a situation where the non-peak reduction components have greater cost than the peak reduction and the peak reduction performance is then sacrificed in order to satisfy sum of the non-peak reduction costs.

Another important aspect concerning the application and deployment (which applies to both MPC and SPC) is that with FDS it is not necessary to have LV substation monitoring infrastructure (which could be costly for the network operator) in order to provide the signal (for SPC) or the feedback (for MCP). Instead of feeder level demand, FDS could be based on the aggregated demand up-scaled from smart meter data (Giasemidis et al. 2017). To provide wider network support, FDS could also be based on a demand forecast at a higher level of the network and coordinate multiple BESS on multiple LV feeders to achieve peak reduction at, for instance, 11 kV feeder or substation.

Finally, due to the natural variability of demand the peak demand occurring on the network would be significantly higher and shorter in duration than the half-hourly average demand used in this chapter. In practice, the schedules presented and discussed in this chapter would be used as half-hourly set-points for a highspeed control system located on the physical BESS, which will be issuing instruction to BESS in near-real-time. Furthermore, the half-hourly demand profiles used in this chapter are aggregates of demand across the three phases of low voltage feeders. Inequality in the number of customers per phase and asynchronous customer behaviour provides an opportunity for further peak reduction by the means of phase-balancing function of the BESS power electronics.

**Acknowledgements** The research work presented in this chapter have been initiated as a part of New Thames Valley Vision (SSET203), a Low Carbon Network Fund project, funded by Ofgem and led by Scottish and Southern Electricity Networks.

## References

Breiman, L. (2001). Random forests. *Machine Learning, 45*(1), 5–32. https://doi.org/10.1023/a:1010933404324.

Breiman, L., Friedman, J. H., Olshen, R. A., & Stone, C. J. (1984). *Classification and regression trees. Statistics/probability series*. Belmont, California, USA: Wadsworth Publishing Company.

Drucker, H., Burges, C. J. C., Kaufman, L., Smola, A. J., & Vapnik, V. (1997). Support vector regression machines. In M. C. Mozer, M. I. Jordan, & T. Petsche (Eds.) *Advances in neural information processing systems* (Vol. 9, pp. 155–161). MIT Press. URL http://papers.nips.cc/paper/1238support-vector-regression-machines.pdf.

Dudek, G. (2015). Short-term load forecasting using Random forests. In *Intelligent systems' 2014. Advances in intelligent systems and computing* (Vol. 323, pp. 821–828).

Evans, G. (2016). New thames valley vision technical impact evaluation impact on DNO network from low carbon promotions, SDRC 9.8b (SSET203). In *Technical Report 3, Scottish and Southern Energy Power Distribution*.

Gajowniczek, K., & Zbkowski, T. (2017). Electricity forecasting on the individual household level enhanced based on activity patterns. *PLOS ONE, 12*(4), 1–26. https://doi.org/10.1371/journal.pone.0174098.

García, C. E., Prett, D. M., & Morari, M. (1989). Model predictive control: Theory and practice-A survey. *Automatica, 25*(3), 335–348. https://doi.org/10.1016/0005-1098(89)90002-2.

Giasemidis, G., Haben, S., Lee, T., Singleton, C., & Grindrod, P. (2017). A genetic algorithm approach for modelling low voltage network demands. *Applied Energy, 203,* 463–473. https://doi.org/10.1016/j.apenergy.2017.06.057.

Haben, S., & Giasemidis, G. (2016). A hybrid model of kernel density estimation and quantile regression for gefcom2014 probabilistic load forecasting. *International Journal of Forecasting, 32,* 1017–1022.

Hida, Y., Yokoyama, R., Shimizukawa, J., Iba, K., Tanaka, K., & Seki, T. (2010). Load following operation of NAS battery by setting statistic margins to avoid risks. *IEEE PES General Meeting, PES, 2010,* 1–5. https://doi.org/10.1109/PES.2010.5588170.

Hong, W. C. (2009). Electric load forecasting by support vector model. *Applied Mathematical Modelling 33*(5), 2444–2454. https://doi.org/10.1016/j.apm.2008.07.010. URL http://www.sciencedirect.com/science/article/pii/S0307904X08001844.

Hu, Z., Bao, Y., & Xiong, T. (2013). Electricity load forecasting using support vector regression with memetic algorithms. *The Scientific World Journal*, 2013.

Joshi, K. A., & Pindoriya, N. M. (2015). Day-ahead dispatch of battery energy storage system for peak load shaving and load leveling in low voltage unbalance distribution networks. In *Power & Energy Society General Meeting,* 2015 *IEEE* (pp. 1–5). https://doi.org/10.1109/pesgm.2015.7285673.

Lahouar, A., & Slama, J. B. H. (2015). Random forests model for one day ahead load forecasting. In *Renewable Energy Congress (IREC),* 2015 *6th International. IEEE, Sousse, Tunisia*.

Megel, O., Mathieu, J. L., & Andersson, G. (2015). Scheduling distributed energy storage units′ to provide multiple services under forecast error. *International Journal of Electrical Power & Energy Systems, 72*, 48–57. https://doi.org/10.1016/j.ijepes.2015.02.010. URL http://linkinghub.elsevier.com/retrieve/pii/S0142061515000939.

Nair, N., & Garimella, N. (2010). Battery energy storage systems: Assessment for small-scale renewable energy integration. *Energy and Buildings, 42,* 2124–2130.

Rowe, M., Yunusov, T., Haben, S., Holderbaum, W., & Potter, B. (2014). The real-time optimisation of DNO owned storage devices on the LV network for peak reduction. *Energies, 7,* 3537–3560.

Sapankevych, N. I., & Sankar, R. (2009). Time series prediction using support vector machines: A survey. *IEEE Computational Intelligence Magazine, 4*(5), 24–38.

Schmidt, O., Hawkes, A., Gambhir, A., & Staffell, I. (2014). The future cost of electrical energy storage based on experience rates. *Nature Energy, 2,* 17110.

Scottish and Southern Energy Power Distribution: New Thames Valley Vision. (2014). URL http://www.thamesvalleyvision.co.uk/.

Tian, L., & Noore, A. (2004). A novel approach for short-term load forecasting using support vector machines. *International Journal of Neural Systems 14*(05), 329–335. https://doi.org/10.1142/s0129065704002078. URL http://www.worldscientific.com/doi/abs/10.1142/S0129065704002078.

Trkay, B. E., & Demren, D. (2011). Electrical load forecasting using support vector machines. In *2011 7th International Conference on Electrical and Electronics Engineering (ELECO)*. IEEE, Bursa, Turkey.

Vapnik, V. (1995). *The nature of statistical learning theory*. New York, USA: Springer.

Yunusov, T., Frame, D., Holderbaum, W., & Potter, B. (2016). The impact of location and type on the performance of low-voltage network connected battery energy storage systems. *Applied Energy, 165*, 202–213 (2016). https://doi.org/10.1016/j.apenergy.2015.12.045. URL http://linkinghub.elsevier.com/retrieve/pii/S0306261915016189.

Yunusov, T., Haben, S., Lee, T., Ziel, F., Holderbaum, W., & Potter, B. (2017). Evaluating the effectiveness of storage control in reducing peak demand on low voltage feeders. In *Proceedings of the 24th International Conference on Electricity Distribution (CIRED).*

# Part III
# Economic Analysis

# Chapter 6
# Modeling and Economic Evaluation of PV Net-Metering and Self-consumption Schemes

**Georgios C. Christoforidis and Ioannis P. Panapakidis**

**Abstract** Due to the high rate of Photovoltaics (PV) installations in many countries, the need for data processing and exploitation is a crucial factor that determines the success of the economic profitability of the installation. Machine learning is a family of tools for information retrieval and knowledge extraction. In the present study, clustering is applied to a set of PV power generation curves that correspond to locational distributed PV installations with the aim of formulating the PV generation profiles and PV clusters. Next, a techno-economic assessment of different policy schemes is applied to selected cluster. The scope is to reduce the need for conducting economic analyses per PV site; grouping PV installations in homogenous clusters can lead to reduced effort in the phase of techno-economic evaluation of the overall operation of the PV technology.

## 6.1 Introduction

The rapid advancement of Smart Grids and Microgrids technologies is evident by numerous researchers, surveys, pilots programs and practical applications (Zhang et al. 2017; Ali et al. 2017). These technologies include small-scaled generation units, storage and flexible loads. Renewable Energy Resources (RES) are considered as an important asset in materializing the distributed generation systems that compose the core of the Smart Grids and Microgrids (Eltigani and Masri 2015). During the last years, the utilization of RES and especially Photovoltaics (PV) have witnessed a vast growth in many countries, making RES power an important contributor in electricity generation across the globe (PVPS 2016). This was due to

G. C. Christoforidis (✉)
Department of Electrical Engineering, Western Macedonia University of Applied Sciences, 50100 Kila Kozanis, Kozani, Greece
e-mail: gchristo@teiwm.gr

I. P. Panapakidis
Department of Electrical Engineering, Technological Educational Institute of Thessaly, 41100 Larisa, Greece

© Springer International Publishing AG, part of Springer Nature 2018

C. Kahraman and G. Kayakutlu (eds.), *Energy Management—Collective and Computational Intelligence with Theory and Applications*, Studies in Systems, Decision and Control 149, https://doi.org/10.1007/978-3-319-75690-5_6

several factors such as government incentives and economic motives, public's environmental awareness, wide variety of size installations, high solar resource availability in many locations, potential of hybridization with other sources and others (Mauleón 2017). Solar technologies are considered as an efficient option to cope with various problems that the global energy sector faces, such as environmental impact, capacity shortage, covering of isolated loads and others. Nevertheless, distributed RES, such as PV systems, have a negative impact on the operation of low and medium voltage grids (Enslin 2014). Furthermore, RES units influence the operational margins of the other generation units and provide obstacles in the secure and reliable energy management of Smart Grids (Martinez-Anido et al. 2016).

In order to motivate consumers and other interested parties to invest in PV installations, several support mechanisms have been successfully implemented in various countries, such as Feed-in-Tariffs (FiTs) (Mabee et al. 2012; Pyrgou et al. 2016; Ye et al. 2017). However, FiTs have exhausted their role in most cases, due to the fact that they indirectly cross-subsidy PV owners at the expense of regular electricity consumers. Thus, they have been disappearing (or considerably reduced) over the past years in most countries. This fact has led to the decline in new installations in Europe since 2013 (EurObserver 2014). Therefore, regulation authorities and market players have been seeking ways to revitalize the PV market. Net-Metering (NEM) is a support mechanism for PV systems specializing mostly in the residential sector, which have been utilized in considerably less countries than the FiT scheme. The consumer is transformed into "prosumer". This policy enables the consumer to hold an active role in competitive energy markets (Ottesen et al. 2016). Under NEM, the electrical grid acts as a virtual storage unit. An offset takes place between the PV generated electricity and the consumption. The difference between the amounts of consumption and generation is called "netting". If netting is positive, i.e. the PV generates an amount that covers the consumption and excess electricity appears, the excess electricity is fed to the grid. Depending of the NEM scheme, this amount can be compensated or not. In the simplest NEM formulation (termed "full NEM"), if the netting is negative, the prosumer is charged only with the extra amount that imports from the grid. Although NEM has the potential to utilize efficiently the PV system in terms of lowering the cost of the electricity payments from the prosumer perspective, the current landscape in EU does not display large adoption. Many questions arise for NEM implementation (Yamamoto 2012; Poullikkas 2013; Satchwell et al. 2015; Bertsch et al. 2017). For instance, will the excess electricity fed to the grid be compensated or transferred to the next billing period and used as credit, what is the compensation price, should the prosumer be charged for the imported electricity to provide a motive for higher self-consumption (a type of partial NEM or net-billing), and others. Another type of support mechanism is self-consumption (Martín-Chivelet and Montero-Gómez 2017). In this case, all the PV electricity is firstly used to cover the consumption, while excess energy is either compensated based on the electricity market prices or not compensated at all. This mechanism is the most favorable for the rest consumers as it does not heavily subsidy PV owners, apart from providing non-monetary incentives (e.g. priority to grid exporting or offering a small premium above real time electricity prices).

This study presents a methodology for the techno-economic evaluation of NEM and self-consumption schemes. The methodology is composed of two stages. The 1st stage refers to the application of clustering in order to form residential PV clusters. The 2nd stages presents a comprehensive economic model for examining the potential of PV support mechanism on a specific time horizon. The scope is to form clusters that include similar PV systems in terms of generation patterns. Instead of applying an economic feasibility analysis for each installation, the analysis can be applied for 1 representative installation per cluster. This approach can be applied to cases with a large number of installations that are geographically distributed. In the present study, the data correspond to actual installation of residential PVs. However, the methodology is applicable in cases where only solar irradiation and consumption data are available. This concept allows the interested party to assess the potential of a PV installation in various regions. Note that apart from the climatic conditions, the various regions may differ in terms of distribution grid characteristics such as topology, number of connection points and others. Hence, if it is foreseen by the legislation of the energy market, region specific network access tariffs can be applied.

## 6.2 Machine Learning Application

### *6.2.1 PV Data Modeling*

In the present study, the data set under study is composed by 11 residential PV installations located in Northern Greece (i.e. the 10 installations) and in Aegean Sea (i.e. 1 installation). The PVs of Northern Greece are installed in the following areas: Agios Petros, Ierissos, Imathia, Panorama, Peraia, Polygiros, Profitis, Serres, Vrasna and Nea Mesimvria. Figure 6.1 shows a map of the locations. Figure 6.2 displays the installation of Ios island, located in the Aegean sea. The average installed capacity of the 11 systems is 9.90 kWp. The available set includes power generation data that cover the period of a complete year. The time interval for data collection is 15 min. The generation data modeling refers to the following steps: Data pre-processing, data representation and information retrieval. Data pre-processing is not obligatory. This step refers to erroneous data removal, such as extremely low values and outliers, data missing data filling and others. In the present set, no pre-processing took place since no erroneous or other problematic data entries were detected.

Data representation refers to the type of mathematical format that is used for data expression. The most common approach is to simulate the data without any further transformation. In the present study we deal with daily power generation curves. The term "pattern" is used to refer to *D*-dimensional vectors that represent the daily

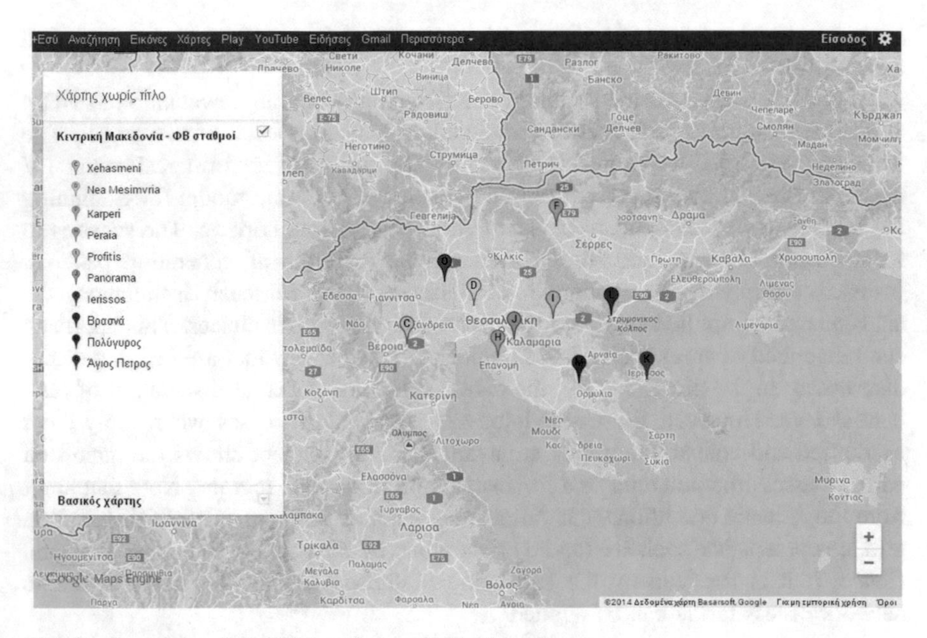


Fig. 6.1 Locations of the PV installations in Northern Greece

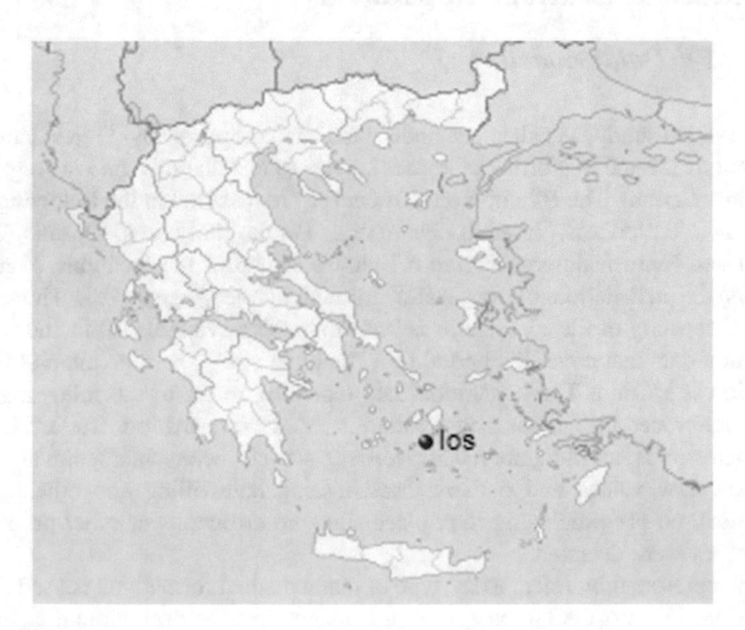


Fig. 6.2 Location of the PV installation in Aegean island

power curves. Let $p = 1, \ldots, P$ with $P = 11$ denote the number of PV installations and $m = 1, \ldots, M$ with $M = 365$ the number of available daily power curves. For each installation, a set of daily power curves is formed denoted as $X^p = \{x_m^p, m = 1, \ldots, M, p = 1, \ldots, P\}$, with:

$$x_m^p = [x_{m1}^p, \ldots, x_{mi}^p]^T \tag{6.1}$$

where $i = 1, \ldots, D$ is the dimension, i.e. the number of elements. In the present study, it is $D = 96$.

The information retrieval step refers to the extraction of exploitable information from the data set for further applications. Here, the application is the techno-economic assessment. Other applications may refer to PV power forecasting, harmonic distortion due to PVs, power flow studies and others. For this step, we utilize an unsupervised machine learning tool and namely, a partitional clustering algorithm. The scope is to formulate a descriptive model of the data by extracting the PV power profiles.

Clustering is a data-driven method. It is suitable in problems where limited or complete absence of prior information about the data structure is available (Xu and Wunsch 2008). A clustering algorithm aims to track the similarities between the patterns and group together patterns of high similarity. The similarity estimation is held into the $D$-dimensional feature space. While clustering operates based on similarities, the patterns magnitude is not relevant. In fact it may cause obstacles in the robustness of the operation. Thus, prior to the clustering, all data should be normalized. For each installation, we use the rated power $P_{PV,rated}^p$ as the basis for the normalization.

By dividing each element of the vector $x_m^p$ with $P_{PV,rated}^p$ we obtain a new set of daily load power curves with normalized values within the [0,1] range, denoted as $Y^p = \{y_m^p, m = 1, .., M, p = 1, .., P\}$, with:

$$y_m^p = [y_{m1}^p, \ldots, y_{mi}^p]^T \tag{6.2}$$

Equation (6.2) refers to the term of "final yield" (de Lima et al. 2017). It refers to yield how many hours within a day the PV system have to operate at its rated power in order to produce the same amount of energy as was recorded.

### 6.2.2 PV Generation Profiles Per Installation

For the purpose of our analysis, the K-means clustering is employed. The operation of the algorithm is illustrated in Fig. 6.3.

The operation of the algorithm is based on a cost function minimization procedure. The algorithm starts by selecting random patterns from the data set and assigns them as the initial centroids. The centroid refers to the average of all patterns that belong to the same cluster. Next, the algorithm distributes the rest patterns into the clusters based on the minimum Euclidean distance between each

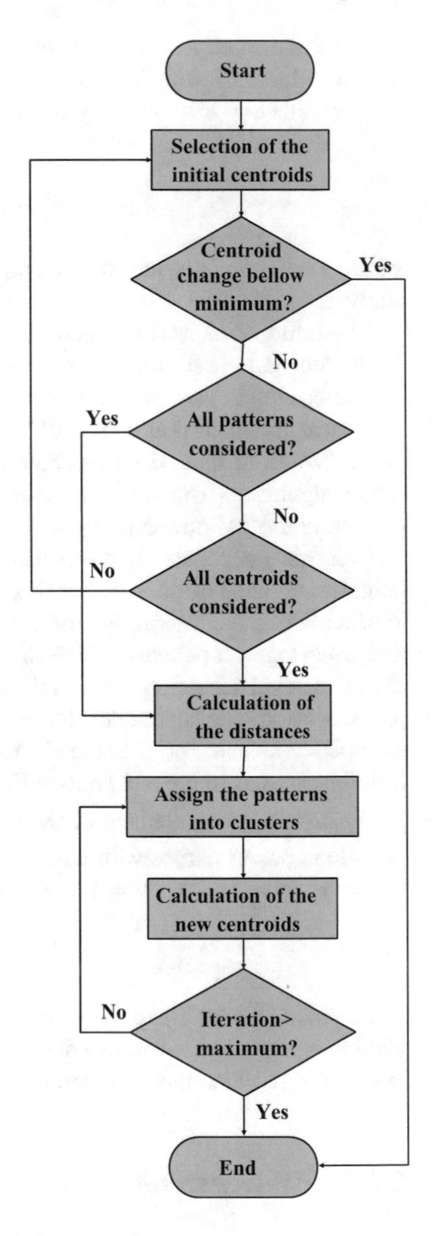


**Fig. 6.3** Fow-chart of the operation of the K-means algorithm

patterns and the set of centroids. A series of repetitions of the aforementioned procedure takes place until the cost function is minimized or the pre-defined number of repetitions is met. For a full mathematical description the reader is referred to (Steinley 2006; Khan and Ahmad 2004).

The K-means is applied separately to each PV installation. Since the number of clusters is unknown, the algorithm is executed for variable number of clusters. Clustering provides a mapping of $M \rightarrow K$, where $K$ is the number of clusters and

$1 \leq K \leq M$. Each cluster is represented by the centroid, which is also expressed by a $D$-dimensional vector:

$$c_k^p = \frac{1}{M_K} \sum_{\substack{m=1 \\ x_m^p \in C_k^p}}^{M} x_m^p \tag{6.3}$$

where $M_K$ is the number of vectors that belong to the cluster $C_k^p$. The set of the clusters is denoted as $C_k^p = \{c_k^p, k = 1, \ldots, K, p = 1, \ldots, P\}$. The K-means algorithm produces the centroids and the patterns' membership in the clusters. For the purpose of evaluating the algorithm's performance and determine the optimal number of clusters, the ratio of Within Cluster Sum of Squares to Between Cluster Variation (WCBCR) indicator is employed:

$$\text{WCBCR} = \frac{\sum_{k=1}^{K} \sum_{m=1}^{M} d^2(c_k^p, x_m^p)}{\sum_{1 \leq s < t}^{M} d^2(c_s^p, c_t^p)} \tag{6.4}$$

K-means is executed for different number of clusters and for each number the value of WCBCR is checked. WCBCR is a measure of both separation and compactness of the produced clusters. The outputs of PV generation clustering per site are the PV power generation profiles per installation.

### 6.2.3 PV Generation Profiles Per Cluster

After the completion of the clustering per installation, the $k$ profiles are extracted. The profiles provide a general view of the PV generation patterns per installation. Next, a selection of a specific profile is done and second clustering takes place. This leads to the formation of the PV power generation profiles per cluster. In this stage, the population of the patterns for clustering is $M = 11$. For the second clustering, the profile of the most populated cluster of the first stage clustering is selected. Again, the number of clusters of the PV installations is unknown; the K-means is executed for variable number of clusters.

## 6.3 Techno-Economic Assessment

### 6.3.1 Overview

A certain PV support mechanism usually requires the determination of various parameters related to the consumption and generation of electrical energy within a time step. Depending on the mechanism's content, a PV installation may require one or two metering devices. The configurations with one meter use a bi-directional meter and are the most common one for full-netting schemes. In partial netting schemes two-meter configurations may be considered. For example, one meter is used to measure the PV generation and the other to measure the incoming electricity. In the present study, the type and cost of the metering system is not considered in the techno-economic analysis. In the following Sections, the self-consumption and the NEM schemes as well as the Scenarios examined in this study are described.

## 6.3.2 *Self-consumption and Utilization Rates*

The volatility of PV systems necessitates the interaction with the grid, unless a combination of storage, demand-side management and energy conservation measures is employed. Generally, the grid is resilient enough to absorb exported energy from PV systems and provide the needed power when PV energy is not sufficient to cover the energy demand of an installation. However, with high PV penetration several problems may emerge, such as reverse power flow, overvoltages, flexibility needs, etc. (Martín-Chivelet and Montero-Gómez 2017). The self-consumption and the utilization (or self-sufficiency) rates are two indicators that describe how PV energy is used on-site and to what extent the grid is utilized. They can be defined for any given time period, although usually they are calculated for a period of 24 h (i.e. daily Self-Consumption Rate). These are defined with the aid of Fig. 6.4, corresponding to a daily Self-Consumption Rate, as follows:

- The Self-Consumption Rate (SCR) is the ratio between the PV energy directly consumed on-site over the total PV energy production, that is: $SCR = \left(\frac{C}{C+B}\right) \cdot 100$

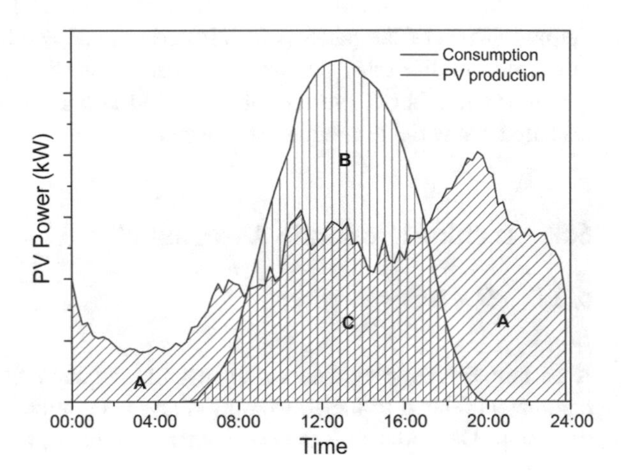


**Fig. 6.4** Sample PV production and consumption profiles of a certain installation

- The Utilization Rate (UR) is a measure of the percentage of PV energy used on-site compared with the total energy demand. This means:

$$UR = \left(\frac{C}{A + C}\right) \cdot 100$$

where $A$, $B$ and $C$ are areas of the PV generation and prosumer's consumption curves.

A high SCR means that PV energy is mostly utilized on-site and excess generated PV energy export to the grid is limited. This, though, may imply that demand-side management is utilized or that the installed PV system is small compared to demand. From a grid technical perspective a high SCR may be beneficial, since high PV injection may lead to overvoltages, while the prosumer may have financial benefits as well, depending on the incentive scheme or general policy.

### 6.3.3 Energy Modelling of Net-Metering and Self-consumption Schemes

Using the above Fig. 6.4 we define the following elements applicable for each time step $\tau$:

- Energy generation $E_{G\tau}$ (kWh): the generated PV energy during time step $\tau$. This can be measured using a dedicated meter or estimated.
- Energy demand $E_{D\tau}$ (kWh): the energy imported from the grid during time step $\tau$. This can be measured by a dedicated meter.
- Energy feed $E_{F\tau}$ (kWh): the excess PV energy injected to the grid during time step $\tau$. Energy demand and feed can be measured with a bi-directional meter or estimated.
- Netted energy $E_{N\tau}$ (kWh): the net energy consumed during time step $\tau$, calculated by subtracting the exported energy from the imported energy i.e. $(E_{D\tau} - E_{F\tau})$.
- Energy consumption $E_{C\tau}$ (kWh): the real energy consumed during time step $\tau$. This can be calculated with the formula $(E_{D\tau} + E_{G\tau}) - E_{F\tau}$.

### 6.3.4 Economic Parameters: Purchased Electricity Cost and Prosumer's Revenue

Within the NEM concept, the prosumer profits is directly related to the cost of purchased electricity from the grid, in a cost-avoidance context. An electricity tariff is composed by many cost elements that, in general, can be divided to the following categories: Electricity production charge, network charge, standing fees, taxes and VAT. From a NEM policy perspective, the significant issue is how a prosumer is charged at each billing period. Therefore, to provide a detailed analysis on the effects of the purchased electricity, a further cost breakdown is held as follows:

- Netted charge $C_N$ (€/kWh): The charge for the prosumer is calculated using the net energy consumed in a billing period.
- Non-netted charge $C_{NN}$ (€/kWh): The charge is calculated using the total energy consumed by the prosumer, i.e. the sum of the imported energy from the grid and the self-consumed PV energy.
- Grid demand charge $C_D$ (€/kWh): The charge is calculated using the total energy imported from the grid.
- Fixed charge $C_F$ (€ or €/kWp or €/kW): This charge may be due to various charge mechanisms that exist, like the standing fees, the power component of transmission and distribution charges, the possible special charge for a NEM prosumer based on the installed power, etc.

Concerning taxes, grid charges or other duties imposed, they can be included in the above categories as well. For example, a full netting scheme implies that all taxes and duties are included in the "Netted charge", which means that they are calculated based on the net consumed energy within a netting period. On the other hand, a partial netting scheme may imply that part or all of taxes and duties are included in the "Non-netted charge" category, whereas the grid charges may be included in the "Grid demand charge". The latter is influenced by the level of self-consumption, since the higher this rate is, the lower the energy imported from the grid and hence the charge for the prosumer. In addition, a policy variation may dictate the selling of the excess produced PV energy to the grid at a certain price. This can be modelled using a revenue element as follows:

- Feed-in repayment $R_F$ (€/kWh): revenue provided to the prosumer for feeding excess PV energy back to the grid in a billing period.

Obviously, when dealing with certain retail electricity tariffs and NEM schemes, a pre-processing is required in order to be able to determine the above cost and revenue elements.

### *6.3.5 Prosumer's Profits*

Within a certain incentive scheme (i.e. NEM or self-consumption), a prosumer may profit by: (a) Directly selling excess PV produced energy to the grid, or (b) indirectly through the avoided electricity cost that would have been charged otherwise. For the purpose of calculating the total economic benefits, both types of profits must be evaluated. It should be noted that the direct profit is straightforward to evaluate. The indirect profit may be more complicated because it contains elements that can be specific to a certain scheme. Before advancing into the calculation procedure, some important definitions related to such schemes are provided. The "Billing Period (BP)" is defined as the period when all the relevant elements are calculated and the prosumer is charged. In case the netted energy is positive (i.e. excess of PV energy generated during the billing period), this energy can be either transferred to the next billing period as "Renewable Energy Credits (RECs)" within a netting period or reimbursed using the Feed-in repayment $R_F$. The "Netting Period (NP)" is a setting that determines for how long the RECs are still valid and taken into account for the prosumer charging. The estimation of the prosumer profits is a multi-step procedure that involves the following 6 steps. It has to be noted that although these steps correspond to the evaluation of NEM schemes, they can be applied to self-consumption schemes as well with minor adjustments. In the latter case, the elements related to netted energy are absent.

Step 1: For each billing period the following energy-related elements are calculated:

- Total PV generated energy, $E_{G,bp}$ (kWh): $E_{G,bp} = \sum_{\tau=1}^{bp} E_{G\tau}$
- Total energy absorbed from the grid, $E_{D,bp}$ (kWh): $E_{D,bp} = \sum_{\tau=1}^{bp} E_{D\tau}$
- Total energy injected to the grid, $E_{F,bp}$ (kWh): $E_{F,bp} = \sum_{\tau=1}^{bp} E_{F\tau}$
- Total netted energy, $E_{N,bp}$ (kWh), adding any RECs from the previous billing period (if applicable): $E_{N,bp} = \sum_{\tau=1}^{bp} E_{N\tau} + REC_{bp-1}$
- Total energy consumption, $E_{C,bp}$ (kWh): $E_{C,bp} = \sum_{\tau=1}^{bp} E_{C\tau}$

Step 2: For each billing period the relative costs and revenues are calculated:

- The total cost for the prosumer has a component related to the netted charge ($C_N$), a component related to the non-netted charge ($C_{NN}$), one related to the grid demand charge ($C_D$) and one to the fixed costs ($C_F$): $C_{NEM,bp} = C_N \cdot E_{N,bp} + C_{NN} \cdot E_{C,bp} + C_D \cdot E_{D,bp} + C_F$
- The total direct revenues for the prosumer (if applicable) for injecting excess PV energy to the grid: $R_{D,bp} = R_F \cdot E_{F,bp}$

Step 3: For each billing period the indirect profit due to avoided electricity cost is calculated:

- Determine the electricity cost for the prosumer without the PV system based on the existing electricity tariff and the total actual consumption (business as usual scenario). This is denoted as $C_{A,bp}$ and generally can be evaluated as: $C_{A,bp} = (C_N + C_{NN} + C_D) \cdot E_{C,bp} + C_F'$, where $C_F'$ denotes the fixed cost as before but without any possible charges related to the examined scheme.
- Evaluate the indirect profit in a billing period by subtracting $C_{NEM,bp}$ from $C_{A,bp}$: $R_{I,bp} = C_{A,bp} - C_{NET,bp}$

Step 4: For each billing period the total profits, $R_{bp,}$ for the prosumer are evaluated:

- Adding the direct and indirect profits in the billing period: $R_{bp} = R_{I,bp} + R_{D,bp}$

Step 5: Repeat Steps 1 to 4 for all billing periods within a netting period:

- Calculate the total profits in the netting period: $R_{np} = \sum_{i=bp,2bp...}^{np} R_i$

Step 6: Repeat Step 5 for all netting periods within a year:

- Calculation of the annual profits: $R_{annual} = \sum_{i=np,2np...}^{year} R_i$

### 6.3.6 Financial Analysis

The indicators used for the economic assessment are the Internal Rate of Return (IRR), the Simple and Discounted Payback Period (SPP, DPP), and the Net Present Value (NPV) (Comello and Reichelstein 2017). These indicators are commonly used in PV projects' evaluation. Generally, a low NPV or IRR value of a PV investment under a specific NEM scheme will indicate that the investment is financially risky. The calculated annual profits from the aforementioned multi-step procedure are used in the financial analysis utilizing annual cash flows. The main financial parameters required for the investment's economic feasibility are the installation/capital costs (€/kWp), the operation & maintenance (O&M) costs (% of capital cost), the DR (%), the investment lifetime (years) and the inflation (%) of both the electricity prices and the O&M costs. Note that other financial parameters may be also used utilized, e.g. the inverter replacement costs at a certain year during the investment's lifetime, the evolution of the feed-in repayment ($R_F$) for the exported energy to the grid, etc. Moreover, parameters related to the reduction of PV generation rate at a certain percentage per-year and the change in the consumption profile or the prosumer may be utilized as well.

### 6.3.7 Scenarios Formulation

For the scope of this study, we assume that the prosumer is charged with the current basic residential tariff (i.e. namely "G1") of the Public Power Corporation (PPC) SA of Greece (Residential "G1" Electricity Tariff 2017). In this Section, the Scenarios considered in this study are described:

- Scenario #1 corresponds to a full NEM scheme with annual netting period. It is the most favorable scenario for the prosumer, since the excess produced PV energy is compensated at retail price (minus standing fees-fixed costs) within a calendar year. This is done by transferring the excess PV energy in the form of RECs to the next billing period, within the annual netting period. These RECs have equal value to the retail price. On the other hand, this is a scheme that utilities favor least, as they lose revenues from the netted network charges.

- Scenario #2 simulates a partial NEM scheme with annual netting period. Now, the network charge and taxes for the prosumer are calculated based on the actual imported energy from the grid, while production and supply charges are fully netted. This means that even in case the produced PV energy and the consumed energy are equal within the netting period, the prosumer has to pay the network charges and the taxes based on his usage of the grid. The transferred RECs within the netting period have a value equal to the generation/production charge of the retail electricity tariff. This scenario rewards those prosumers that manage to keep a high self-consumption ratio and penalizes those with a low SCR, through the partial netted network charges and taxes. The utilities are less affected in this way.
- Scenario #3 is a pure self-consumption scheme, in which any excess produced PV energy is not compensated or netted in any way. In such a scheme, the prosumer aims to increase his SCR and limit as much as possible the exporting of PV energy to the grid, since he receives no compensation for that whatsoever. In order to achieve that, a prosumer under a pure self-consumption scheme may consider additionally installing an energy storage system, or applying demand-side management techniques.
- Scenario #4 is a self-consumption scheme where the excess produced PV energy is compensated based on a certain tariff. In this example, this tariff is equal to the average System Marginal Price (SMP) of the Greek interconnected system of 2013 (i.e. 0.04 €/kWh approximately). This scheme is more profitable for the prosumer compared to Scenario #3, but less profitable than Scenarios #1 and #2. It may be considered fairer, as the prosumer is compensated for providing energy to the grid and at the average generation cost of the overall system. Another formulation of this scheme includes the compensation of the prosumer with the actual SMP for each hour of the day, or the addition on top of the SMP of a premium (e.g. 10%) to cover the supply costs.
- Scenario #5 corresponds to a full NEM scheme, but now with an hourly netting period and without any compensation for excess generated PV energy. This is a NEM scheme that results in financial benefits similar to that of a pure self-consumption scheme, as shown in Fig. 6.4. However, in case the SCR of the prosumer changes considerably within an hour, then this scheme may in reality result in higher profits compared to the pure self-consumption scheme.

Recall that the aim of the analysis is to evaluate the economic attractiveness of investing in a PV system under different NEM and self-consumption schemes.

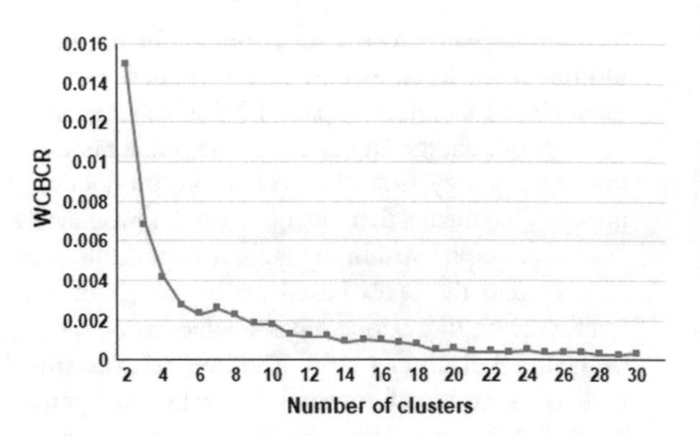


**Fig. 6.5** WCBCR indicator curve for the PV system located in Ios island

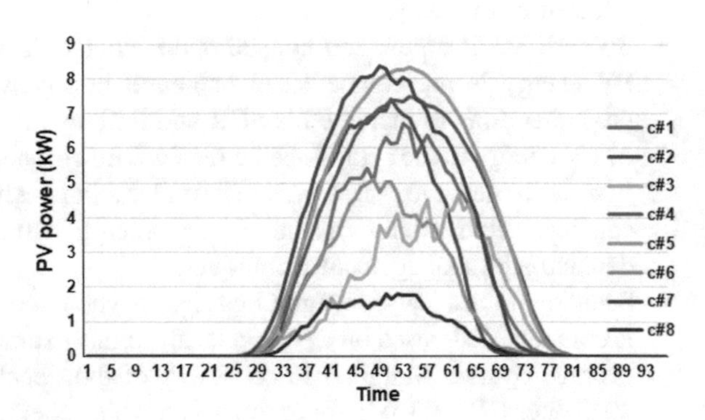


**Fig. 6.6** PV generation profiles

## 6.4 Results

### *6.4.1 Clustering Per PV Installation—1st Stage Clustering*

The K-means algorithm is executed for 2–30 clusters with an increasing step equal to 1. Due to space limitations, only the results of 1 system will be presented. Figure 6.5 shows the WCBCR indicator for the PV system installed in Ios island. While the number of clusters is increasing, the WCBCR receives in general terms lower values, i.e. clustering is more robust. By utilizing the “knee” point detection method on the WCBCR curve, the optimal number of clusters is drawn (Panapakidis et al. 2015).

For the PV system in Ios, the optimal number is c = 8. The optimal number of clusters for the remaining system varies between 5 and 9. The PV generation profiles of the PV system are presented in Fig. 6.6. It is noticeable that there is diversity in the profiles, indicated as c#*i*, *i* = 1, …, 8.

### 6.4.2 PV Systems Clusters—2nd Stage Clustering

After the profile of the most populated cluster is selected, the K-means is applied again and the PV system clusters are drawn. The algorithm is executed for 2–11 clusters. The optimal number of clusters is 4. The 1st cluster includes the following PV systems: Nea Mesimvria, Imathia, Peraia and Agios Petros, The 2nd cluster includes the following PV systems: Panorama, Ierissos and Profitis. The 3rd cluster includes the following PV systems: Serres and Vrasna. The 4th cluster includes the following PV systems: Ios and Poligyros.

### 6.4.3 Scenarios Comparison

The 2nd stage clustering leads to PV system clusters. Instead of applying a techno-economic analysis per park, the analysis can be limited to 1 park per cluster. Recall that the scope is to reduce the computational effort for applying system specific economic analyses. Since the analysis takes into account the consumption, the clustering can be applied also to the load data. While, residential consumers in Greece present in general similar patterns, in the present study the categorization of the PV systems is held only by using the PV generation profiles since they correspond to different climatic conditions (i.e, different solar irradiation patterns).

The analysis will be restricted to 1 system and namely the one in Ios island. In addition, the analysis will only regard average weekday and weekend daily load curves. Figure 6.7 shows the average daily load curve of the prosumer in Ios island for specific months. For the examination of the above Scenarios we consider the following inputs:

- Average annual PV produced energy: 1470 kWh/kWp of installed PV capacity
- System losses: 10%
- Annual degradation of PV panels: 1%
- Allowed PV system capacity: 1–10 kWp
- Production and supply charges: 0.1025 €/kWh
- Network charges: 0.0266 €/kWh
- Taxes: 0.0727 €/kWh
- The VAT is not included in any of the calculations
- PV system costs: Installation costs—1400 €/kWp, Connection costs—300€, Operation & Maintenance (including insurance) costs—1.5%/year of the installation costs
- Other financial parameters: Discount rate—5%, Inflation (applies to both the electricity tariff and O & M costs): 2%

The 5 Scenarios are compared in Figs. 6.8, 6.9 and 6.10 for different installed capacities. It should be noted that the comparison can take place in terms of

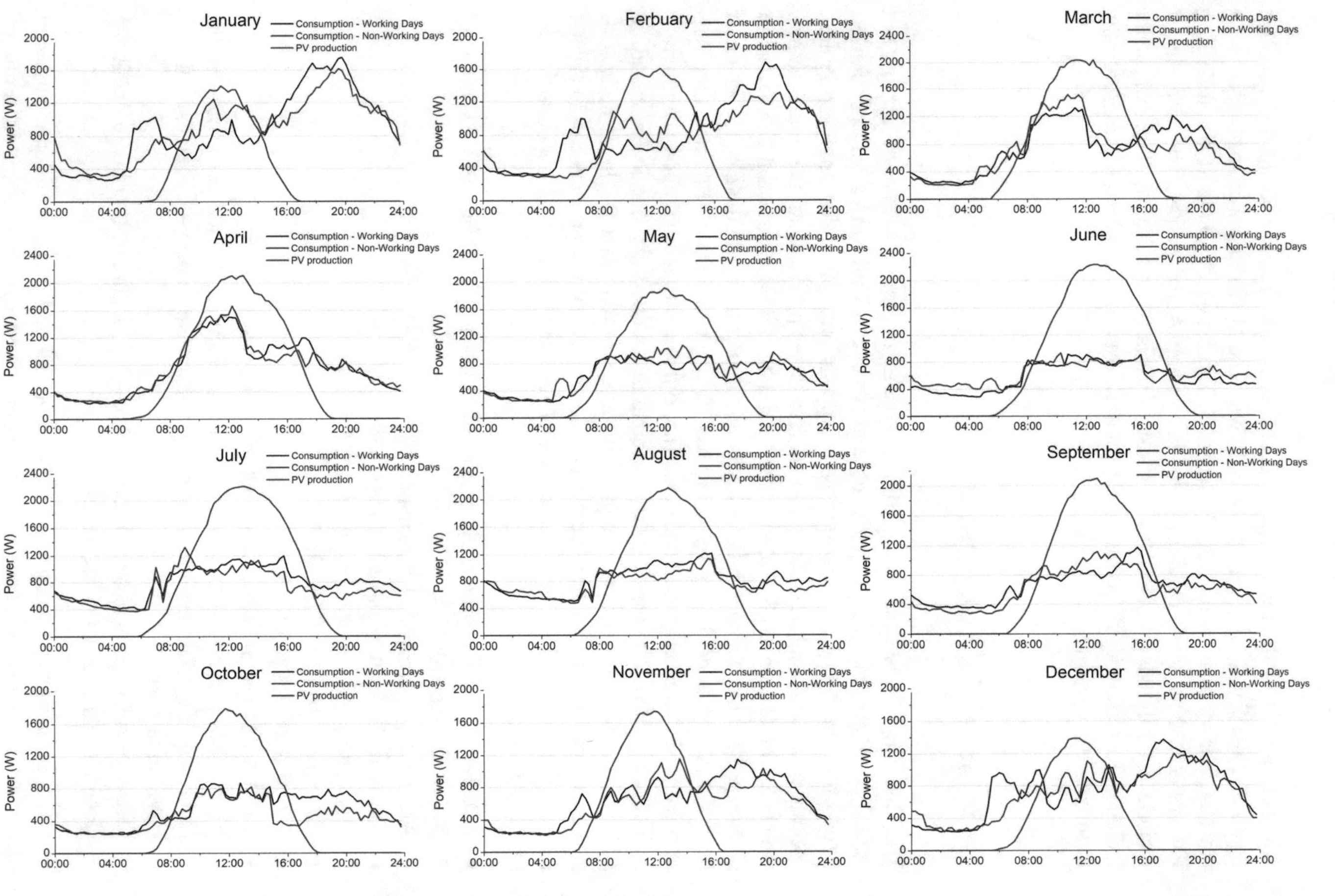


**Fig. 6.7** Averaged weekday and weekend load curves

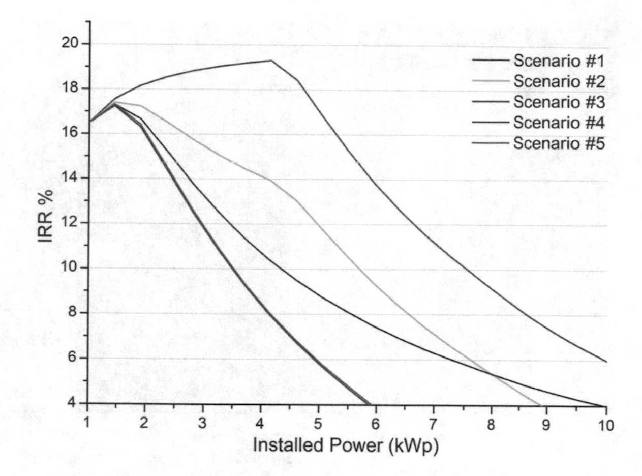


**Fig. 6.8** Variation of the IRR for different PV installed capacities

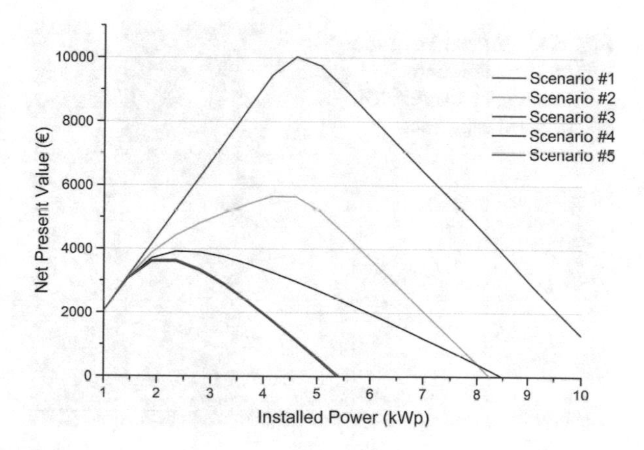


**Fig. 6.9** Variation of the NPV for different PV installed capacities

variations in the installation cost, discount rate values and others. According to Fig. 6.8, the IRR presents a similar trend in all Scenarios, i.e. it receives lower values while the installed capacity increases. This is less evident in Scenario #1 where it reaches its maximum in 4.50 kWp. Scenario #3 and Scenario #5 present almost similar IRR values.

The less favorable scenario for the prosumer is Scenario #3. No considerable economic benefits can be reached if there is no compensated PV generation. However, an important factor is the PV generated electricity selling value to the grid. If the electricity is sold at a constant value, the prosumer is immune to fluctuations of real time electricity prices, but will also not profit when this price increases.

Similar conclusions are drawn by examining the NPV indicator. The optimal installed capacities range for the examined prosumer lie between 4 and 5.50 kWp. If the support mechanism is self-consumption, the prosumer should

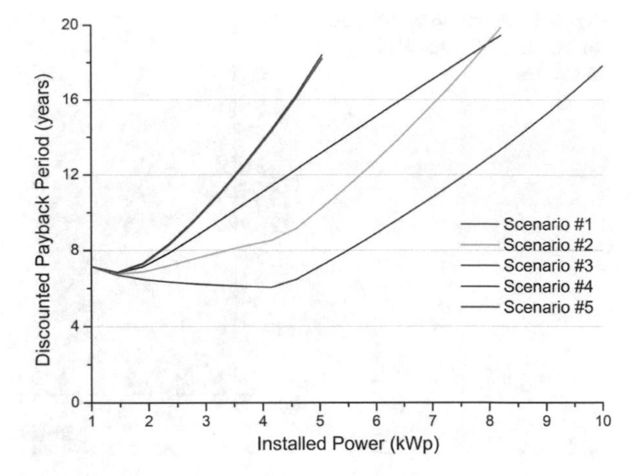


**Fig. 6.10** Variation of DPP for different PV installed capacities

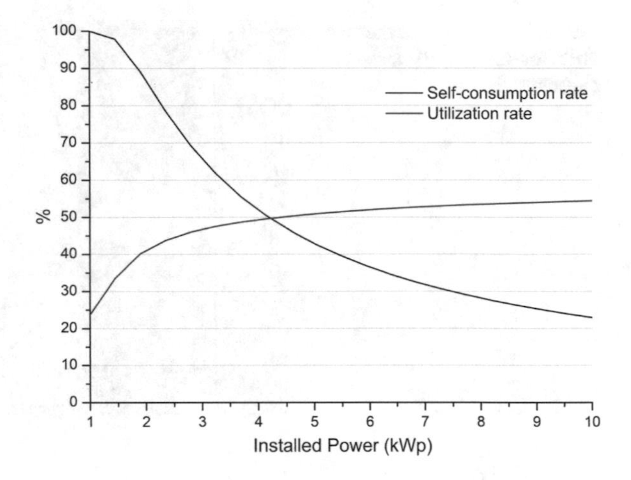


**Fig. 6.11** Variations of the self-consumption and utilization rates for different PV installed capacities

select a lower installed capacity in order to maximize his economic benefits. According to Fig. 6.9, the taxes and other costs not directly associated with the amount of purchased electricity (i.e. as represented by Scenario #2) have a visible impact on the profits. The results presented in Fig. 6.10 show that no scheme can lead to payback less than 4 years period. As the installed capacity increases, the payback period also increases. The DPP curves of Scenario #2 and Scenario #4 cross at 8 kWp. This means that if the installation capacity of the system is 8 kWp, the two scenarios will result in the same payback period. Finally, Fig. 6.11 presents the variations of the self-consumption and utilization rates. It can been noticed that the most favorable capacity in the case of self-consumption is 4.50 kWp. After this value, further increase of the capacity do not considerably influence the utilization rate.

## 6.5 Concluding Remarks

While FiTs lead to installation increments across the globe, many experts believe that they have reached their full potential. Alternate support schemes such as NEM and self-production may aid to the revitalization of the PV markets in countries where storage is still not competitive. It should be noted that compared to FiTs, self-consumption and NEM are more complicated since they have to deal with more issues, such as excess electricity compensation policy, selling price to the grid, network access costs and others. Therefore, they are more country specific.

The present study provides a methodology of techno-economic assessment of various NEM and self-consumption schemes. The methodology is not restricted by the location or the installation capacity, a fact that makes it suitable for PV feasibility studies prior to their implementation. The clustering tool is employed to derive the profiles per PV system and PV clusters. The profiles per PV system can aid in the examination of the generation patterns. For instance, outliers may be detected and isolated, seasonalities may be analyzed, etc. Also, in cases of large data sets, clustering can provide a data reduction approach; the necessity of continuous gathering of data would be the case, since the PV system generation trend can be represented by a reduced set of curves, i.e. the profiles. Furthermore, clustering provides the potential to group together PV systems with similar behavior and thus, lowering the need for techno-economic studies per system. It should be clarified that important factors that influence the economic attractiveness of any NEM and self-consumption scheme are the consumption of the prosumer and the compensation price.

## References

Ali, A., Li, W., Hussain, R., He, X., Williams, B. W., & Memon, A. H. (2017). Overview of current microgrid policies, incentives and barriers in the European Union, United States and China. *Sustainability, 9*(7), 1146.

Bertsch, V., Geldermann, J., & Lühn, T. (2017). What drives the profitability of household PV investments, self-consumption and self-sufficiency? *Applied Energy, 204,* 1–15.

Comello, S., & Reichelstein, S. (2017). Cost competitiveness of residential solar PV: The impact of net metering restrictions. *Renewable and Sustainable Energy Reviews, 75,* 46–57.

de Lima, L. C., de Araújo Ferreira, L., & de Lima Morais, F. H. B. (2017). Performance analysis of a grid connected photovoltaic system in northeastern Brazil. *Energy for Sustainable Development, 37,* 79–85.

Eltigani, D., & Masri, S. (2015). Challenges of integrating renewable energy sources to smart grids: A review. *Renewable and Sustainable Energy Reviews, 52,* 770–780.

Enslin, J. (2014). Integration of photovoltaic solar power-the quest towards dispatchability. *IEEE Instrumentation and Measurement Magazine, 17*(2), 21–26.

EurObserver. (2014). *Photovoltaic barometer*. http://www.energies-renouvelables.org.

Khan, S. S., & Ahmad, A. (2004). Cluster center initialization algorithm for K-means clustering. *Pattern Recognition Letters, 25*(11), 1293–1302.

Mabee, W. E., Mannion, J., & Carpenter, T. (2012). Comparing the feed-in tariff incentives for renewable electricity in Ontario and Germany. *Energy Policy, 40,* 480–489.

Martín-Chivelet, N., & Montero-Gómez, D. (2017). Optimizing photovoltaic self-consumption in office buildings. Energy and Buildings.

Martinez-Anido, C. B., Botor, B., Florita, A. R., Draxl, C., Lu, S., Hamann, H. F., et al. (2016). The value of day-ahead solar power forecasting improvement. *Solar Energy, 129,* 192–203.

Mauleón, I. (2017). Photovoltaic investment roadmaps and sustainable development. *Journal of Cleaner Production, 167,* 1112–1121.

Ottesen, S. Ø., Tomasgard, A., & Fleten, S. E. (2016). Prosumer bidding and scheduling in electricity markets. *Energy, 94,* 828–843.

Panapakidis, I., Alexiadis, M., & Papagiannis, G. (2015). Evaluation of the performance of clustering algorithms for a high voltage industrial consumer. *Engineering Applications of Artificial Intelligence, 38,* 1–13.

Poullikkas, A. (2013). A comparative assessment of net metering and feed in tariff schemes for residential PV systems. *Sustainable Energy Technologies and Assessments, 3,* 1–8.

PVPS, I. (2016). *Annual report 2015 [EB/OL].* http://www.iea-pvps.org.

Pyrgou, A., Kylili, A., & Fokaides, P. A. (2016). The future of the Feed-in Tariff (FiT) scheme in Europe: The case of photovoltaics. *Energy Policy, 95,* 94–102.

Residential "G1" Electricity Tariff. (2017). *Public Power Corporation Sa*. http://www.dei.gr.

Satchwell, A., Mills, A., & Barbose, G. (2015). Quantifying the financial impacts of net-metered PV on utilities and ratepayers. *Energy Policy, 80,* 133–144.

Steinley, D. (2006). K-means clustering: a half-century synthesis. *British Journal of Mathematical and Statistical Psychology, 59*(1), 1–34.

Xu, R., & Wunsch, D. (2008). *Clustering*. NewYork: Wiley.

Yamamoto, Y. (2012). Pricing electricity from residential photovoltaic systems: A comparison of feed-in tariffs, net metering, and net purchase and sale. *Solar Energy, 86*(9), 2678–2685.

Ye, L. C., Rodrigues, J. F., & Lin, H. X. (2017). Analysis of feed-in tariff policies for solar photovoltaic in China 2011–2016. *Applied Energy, 203,* 496–505.

Zhang, Y., Chen, W., & Gao, W. (2017). A survey on the development status and challenges of smart grids in main driver countries. *Renewable and Sustainable Energy Reviews, 79,* 137–147.

# Chapter 7
# Life Long Economic Analysis for Industrial Microgrids: A Case Study in Turkey

**Cagri Ozturk, Irem Duzdar Argun and M. Özgür Kayalica**

**Abstract** Microgrids are used prevalently in isolated sites as a solution for multiple resource usage and distributed energy generation. Industrial Zones are constructed as isolated sites, where expectations include reducing the energy costs, providing local energy supply with fewer fluctuations and reducing greenhouse gas emissions. To encourage the microgrids in a developing country of Small and Medium-sized Enterprises (SMEs) placed in industrial zones, pre-investment studies are to be run. This article aims at minimizing the total energy costs of an organized industrial zone in parallel with mitigation of emission for climate change. The costs depend on the number and power of the Wind Turbines (WT) and the capacity of Photovoltaic (PV) panels when renewable energy sources and power storage construct the resources. A Mixed Integer Nonlinear Programming (MINLP) model is proposed to optimize the number of installations to satisfy the current demand. Lifelong carbon emission and cost analysis are performed to minimize the total cost of ownership. In this initial study, uncertainties caused by the renewable energy supply are smoothed by limited use of one gas tribune and grid connection. A case study of the model is implemented for Gebze Industrial Zone. This project will contribute to the researches on microgrids for a long term optimization model.

## 7.1 Introduction

Global warming, rising energy demand due to industrial and technological developments, and depleted fossil fuels force countries to use new energy systems based on renewable energy sources. Air pollution mitigation efforts point to the specific renewable energy sources as wind turbines and solar panels. Microgrids are smart systems providing power with Distributed Energy Resources (DER) and they are

C. Ozturk (✉) · M. Ö. Kayalica
Istanbul Technical University, Macka, Istanbul, Turkey
e-mail: ozturkcagr@itu.edu.tr

I. D. Argun
Duzce University, Duzce, Turkey

© Springer International Publishing AG, part of Springer Nature 2018

C. Kahraman and G. Kayakutlu (eds.), *Energy Management—Collective and Computational Intelligence with Theory and Applications*, Studies in Systems, Decision and Control 149, https://doi.org/10.1007/978-3-319-75690-5_7

controlled independently. They can operate in two different modes; in grid-connected mode, they allow import power from external grid, but in island mode they are isolated from the external grid depending on reliability of responding the demand. Microgrids have efficient distribution systems due to DER located close to the consumers. They form a safe and reliable power supply system based on consumer preferences and power quality requirements. Thus, they have a grid structure with sufficient power generation and balancing resources to operate autonomously and independently from the external grid during interruptions.

This study aims to realize two tandem objectives: first one is minimizing the long term total energy cost of an industrial zone, and the second one is mitigation of emissions. The costs depend on the number and power of the Wind Turbines (WT) and the capacity of Photovoltaic (PV) panels, while renewable energy sources and power storage construct the resources. This article will propose an optimization model that depends on the existing demand and availabilities using Mixed Integer Nonlinear Programming (MINLP) method. This model considers the capacities and the constraints of the zone to optimize size and number of WT and PV and capacity of limited gas turbines, as well as external grid usage in line with the local constraints. Energy supply is observed hourly because of the industrial demand and renewable resource fluctuations. In order to stabilize the energy supply, connecting to the gas turbine and/or external grid is used to lower the demand peaks.

The case study is applied in Gebze Industrial Zone (GIZ) Power consumption data and constraints of the region are considered and demand is responded through a microgrid system.

This paper is organized that the following section gives a review of previous researches related to microgrid systems. Section 7.3 represents using methodology in this study. Section 7.4 presents the proposed model constructed using the designed constraints and net present value calculation. The case study and results are presented in Sect. 7.5 and finally, conclusion and recommendations are given.

This project will contribute to the research on microgrids with a pre investment analysis of a long-term optimization model.

## 7.2 Literature Review

Microgrids are widely considered as a solution for both climate change and peak shaving in recent years. Energy management solutions for microgrids have some advantages and disadvantages in combining the environmentalist and ecologic views as examined by Li et al. (2016). The economic assessment of the microgrids is still stimulating. Dicorato et al. (2009) realized that economic factors influencing the application of microgrids have been accrued. They studied a model to reduce the costs of microgrid application under various limitations. Xie et al. (2015) studied the economic evaluation of microgrids on the basis of discount rates, unit costs of power, and the government subsidy. In this study Xie et al. (2015)

proposed an approach by integrating the economic analysis with optimizing allocation of microgrids.

The optimization effort is observed by an objective to reduce the annual costs, considering the capacity and power constraints. The main intent of Costa and Matos (2006) is to build a model defining all of the cost factors (investment, maintenance, operation, etc.) and the benefits. This model is regards the legal conditions necessary to define the benefits, where, there is a way to evaluate the risks caused by data variations and system parameters. A methodology is proposed by Asanol and Bandol (2007) in order to optimize the operation and economic design for application of microgrids to the renewable energy sources. Asona and Bando (2008) presented a similar evaluation using a methodology based on the economic scale of a gasoline engine's partial load efficiency with the optimal capacity and number of equipments used. The annual work program is derived as the result of optimal planning technique to minimize the annual expenses. This study estimates the required reserves as well.

Microgrid energy management systems and cost analyses are evaluated for many different countries and a large span of scenarios. Various types of algorithms and models are used for finding the optimum model for energy management especially in local scale for decreasing dependency of the external grid and decreasing carbon emission rate are planned with using various algorithms and optimization models. A study run by Yu et al. (2016) proposes economical and green solutions at the design phase of the microgrid system with various energy resources. The solution is achieved by the considering multiple operation inconstancies and different expectations of stakeholders. Hence, a multi objective optimization model to plan the microgrid operation, based on an economic robustness is proposed. This model implies the balance between the investment expenditures and the environmental benefits as well.

Baghaee et al. (2016) investigated minimization of power loss, load loss and annual cost for hydrogen based storage microgrid systems for 20-year period. Multi-objective Particle Swarm Optimization (MOPSO) algorithm is tested for fuel cell, hydrogen storage, PV, WT and direct current. Specific target was to show the benefits and drawbacks of hydrogen storage systems in the long term. Two-Stage Stochastic-Programming is proposed by Hu et al. (2016) to find the optimum working conditions of renewable energy sources in microgrids. In this research minimization of power trading among the microgrids is analyzed as a function of battery usage and capacity. In this paper, energy demand and supply uncertainties are balanced in parallel with providing depreciation of greenhouse gas emissions.

A model is proposed by Mao et al. (2013) to examine and judge economically and to allocate the industrial PV microgrids optimally. Particle Swarm Optimization method is employed to minimize the annual energy expenses, besides reducing the emission for the benefits of industrial users. The verification is done by simulation of three months data gathered from 500 kWh industrial microgrid established in Dongguan City of China. Another model has been proposed by Chen et al. (2011) to define the optimal size and to make economic survey for the energy accumulation system in terms of present net values. Genetic algorithm is employed to allocate and operate the energy storage systems for low voltage microgrid applications. The purpose of the paper prepared by Biczel and Koniak (2011) is to develop a model simulating the power production and designing the storage system capacity. This model computes the required data to optimize the microgrid technically and economically. Another survey is done at Savona Campus of Genoa University describing the basic parts of the Smart Polygeneration Microgrid (SPM). The main idea of this study is to design a model to optimally direct the SPM to minimize the daily operating costs. The reported results exhibit the optimality, and the reduced $CO_2$ emission. The results also show that the energy is saved by optimal operation of SPM after having investigated the traditional conditions by Bracco et al. (2014). Objective of the model is to minimize the expected expenses of correcting operations. Similar to that, a stochastic optimization problem, converted to a nonlinear programming problem is analyzed by Parisioa et al. (2016).

## 7.3 Methodology

### *7.3.1 Assumptions*

Power purchasing/selling prices of external grid, wind speed and solar radiation parameters generate uncertainties. Wind speed and solar radiation are changing with ambient and weather condition thus 1-year data is used.

Energy losses due to cable use and electrical conduction have been neglected. Economic lives of wind turbine and solar PV chosen are taken as 25 years.

### *7.3.2 Power Generation Functions*

Another assumption is that, according to weather reports below equations are used to find wind turbine and solar panel power generation amounts. Wind speed is the major factor in producing power according to (7.1) where, $A$ represents the area of wind turbine, $\rho$ represents to density of air, and represents wind speed (Breeze 2016).

$$f_{WT}(v) = \frac{1}{2}\rho A v^3 \tag{7.1}$$

According to Moradi et al. (2013) photo voltaic generators use maximum solar radiation as main variable as in (7.2). In this equation, other variables are maximum power, temperature coefficient, cell temperature, and reference temperature are $P_{max}$, $I$, $I_{max}$, $k$, $T_c$, and $T_r$.

$$f_{PV}(I) = P_{max}\frac{I}{I_{max}}(1+k(T_c - T_r)) \tag{7.2}$$

### 7.3.3 Net Present Value (NPV)

NPV is used to calculate the current investment value. Whole investment period income and expenses are considered according to risk level of the investment. It is a method frequently used for life cycle assessments. According to Shaffie and Jaaman (2016), the NPV value uses the function in (7.3), where $C_0$, $C_{Fi}$, $r$ consecutively represent capital cost, cash flows, and discount rate.

$$NPV = -C_0 + \sum_{i=1}^{n}\frac{C_{Fi}}{(1+r)^i} \tag{7.3}$$

### 7.3.4 Levelized Cost of Electricity

Energy generation cost per kWh is calculated according to the Levelized Cost of Electricity (LCOE) generation as in (7.4). LCOE values vary by technology, country, capital and operating cost and efficiencies. The financial account of such renewable energy technologies is based on discounting financial flows. In this equation, $I_t$ represents investment expenditures, $M_t$ is the operations and maintenance expenditures, $F_t$ shows resource expenditures, $E_t$ is the power amount generated generation, and r is discount rate, whereas n is the economic life of the system.

$$\mathrm{LCOE} = \frac{\sum_{t=1}^{N}\frac{I_t+M_t+F_t}{(1+r)^t}}{\sum_{t=1}^{N}\frac{E_t}{(1+r)^t}} \tag{7.4}$$

Proposed renewable energy sources' economic life are accepted as 25 years and discount rate is accepted as 7%. Renewable energy sources resource expenditures were given as zero. Table 7.1 shows levelized cost values of various energy sources.

**Table 7.1** Levelized cost of energy sources

| Energy source | Unit production cost ($/kWh) |
|---|---|
| Wind turbine | 0.0394 |
| Solar PV panel | 0.0450 |
| Gas turbine | 0.0486 |

Table 7.2 Purchasing electricity prices of external grid

| Energy source | Unit production cost ($/kWh) |
|---|---|
| Day rate | 0.0554 |
| Peak | 0.0981 |
| Off peak | 0.0240 |

In this study, three tariff grid is used and purchase prices of three tariff external grid can be observed in Table 7.2.

### 7.3.5 *Mixed Integer Nonlinear Programming*

Mixed Integer Modeling can be both linear and Non-Linear. According to Bussieck and Pruessner (2003) these are models with a minimizing objective function and/or several constraints using both discrete (integer) and continuous decision variables.

## 7.4 Proposed Model

This Section begins with formulating a MINLP model for energy management in microgrids. In this model wind, solar, natural gas and demand data are used for specified period in a certain location.

In the first stage, the decision of investments for wind turbine, photovoltaic panel, and gas turbine are determined for minimizing total cost. After that capital costs, replacement costs, operation and maintaining costs per power plant are calculated through these data, thus a lifecycle analysis can be provided with multiple objectives.

t denotes time period hourly; $X_{WT}$, $X_{PV}$ and $X_{NG}$ are decision variables representing the number of gas turbines, wind turbines, and photo voltaic panels; $C_{G,t}$ represents hourly purchasing electricity cost of external grid per kWh these values are assumed as real hourly purchasing electricity prices of grid, and $C_{NG}$, $C_{WT}$, $C_{PV}$ represents natural gas turbine, wind turbine, and photovoltaic panel electricity generation cost per kWh. $E_{G,t}$ is the decision variable that represents purchasing power at time t from external grid, $E_{NG,t}$, $E_{WT,t}$ and $E_{PV,t}$ are decision variables for amount of power generated at time t as kWh per wind turbine, photo voltaic panels and gas turbine. Main objective in this model is that minimizing total cost of power plants and external grid purchasing energy are calculated in (7.5).

$$\text{Min } Z = \sum_{t=1}^{8760} C_{G,t}\, E_{G,t} + C_{NG}\, E_{NG,t}\, X_{NG} + C_{WT}\, E_{WT,t}\, X_{WT} + C_{PV}\, E_{PV,t}\, X_{PV} \tag{7.5}$$

$X_{WT}$, $X_{PV}$, and $X_{NG}$ are the number of wind power, photo voltaic and natural gas generators, respectively; $E_{G,t}$, $E_{NG,t}X_{NG}$, $E_{WT,t}X_{WT}$, $E_{PV,t}X_{PV}$ are used to represent the amounts of purchasing electricity from external grid and generated electricity using wind, photo voltaic, and natural gas generators. Hourly demand of an industrial zone is $D_t$ and hourly demand constraint is as follows:

$$E_{G,t} + E_{NG,t}X_{NG} + E_{WT,t}X_{WT} + E_{PV,t}X_{PV} = D_t \quad \forall\, t \tag{7.6}$$

The probability density function of wind speed $f_{WT}$ (ν) is used for determination of wind turbine capacity will be multiplied with efficiency coefficient of wind turbine $\eta_{WT}$. In this function ν is wind speed and as a result of uncertainty of wind speed, it has empirical distribution. For determination of PV panels electricity production amounts according to solar irradiance values, $f_{PV}$ *(I)* is used and in this function I denotes solar irradiance. Using efficiency of gas turbine coefficient identifies capacity of gas turbine and production capacity in hour as $\eta_{NG}$ and $K_{NG}$. All of the capacity constraints are identified in (7.7)–(7.9).

$$E_{WT,t} \quad \forall\, t \leq f_{WT}(\nu)\eta_{WT} \quad \forall\, t \tag{7.7}$$

$$E_{PV,t} \leq f(I) \quad \forall\, t \tag{7.8}$$

$$E_{NG,t} \leq \eta_{NG} K_{NG} \quad \forall\, t \tag{7.9}$$

Any Industrial Zone has limited area for construction of wind turbines, photo voltaic panels, and gas turbine. For installation of PV panels, roofs, building sides, and specified area in GIZ are used. Total installed capacities of wind turbine generators, photo voltaic generators, and gas turbine generators are $X_{WT}$ * $I_{WT}$, $X_{PV}$ * $I_{PV}$, and $X_{NG}$ * $I_{NG}$, where I represents footprint of power generator. The available area limits are shown by $A_{WT}$, $A_{PV}$, $A_{NG}$. Finally, the capacity constraints for the wind turbine generators, photo voltaic generators, and gas turbine generators can be shown as in functions (7.10)–(7.12).

$$X_{WT}\, I_{WT} \leq A_{WT} \tag{7.10}$$

$$X_{PV}\, I_{PV} \leq A_{PV} \tag{7.11}$$

$$X_{NG}\, I_{NG} \leq A_{NG} \tag{7.12}$$

Last constraint of the model is limited carbon mitigation according to European Carbon Emission Standards, which is 80% of current $CO_2$ emission. The emission constraint is defined in (7.13), where, κ represents carbon limit, which is produced by power plants. Total carbon emission rates are calculated for wind turbine generator, photovoltaic generator, and gas turbine generator. $P_{WT}$, $P_{PV}$, and $P_{NG}$ represents carbon emission coefficient of electricity producing for gas turbine, wind turbine, and photovoltaic panel.

**Table 7.3** Carbon emission amount of energy sources

| Energy source | Carbon emission amount (kg/kWh) |
|---|---|
| Wind turbine | 0.02 |
| Solar PV panel | 0.19 |
| Gas turbine | 0.454 |
| External grid | 0.4 |

$$\sum_{t=1}^{8760} P_{NG}\, E_{NG,t} X_{NG} + P_{WT}\, E_{WT,t} X_{WT} + P_{PV}\, E_{PV,t}\, X_{PV} \leq \kappa \tag{7.13}$$

Table 7.3 shows carbon emission amount as kg per kWh of different type of energy sources. Carbon emission value of the grid is assumed as 0.4 kg in Turkey, since 80% of power generation is still using fuel, coal, and natural gas.

According to designed microgrid's optimization model integer variable constraints and non-negativity constraints are listed in (7.14) and (7.15).

$$X_{WT}, X_{PV}, \text{ and } X_{NG} \text{ are integer variables} \tag{7.14}$$

$$\begin{aligned} &C_{WT},\ CPV,\ CNG,\ CG{,}t,\ Dt \geq 0 \quad \forall\, t \\ &E_{WT,t}, E_{PV,t}, E_{NG,t}, E_{G,t} \geq 0 \quad \forall\, t \\ &f_{WT}(v), f_{PV}(I), K_{NG} \geq 0 \\ &A_{WT}, I_{WT}, A_{PV}, I_{PV}, A_{NG}, I_{NG} \geq 0 \\ &P_{WT},\ P_{PV}, P_{NG} \geq 0 \end{aligned} \tag{7.15}$$

In the second stage previously mentioned, the NPV function in (7.3) uses the results of this optimization problem.

## 7.5 Case Results and Discussions

Gebze Industrial Zone has 5,160,000 $m^2$ area. There are 224 companies in GIZ, which includes many industries such as metal, plastic, paint, food and chemical industry. GIZ consumes averagely 546,685 MW of electricity in a year. Its electricity expense is 29,814,324.57 $ for a year.

Mixed Integer Nonlinear Programming Model and then the NPV are run to optimize the energy use of GIZ.

The optimum number of power plants are assumed as 14 wind turbines with 3.45 MW capacity each, which are totally 48.3 MW; moreover, 6 MW capacity PV panels, and one gas turbine with 40 MW capacity energy sources usage are proposed in designed microgrid. In the proposed microgrid, the storage is not considered since there is still no used batteries which would make battery storage cheaper for GIZ. When the external grid is used it is much cheaper than the storage.

Implementing the mixed integer programming, total carbon emission is decreased from 248,195,371 to 197,460,075 kg which is approximately 20.5%. Total cost is decreased from 29,814,324.57 to 22,388,815.38 $ which is approximately 25%.

Not only cost but also carbon emission amounts were decreased significantly even carbon tax is not applied in Turkey. If carbon tax is applied, cost reduction is expected to be much higher. The main issue is that investment cost of the wind turbines is very high but total cost of ownership is reduced significantly in the long term.

## 7.6 Conclusion and Future Work

This paper considers the global importance of microgrids and presents a brief literature review to demonstrate the role of microgrids for industrial zones. Our aim is to suggest an economical model for using microgrids in high energy consuming industrial zone, which combines Small and Medium-sized Enterprises (SMEs) and the large companies. That is why, the study is based on reducing both costs and carbon emissions.

The proposed model uses both Net Present Value to calculate the lifelong investment value and mixed integer nonlinear model for minimizing operational costs. The analysis aims to find the most economic structure of a micro grid operation using demand, available space, available wind and illuminations.

The case is run for Gebze Industrial Zone, where chemical and metal production sites exist with the small manufacturing companies side by side. The annual power demand of the zone is one of the highest among hundreds of industrial zones in Turkey.

Results achieved by the case implementation, gives the most economic number of wind, solar and natural gas investments without considering the uncertainties of renewables.

In the extension of this study, uncertainties will also be handled by considering the different sizes of turbines and different conditions. Further studies are recommended to compare the same model for different regional conditions and applying a robust optimization with stochastic variables.

## References

Asano, H., & Bando, S. (2008, July). Economic evaluation of microgrids. In *Power and Energy Society General Meeting-Conversion and Delivery of Electrical Energy in the 21st Century* (pp. 1–6). IEEE.

Asanol, H., & Bandol, S. (2007, April). Economic analysis of microgrids. In *Power Conversion Conference-Nagoya PCC'07* (pp. 654–658). IEEE.

Baghaee, H. R., Mirsalim, M., Gharehpetian, G. B., & Talebi, H. A. (2016). Reliability/cost-based multi-objective Pareto optimal design of stand-alone wind/PV/FC generation microgrid system. *Energy, 115,* 1022–1041.

Biczel, P., & Koniak, M. (2011). Design of power plant capacity in DC hybrid system and microgrid. *COMPEL—The International Journal for Computation and Mathematics in Electrical and Electronic Engineering, 30*(1), 336–350.

Bracco, S., Delfino, F., Pampararo, F., Robba, M., & Rossi, M. (2014). A mathematical model for the optimal operation of the University of Genoa Smart Polygeneration Microgrid: Evaluation of technical, economic and environmental performance indicators. *Energy, 64,* 912–922.

Breeze, P. (2016). *Wind power generation* (pp. 67–73).

Bussieck, M. R., & Pruessner, A. (2003). *Mixed-integer nonlinear programming* (pp. 1–2).

Chen, C., Duan, S., Cai, T., Liu, B., & Hu, G. (2011). Optimal allocation and economic analysis of energy storage system in microgrids. *IEEE Transactions on Power Electronics, 26*(10), 2762–2773.

Costa, P. M., & Matos, M. A. (2006, June). Economic analysis of microgrids including reliability aspects. In *International Conference on Probabilistic Methods Applied to Power Systems (PMAPS 2006)* (pp. 1–8). IEEE.

Dicorato, M., Forte, G., & Trovato, M. (2009, June). A procedure for evaluating microgrids technical and economic feasibility issues. In *PowerTech, Bucharest* (pp. 1–6). IEEE.

Hu, M. C., Lu, S. Y., & Chen, Y. H. (2016). Stochastic programming and market equilibrium analysis of microgrids energy management systems. *Energy, 113,* 662–670.

Li, M., Zhang, X., Li, G., & Jiang, C. (2016). A feasibility study of microgrids for reducing energy use and GHG emissions in an industrial application. *Applied Energy, 176,* 138–148.

Mao, M., Jin, P., Zhao, Y., Chen, F., & Chang, L. (2013, September). Optimal allocation and economic evaluation for industrial PV microgrid. In *Energy Conversion Congress and Exposition (ECCE)* (pp. 4595–4602). IEEE.

Moradi, M. H., Hajinazari, M., Jamasb, S., & Paripour, M. (2013). An energy management system (EMS) strategy for combined heat and power (CHP) systems based on a hybrid optimization method employing fuzzy programming. *Energy, 49,* 86–101.

Parisio, A., Rikos, E., & Glielmo, L. (2016). Stochastic model predictive control for economic/environmental operation management of microgrids: An experimental case study. *Journal of Process Control, 43,* 24–37.

Shaffie, S. S., & Jaaman, S. H. (2016). Monte Carlo on net present value for capital investment in Malaysia. *Procedia-Social and Behavioral Sciences, 219,* 688–693.

Xie, D., Du, Z., Ding, H., Zhang, J., Ma, L., Zhang, S. (2015, July). An integrated configuration optimization and economic evaluation approach for microgrids. In *34th ChineseControl Conference (CCC)* (pp. 7877–7882). IEEE.

Yu, N., Kang, J. S., Chang, C. C., Lee, T. Y., & Lee, D. Y. (2016). Robust economic optimization and environmental policy analysis for microgrid planning: An application to Taichung Industrial Park, Taiwan. *Energy, 113,* 671–682.

# Chapter 8
# Wind Energy Investment Analyses Based on Fuzzy Sets

**Cengiz Kahraman, Sezi Çevik Onar, Başar Öztayşi, İrem Uçal Sarı and Esra İlbahar**

**Abstract** Engineering economics deals with the investment decisions, where the investment parameters are very hard to estimate exactly. In the cases where we do not have the required data for parameter estimation, possibilistic approaches may be used. In this chapter, a brief literature review on wind energy investments is first presented. Later, the chapter gives present worth analysis (PWA) methods extended to fuzzy sets. The chapter introduces ordinary fuzzy PWA, type-2 fuzzy PWA, intuitionistic fuzzy PWA, and hesitant fuzzy PWA. A numerical application for each extension is presented.

## 8.1 Introduction

There is an increasing energy need in the world and carbon-based fuels are the main sources for fulfilling this need. Yet, these carbon-based energy sources damage the ecological environment and they are limited sources. Renewable energy sources are the best alternatives for carbon-based fuels since they are eco-friendly and can provide energy unlimitedly.

Wind energy can become an efficient energy source for many regions. The uncertainty in electricity prices and energy production levels of wind turbines limits the wind energy investments. Especially, the costs and benefits of the long-term wind energy investments are hard to calculate with the traditional engineering economic analysis since they need precise values of investment parameters (Cevik Onar and Kilavuz 2015).

Ordinary fuzzy sets and their extensions such as type-2 fuzzy sets, intuitionistic fuzzy sets, and hesitant fuzzy sets are exceptional tools for dealing with uncertainty in human thoughts and perceptions (Kahraman et al. 2016b). Ordinary fuzzy sets (Zadeh 1965) use membership degrees for representing vagueness and imprecise-

C. Kahraman (✉) · S. Çevik Onar · B. Öztayşi · İ.U. Sarı · E. İlbahar
Industrial Engineering Department, Istanbul Technical University,
Macka, Istanbul, Turkey
e-mail: kahraman@itu.edu.tr

© Springer International Publishing AG, part of Springer Nature 2018

C. Kahraman and G. Kayakutlu (eds.), *Energy Management—Collective and Computational Intelligence with Theory and Applications*, Studies in Systems, Decision and Control 149, https://doi.org/10.1007/978-3-319-75690-5_8

ness. Type-2 fuzzy sets introduced by Zadeh (1975) employ three dimensional membership functions. Type-2 fuzzy sets have grades of membership that are themselves fuzzy. Intuitionistic fuzzy sets introduced by Atanassov (1986) employ both membership and non-membership degrees for defining uncertainty. Hesitant fuzzy sets developed by Torra (2010) represent the hesitancies in decision makers mind. Fuzzy net present worth analysis enables evaluating investment alternatives under vague and incomplete information. The extensions of fuzzy sets enable better defining the uncertainties inherent in investment parameters through their membership functions.

The wind energy investments involve uncertain, vague and incomplete parameters. Therefore, applying classical present worth analyses may create unrealistic results. Calculating present worth with vague and incomplete data may produce incorrect and misleading decisions. Therefore, this chapter shows the calculation of the fuzzy PW of a wind energy investment based on fuzzy parameters. Ordinary fuzzy PW, intuitionistic fuzzy PW and hesitant fuzzy PW are employed in wind energy investment problems.

The rest of the chapter is organized as follows: Sect. 8.2 summarizes the literature on wind energy investments. Section 8.3 presents the fuzzy present worth analyses based on extensions of fuzzy sets. In Sect. 8.4, a wind energy investment problem is analyzed with ordinary fuzzy PW, Intuitionistic fuzzy PW and hesitant fuzzy PW. Section 8.5 concludes the chapter.

## 8.2 Wind Energy Investments: A Literature Review

Much research on wind energy investments exists in the literature. The recent studies in this field will be further examined under two categories as classical techniques and fuzzy techniques.

### *8.2.1 Classical Techniques*

Caralis et al. (2014) investigated the profitability of wind energy investments by employing a Monte Carlo approach to deal with the uncertainties. In their study, Monte Carlo simulation and a typical financial model were integrated to examine different cases of wind energy development. Uncertain parameters considered in the study of Caralis et al. (2014) are wind capacity factor, investment cost, interest rate, feed-in-tariff, absorption rate, grid accessibility. Kucukali (2016) utilized a scoring technique for the assessment of an onshore wind energy project. The proposed method enables decision makers to determine the most appropriate wind energy project by examining the risks of the alternatives. Site geology, land use and permits, environmental impact, grid connection, social acceptance, macroeconomic, natural hazards, change of laws, access road, and revenue are the risks considered in the study of Kucukali (2016). Liu and Zeng (2017) used system dynamics approach to evaluate renewable energy investment risk, particularly wind power projects. After risks in renewable energy investment were analyzed in three categories as technical risk, policy risk and market risk, causal loop diagram for investment risk assessment was formed. The simulation results which are obtained using VENSIM software indicated that policy risk is more crucial in early stage of an investment whereas market risks become more significant with technological advancements and incentive policies improvement (Liu and Zeng 2017). Fazelpour et al. (2017) examined the wind resource and economic feasibility to assess investment risks. The Weibull distribution function was utilized to estimate the wind power and energy density. Windographer software was used to examine the wind direction. For the economic assessment, four types of wind turbines were taken into consideration. These wind turbines are different with respect to rotor diameter, variable rotor speed, nominal power output, cut-in wind speed, rated wind speed, cut-out wind speed, survival wind speed. Monthly capacity factor, energy output and cost of energy of the alternatives with these wind turbines were evaluated. Al-Sharafi et al. (2017) investigated the feasibility of solar and wind energy systems for power generation and hydrogen production and performed an economic analysis by using simulation software, Hybrid Optimization of Multiple Energy Resources (HOMER). Aquila et al. (2017) investigated wind power feasibility under uncertainty by employing Monte Carlo simulation and Value at Risk technique. The proposed framework is quite useful for potential investors because it is able to show the influence of the uncertainty on wind power and electricity prices. Kitzing et al. (2017) proposed a real options model to assess wind energy investments. The proposed model involves an upper capacity limit by considering investment timing and continuous sizing. Moreover, several uncertainty factors such as power price and wind speed are taken into consideration in a stochastic process in the study of Kitzing et al. (2017).

### *8.2.2 Fuzzy Techniques*

Shamshirband et al. (2014) employed adaptive neuro-fuzzy optimization to maximize the net profit of a wind farm. While applying an intelligent optimization method based on the adaptive neuro-fuzzy inference system, net present value and interest rate of return were considered as the measures of net profit. Interest rate per year and unit sale price of electricity were utilized as inputs of optimization scheme whereas output was the optimal number of turbines which is an indicator of maximal net profit. In this study, while determining the optimal number of wind turbines, aerodynamic interactions between the turbines, as well as cost factors, are taken into consideration. In this way, both optimal solution with respect to the maximum net profit and the optimal layout for wind turbines were achieved (Shamshirband et al. 2014). Wu et al. (2014) investigated evaluation criteria considered in the process of wind farm project plan selection and proposed a framework to select the best wind farm project. Criteria considered in this study are construction, resource, wind turbine, financial analysis, social risk, policy risk, technological risk, good influence, bad influence, the influence of project to the local society and stabilization, the influence of project to the local economy and employment, and the influence of project to resource utilization. Wu et al. (2014) employed intuitionistic fuzzy numbers, intuitionistic fuzzy Choquet operator, and generalized intuitionistic fuzzy ordered geometric averaging operator to reduce the probability of information loss and to stay away the independent assumption of multi-criteria decision making methods (Wu et al. 2014). Onar et al. (2015) utilized interval-valued intuitionistic fuzzy sets for the assessment of wind energy investments. Interval-valued intuitionistic fuzzy sets are employed because of its ability to cope with vagueness and impreciseness in a more comprehensive manner. The proposed approach provides an overall performance measurement for wind energy technology alternatives by considering the following criteria: reliability, cooperation, domesticity, performance, cost factors, availability, maintenance, and technical characteristics (Onar et al. 2015). Shafiee (2015) utilized fuzzy analytic network process to determine the most appropriate risk mitigation strategy for offshore wind farms by employing safety, added value, cost and feasibility criteria. Variation of offshore site layout, improvement of maintenance services, upgrading the monitoring systems, and modification in design of wind turbines are the alternatives considered in the study of Shafiee (2015). Petković et al. (2016) investigated the most influential factors on the net present value of a wind farm using adaptive neuro-fuzzy inference system. In their study, seven inputs, number of turbines, power production, cost per power unit, cost, efficiency, interest rate per year, unit sale price of electricity, are selected to analyze the wind farm net present value (Petković et al. 2016). Wu et al. (2016) proposed an inexact fixed-mix fuzzy-stochastic programming method for heat supply management in wind power heating system under uncertainty. In their study, uncertainties are presented as interval values, random variables and fuzzy sets. The proposed approach is a combination of interval-parameter programming, fixed-mix stochastic programming and fuzzy mathematical programming. The proposed approach enables decision makers to observe interval solutions and plausibility degrees of constraint violation in order to determine the best heat supply management strategies (Wu et al. 2016). Gumus et al. (2016) introduced a multi-criteria decision making method consisting of an intuitionistic fuzzy entropy method, an intuitionistic fuzzy weighted geometric averaging operator and intuitionistic fuzzy weighted arithmetic averaging operator for sustainable energy problems. The selection of V80 and V90 onshore and offshore wind turbines was investigated using the proposed method (Gumus et al. 2016). Cunico et al. (2017) proposed a mathematical model taking several uncertain parameters into consideration to analyze investments in the energy sector. It is aimed at covering both pessimistic and optimistic scenarios by integrating uncertain parameters in their decision making model. Therefore, a fuzzy approach and a set of possibilistic techniques were employed to handle the problem. The uncertain parameters considered in their study are uncertainty in the price of fossil resources, the trend in the growing demand and the variation in the availability of fossil reserves (Cunico et al. 2017). Chang (2017) introduced a fuzzy score technique to optimally locate wind turbines. In this study, the proposed technique was utilized to measure the Euclidean distance between the achievement function and their aspirations (Chang 2017). Morshedizadeh et al. (2017) investigated the utilization of imputation techniques and adaptive neuro-fuzzy inference system to predict wind turbine power production. It was revealed that appropriate combinations of decision tree and mean value for imputation might enhance the prediction performance (Morshedizadeh et al. 2017).

There are various studies in the literature on wind energy investments. These studies have different objectives such as analyzing wind energy technology investments, maximizing investment profit, identifying optimal investment decisions, investigating suitability of a region, predicting energy output of a wind farm, and selecting a suitable site for investment. These studies utilize different methods such as Benefit/Cost analysis, real option analysis, adaptive neuro-fuzzy inference system, optimization, and AHP to achieve these objectives. Moreover, evaluation criteria or employed parameters may change with respect to the objective of the study. Table 8.1 shows some representative studies on wind energy investments in the literature.

## 8.3 Fuzzy Present Worth Analysis

Fuzzy logic is used to determine uncertainty occurred from linguistic assumptions. It is possible to represent linguistic definitions in a mathematical form using fuzzy sets. Fuzzy numbers have different types which determine the linguistic terms in different ways. In this section, present worth analysis is constructed using different types of fuzzy numbers such as ordinary fuzzy numbers, intuitionistic fuzzy numbers, type-2 fuzzy numbers and hesitant fuzzy numbers.

Especially in public sector projects such as highways, infrastructure, power generation facilities, project alternatives have very long expected useful lives. In such kind of projects, planning horizon could be taken as infinite to be effective. In this section, the present worth analysis for infinite time horizon is proposed using different types of fuzzy numbers.

### *8.3.1 Ordinary Fuzzy Present Worth Analysis*

There are different types of ordinary fuzzy numbers such as triangular fuzzy numbers, trapezoidal fuzzy numbers, L-R type fuzzy numbers etc. The most used ordinary fuzzy numbers are triangular fuzzy numbers due to their easy calculations. Chiu and Park (1994) defined triangular fuzzy net present value $\left(\widetilde{NPV}\right)$ formula as

**Table 8.1** Some representative studies on wind energy investments

| Authors | MCDM method | Classical method | Fuzzy | Objective | Evaluation criteria/parameters |
|---|---|---|---|---|---|
| Kahraman et al. (2016a) | None | Benefit/cost analysis | Interval-valued intuitionistic | To analyze wind energy technology investments | Present worth, annual worth |
| Ashkaboosi et al. (2016) | None | Bi-level optimization technique | None | To maximize the profit of investment and market clearing for the wind power | None |
| Petković (2015) | None | Adaptive neuro-fuzzy inference system | Type I | To propose a model to provide economically optimal layouts for wind farm | Wake effect, wind regime, cost factors |
| Panduru et al. (2014) | None | Fuzzy logic controller | Type I | To monitor the energy generated by a wind turbine, and to control its distribution | None |
| Sheen (2014) | None | Real option analysis | Type I | To evaluate economic effectiveness of wind power investment projects | Net present value |
| Yeh and Huang (2014) | DEMATEL, ANP | Goal/question/metric | Type I | To examine the important factors considered in selection of wind farm location | Safety and quality, economy and benefit, social impression, environment and ecology, regulation, policy |
| Baringo and Conejo (2013) | None | Multi-stage stochastic programming | None | To identify optimal investment decisions on wind power facilities | None |
| Ersoz et al. (2013) | None | Adaptive-network based fuzzy inference systems | Type I | To investigate the suitability of a region for wind plant investment | Average moisture,temperature and pressure |
| Soroudi (2012) | None | Scenario-based approach | Type I | To quantify the impact of distributed generation units on active loss and voltage profile | Active loss, technical risk |

(continued)

**Table 8.1** (continued)

| Authors | MCDM method | Classical method | Fuzzy | Objective | Evaluation criteria/parameters |
|---|---|---|---|---|---|
| Lee (2011) | None | Real option analysis | None | To assess the value of wind energy investment opportunities | Underlying price, exercise price, time to maturity, risk-free rate, volatility |
| Madlener et al. (2011) | None | Fuzzy portfolio optimization with SMAD as risk measure | Type I | To show the implementation of fuzzy portfolio optimization on onshore wind power plants | Technical characteristics, economic characteristics |
| Onat and Ersoz (2011) | None | Adaptive-network-based fuzzy inference system | Type I | To analyze wind climate and wind energy potential | Average pressure, temperature, humidity |
| Aydin et al. (2010) | "and", "or", OWA | GIS | Type I | To proivde a decision support system for site selection of wind turbines | Sufficient potential for wind energy generation, satisfaction of most of the environmental objectives |
| Lee et al. (2009) | AHP | Benefits, opportunities, costs and risks (BOCR) | None | To choose a suitable wind farm project | Wind availability, Site advantage, WEG functions, financial schemes, policy support, advanced technologies, Wind turbine, connection, foundation, concept conflict, technical risks, uncertainty of land |
| Sheen (2009) | None | Geometric moment fuzzy ranking algorithm,and cost-benefit analysis | Type I | To assess the feasibility of wind generation investment | interest rate, inflation rate, investment, and operating revenue and/or cost |
| Cavallaro and Ciraolo (2005) | NAIADE method | None | Type I | To investigate the feasibility of installing wind energy turbines | Investment costs, operating and maintenance costs, energy production capacity, savings of finite energy sources, maturity of technology, realization time, $CO_2$ emissions avoided, visual impact, acoustic noise, Impact on ecosystems, social acceptability |

(continued)

**Table 8.1** (continued)

| Authors | MCDM method | Classical method | Fuzzy | Objective | Evaluation criteria/parameters |
|---|---|---|---|---|---|
| Celik (2003) | None | Weibull wind speed distribution model | None | To derive the wind energy output for small-scale wind power generators | Climate, topography, wind speed measurement |
| Pinson and Kariniotakis (2003) | None | Adaptive fuzzy neural networks | Type I | To predict the power production of wind farms | Online SCADA measurement, numerical weather predictions |
| Sen and Sahin (1997) | None | Cumulative semivariogram method | None | To evaluate regional wind power | None |

given in Eq. 8.1, where $\widetilde{F}_i = (f_{t_l}; f_{t_m}; f_{t_r})$, denotes net cash flows occurred in time period $t$ and $\tilde{i}_t = (i_{t_l}; i_{t_m}; i_{t_r})$ denotes the fuzzy interest rate.

$$\widetilde{NPV} = \left( \sum_{t=0}^{n} \left( \frac{\max(f_{t_l};\ 0)}{\prod_{t'=0}^{t}(1+i_{t'_r})} + \frac{\min(f_{t_l};\ 0)}{\prod_{t'=0}^{t}\left(1+i_{t'_l}\right)} \right);\ \sum_{t=0}^{n} \frac{f_{t_m}}{\prod_{t'=0}^{t}(1+i_{t'_m})};\ \sum_{t=0}^{n} \left( \frac{\max(f_{t_r};\ 0)}{\prod_{t'=0}^{t}\left(1+i_{t'_l}\right)} + \frac{\min(f_{t_r};\ 0)}{\prod_{t'=0}^{t}(1+i_{t'_r})} \right) \right) \quad (8.1)$$

When the time horizon is infinite the fuzzy net present worth is calculated by Eq. 8.2:

$$\widetilde{NPV} = \left( \sum_{t=0}^{n} \left( \frac{\max(f_{t_l}; 0)}{i_{t'_r}} + \frac{\min(f_{t_l}; 0)}{i_{t'_l}} \right); \sum_{t=0}^{n} \frac{f_{t_m}}{i_{t'_m}}; \sum_{t=0}^{n} \left( \frac{\max(f_{t_r}; 0)}{i_{t'_l}} + \frac{\min(f_{t_r}; 0)}{i_{t'_r}} \right) \right) \quad (8.2)$$

In this chapter Eq. 8.3 is used for the defuzzifiction of ordinary fuzzy sets:

$$Def\left(\widetilde{F}\right) = \frac{f_l + 2f_m + f_u}{4} \quad (8.3)$$

### 8.3.2 *Type-2 Fuzzy Present Worth Analysis*

The concept of a type-2 fuzzy set was introduced by Zadeh as an extension of the concept of an ordinary fuzzy set called an ordinary fuzzy set (Zadeh 1974). A type-2 fuzzy set $\tilde{\tilde{A}}$ in the universe of discourse $X$ can be represented by a type-2 membership function $\mu_{\tilde{\tilde{A}}}$, shown as follows (Zadeh 1975):

$$\tilde{\tilde{A}} = \left\{ (x, u), \mu_{\tilde{\tilde{A}}}(x, u) \middle| \quad \forall\, x \in X, \forall\, u \in J_x \subseteq [0, 1], 0 \le \mu_{\tilde{\tilde{A}}}(x, u) \le 1 \right\} \quad (8.4)$$

where $J_x$ denotes an interval [0,1]. In the literature review, it is seen that triangular interval type-2 fuzzy sets are the most preferred interval type-2 fuzzy sets. A triangular interval type-2 fuzzy set is represented as $\tilde{\tilde{A}}_i = \left( (a_{il}^U, a_{im}^U, a_{ir}^U; H(\widetilde{A}_i^U)\right), \left(a_{il}^L, a_{im}^L, a_{ir}^L; H(\widetilde{A}_i^L)\right)$ where $\widetilde{A}_i^L$ and $\widetilde{A}_i^U$ are ordinary fuzzy sets, $a_{il}^U, a_{im}^U, a_{ir}^U, a_{il}^L, a_{im}^L$ and $a_{ir}^L$ are the references points of the interval type-2 fuzzy set $\tilde{\tilde{A}}_i$,$H(\tilde{A}_i^U)$ denotes the membership value of the element $a_i^U$ in the upper triangular membership function $\tilde{A}_i^U$, $H(\tilde{A}_i^L)$ denotes the membership value of the element $a_i^L$ in the lower triangular membership function $\tilde{A}_i^L$, $H(\tilde{A}_i^U) \in [0,1], H(\tilde{A}_i^L) \in [0,1]$ and $1 \leq i \leq 2$. Kuo- Ping (2011) gives detailed information on the basic algebraic operations of type-2 fuzzy sets.

Ucal Sari and Kahraman (2015) introduced type-2 fuzzy net present worth method. Triangular interval type-2 fuzzy net present value ($N\tilde{\tilde{P}}V$) is formulized in Eq. 8.5 where $\tilde{\tilde{F}}_t = (f_{tl}^U, f_{tm}^U, f_{tr}^U; H(\tilde{f}_t^U)), (f_{tl}^L, f_{tm}^L, f_{tr}^L; H(\tilde{f}_t^L))$ denotes the cash flow occurred at time $t$ and $\tilde{\tilde{i}}_t = (i_{tl}^U, i_{tm}^U, i_{tr}^U; H(\tilde{i}_t^U)),(i_{tl}^L, i_{tm}^L, i_{tr}^L; H(\tilde{i}_t^L)), \forall \tilde{i} > 0$ denotes the discount rate at time $t$:

$$N\tilde{\tilde{P}}V = \left( \left( \sum_{t=0}^{n} \frac{f_{tl}^U}{\prod_{t'=0}^{t}(1+i_{t'r}^U)}, \sum_{t=0}^{n} \frac{f_{tm}^U}{\prod_{t'=0}^{t}(1+i_{t'm}^U)}, \sum_{t=0}^{n} \frac{f_{tr}^U}{\prod_{t'=0}^{t}(1+i_{t'l}^U)}; \min\left(H(\tilde{f}_t^U), H(\tilde{i}_t^U)\right) \right) \right.$$
$$\left. , \left( \sum_{t=0}^{n} \frac{f_{tl}^L}{\prod_{t'=0}^{t}(1+i_{t'r}^L)}, \sum_{t=0}^{n} \frac{f_{tm}^L}{\prod_{t'=0}^{t}(1+i_{t'm}^L)}, \sum_{t=0}^{n} \frac{f_{tr}^L}{\prod_{t'=0}^{t}(1+i_{t'l}^L)}; \min\left(H(\tilde{f}_t^L), H(\tilde{i}_t^L)\right) \right) \right) \tag{8.5}$$

When the time horizon is infinite triangular interval type-2 fuzzy net present worth is calculated by Eq. 8.6:

$$N\tilde{\tilde{P}}V = \left( \left( \sum_{t=0}^{n} \frac{f_{tl}^U}{i_{t'r}^U}, \sum_{t=0}^{n} \frac{f_{tm}^U}{i_{t'm}^U}, \sum_{t=0}^{n} \frac{f_{tr}^U}{i_{t'l}^U}; \min\left(H(\tilde{f}_t^U), H(\tilde{i}_t^U)\right) \right), \left( \sum_{t=0}^{n} \frac{f_{tl}^L}{i_{t'r}^L}, \sum_{t=0}^{n} \frac{f_{tm}^L}{i_{t'm}^L}, \sum_{t=0}^{n} \frac{f_{tr}^L}{i_{t'l}^L}; \min\left(H(\tilde{f}_t^L), H(\tilde{i}_t^L)\right) \right) \right) \tag{8.6}$$

In this chapter, realistic type reduction indices are used for the defuzzification of type 2 fuzzy sets. Realistic type reduction indices is calculated by Eq. 8.7 which transforms $\tilde{\tilde{A}}$ into an ordinary fuzzy set where $\underline{\mu}_{\tilde{A}}(x)$ and $\overline{\mu}_{\tilde{A}}(x)$ are lower and upper membership functions of the $\tilde{\tilde{A}}$ (Niewiadomski et al. 2006).

$$TR_{re}(\tilde{\tilde{A}}) = \frac{\underline{\mu}_{\tilde{A}}(x) + \overline{\mu}_{\tilde{A}}(x)}{2}, \quad x \in X \tag{8.7}$$

Equation 8.3 can be used to rank the ordinary fuzzy set which is obtained by type reduction indices method.

### 8.3.3 Intuitionistic Fuzzy Present Worth

Atanassov (1986) introduced triangular intuitionistic fuzzy numbers (TIFN) $\widetilde{A}$. TIFN utilizes both membership value and non-membership value of a fuzzy number. Formulas of membership function $\left(\mu_{\tilde{A}}(x)\right)$ and non-membership function $\left(v_{\tilde{A}}(x)\right)$ are as follows:

$$\mu_{\tilde{A}}(x) = \begin{cases} \frac{x-l}{m-l}, & for\ l \leq x \leq m \\ \frac{u-x}{u-m}, & for\ m \leq x \leq u \\ 0, & otherwise \end{cases} \tag{8.8}$$

and

$$v_{\tilde{A}}(x) = \begin{cases} \frac{m-x}{m-\acute{l}}, & for\ \acute{l} \leq x \leq \acute{m} \\ \frac{x-\acute{m}}{\acute{u}-\acute{m}}, & for\ \acute{m} \leq x \leq \acute{u} \\ 1, & otherwise \end{cases} \tag{8.9}$$

where $l \leq m \leq u$, $\acute{l} \leq \acute{m} \leq \acute{u}$, $0 \leq \mu_{\tilde{A}}(x) + v_{\tilde{A}}(x) \leq 1$ and it is denoted by

$$\widetilde{A}_{TIFN} = \left((l, m, u), \left(\acute{l}, \acute{m}, \acute{u}\right)\right). \tag{8.10}$$

The sum of membership and non-membership values should be less than or equal to 1. The basic algebraic operations are determined by Mapatra and Roy (2009), Atasannov (2012) and Kumar and Hussein (2014).

In this chapter, TIFNs are ranked using the deffuzzification method which is proposed by Kahraman et al. (2015).

The rank of a TIFN $\widetilde{A} = \left((l, m, u), \left(\acute{l}, \acute{m}, \acute{u}\right)\right)$ is determined as follows:

$$R\left(\widetilde{A}\right) = \frac{1}{2}\left(\frac{l + 2m + u}{4} + \frac{\acute{l} + 2\acute{m} + \acute{u}}{4}\right) = \frac{l + \acute{l} + 2m + 2\acute{m} + u + \acute{u}}{8} \tag{8.11}$$

Triangular fuzzy number intuitionistic fuzzy weighted geometric $(TFNIFWG_w)$ operator is used to aggregate triangular intuitionistic fuzzy sets (Chen et al. 2010):

$$
\begin{aligned}
&TFNIFWG_w\left(\widetilde{A}_1, \widetilde{A}_2, \ldots, \widetilde{A}_n\right) = \\
&\quad \left( \left(1-\prod_{j=1}^{n}(1-l_j)^{w_j}, 1-\prod_{j=1}^{n}(1-m_j)^{w_j}, 1-\prod_{j=1}^{n}(1-u_j)^{w_j}\right), \right. \\
&\quad \left. \left(1-\prod_{j=1}^{n}(1-l_j)^{w_j}, 1-\prod_{j=1}^{n}(1-\acute{m}_j)^{w_j}, 1-\prod_{j=1}^{n}(1-u_j)^{w_j}\right) \right)
\end{aligned}
\tag{8.12}
$$

Kahraman et al. (2015) introduced intuitionistic fuzzy net present worth and intuitionistic fuzzy annual worth methods. The parameters used in the calculations are expressed by TFIN in Eqs. 8.13–8.18 where $m$ evaluations are made for each of the parameter.

$$
\widetilde{FC}_{T,I} = \left\{ \begin{array}{c} \left\langle fc_1, \left(TFN_1, T\acute{F}N_1\right), \ldots, \left(TFN_m, T\acute{F}N_m\right) \right\rangle, \\ \left\langle fc_2, \left(TFN_1, T\acute{F}N_1\right), \ldots, \left(TFN_m, T\acute{F}N_m\right) \right\rangle \\ , \ldots, \\ \left\langle fc_k, \left(TFN_1, T\acute{F}N_1\right), \ldots, \left(TFN_m, T\acute{F}N_m\right) \right\rangle \end{array} \right\} \tag{8.13}
$$

$$
\widetilde{UAC}_{T,I} = \left\{ \begin{array}{c} \left\langle uac_1, \left(TFN_1, T\acute{F}N_1\right), \ldots, \left(TFN_m, T\acute{F}N_m\right) \right\rangle, \\ \left\langle uac_2, \left(TFN_1, T\acute{F}N_1\right), \ldots, \left(TFN_m, T\acute{F}N_m\right) \right\rangle \\ , \ldots, \\ \left\langle uac_k, \left(TFN_1, T\acute{F}N_1\right), \ldots, \left(TFN_m, T\acute{F}N_m\right) \right\rangle \end{array} \right\} \tag{8.14}
$$

$$
\widetilde{UAB}_{T,I} = \left\{ \begin{array}{c} \left\langle uab_1, \left(TFN_1, T\acute{F}N_1\right), \ldots, \left(TFN_m, T\acute{F}N_m\right) \right\rangle, \\ \left\langle uab_2, \left(TFN_1, T\acute{F}N_1\right), \ldots, \left(TFN_m, T\acute{F}N_m\right) \right\rangle \\ , \ldots, \\ \left\langle uab_k, \left(TFN_1, T\acute{F}N_1\right), \ldots, \left(TFN_m, T\acute{F}N_m\right) \right\rangle \end{array} \right\} \tag{8.15}
$$

$$
\widetilde{SV}_{T,I} = \left\{ \begin{array}{c} \left\langle sv_1, \left(TFN_1, T\acute{F}N_1\right), \ldots, \left(TFN_m, T\acute{F}N_m\right) \right\rangle, \\ \left\langle sv_2, \left(TFN_1, T\acute{F}N_1\right), \ldots, \left(TFN_m, T\acute{F}N_m\right) \right\rangle, \\ , \ldots, \\ \left\langle sv_k, \left(TFN_1, T\acute{F}N_1\right), \ldots, \left(TFN_m, T\acute{F}N_m\right) \right\rangle \end{array} \right\} \tag{8.16}
$$

$$
\tilde{i}_{T,I} = \left\{ \begin{array}{c} \left\langle i_1, \left(TFN_1, T\acute{F}N_1\right), \ldots, \left(TFN_m, T\acute{F}N_m\right) \right\rangle, \\ \left\langle i_2, \left(TFN_1, T\acute{F}N_1\right), \ldots, \left(TFN_m, T\acute{F}N_m\right) \right\rangle \\ , \ldots, \\ \ldots, \left\langle i_k, \left(TFN_1, T\acute{F}N_1\right), \ldots, \left(TFN_m, T\acute{F}N_m\right) \right\rangle \end{array} \right\} \tag{8.17}
$$

$$\tilde{n}_{T,I} = \left\{ \begin{array}{c} \langle n_1, (TFN_1, T\acute{F}N_1), \ldots, (TFN_m, T\acute{F}N_m) \rangle, \\ \langle n_2, (TFN_1, T\acute{F}N_1), \ldots, (TFN_m, T\acute{F}N_m) \rangle \\ , \ldots, \\ , \langle n_k, (TFN_1, T\acute{F}N_1), \ldots, (TFN_m, T\acute{F}N_m) \rangle \end{array} \right\} \quad (8.18)$$

where FC represents the first cost of the alternative, UAC represents uniform annual cost of the alternative, UAB represents uniform annual benefit, n represents project life, i represents interest rate, and SV represents salvage value.

The intuitionistic fuzzy present worth $\left(\widetilde{PW}_{T,I}\right)$ of an investment alternative can be calculated by Eq. 8.19 or Eq. 8.20:

$$\begin{aligned} \widetilde{PW}_{T,I} = & -\widetilde{FC}_{T,I} - \widetilde{UAC}_{T,I}\left(\frac{P}{A}, \tilde{i}_{T,I}, \tilde{n}_{T,I}\right) \\ & + \widetilde{UAB}_h\left(\frac{P}{A}, \tilde{i}_{T,I}, \tilde{n}_{T,I}\right) + \widetilde{SV}_h\left(\frac{P}{F}, \tilde{i}_{T,I}, \tilde{n}_{T,I}\right) \end{aligned} \quad (8.19)$$

or

$$\begin{aligned} \widetilde{PW}_{T,I} = & -\widetilde{FC}_{T,I} - \widetilde{UAC}_{T,I}\left[\frac{\left(1+\tilde{i}_{T,I}\right)^{\tilde{n}_{T,I}} - 1}{\tilde{i}_{T,I}\left(1+\tilde{i}_{T,I}\right)^{\tilde{n}_{T,I}}}\right] \\ & + \widetilde{UAB}_{T,I}\left[\frac{\left(1+\tilde{i}_{T,I}\right)^{\tilde{n}_{T,I}} - 1}{\tilde{i}_{T,I}\left(1+\tilde{i}_{T,I}\right)^{\tilde{n}_{T,I}}}\right] + \widetilde{SV}_{T,I}\left(1+\tilde{i}_{T,I}\right)^{-\tilde{n}_{T,I}} \end{aligned} \quad (8.20)$$

where

$$\widetilde{FC}_{T,I} = \bigcup_{j=1}^{k} TFNIFWG_w\left(\left\langle \begin{array}{c} fc_j, (TFN_1, T\acute{F}N_1), \\ \ldots, (TFN_m, T\acute{F}N_m) \end{array} \right\rangle\right)$$

$$\widetilde{UAC}_{T,I} = \bigcup_{j=1}^{k} TFNIFWG_w\left(\left\langle \begin{array}{c} uac_j, (TFN_1, T\acute{F}N_1), \\ \ldots, (TFN_m, T\acute{F}N_m) \end{array} \right\rangle\right),$$

$$\widetilde{UAB}_{T,I} = \bigcup_{j=1}^{k} TFNIFWG_w\left(\left\langle \begin{array}{c} uab_j, (TFN_1, T\acute{F}N_1), \\ \ldots, (TFN_m, T\acute{F}N_m) \end{array} \right\rangle\right),$$

$$\widetilde{SV}_{T,I} = \bigcup_{j=1}^{k} TFNIFWG_w\left(\left\langle \begin{array}{c} sv_j, (TFN_1, T\acute{F}N_1), \\ \ldots, (TFN_m, T\acute{F}N_m) \end{array} \right\rangle\right),$$

$$\tilde{i}_{T,I} = \bigcup_{j=1}^{k} TFNIFWG_w\left(\left\langle \begin{array}{c} i_j, (TFN_1, T\acute{F}N_1), \\ \ldots, (TFN_m, T\acute{F}N_m) \end{array} \right\rangle\right),$$

$$\tilde{n}_{T,I} = \bigcup_{j=1}^{k} TFNIFWG_w\left(\left\langle \begin{array}{c} n_j, (TFN_1, T\acute{F}N_1), \\ \ldots, (TFN_m, T\acute{F}N_m) \end{array} \right\rangle\right).$$

When the time horizon is infinite, triangular intuitionistic fuzzy present worth is calculated by Eq. 8.21:

$$\widetilde{PW}_{T,I} = -\widetilde{FC}_{T,I} - \left(\frac{\widetilde{UAC}_{T,I}}{\tilde{i}_{T,I}}\right) + \left(\frac{\widetilde{UAB}_{T,I}}{\tilde{i}_{T,I}}\right) \tag{8.21}$$

The defuzzified values of these parameters are needed for further calculations. For instance, the defuzzified value of $\widetilde{FC}_{T,I}$ is obtained by the following process:

$$TFNIFWG_w\left(\left\langle \begin{matrix} fc_j, (TFN_1, T\acute{F}N_1), \\ \ldots, (TFN_m, T\acute{F}N_m) \end{matrix} \right\rangle\right) = \tilde{\mu}_{fc_j}$$
$$= \left(\left(\mu_{fc_{jl}}, \mu_{fc_{jm}}, \mu_{fc_{ju}}\right), \left(\acute{\mu}_{fc_{jl}}, \acute{\mu}_{fc_{jm}}, \acute{\mu}_{fc_{ju}}\right)\right),$$
$$j = 1, \ldots, k \tag{8.22}$$

Defuzzified value of $\left(\left(\mu_{fc_{jl}}, \mu_{fc_{jm}}, \mu_{fc_{ju}}\right), \left(\acute{\mu}_{fc_{jl}}, \acute{\mu}_{fc_{jm}}, \acute{\mu}_{fc_{ju}}\right)\right)$ is $Def\left(\tilde{\mu}_{fc_j}\right)$ which is obtained by Eq. 8.11. The defuzzified value of $\widetilde{FC}_{T,I}$ is obtained by Eq. 8.23:

$$Def\widetilde{FC}_{T,I} = \frac{\sum_{j=1}^{k} fc_j \left(Def\left(\tilde{\mu}_{fc_j}\right)\right)^2}{\sum_{j=1}^{k} \left(Def\left(\tilde{\mu}_{fc_j}\right)\right)^2} \tag{8.23}$$

Other parameters could be deffuzzified in a similar way.

### 8.3.4 Hesitant Fuzzy Environmental Economics Methods

Kahraman et al. (2015) introduced hesitant fuzzy net present worth and hesitant fuzzy annual worth methods. A hesitant fuzzy set (HFS) is another extension of fuzzy sets that aims to model the uncertainty originated by the hesitation that might arise in the assignment of membership degrees of the elements to a fuzzy set (Kahraman et al. 2017).

Triangular Fuzzy Hesitant Fuzzy Sets (TFHFS) are proposed in 2013 by Yu. In TFHFS several triangular fuzzy numbers are used to express the membership degree of an element.

A TFHFS $\widetilde{E}$ on a fixed set X is defined in terms of a function $\tilde{f}_{\tilde{E}}(x)$ that returns several triangular fuzzy values,

$$\tilde{E} = \left\{ \langle x, \tilde{f}_{\tilde{E}}(x) \rangle \middle| x \epsilon X \right\} \tag{8.24}$$

where $\tilde{f}_{\tilde{E}}(x)$ is a set of several triangular fuzzy numbers which express the possible membership degrees of an element $x \in X$ to a set $\tilde{E}$.

For a Triangular Fuzzy Hesitant Fuzzy Set (TFHFS), $\tilde{f}$, $s(\tilde{f}) = \frac{1}{l_{\tilde{f}}} \sum_{\widetilde{TFN} \in \tilde{f}} \bar{X}\left(\widetilde{TFN}\right)$ is called the score function of $\tilde{f}$ with $l_{\tilde{f}}$ being the number of TFNs in $\tilde{f}$ (Yu 2013). $h(\tilde{f}) = \frac{1}{l_{\tilde{f}}} \sum_{\widetilde{TFN} \in \tilde{f}} \sigma\left(\widetilde{TFN}\right)$ is called the deviation function of $\tilde{f}$. For $\tilde{f}_1$ *and* $\tilde{f}_2$,

$$\begin{aligned}
&\text{If } s(\tilde{f}_1) > s(\tilde{f}_2), \quad \text{then } \tilde{f}_1 \geq \tilde{f}_2 \\
&\text{If } s(\tilde{f}_1) = s(\tilde{f}_2), h(\tilde{f}_1) = h(\tilde{f}_2), \quad \text{then } \tilde{f}_1 = \tilde{f}_2 \\
&\text{If } s(\tilde{f}_1) = s(\tilde{f}_2), h(\tilde{f}_1) > h(\tilde{f}_2), \quad \text{then } \tilde{f}_1 < \tilde{f}_2 \\
&\text{If} s(\tilde{f}_1) = s(\tilde{f}_2), h(\tilde{f}_1) > h(\tilde{f}_2), \quad \text{then } \tilde{f}_1 > \tilde{f}_2
\end{aligned}$$

Let $\tilde{f}_1$ *and* $\tilde{f}_2$ be two THHFEs, then

$$\tilde{f}_1 \oplus \tilde{f}_2 = \left\{ (l_1 + l_2 - l_1.l_2, m_1 + m_2 - m_1.m_2, u_1 + u_2 - u_1.u_2) \middle| \widetilde{TFN}_1 \in \tilde{f}_1, \widetilde{TFN}_2 \in \tilde{f}_2 \right\} \tag{8.25}$$

$$\tilde{f}_1 \otimes \tilde{f}_2 = \left\{ l_1.l_2, m_1.m_2, u_1.u_2 \middle| \widetilde{TFN}_1 \in \tilde{f}_1, \widetilde{TFN}_2 \in \tilde{f}_2 \right\} \tag{8.26}$$

$$\tilde{f}^{\lambda} = \left\{ (l)^{\lambda}, (m)^{\lambda}, (u)^{\lambda} \middle| \widetilde{TFN} \in \tilde{f} \right\}, \quad \lambda > 0 \tag{8.27}$$

$$\lambda\tilde{f} = \left\{ 1 - (1 - l)^{\lambda}, 1 - (1 - m)^{\lambda}, 1 - (1 - u)^{\lambda} \middle| \widetilde{TFN} \in \tilde{f} \right\}, \quad \lambda > 0 \tag{8.28}$$

where $\widetilde{TFN}_1 = (l_1, m_1, u_1)$ and $\widetilde{TFN}_2 = (l_2, m_2, u_2)$.

For aggregating triangular fuzzy hesitant fuzzy sets, Triangular Fuzzy Hesitant Fuzzy Weighted Averaging (TFHFWA) operator is used. Let $\tilde{f}_j (j = 1, 2, \ldots, n)$ be a collection of TFHFEs. $w = (w_1, w_2, \ldots, w_n)^T$ is the weight vector of $\tilde{f}_j (j = 1, 2, \ldots, n)$ with $w_j \epsilon [0, 1]$ and $\sum_{j=1}^{n} w_j = 1$, then a TFHFWA operator is a mapping TFHFWA: $F^n \rightarrow \bar{F}$ such that

$$TFHFWA(\tilde{f}_1,\tilde{f}_2,\ldots,\tilde{f}_n) = \oplus_{j=1}^{n}(w_i\tilde{f}_j)$$
$$\left\{1-\prod_{j=1}^{n}(1-L_j)^{w_j}, 1-\prod_{j=1}^{n}(1-M_j)^{w_j},\right.$$
$$\left. 1-\prod_{j=1}^{n}(1-U_j)^{w_j} | \widetilde{TFN}_1 \in \tilde{f}_1, \widetilde{TFN}_1 \in \tilde{f}_1, \ldots, \widetilde{TFN}_n \in \tilde{f}_n\right\} \tag{8.29}$$

For the defuzzification of triangular hesitant fuzzy sets, the defuzzified value of a hesitant $\widetilde{TFN} = (l, m, u)$ can be defined as follows:

$$Def\left(\widetilde{TFN}\right) = \frac{l+2m+u}{4} \tag{8.30}$$

In the hesitant fuzzy present worth analysis, investment parameters are expressed using triangular fuzzy hesitant fuzzy sets. The parameters used in the calculations are expressed by TFHFS in Eqs. 8.31–8.36 where $m$ evaluations are made for each of the parameter.

$$\widetilde{FC}_{T,h} = \left\{\begin{array}{c}\langle fc_1, TFN_1,\ldots, TFN_m\rangle, \langle fc_2, TFN_1,\ldots, TFN_m\rangle,\\ \ldots, \langle fc_k, TFN_1,\ldots, TFN_m\rangle\end{array}\right\} \tag{8.31}$$

$$\widetilde{UAC}_{T,h} = \left\{\begin{array}{c}\langle uac_1, TFN_1,\ldots, TFN_m\rangle,\\ \langle uac_2, TFN_1,\ldots, TFN_m\rangle,\\ \ldots, \langle uac_k, TFN_1,\ldots, TFN_m\rangle\end{array}\right\} \tag{8.32}$$

$$\widetilde{UAB}_{T,h} = \left\{\begin{array}{c}\langle uab_1, TFN_1,\ldots, TFN_m\rangle,\\ \langle uab_2, TFN_1,\ldots, TFN_m\rangle,\\ \ldots, \langle uab_k, TFN_1,\ldots, TFN_m\rangle\end{array}\right\} \tag{8.33}$$

$$\widetilde{SV}_{T,h} = \left\{\begin{array}{c}\langle sv_1, TFN_1,\ldots, TFN_m\rangle, \langle sv_2, TFN_1,\ldots, TFN_m\rangle,\\ \ldots, \langle sv_k, TFN_1,\ldots, TFN_m\rangle\end{array}\right\} \tag{8.34}$$

$$\tilde{i}_{T,h} = \left\{\begin{array}{c}\langle i_1, TFN_1,\ldots, TFN_m\rangle, \langle i_2, TFN_1,\ldots, TFN_m\rangle,\\ \ldots, \langle i_k, TFN_1,\ldots, TFN_m\rangle\end{array}\right\} \tag{8.35}$$

$$\tilde{n}_{T,h} = \left\{\begin{array}{c}\langle n_1, TFN_1,\ldots, TFN_m, n_2, TFN_1,\ldots, TFN_m\rangle,\\ \ldots, \langle n_k, TFN_1,\ldots, TFN_m\rangle\end{array}\right\} \tag{8.36}$$

where FC represents the first cost of the alternative, UAC represents uniform annual cost of the alternative, UAB represents uniform annual benefit, $n$ represents project life, $i$ represents interest rate, and SV represents salvage value.

The hesitant fuzzy present worth $\left(\widetilde{PW}_{T,h}\right)$ of an investment alternative can be calculated by Eq. 8.37 or Eq. 8.38:

$$\begin{aligned}\widetilde{PW}_{T,h} = & -\widetilde{FC}_{T,h} - \widetilde{UAC}_{T,h}\left(\frac{P}{A}, \tilde{i}_{T,h}, \tilde{n}_{T,h}\right) \\ & + \widetilde{UAB}_{h}\left(\frac{P}{A}, \tilde{i}_{T,h}, \tilde{n}_{T,h}\right) + \widetilde{SV}_{h}\left(\frac{P}{F}, \tilde{i}_{T,h}, \tilde{n}_{T,h}\right)\end{aligned} \tag{8.37}$$

or

$$\begin{aligned}\widetilde{PW}_{T,h} = & -\widetilde{FC}_{T,h} - \widetilde{UAC}_{T,h}\left[\frac{\left(1+\tilde{i}_{T,h}\right)^{\tilde{n}_{T,h}}-1}{\tilde{i}_{T,h}\left(1+\tilde{i}_{T,h}\right)^{\tilde{n}_{T,h}}}\right] \\ & + \widetilde{UAB}_{T,h}\left[\frac{\left(1+\tilde{i}_{T,h}\right)^{\tilde{n}_{T,h}}-1}{\tilde{i}_{T,h}\left(1+\tilde{i}_{T,h}\right)^{\tilde{n}_{T,h}}}\right] + \widetilde{SV}_{T,h}\left(1+\tilde{i}_{T,h}\right)^{-\tilde{n}_{T,h}}\end{aligned} \tag{8.38}$$

where

$$\widetilde{FC}_{T,h} = \bigcup_{j=1}^{k} TFHFWA\left(\left\langle fc_j, TFN_1, \ldots, TFN_m\right\rangle\right)$$

$$\widetilde{UAC}_{T,h} = \bigcup_{j=1}^{k} TFHFWA\left(\left\langle uac_j, TFN_1, \ldots, TFN_m\right\rangle\right),$$

$$\widetilde{UAB}_{T,h} = \bigcup_{j=1}^{k} TFHFWA\left(\left\langle uab_j, TFN_1, \ldots, TFN_m\right\rangle\right),$$

$$\widetilde{SV}_{T,h} = \bigcup_{j=1}^{k} TFHFWA\left(\left\langle sv_j, TFN_1, \ldots, TFN_m\right\rangle\right),$$

$$\tilde{i}_{T,h} = \bigcup_{j=1}^{k} TFHFWA\left(\left\langle i_j, TFN_1, \ldots, TFN_m\right\rangle\right),$$

$$\tilde{n}_{T,h} = \bigcup_{j=1}^{k} TFHFWA\left(\left\langle n_j, TFN_1, \ldots, TFN_m\right\rangle\right).$$

When the time horizon is infinite, triangular intuitionistic fuzzy present worth is calculated by Eq. 8.39:

$$\widetilde{PW}_{T,h} = -\widetilde{FC}_{T,h} - \left(\frac{\widetilde{UAC}_{T,h}}{\tilde{i}_{T,h}}\right) + \left(\frac{\widetilde{UAB}_{T,h}}{\tilde{i}_{T,h}}\right) \tag{8.39}$$

For the defuzzification of triangular hesitant fuzzy sets, the defuzzified value of $\widetilde{FC}_{T,h}$ is obtained as follows:

$$TFHFWA(\langle fc_j, TFN_1, \ldots, TFN_m \rangle) = \tilde{\mu}_{\mathrm{fc_j}} = \left(\mu_{fc_{jl}}, \mu_{fc_{jm}}, \mu_{fc_{ju}}\right), \quad j = 1, \ldots, k \tag{8.40}$$

Defuzzified value of $\left(\mu_{\mathrm{fc_{jl}}}, \mu_{\mathrm{fc_{jm}}}, \mu_{\mathrm{fc_{ju}}}\right)$ is $Def\left(\tilde{\mu}_{\mathrm{fc_j}}\right)$ which is obtained by Eq. 8.30. Other parameters could be deffuzzified in a similar way.

## 8.4 An Application

Wind turbines have two major types based on their axis; the horizontal axis wind turbine (HAWT) and the vertical axis wind turbine (VAWT). In general HAWTs have greater capacities than VAWTs. Therefore, HAWTs are preferred for the industrial energy production. Mostly the useful life of HAWT is considered as 20 years. However the useful life of a wind turbine could increase by regular maintenances. In this chapter, a HAWT type wind turbine is analyzed for two scenarios that are (1) using the turbine without additional maintenances and reinvest at the end of its useful life, (2) using turbine with routine maintenances and take its useful life as infinite.

The economic parameter values of two alternatives are represented by different types of fuzzy numbers in Tables 8.2, 8.3, 8.4, 8.5 and 8.6.

**Table 8.2** Parameters defined by ordinary fuzzy sets

| Parameter | Scenerio I possible cash flows (1000€) | Scenerio II possible cash flows (1000€) |
|---|---|---|
| $\widetilde{FC}$ | (630,650,670) | (630,650,670) |
| $\widetilde{UAC}$ | (40,45,50) | (40,45,50) |
| $\widetilde{UAB}$ | (300,350,400) | (300,350,400) |
| $\widetilde{MC}$ (once in each five years) | – | (50,80,110) |
| $\widetilde{SV}$ | (100,130,150) | |
| $\tilde{i}\%$ | (7,8,9) | (7,8,9) |
| $nn$ | 20 | infinite |

**Table 8.3** Parameters defined by type 2 fuzzy sets

| Parameter | Scenerio I possible cash flows (1000€) | Scenerio II possible cash flows (1000€) |
|---|---|---|
| $\widetilde{\widetilde{FC}}$ | (630,650,670;1) (640,650,660;0.9) | (630,650,670;1) (640,650,660;0.9) |
| $\widetilde{\widetilde{UAC}}$ | (40,45,50;1)(42,45,48;0.9) | (40,45,50;1)(42,45,48;0.9) |
| $\widetilde{\widetilde{UAB}}$ | (300,350,400;1)(310,350,390,0.9) | (300,350,400;1) (310,350,390,0.9) |
| $\widetilde{\widetilde{MC}}$ (once in each five years) | – | (50,80,110;1)(60,80,100;0.9) |
| $\widetilde{\widetilde{SV}}$ | (100,130,150;1) (110,130,140;0.9) | |
| $\tilde{\tilde{i}}\%$ | (7,8,9;1)(7.5,8,8.5;0.9) | (7,8,9;1)(7.5,8,8.5;0.9) |
| $n$ | 20 | Infinite |

**Table 8.4** Experts' compromised membership degrees based on IVIFS

| Parameter | Possible values | Experts' weights | | |
|---|---|---|---|---|
| | | E1 | E2 | E3 |
| | | 0.3 | 0.4 | 0.3 |
| FC | $630,000 | ([0.3,0.6][0.2,0.4]) | ([0.4,0.6][0.2,0.4]) | ([0.3,0.5][0.2,0.45]) |
| | $650,000 | ([0.4,0.5][0.1,0.4]) | ([0.3,0.5][0.3,0.5]) | ([0.4,0.6][0.1,0.3]) |
| | $670,000 | ([0.3,0.7][0.2,0.25]) | ([0.2,0.6][0.2,0.4]) | ([0.4,0.5][0.3,0.5]) |
| UAC | $40.000 | ([0.2,0.5][0.3,0.5]) | ([0.4,0.7][0.1,0.3]) | ([0.7,0.8][0.05,0.1]) |
| | $45,000 | ([0.4,0.7][0.1,0.3]) | ([0.5,0.7][0.1,0.3]) | ([0.3,0.6][0.1,0.3]) |
| | $50,000 | ([0.5,0.7][0.1,0.3]) | ([0.5,0.7][0.1,0.2]) | ([0.3,0.5][0.3,0.4]) |
| UAB | $300,000 | ([0.4,0.5][0.3,0.4]) | ([0.6,0.8][0.1,0.2]) | ([0.5,0.8][0.1,0.2]) |
| | $350,000 | ([0.4,0.7][0.1,0.1]) | ([0.5,0.7][0.1,0.3]) | ([0.4,0.7][0.1,0.3]) |
| | $400,000 | ([0.6,0.8][0.05,0.1]) | ([0.4,0.6][0.2,0.4]) | ([0.3,0.6][0.2,0.4]) |
| MC | $50,000 | ([0.1,0.3][0.5,0.6]) | ([0.4,0.7][0.1,0.2]) | ([0.5,0.6][0.2,0.4]) |
| | $80,000 | ([0.4,0.6][0.1,0.2]) | ([0.4,0.8][0,0.1]) | ([0.4,0.6][0.1,0.3]) |
| | $110,000 | ([0.6,0.8][0,0.1]) | ([0.5,0.6][0.1,0.3]) | ([0.3,0.4][0.2,0.5]) |
| SV | $100,000 | ([0.3,0.5][0.2,0.4]) | ([0.5,0.7][0.1,0.3]) | ([0.4,0.5][0.2,0.5]) |
| | $130,000 | ([0.5,0.7][0.1,0.2]) | ([0.4,0.6][0.2,0.4]) | ([0.3,0.5][0.3,0.5]) |
| | $150,000 | ([0.6,0.7][0.05,0.1]) | ([0.3,0.4][0.3,0.5]) | ([0.1,0.2][0.5,0.7]) |
| i | 7% | ([0.4,0.7][0.1,0.2]) | ([0.6,0.7][0.1,0.3]) | ([0.3,0.5][0.4,0.5]) |
| | 8% | ([0.2,0.5][0.3,0.4]) | ([0.5,0.7][0.1,0.3]) | ([0.6,0.8][0,0.1]) |
| | 9% | ([0.6,0.8][0,0.1]) | ([0.4,0.5][0.2,0.4]) | ([0.2,0.4][0.4,0.5]) |

**Table 8.5** Aggregated and defuzzified matrix for IVIFS

| Parameter | Possible values | Aggregated value | Defuzzified Value of membership | Defuzzified value of parameter |
|---|---|---|---|---|
| FC | $630,000 | ([0.341,0.572][0.2,0.415]) | 0.803 | 649,867 |
| | $650,000 | ([0.361,0.532][0.186,0.446]) | 0.788 | |
| | $670,000 | ([0.294,0.607][0.231,0.392]) | 0.795 | |
| UAC | $40.000 | ([0.468,0.69][0.151,0.31]) | 0.962 | 45,074 |
| | $45,000 | ([0.415,0.672][0.1,0.3]) | 0.944 | |
| | $50,000 | ([0.494,0.702][0.165,0.294]) | 0.962 | |
| UAB | $300,000 | ([0.471,0.69][0.165,0.266]) | 0.973 | 348,840 |
| | $350,000 | ([0.442,0. 7][0.1,0.245]) | 0.984 | |
| | $400,000 | ([0.443,0.675][0.157,0.322]) | 0.939 | |
| MC | $50,000 | [0.358,0.578][0.271,0.403] | 0.799 | 83,199 |
| | $80,000 | ([0.4,0.696][0.06,0.194] | 0.984 | |
| | $110,000 | ([0.482,0.633][0.103,0.317]) | 0.952 | |
| SV | $100,000 | ([0.415,0.592][0.161,0.395) | 0.864 | 123,716 |
| | $130,000 | ([0.405,0.607][0.203,0.380]) | 0.860 | |
| | $150,000 | ([0.361,0.468][0.306,0.532]) | 0.705 | |
| i | 7% | ([0.465,0.65][0.203,0.341]) | 0.921 | 796 |
| | 8% | ([0.461,0.69][0.138,0.279]) | 0.971 | |
| | 9% | ([0.42,0.598][0.215,0.358]) | 0.866 | |

**Table 8.6** Experts' compromised membership degrees based on triangular HFS

| Parameter | Possible values | Experts' weights | | |
|---|---|---|---|---|
| | | E1 | E2 | E3 |
| | | 0.3 | 0.4 | 0.3 |
| FC | $630,000 | (0.3,0.4,0.6) | (0.4,0.5,0.6) | (0.3,0.4,0.5) |
| | $650,000 | (0.4,0.5,0.6) | (0.3,0.4,0.5) | (0.4,0.5,0.6) |
| | $670,000 | (0.3,0.5,0.7) | (0.2,0.5,0.6) | (0.4,0.4,0.5) |
| UAC | $40.000 | (0.2,0.3,0.5) | (0.4,0.5,0.7) | (0.7,0.8,0.9) |
| | $45,000 | (0.4,0.6,0.7) | (0.5,0.6,0.7) | (0.4,0.6,0.7) |
| | $50,000 | (0.5,0.6,0.7) | (0.6,0.7,0.8) | (0.3,0.4,0.5) |
| UAB | $300,000 | (0.4,0.5,0.6) | (0.6,0.7,0.8) | (0.5,0.7,0.8) |
| | $350,000 | (0.5,0.6,0.8) | (0.5,0.6,0.7) | (0.4,0.5,0.7) |
| | $400,000 | (0.6,0.7,0.9) | (0.4,0.5,0.6) | (0.3,0.5,0.6) |
| MC | $50,000 | (0.1,0.2,0.3) | (0.4,0.5,0.7) | (0.5,0.5,0.6) |
| | $80,000 | (0.5,0.7,0.8) | (0.7,0.8,0.9) | (0.6,0.7,0.8) |
| | $110,000 | (0.6,0.8,0.9) | (0.5,0.6,0.7) | (0.3,0.4,0.5) |
| SV | $100,000 | (0.4,0.5,0.6) | (0.5,0.6,0.7) | (0.4,0.5,0.5) |
| | $130,000 | (0.6,0.7,0.8) | (0.4,0.5,0.6) | (0.3,0.4,0.5) |
| | $150,000 | (0.5,0.7,0.8) | (0.3,0.4,0.5) | (0.1,0.1,0.2) |
| i | 7% | (0.4,0.5,0.7) | (0.6,0.7,0.7) | (0.3,0.4,0.5) |
| | 8% | (0.2,0.4,0.5) | (0.5,0.6,0.7) | (0.6,0.8,0.9) |
| | 9% | (0.6,0.7,0.9) | (0.4,0.4,0.5) | (0.2,0.3,0.4) |

### 8.4.1 Evaluation Using Ordinary Fuzzy Present Worth

Table 8.2 shows the values of parameters using ordinary triangular fuzzy sets.

In the present worth analysis, period is defined as least common multiples of the useful alternative lives. Therefore, in our analysis, the analysis period is taken as infinite. In scenario 1, there will be cash inflow series from the salvage values and cash outflow series from the reinvestment costs which occur once in each 20 years period. To calculate the fuzzy present worth for scenario 1, first effective interest rate for 20 years should be calculated as follows:

$$\begin{aligned}\tilde{i}_{20} &= \left(1+\tilde{i}_1\right)^{20}-1 = \left((1+i_l)^{20}-1,(1+i_m)^{20}-1,(1+i_r)^{20}-1\right)\\ &= \left((1+0.07)^{20}-1,(1+0.08)^{20}-1,(1+0.09)^{20}-1\right)\\ &= (2.8697, 3.6609, 4.6044)\end{aligned}$$

Fuzzy net present worth of scenario 1 is calculated using Eq. 8.2 as follows:

$$\begin{aligned}\widetilde{NPV} &= -\widetilde{FC} - \frac{\widetilde{UAC}}{\tilde{i}} + \frac{\widetilde{UAB}}{\tilde{i}} + \frac{\widetilde{SV}}{\tilde{i}_{20}} - \frac{\widetilde{FC}}{\tilde{i}_{20}}\\ NPV_l &= -FC_r - \frac{UAC_r}{i_r} + \frac{UAB_l}{i_r} + \frac{SV_l}{\tilde{i}_{20r}} - \frac{FC_r}{\tilde{i}_{20r}}\\ NPV_m &= -FC_m - \frac{UAC_m}{i_m} + \frac{UAB_m}{i_m} + \frac{SV_m}{\tilde{i}_{20m}} - \frac{FC_m}{\tilde{i}_{20m}}\\ NPV_r &= -FC_l - \frac{UAC_l}{i_l} + \frac{UAB_r}{i_l} + \frac{SV_r}{\tilde{i}_{20l}} - \frac{FC_l}{\tilde{i}_{20l}}\end{aligned}$$

Using the formulas given above $\widetilde{NPV}$ is calculated as $(2702.8, 3020.46, 4345.59)$ for scenario 1.

To calculate the fuzzy present worth for scenario 2, the effective interest rate for 5 years should be calculated as follows:

$$\begin{aligned}\tilde{i}_5 &= \left(1+\tilde{i}_1\right)^5-1 = \left((1+i_l)^5-1,(1+i_m)^5-1,(1+i_r)^5-1\right)\\ &= \left((1+0.07)^5-1,(1+0.08)^5-1,(1+0.09)^5-1\right)\\ &= (0.4025, 0.4693, 0.5386)\end{aligned}$$

Fuzzy net present worth of scenario 2 is calculated using the following equation:

$$\widetilde{NPV} = -\widetilde{FC} - \frac{\widetilde{UAC}}{\tilde{i}} + \frac{\widetilde{UAB}}{\tilde{i}} + \frac{\widetilde{MC}}{\tilde{i}_5}$$

$\widetilde{NPV}$ is calculated as $(1903.545, 2992.033, 4388.634)$ for scenario 2.

Defuzzified values of $\widetilde{NPV}$ for scenario 1 and 2 are calculated using Eq. 8.3 as 3272.328 and 3069.061, respectively.

### 8.4.2 Evaluation Using Type 2 Fuzzy Present Work

Table 8.3 shows the values of parameters using triangular interval type 2 fuzzy sets.

Effective interest rates for 5 and 20 years are calculated as follows:

$$\begin{aligned}
\tilde{\tilde{i}}_5 &= \left(1+\tilde{\tilde{i}}_1\right)^5 - 1 = \left(\left(1+i_l^U\right)^5 - 1, \left(1+i_m^U\right)^5 - 1, \left(1+i_r^U\right)^5 - 1\right); 1 \\
&\quad \left(\left(1+i_l^L\right)^5 - 1, \left(1+i_m^L\right)^5 - 1, \left(1+i_r^L\right)^5 - 1\right); 0.9 \\
&= \left((1+0.07)^5 - 1, (1+0.08)^5 - 1, (1+0.09)^5 - 1\right); 1 \\
&\quad \left((1+0.075)^5 - 1, (1+0.08)^5 - 1, (1+0.085)^5 - 1\right); 0.9 \\
&= (0.4025, 0.4693, 0.5386; 1)(0.4356, 0.4693, 0.5036; 0.9)
\end{aligned}$$

$$\begin{aligned}
\tilde{\tilde{i}}_{20} &= \left(1+\tilde{\tilde{i}}_1\right)^{20} - 1 \\
&= \left(\left(1+i_l^U\right)^{20} - 1, \left(1+i_m^U\right)^{20} - 1, \left(1+i_r^U\right)^{20} - 1\right); 1 \\
&\quad \left(\left(1+i_l^L\right)^{20} - 1, \left(1+i_m^L\right)^{20} - 1, \left(1+i_r^L\right)^{20} - 1\right); 0.9 \\
&= \left((1+0.07)^{20} - 1, (1+0.08)^{20} - 1, (1+0.09)^{20} - 1\right); 1 \\
&\quad \left((1+0.075)^{20} - 1, (1+0.08)^{20} - 1, (1+0.085)^{20} - 1\right); 0.9 \\
&= (2.8697, 3.6609, 4.6044; 1)(3.2478, 3.6609, 4.1120; 0.9)
\end{aligned}$$

$\tilde{\tilde{NPV}}$ s are calculated using Eq. 8.6, as $(1983.983, 3020.458, 4345.592; 1)$

$(2288.598, 3020.458, 3846.05; 0.9)$ and $(1903.544, 2992.033, 4388.634; 1)$

$(2223.783, 2992.033, 3862.259; 0.9)$ for scenario 1 and 2, respectively.

Defuzzified values are calculated using Eqs. 8.7 and 8.3 as 3,068.257 and 3043.294 respectively.

Table 8.7 Aggregated and defuzzified values of triangular HFS

| Parameter | Possible values | Aggregated value | Defuzzified Value of membership | Defuzzified value of parameter |
|---|---|---|---|---|
| FC | $630,000 | (0.341,0.442,0.572) | 0.449 | 650,347 |
| | $650,000 | (0.361,0.462,0.562) | 0.462 | |
| | $670,000 | (0.294,0.471,0.607) | 0.461 | |
| UAC | $40.000 | (0.468,0.579,0.748) | 0.594 | 45,020 |
| | $45,000 | (0.442,0.6,0.7) | 0.585 | |
| | $50,000 | (0.494,0.597,0.702) | 0.597 | |
| UAB | $300,000 | (0.516,0.650,0.753) | 0.642 | 346,515 |
| | $350,000 | (0.471,0.572,0.734) | 0.587 | |
| | $400,000 | (0.443,0.571,0.736) | 0.580 | |
| MC | $50,000 | (0.358,0.424,0.578) | 0.446 | 85,035 |
| | $80,000 | (0.618,0.744,0.848) | 0.739 | |
| | $110,000 | (0.482,0.633,0.748) | 0.624 | |
| SV | $100,000 | (0.442,0.542,0.618) | 0.536 | 124,043 |
| | $130,000 | (0.443,0.546,0.652) | 0.547 | |
| | $150,000 | (0.317,0.449,0.562) | 0.444 | |
| i | 7% | (0.465,0.569,0.650) | 0.563 | 794 |
| | 8% | (0.461,0.633,0.748) | 0.619 | |
| | 9% | (0.420,0.489,0.674) | 0.518 | |

### 8.4.3 Evaluation Using Intuitionistic Fuzzy Present Worth

Table 8.4 shows the values of parameters using interval valued intuitionistic fuzzy sets.

Table 8.5 shows the aggregated and defuzzified values for IVIFS based on Eqs. (8.11) and (8.12).

Using the data shown in Table 8.4 NPVs for Scenario 1 and 2 are calculated as $3,020.785 and $2.987,552, respectively.

### 8.4.4 Evaluation Hesitant Fuzzy Annual Worth

Possible values of the parameters and their corresponding compromised membership degrees are given in Table 8.6 using triangular HFS.

Table 8.7 shows the aggregated and defuzzified values for Triangular HFS based on Eqs. (8.29) and (8.30).

Using the data shown in Table 8.6 NPVs for Scenarios 1 and 2 are calculated as $2,996.775 and $2,959.74, respectively.

## 8.5 Conclusions

Wind energy investments involve several uncertain parameters; each can be represented by linguistic terms or fuzzy numbers. The cost and benefit parameters of wind energy investments can be better represented by fuzzy sets. Thus, an investment decision report can be presented to the investor with a list of possible results and their membership degrees.

PW analysis is the most used investment analysis technique. However, applying classical PW analysis under vagueness may produce unrealistic suggestions. Taking all possibilities into consideration before an investment decision is given is extremely important. Fuzzy PW analysis exhibits all possibilities regarding the investment outcomes together with their membership degrees.

For further research, other extensions of fuzzy sets such as Pythagorean fuzzy sets can be used for analyzing the wind energy investments. Other renewable energy alternatives can be also examined such as biomass energy, solar energy, geothermal energy, hydroelectric energy, ocean energy, or hydrogen energy under fuzziness. Types of fuzzy numbers can be changed alternatively such as trapezoidal fuzzy numbers or LR-type fuzzy numbers.

## References


Al-Sharafi, A., Sahin, A. Z., Ayar, T., & Yilbas, B. S. (2017). Techno-economic analysis and optimization of solar and wind energy systems for power generation and hydrogen production in Saudi Arabia. *Renewable and Sustainable Energy Reviews, 69,* 33–49.

Aquila, G., Junior, P. R., de Oliveira Pamplona, E., & de Queiroz, A. R. (2017). Wind power feasibility analysis under uncertainty in the Brazilian electricity market. *Energy Economics, 65,* 127–136.

Ashkaboosi, M., Nourani, S. M., Khazaei, P., Dabbaghjamanesh, M., & Moeini, A. (2016). An optimization technique based on profit of investment and market clearing in wind power systems. *American Journal of Electrical and Electronic Engineering, 4*(3), 85–91.

Atanassov, K. (2012). *On Intuitionistic Fuzzy Sets Theory*. Berlin, Heidelberg: Springer.

Atanassov, K.T. (1986), Intuitionistic fuzzy sets. *Fuzzy Sets and Systems, 20,* 87–96 (1986).

Aydin, N. Y., Kentel, E., & Duzgun, S. (2010). GIS-based environmental assessment of wind energy systems for spatial planning: A case study from Western Turkey. *Renewable and Sustainable Energy Reviews, 14*(1), 364–373.

Baringo, L., & Conejo, A. J. (2013). Risk-constrained multi-stage wind power investment. *IEEE Transactions on Power Systems, 28*(1), 401–411.

Caralis, G., Diakoulaki, D., Yang, P., Gao, Z., Zervos, A., & Rados, K. (2014). Profitability of wind energy investments in China using a Monte Carlo approach for the treatment of uncertainties. *Renewable and Sustainable Energy Reviews, 40,* 224–236.

Cavallaro, F., & Ciraolo, L. (2005). A multicriteria approach to evaluate wind energy plants on an Italian island. *Energy Policy, 33*(2), 235–244.

Celik, A. N. (2003). Energy output estimation for small-scale wind power generators using Weibull-representative wind data. *Journal of Wind Engineering and Industrial Aerodynamics, 91*(5), 693–707.

Chang, C. T. (2017). Fuzzy score technique for the optimal location of wind turbines installations. *Applied Mathematical Modelling, 44,* 576–587.

Chen, D., Zhang, L., & Jiao, J. (2010). Triangle fuzzy number intuitionistic fuzzy aggregation operators and their application to group decision making. In *International Conference on Artificial Intelligence and Computational Intelligence AICI 2010* (pp. 350–357).

Chiu, C. Y., & Park, C. S. (1994). Fuzzy cash flow analysis using present worth criterion. *The Engineering Economist, 39*(2), 113–138.

Cunico, M. L., Flores, J. R., & Vecchietti, A. (2017). Investment in the energy sector: An optimization model that contemplates several uncertain parameters. *Energy, 138,* 831–845.

Ersoz, S., Akinci, T. C., Nogay, H. S., & Dogan, G. (2013). Determination of wind energy potential in Kirklareli-Turkey. *International Journal of Green Energy, 10*(1), 103–116.

Fazelpour, F., Markarian, E., & Soltani, N. (2017). Wind energy potential and economic assessment of four locations in Sistan and Balouchestan province in Iran. *Renewable Energy, 109,* 646–667.

Gumus, S., Kucukvar, M., & Tatari, O. (2016). Intuitionistic fuzzy multi-criteria decision making framework based on life cycle environmental, economic and social impacts: The case of US wind energy. *Sustainable Production and Consumption, 8,* 78–92.

Kahraman, C., Çevik Onar, S., & Oztaysi, B. (2015). Engineering economic analyses using intuitionistic and hesitant fuzzy sets. *Journal of Intelligent & Fuzzy Systems, 29*(3), 1151–1168.

Kahraman, C., Cevik Onar, S., & Oztaysi, B. (2016a). A comparison of wind energy investment alternatives using interval-valued intuitionistic fuzzy benefit/cost analysis. *Sustainability, 8*(2), 118.

Kahraman, C., Oztaysi, B., & Cevik Onar, S. (2016b). A comprehensive literature review of 50 years of fuzzy set theory. *International Journal of Computational Intelligence Systems, 9* (sup1), 3–24.

Kahraman, C., Sarı, İ. U., Onar, S. C., & Oztaysi, B. (2017). Fuzzy Economic analysis methods for environmental economics. In *Intelligence Systems in Environmental Management: Theory and Applications* (pp. 315–346). Berlin: Springer.

Kitzing, L., Juul, N., Drud, M., & Boomsma, T. K. (2017). A real options approach to analyse wind energy investments under different support schemes. *Applied Energy, 188,* 83–96.

Kucukali, S. (2016). Risk scorecard concept in wind energy projects: An integrated approach. *Renewable and Sustainable Energy Reviews, 56,* 975–987.

Kumar, P. S., & Hussain, R. J. (2014). A method for solving balanced intuitionistic fuzzy assignment problem. *International Journal of Engineering Research and Applications, 4*(3), 897–903.

Kuo Ping, C. (2011). Multiple criteria group decision making with triangular interval type-2 fuzzy sets. In *Proceedings of 2011 IEEE International Conference on Fuzzy Systems (FUZZ), Tapei* (pp. 1098-7584), June 27–30, 2011.

Lee, A. H., Chen, H. H., & Kang, H. Y. (2009). Multi-criteria decision making on strategic selection of wind farms. *Renewable Energy, 34*(1), 120–126.

Lee, S. C. (2011). Using real option analysis for highly uncertain technology investments: The case of wind energy technology. *Renewable and Sustainable Energy Reviews, 15*(9), 4443–4450.

Liu, X., & Zeng, M. (2017). Renewable energy investment risk evaluation model based on system dynamics. *Renewable and Sustainable Energy Reviews, 73,* 782–788.

Madlener, R., Glensk, B., & Weber, V. (2011). Fuzzy portfolio optimization of onshore wind power plants. FCN Working Papers 10/2011, E.ON Energy Research Center, Future Energy Consumer Needs and Behavior (FCN), Revised Jul 2014.

Mahapatra, G. S., & Roy, T. K. (2009). Reliability evaluation using triangular intuitionistic fuzzy numbers arithmetic operations. *World Academy of Science, Engineering and Technology, 3*(2), 422–429.

Morshedizadeh, M., Kordestani, M., Carriveau, R., Ting, D. S.-K., & Saif, M. (2017). Application of imputation techniques and Adaptive Neuro-Fuzzy Inference System to predict wind turbine power production. *Energy, 138*(C), 394–404. Elsevier.

Niewiadomski, A., Ochelska, J., & Szczepaniak, P. S. (2006). Interval-valued linguistic summaries of databases. *Control and Cybernetics, 35*(2), 415–443.

Onar, S. C., & Kilavuz, T. N. (2015). Risk analysis of wind energy investments in Turkey. *Human and Ecological Risk Assessment*, *21*, 1230–1245.

Onar, S. C., Oztaysi, B., Otay, İ., & Kahraman, C. (2015). Multi-expert wind energy technology selection using interval-valued intuitionistic fuzzy sets. *Energy, 90,* 274–285.

Onat, N., & Ersoz, S. (2011). Analysis of wind climate and wind energy potential of regions in Turkey. *Energy, 36*(1), 148–156.

Panduru, K. K., Riordan, D., & Walsh, J. (2014). Fuzzy logic based intelligent energy monitoring and control for renewable energy. In *Irish Signals & Systems Conference 2014 and 2014 China-Ireland International Conference on Information and Communications Technologies (ISSC 2014/CIICT 2014).*

Petković, D. (2015). Adaptive neuro-fuzzy optimization of the net present value and internal rate of return of a wind farm project under wake effect. *Journal of CENTRUM Cathedra: The Business and Economics Research Journal*, *8*(1), 11–28.

Petković, D., Shamshirband, S., Kamsin, A., Lee, M., Anicic, O., & Nikolić, V. (2016). Survey of the most influential parameters on the wind farm net present value (NPV) by adaptive neuro-fuzzy approach. *Renewable and Sustainable Energy Reviews, 57,* 1270–1278.

Pinson, P., & Kariniotakis, G. N. (2003, June). Wind power forecasting using fuzzy neural networks enhanced with on-line prediction risk assessment. In *Power Tech Conference Proceedings, 2003 IEEE Bologna* (Vol. 2, 8 pp). IEEE.

Şen, Z., & Şahin, A. D. (1997). Regional assessment of wind power in western Turkey by the cumulative semivariogram method. *Renewable Energy, 12*(2), 169–177.

Shafiee, M. (2015). A fuzzy analytic network process model to mitigate the risks associated with offshore wind farms. *Expert Systems with Applications, 42*(4), 2143–2152.

Shamshirband, S., Petković, D., Ćojbašić, Ž., Nikolić, V., Anuar, N. B., Shuib, N. L. M., et al. (2014). Adaptive neuro-fuzzy optimization of wind farm project net profit. *Energy Conversion and Management, 80*, 229–237.

Sheen, J. N. (2009). Applying fuzzy engineering economics to evaluate project investment feasibility of wind generation. *WSEAS Transactions on Systems, 8*(4), 501–510.

Sheen, J. N. (2014). Real option analysis for renewable energy investment under uncertainty. In *Proceedings of the 2nd International Conference on Intelligent Technologies and Engineering Systems (ICITES2013)* (pp. 283–289). Cham: Springer.

Soroudi, A. (2012). Possibilistic-scenario model for DG impact assessment on distribution networks in an uncertain environment. *IEEE Transactions on Power Systems, 27*(3), 1283–1293.

Torra, V. (2010). Hesitant fuzzy sets. *International Journal of Intelligent Systems, 25*(6), 529–539.

Ucal Sari, I., & Kahraman, C. (2015). Interval Type-2 fuzzy capital budgeting. *International Journal of Fuzzy Systems, 17*(4), 635–646.

Wu, C. B., Huang, G. H., Li, W., Zhen, J. L., & Ji, L. (2016). An inexact fixed-mix fuzzy-stochastic programming model for heat supply management in wind power heating system under uncertainty. *Journal of Cleaner Production, 112,* 1717–1728.

Wu, Y., Geng, S., Xu, H., & Zhang, H. (2014). Study of decision framework of wind farm project plan selection under intuitionistic fuzzy set and fuzzy measure environment. *Energy Conversion and Management, 87,* 274–284.

Yeh, T. M., & Huang, Y. L. (2014). Factors in determining wind farm location: Integrating GQM, fuzzy DEMATEL, and ANP. *Renewable Energy, 66,* 159–169.

Yu, D. (2013). Triangular hesitant fuzzy set and its application to teaching quality evaluation. *Journal of Information & Computational Science, 10*(7), 1925–1934.

Zadeh, L. A. (1965). Fuzzy sets. *Information and Control, 8,* 338–353.

Zadeh, L. A. (1974). Fuzzy logic and its application to approximate reasoning. *Information Processing, 74,* 591–594.

Zadeh, L. A. (1975). The concept of a linguistic variable and its application to approximate reasoning—I. *Information Sciences, 8*(3), 199–249.

# Chapter 9
# Assessment of Fuel Tax Policies to Tackle Carbon Emissions from Road Transport—An Application of the Value-Based DEA Method Including Robustness Analysis

**Maria do Castelo Gouveia and Isabel Clímaco**

**Abstract** The transport sector has increased GHG emissions making it the second largest emitter in the EU after the energy generation sector. Given its share in total GHG emissions, the transport sector plays a critical role in the mitigation efforts required by the Paris Agreement on Climate Change. Fuel taxation can be used to internalize externalities, including those linked to fuel use as GHG emissions and local air pollution. Road transport policies have relied on fuel efficiency standards. A major outcome of this option was the prevailing preferential tax treatment for diesel fuel. This paper aims to assess the potential of fuel tax reforms to deal with carbon emissions from road transport in some EU countries. For this purpose, the Value-Based Data Envelopment Analysis method is used to obtain robust conclusions in face of sources of uncertainty. The adjustment of diesel excise tax levels towards gasoline taxation levels as well as the potential effects of introducing a carbon content-based tax on both diesel and gasoline are studied. The performance evaluation identifies the countries exhibiting the best practices. This approach offers decision makers the possibility to incorporate their priorities in appraising fuel tax policies considering uncertain factors to obtain robust conclusions.

---

M. do Castelo Gouveia (✉) · I. Clímaco
ISCA—Coimbra Business School, Polytechnic Institute of Coimbra, Coimbra, Portugal
e-mail: mgouveia@iscac.pt

M. do Castelo Gouveia
INESC Coimbra, Coimbra, Portugal

I. Clímaco
Centre for Health Studies and Research, University of Coimbra, Coimbra, Portugal

© Springer International Publishing AG, part of Springer Nature 2018

C. Kahraman and G. Kayakutlu (eds.), *Energy Management—Collective and Computational Intelligence with Theory and Applications*, Studies in Systems, Decision and Control 149, https://doi.org/10.1007/978-3-319-75690-5_9

## 9.1 Introduction

### *9.1.1 Motivation and Interest of the Study*

Preventing climate change is a strategic priority for the EU. Policies to reduce energy consumption in the personal transport sector are likewise an essential component of European energy policies with impact on hazardous emissions. According to its climate and energy targets, the EU aims to curb its greenhouse gas (GHG) emissions by 40% from 1990 levels by 2030, and continue this trend to achieve 80–95% reduction by 2050. Responsible for about a quarter of EU's GHG emissions, the transport sector is the only sector where emissions have grown over the past 25 years, due to a strong dependency on oil. Given its share in total GHG emissions, the transport sector should make a large contribution towards the mitigation efforts established by the Paris Agreement on Climate Change signed in December 2015. In order to achieve the 2 °C target, the European Commission estimates that GHG emissions from EU's transport sector need to be reduced by 70% below 2008 levels by 2050 (European Commission 2011).

Fuel taxation can be used to internalize a wide range of externalities, including those directly linked to fuel use, such as GHG emissions and local air pollution. In Europe, fuel taxes were not originally designed as an economic instrument, but rather as a fiscal instrument to raise revenues and finance government expenditure. Since $CO_2$ emissions are closely related to fuel consumption, fuel taxation stands as a relevant tool to internalize the associated external costs (Newbery 1992, 2001; Parry and Small 2005; Parry 2007; Sterner 2007). In fact, many governments defend their fuel duties on environmental grounds (Newbery 2005).

The profile of passenger car fleets has been deeply transformed in Europe during the last decades. The percentage of diesel cars increased from 11% in 1991 to almost 38% in 2011, a process referred to as dieselization. The advantage in fuel efficiency of diesel motor cars and the practice of a diesel incentive fuel tax policy in most European countries may help to explain this dieselization process. Nevertheless, the success of dieselization as a measure to control $CO_2$ emissions has been recently called into question by many authors in the transport literature (Schipper and Fulton 2013; González and Marrero 2012). As has been pointed out in a previous study, the higher carbon content per liter of diesel partially offsets the fuel efficiency of diesel powered cars (United States Environmental Protection Agency 2011). Moreover, a replacement of gasoline by diesel vehicles generates a rebound impact on kilometers travelled, caused by the effect of using more efficient motor cars (in terms of liters per km driven) and a lower diesel price (tax included) (Rodríguez-López et al. 2015).

The Energy Tax Directive (European Commission 2003) establishes the values of EUR 0.359 per liter of unleaded gasoline and 0.330 for diesel as the minimum taxation rates. With the exception of the UK, where the taxes on gasoline and diesel are equal, all EU countries have considerably lower taxes for diesel than for gasoline. However, the lower tax rates applied to diesel are not consistent with environmental and other social externalities associated with diesel usage. Tax advantages for diesel are not justified from air pollution and impact of climate change perspectives. In fact, as taxation is imposed per liter, the fuel efficiency advantage of diesel cars per km driven is a benefit that is internalized by the vehicle user, contrarily to the cost of $CO_2$ emissions (Harding 2014; Zimmer and Koch 2016). The promotion of more fuel-efficient cars through fiscal measures is one of the pillars of the EU's strategy to incentivize reductions of GHG emissions of light-duty vehicles (European Commission 2007). In 2005, the European Commission presented a proposal for a directive aimed at restructuring motor vehicle taxes taking $CO_2$ emissions into account. Neither this proposal nor its reiteration proposed in 2012 were approved.

The necessary transition to a low-carbon economy requires the commitment of all EU member states to the reduction of GHG. The Energy Tax Directive from 2003 is the only common framework that constraints the fuel tax policies of member states. Each country has different economic and political agendas, which constrain their individual contribution to tackle carbon emissions from road transport.

This paper aims at assessing the potential of fuel tax reforms to deal with carbon emissions from road transport in some EU countries using the Value-Based Data Envelopment Analysis (DEA) method. The need to obtain robust conclusions in face of several sources of uncertainty is taken into account. The Value-based DEA method combines DEA and multi-criteria decision analysis, encompassing managerial preference information. This study considers fuel pricing policy scenarios to reduce GHG emissions in a sample of EU countries. The adjustment of diesel excise tax levels towards gasoline taxation levels, as well as the potential effects of introducing a carbon content-based tax of 50 € per ton of $CO_2$ on both diesel and gasoline, are studied.

Data Envelopment Analysis (DEA) is a non-parametric method for evaluating the relative efficiency of decision making units (DMUs), with some degree of autonomy operating in a relatively homogeneous environment that use multiple inputs to produce multiple outputs (Charnes et al. 1978). DEA models compare the performances of DMUs and determine the ones exhibiting the best practices, which form the (Pareto-Koopmans) efficient frontier, also enabling to measure the gaps to best practices and to identify benchmarks against which such inefficient DMUs should be compared. DEA models have been widely used for performance evaluation in different domains.

The Value-Based DEA method (Gouveia et al. 2008) exploits the links between DEA and Multi-criteria Decision Analysis (MCDA), in which DMUs are compared being the input (criteria being minimized) and output factors (criteria being maximized) converted into value functions constructed using managerial/policy preferences. Additive value functions (which may be non-linear or piecewise linear) are used to aggregate the values associated with each DMU, based on Multi-Attribute Utility Theory (MAUT) (von Winterfeldt and Edwards 1986). Preferences (which may derive from managerial focus or policy perspectives) are used to restrict the weights associated with those functions so that the evaluation can make more useful insights to emerge. The Value-Based DEA method produces an efficiency measure assigned to each DMU that has a "min-max regret" (loss of value) interpretation. This method has been modified to include the concept of super-efficiency to assess the robustness of DMUs in face of uncertain information (Gouveia et al. 2013). Uncertainty is incorporated by considering the DMU performances in each factor as intervals (i.e. uncertain but bounded). The robustness analysis assesses whether each DMU is surely efficient, potentially efficient, or surely inefficient.

This paper is aimed at assessing the potential of fuel tax reforms to deal with carbon emissions from road transport in some EU countries. For this purpose, the Value-Based DEA method is used, taking into account the need to obtain robust conclusions in face of several sources of uncertainty. The study considers fuel pricing policy scenarios to reduce GHG emissions in a set of EU countries. The adjustment of diesel excise tax levels towards gasoline taxation levels as well as the potential effects of introducing a carbon content-based tax of 50 € per ton of $CO_2$ on both diesel and gasoline are studied. The performance evaluation by means of the Value-Based DEA method enables to identify the countries that exhibit the best practices defining an efficiency frontier. The gaps to best practices of non-frontier countries are measured and benchmarks against which those inefficient countries should be compared with are identified. This approach offers decision makers the possibility to incorporate their priorities in appraising fuel tax policies considering the uncertain factors at stake in order to obtain robust conclusions, i.e. recommendations that are somehow immune to plausible ranges of variation of some input information.

The chapter is structured as follows: in the next section, a brief review of the literature is presented. Section 9.2 synopsizes the Value-Based DEA method. In Sect. 9.3, a discussion of how to perform the robustness analysis in the context of the approach presented in the previous section is presented. Section 9.4 is devoted to the quantitative data (indicators and data sources) regarding fuel excises taxes for European countries according to the assessment criteria and the elicitation protocols used to obtain the value functions. Section 9.5 analyzes of the results obtained with the Value-Based DEA method for assessing fuel tax reforms' potential to curb carbon emissions from road transport in terms of economic and environmental criteria. In Sect. 9.6, the ranges of efficiency are computed and the robustness of each country is analyzed. Concluding remarks are presented in Sect. 9.7.

### *9.1.2 Brief Literature Review*

Diesel cars have been gaining a growing market share over gasoline cars, namely due the policy focus on fuel savings as well as the expected reduction of emissions due to higher fuel efficiency of diesel engines, so that diesel cars had over 55% of the new vehicle market in the EU in 2009 (Schipper and Fulton 2013). However, a greater travel rebound effect has been witnessed and an energy consumption increase is expected due to lower prices (Ajanovic and Haas 2012).

Fuel taxes have been recognized as relevant instruments to influence total consumption by changing individual driving behavior, create incentives for companies to develop advanced fuel saving technologies, contribute to tackling climate change and other environmental problems (Sterner 2007; Bonilla 2009; Ryan et al. 2009; Kloess and Müller 2011; Carreno et al. 2014; Sterner and Köhlin 2015; Coria 2012).

The economic and environmental consequences resulting from energy taxation have been addressed in several studies (Gago et al. 2014), both by means of ex-ante simulations based on tax policy proposals and ex-post empirical analysis on actual energy tax implementations. Arbolino and Romano (2014) evaluate Environmental Tax Reforms at the European level using Hierarchical Cluster Analysis and Quantitative SWOT analysis. Filipović and Golušin (2015) propose a methodological approach based on an Environmental Taxation Efficiency (ETE) indicator, which encompasses environmental taxation effects per capita. Zimmer and Koch (2016) assess the potential of fuel tax reforms to curb harmful air pollutants and carbon emissions from road transport in Europe to evaluate the potential of two fuel pricing policy scenarios to reduce $CO_2$, PM2.5 and $NO_x$ emissions. Schipper and Fulton (2013) found that the shift to diesel cars played little role in the reductions in aggregate new vehicle $CO_2$ emissions intensity for the EU12. Santos (2017) conclude that road transport externalities in the 22 countries analyzed in her study are not being internalized, thus leading to the recommendation that instruments should be developed for this purpose in a first step by increasing fuel taxes.

A review of DEA approaches in energy efficiency, including the eco-efficiency of transportation modes, is made in Mardani et al. (2017).

## 9.2 The Value-Based DEA Method

Data Envelopment Analysis is a quantitative, empirical and non-parametric (data-driven) method, which measures the relative performance of similar (comparable) organizational units (Decision Making Units, DMUs), generating a relative efficiency score for each DMU under evaluation. Each DMU is free to choose the weighted ratio between outputs and inputs in order to maximize its relative efficiency. Those DMUs with maximum performance will form the empirical efficient frontier, or "envelopment surface". This allows identifying reference units whose performance ratings serve as a benchmark for other units, enveloped by the efficiency frontier.

The Value-Based DEA method developed by Gouveia et al. (2008) is a variant of the additive DEA model (Charnes et al. 1985) with oriented projections (Ali et al. 1995), incorporating managerial/policy preferences using concepts from MAUT under imprecise information. The Value-Based DEA method considers that "the inputs are usually the "less-the-better" type of performance measures and the outputs are usually the "more-the-better" type of performance measures" (Cook et al. 2014), where the identification of a production process is not relevant.

The Value-based DEA method deals with the DMUs as alternatives of a multi-criteria evaluation model, where the criteria (the ones to be minimized and the ones to be maximized) are converted into value functions according to preference information provided by managers/policy makers (henceforth designated as decision maker, DM). This overcomes the scale-dependence problem of the additive DEA model, since all the input and output measures are translated into value units.

The set of n DMUs to be evaluated is: $\{DMU_j: j = 1, \ldots, n\}$. Each $DMU_j$ is evaluated on m factors to be minimized $x_{ij}(i = 1, \ldots, m)$ and p factors to be maximized $y_{rj}(r = 1, \ldots, p)$.

The measure of performance on criterion c is: $\{v_c(DMU_j), c = 1, \ldots, q,\ with\ q = m + p, j = 1, \ldots, n\}$ based on a value function (or utility function) $v_c(.)$. This measure is established using the preferences expressed by the DM. Considering that $p_{cj}$ is the performance of DMU j in factor c, the value functions must be defined such that for each factor c the worst $p_{cj}, j = 1, \ldots, n$, has the value 0 and the best $p_{cj}, j = 1, \ldots, n$, has the value 1, resulting in maximization of all factors.

A preliminary phase of Value-Based DEA method comprises the assessment of marginal (partial) value functions (scored between 0 and 1) on each criterion to establish a global value function. According to the additive MAUT model, the value obtained is $V(DMU_j) = \sum_{c=1}^{q} w_c v_c(DMU_j)$, where $w_c \geq 0, \forall c = 1, \ldots, q$ and $\sum_{c=1}^{q} w_c = 1$ (by convention). The weights $w_1, \ldots, w_q$ considered in the aggregation are the scale coefficients of the value functions and reflect the value trade-offs of the DM. Furthermore, the efficiency measure of each DMU gains the meaning of the "min-max regret" (value loss) measure.

After the preliminary phase in which the factors (to be minimized and to be maximized) are converted into value scales, the Value-Based DEA method can be described in two phases:

Phase 1: Compute the efficiency measure, $d_k^*$, for each DMU, $k = 1, \ldots, n$, and the corresponding weighting vector $w_k^*$ *by* solving the linear problem (9.1).

Phase 2: If $d_k^* \geq 0$ then solve the "weighted additive" problem (9.2), using the optimal weighting vector resulting from Phase 1, $w_k^*$, and determine the corresponding projected point of the DMU under evaluation.

Formulation (9.1) considers the super-efficiency concept (Andersen and Petersen 1993), which allows the discrimination of the efficient units, when assessing the *k*-th DMU (Gouveia et al. 2013):

$$
\begin{aligned}
&\min_{d_k, w} d_k \\
&\text{s.t.} \sum_{c=1}^{q} w_c v_c(DMU_j) \\
&- \sum_{c=1}^{q} w_c v_c(DMU_k) \le d_k, \quad j = 1, \ldots, n;\ j \ne k \\
&\sum_{c=1}^{q} w_c = 1 \\
&w_c \ge 0, \quad \forall\, c = 1, \ldots, q
\end{aligned}
\tag{9.1}
$$

The efficiency measure, $d_k^*$, for each DMU $k$ ($k$ = 1, …, $n$) and the corresponding weighting vector are computed via formulation (9.1). The score $d_k^*$ is the distance defined by the value difference to the best of all DMUs (note that the best DMU will also depend on $w$), excluding itself from the reference set. If the optimal value $d_k^*$ of the objective function in (9.1) is not positive, then the $DMU_k$ under evaluation is efficient, otherwise it is inefficient.

In case the DMU is inefficient, Phase 2 finds an efficient target by solving the linear problem (9.2):

$$
\begin{aligned}
&\min_{\lambda, s} z_k = -\sum_{c=1}^{q} w_c^* s_c \\
&\text{s.t.} \sum_{j=1, j \ne k}^{n} \lambda_j v_c(DMU_j) - s_c \\
&= v_c(DMU_k), c = 1, \ldots, q \\
&\sum_{j=1, j \ne k}^{n} \lambda_j = 1 \\
&\lambda_j, s_c \ge 0, j = 1, \ldots, k-1,\ k+1, \ldots, n;\ c = 1, \ldots, q
\end{aligned}
\tag{9.2}
$$

The variables $\lambda_j$, $j$ = 1, …, $k$ − 1, $k$ + 1, …, $n$ define a convex combination of the value score vectors associated with the $n$ − 1 DMUs. The set of efficient DMUs (possibly a single one) defining the convex combination with $\lambda_j > 0$ are called the "peers" of $DMU_k$ under evaluation. The convex combination corresponds to a point on the efficient frontier that is better than $DMU_k$ by a difference of value of $s_c$ (slack) in each criterion $c$.

The case study (Sect. 9.4) requires the expression of managerial/policy preferences, which are made operations through weights restrictions. So, the set $W$ denote the set of weight vectors necessary to be added to problem (9.1), which leads to a new formulation with $(w_1, \ldots, w_q) \in W$.

## 9.3 A Robustness Assessment Based on the Computation of Stability Intervals

Several researchers have paid attention to the sensitivity of the results to perturbations in data and the robustness of the efficiency scores resulting from these perturbations, based on super-efficiency DEA approaches (Seiford and Zhu 1998a, b; Zhu 1996; 2001; 2003). In our approach to deal with uncertainty, we consider that the perturbations in the coefficients in each factor (input or output) are captured through interval coefficients and then converted into utility scales. According to Gouveia et al. (2013), an optimistic efficiency measure and a pessimistic efficiency measure are computed. Using these two efficiency measures we can classify each DMU as surely efficient, potentially efficient, or surely inefficient for a given tolerance value. Unlike the standard super-efficiency approach for oriented models, no infeasibility concerns arise in our approach.

Let us consider that the value $p_{cj}$ (performance of DMU $j$ in factor $c$) is uncertain but bounded within the range $p_{cj}^{L} \leq p_{cj} \leq p_{cj}^{U}$. For this work, we consider that all performances are applied a common tolerance $\delta$, such that $p_{cj}^{L} = p_{cj}(1-\delta) \leq p_{cj} \leq p_{cj}(1+\delta) = p_{cj}^{U}$. Assuming $v_c$ is monotonic, the previous inequalities imply $v_c^{L}(DMU_j) \leq v_c(DMU_j) \leq v_c^{U}(DMU_j)$, if factor $c$ is to be maximized, or $v_c^{L}(DMU_j) \geq v_c(DMU_j) \geq v_c^{U}(DMU_j)$, if factor $c$ is to be minimized.

To compute the optimistic efficiency measure we consider the best value of the intervals for the DMU being evaluated and the worst value of the intervals for all other DMUs. The reverse is considered to compute the pessimistic efficiency measure. We compute the optimistic efficiency measure $d_k^{opt*}$ for DMU $k$ solving the following linear problem:

$$
\begin{aligned}
&\min_{d_k, w} d_k^{opt} \\
&\text{s.t.} \sum_{c=1}^{q} w_c v_c^{L}(DMU_j) \\
&\quad - \sum_{c=1}^{q} w_c v_c^{U}(DMU_k) \leq d_k^{opt}, \; j = 1, \ldots, n; \; j \neq k \\
&\sum_{c=1}^{q} w_c = 1 \\
&w_c \geq 0, \forall c = 1, \ldots, q \\
&(w_1, \ldots, w_q) \in W
\end{aligned}
\tag{9.3}
$$

We compute the pessimistic efficiency measure $d_k^{pes*}$ for DMU $k$ by formulation (9.4).

$$
\begin{aligned}
&\min_{d_k, w} d_k^{pes} \\
&\text{s.t.} \sum_{c=1}^{q} w_c v_c^U \left(DMU_j\right) \\
&\qquad - \sum_{c=1}^{q} w_c v_c^L (DMU_k) \leq d_k^{pes}, j = 1, \ldots, n; j \neq k \\
&\sum_{cc=1}^{q} w_c = 1 \\
&w_c \geq 0, \forall c = 1, \ldots, q \\
&\left(w_1, \ldots, w_q\right) \in W
\end{aligned}
\tag{9.4}
$$

A DMU is classified as robust to changes in its factors if it remains efficient (or inefficient). In such case, the DMU can be declared robustly efficient (or robustly inefficient) for the tolerance considered.

## 9.4 Case Study

We consider two distinct scenarios as in Zimmer and Koch (2016): an abolition of the favored tax treatment for diesel and an establishment of a carbon content-based tax. In turn, this led to the development of two models. These models determine the relative changes in $CO_2$ emissions in 2020 in relation to the status quo (2013) for two tax reform scenarios: (i) the adaptation of diesel excise tax levels to the levels of gasoline taxation in 2013 and (ii) the introduction of carbon content-based tax of 50 €/t$CO_2$ on both diesel and gasoline in 2013.[1]

### *9.4.1 Problem and Proposed Model*

Data have been collected from a variety of publicly available sources. The excise duties for gasoline[2] and diesel[3] were taken from OECD studies. The consumption of gasoline and diesel originate from the EU data[4] (Table 9.1).

---

[1]The Fifth Assessment Report of the IPCC documents estimates around 241 USD per ton of carbon for studies published after 2007, corresponding to around 65 USD/t$CO_2$ or 50€/t$CO_2$ (Zimmer and Koch 2016).

[2]https://www.oecd.org/ctp/consumption/Table-4.A4.5-Taxation-of-premium-unleaded-(94-96%20RON)-gasoline-(per%20L)-2013-Dec-2014.xls.

[3]https://www.oecd.org/ctp/consumption/Table-4.A4.6-Taxation-of-automotive-diesel-(per%20L)-2013-Dec-2014.xls.

[4]https://ec.europa.eu/energy/en/data-analysis/weekly-oil-bulletin.

**Table 9.1** Data regarding gasoline and diesel in 2013

| | Gasoline | | | | Diesel | | | |
|---|---|---|---|---|---|---|---|---|
| | Excise (€/l) | Consumption (kt) | Consumption (Ml) | Fiscal revenue (M€) | Excise (€/l) | Consumption (kt) | Consumption (Ml) | Fiscal revenue (M€) |
| Austria | 0.49 | 1603 | 2,180,080 | 1075 | 0.41 | 6448 | 7,621,536 | 3117 |
| Belgium | 0.61 | 996 | 1,354,560 | 832 | 0.43 | 6680 | 7,895,760 | 3379 |
| Czech Republic | 0.49 | 1537 | 2,090,320 | 1033 | 0.42 | 4144 | 4,898,208 | 2065 |
| Finland | 0.65 | 890 | 1,210,400 | 788 | 0.47 | 2436 | 2,879,352 | 1353 |
| France | 0.61 | 3706 | 5,039,480 | 3089 | 0.44 | 34,154 | 40,370,028 | 17,722 |
| Germany | 0.66 | 14,597 | 19851920 | 13,003 | 0.47 | 34,840 | 41,180,880 | 19,355 |
| Hungary | 0.42 | 1171 | 1,592,560 | 661 | 0.38 | 2603 | 3,076,746 | 1177 |
| Italy | 0.73 | 8025 | 10,914,000 | 7945 | 0.62 | 22,353 | 26,421,825 | 16,302 |
| Netherlands | 0.75 | 3958 | 5382880 | 4053 | 0.44 | 6844 | 8,089,608 | 3559 |
| Poland | 0.40 | 3635 | 4943600 | 1965 | 0.35 | 11,719 | 13,851,858 | 4806 |
| Spain | 0.46 | 4336 | 5,896,960 | 2736 | 0.37 | 24,238 | 28,649,316 | 10,600 |
| United Kingdom | 0.68 | 12,173 | 16,555,280 | 11,297 | 0.68 | 21,926 | 25,916,532 | 17,686 |

In order to evaluate the gasoline and diesel taxation in light of these principles, the following perspectives are considered:

(i) Potential of fiscal revenue—the potential gain of fiscal revenue in 2020 as a percentage of total fiscal revenue in 2013 is considered as proxy.
(ii) Contribution to correct external costs (relative changes in $CO_2$ emissions).
(iii) Distributional effects or fiscal equity.

The model includes two distinct policy scenarios (Model 1 and Model 2) of fuel tax reforms regarding 2020:

Model 1—Adjustment of diesel excise taxes to gasoline taxation levels.
Model 2—Introduction of a $CO_2$ content-based tax of 50 €/t$CO_2$.

The aforementioned proposals were also analyzed by Zimmer and Koch (2016) in order to assess how tax reforms contribute to climate change goals, including the inference of a carbon content-based tax per liter of diesel and gasoline of 50 €/t$CO_2$, in addition to the pre-existing tax level. These authors make the assumption that this $CO_2$ tax falls within the range of social costs of carbon estimates (for which there is not a consensus).

Table 9.2 displays a summary of the factors (criteria) used in each model with the respective description and metrics.

Data concerning the potential fiscal revenue and the contribution to correct the external costs for the two distinct fuel policy scenarios considered (Tables 9.3 and 9.4) were gathered from Zimmer and Koch (2016).

**Table 9.2** Description of factors in models 1 and 2

| Factors | Description | Direction of optimization |
|---|---|---|
| $x_{CO_2Red\_M1}$ | Relative changes in $CO_2$ emissions compared to the status quo (2013) for diesel excise tax adjustment (Model 1) | Minimize the performances in column 2 of Table 9.4 |
| $y_{PGR\_M1}$ | Potential gain of revenue in 2020 as a percentage of total fiscal revenue in 2013 (Model 1) | Maximize the performances resulting from column 2 of Table 9.3 divided by the sum of column 5 and column 9 of Table 9.1 |
| $y_{ATFT}$ | Average transport fuel taxes as a percentage of pre-tax expenditure (Model 1 and Model 2) | Maximize the performances in column 6 of Table 9.5 |
| $x_{CO_2Red\_M2}$ | Relative changes in $CO_2$ emissions compared to the status quo (2013) with the introduction of a carbon content-based tax of 50 €/t$CO_2$ (Model 2) | Minimize the performances in column 3 of Table 9.4 |
| $y_{PGR\_M2}$ | Potential gain of revenue in 2020 as a percentage of total fiscal revenue in 2013, with 50 €/t$CO_2$ (Model 2) | Maximize the performances resulting from column 3 of Table 9.3 divided by the sum of column 5 and column 9 of Table 9.1 |

**Table 9.3** Fiscal revenue with the adoption of the distinct fuel policies

| | Increase in fiscal revenue in 2020 with the adjustment of excise taxes (M€) | Increase in fiscal revenue in 2020 with the adoption of a carbon content-based tax (M€) |
|---|---|---|
| Austria | 409 | 799 |
| Belgium | 918 | 857 |
| Czech Republic | 212 | 573 |
| Finland | 314 | 421 |
| France | 4046 | 4089 |
| Germany | 4131 | 5470 |
| Hungary | 58 | 373 |
| Italy | 1735 | 3143 |
| Netherlands | 1218 | 1133 |
| Poland | 383 | 1424 |
| Spain | 1510 | 2760 |
| United Kingdom | 0 | 3800 |

*Source* Data from Zimmer and Koch (2016)

**Table 9.4** Relative changes in $CO_2$ emissions compared to the status quo (2013) for both policy scenarios

| $\Delta CO_2$ in 2020 (w.r.t. to 2013) | | |
|---|---|---|
| | Model 1 (%) | Model 2 (%) |
| Austria | −5.8 | −9.6 |
| Belgium | −11.8 | −9.3 |
| Czech Republic | −4.1 | −9.0 |
| Finland | −8.7 | −8.0 |
| France | −11.9 | −10.0 |
| Germany | −9.0 | −8.1 |
| Hungary | −1.7 | −8.9 |
| Italy | −5.4 | −7.6 |
| Netherlands | −14.7 | −8.0 |
| Poland | −3.2 | −9.9 |
| Spain | −6.1 | −9.9 |
| United Kingdom | 0.0 | −7.3 |

*Source* Data from Zimmer and Koch (2016)
*Note* Emissions refer to calculated emissions from diesel and gasoline consumption for road transportation based on elasticity estimates and emission factors per litre. Reported changes assume that GDP per capita and vehicle stocks per driver remain unchanged

Table 9.4 shows the predicted reductions in $CO_2$ emissions for Model 1 (diesel excise tax adjustment) and Model 2 (introducing a 50 €/t$CO_2$ carbon tax) in relation to the $CO_2$ emission levels from use of diesel and gasoline in 2013. Accordingly,

**Table 9.5** Average transport fuel taxes as percentage of pre-tax expenditure by expenditure decile

| Deciles | 1 (Poorest) (%) | 2 (%) | 9 (%) | 10 (Richest) (%) | (9 + 10)/(1 + 2) (%) |
|---|---|---|---|---|---|
| Austria | 1.40 | 1.80 | 1.70 | 1.40 | 96.88 |
| Belgium | 1.10 | 1.50 | 1.30 | 1.10 | 92.31 |
| Czech Republic | 1.30 | 1.70 | 2.00 | 2.10 | 136.67 |
| Finland | 1.50 | 1.90 | 1.80 | 1.40 | 94.12 |
| France | 1.00 | 1.30 | 1.90 | 1.40 | 84.38 |
| Germany | 1.50 | 1.80 | 1.60 | 1.50 | 143.48 |
| Hungary | 0.60 | 1.10 | 2.20 | 2.30 | 93.94 |
| Italy | 2.30 | 2.40 | 2.00 | 1.60 | 113.95 |
| Netherlands | 1.70 | 1.70 | 1.90 | 1.70 | 264.71 |
| Poland | 1.30 | 1.50 | 2.10 | 1.80 | 76.60 |
| Spain | 1.60 | 1.60 | 1.50 | 1.20 | 105.88 |
| United Kingdom | 1.60 | 2.70 | 2.80 | 2.10 | 139.29 |

*Source* Data from Flues and Thomas (2015)

the more negative the better, since the countries with more negative performances in this criterion will experience the highest reduction in carbon emissions in 2020. It should be noted that the changes in emissions displayed pertain to annual emissions, i.e. in 2020, and not to accumulated emission reductions up to 2020 (Zimmer and Koch 2016).

To better evaluate the distributional (equity) effects of fuel taxes, we considered the average transport fuel taxes as a percentage of pre-tax expenditure by expenditure decile taken from an OECD study (Flues and Thomas 2015). The major advantage of the use of expenditure as the basis for the measurement of tax burdens when assessing consumption taxes' distributional effects is that expenditure is a better measure of current and lifetime well-being than current income. Expenditure will vary to a lesser extent than income over a lifetime, since households generally present a given level of consumption smoothing due to varying consumption needing a longer time to occur.

Taking these data into account, it was necessary to unveil three different patterns across countries. The first one pertains to the progressive effect throughout the whole expenditure distribution, the second considers the approximately proportional impact of transport fuel taxes on households across deciles in the expenditure distribution, and the third refers to acknowledging that households in the middle of the expenditure distribution face the highest transport fuel tax burden (Flues and Thomas 2015). For this end, we considered the four extreme deciles (1 + 2 and 9 + 10) (see Table 9.5) and calculated the ratio (9 + 10)/(1 + 2).

**Table 9.6** Performances of DMUs in original scales and in value scales

| DMUs | Factors in original scales | | | | | Factors in value scales | | | | |
|---|---|---|---|---|---|---|---|---|---|---|
| | $x_{CO_2Red\_M1}$ (%) | $x_{CO_2Red\_M2}$ (%) | $y_{PGR\_M1}$ (%) | $y_{PGR\_M2}$ (%) | $y_{ATFT}$ (%) | $v_{CO_2Red\_M1}$ (%) | $v_{CO_2Red\_M2}$ (%) | $v_{PGR\_M1}$ (%) | $v_{PGR\_M2}$ (%) | $v_{TFT}$ (%) |
| AT | −5.80 | −9.60 | 9.76 | 19.06 | 96.88 | 0.242 | 0.400 | 0.903 | 0.273 | 0.108 |
| BE | −11.80 | −9.30 | 21.80 | 20.35 | 92.31 | 0.567 | 0.325 | 0.990 | 0.307 | 0.084 |
| CZ | −4.10 | −9.00 | 6.84 | 18.50 | 136.67 | 0.171 | 0.250 | 0.864 | 0.258 | 0.319 |
| FI | −8.70 | −8.00 | 14.66 | 19.66 | 143.48 | 0.385 | 0.150 | 0.947 | 0.289 | 0.356 |
| FR | −11.90 | −10.00 | 19.44 | 19.65 | 93.94 | 0.575 | 0.500 | 0.977 | 0.289 | 0.092 |
| DE | −9.00 | −8.10 | 12.77 | 16.90 | 94.12 | 0.400 | 0.160 | 0.932 | 0.213 | 0.093 |
| HU | −1.70 | −8.90 | 3.16 | 20.29 | 264.71 | 0.071 | 0.240 | 0.780 | 0.305 | 1.000 |
| IT | −5.40 | −7.60 | 7.16 | 12.96 | 76.60 | 0.225 | 0.110 | 0.869 | 0.078 | 0.000 |
| NL | −14.70 | −8.00 | 16.00 | 14.88 | 105.88 | 0.830 | 0.150 | 0.956 | 0.148 | 0.156 |
| PL | −3.20 | −9.90 | 19.49 | 72.47 | 139.29 | 0.133 | 0.475 | 0.977 | 0.950 | 0.333 |
| ES | −6.10 | −9.90 | 11.32 | 20.70 | 84.38 | 0.255 | 0.475 | 0.919 | 0.315 | 0.041 |
| UK | 0.00 | −7.30 | 0.00 | 13.11 | 113.95 | 0.000 | 0.080 | 0.000 | 0.084 | 0.199 |

**Table 9.7** Results from Phase 1 and Phase 2 with weight restrictions

| DMUs | Phase 1 | | | | | Phase 2 | | | | | | |
|---|---|---|---|---|---|---|---|---|---|---|---|---|
| | $d^*$ | $w^*_{CO_2Red\_M1}$ | $w^*_{CO_2Red\_M2}$ | $w^*_{PGR\_M1}$ | $w^*_{PGR\_M2}$ | $w^*_{ATFT}$ | $s^*_{CO_2Red\_M1}$ | $s^*_{CO_2Red\_M2}$ | $s^*_{PGR\_M1}$ | $s^*_{PGR\_M2}$ | $s^*_{ATFT}$ | Peers $\lambda$ |
| PL | **−0.317** | 0.000 | 0.500 | 0.000 | 0.500 | 0.000 | | | | | | |
| HU | **−0.232** | 0.000 | 0.000 | 0.000 | 0.000 | 1.000 | | | | | | |
| NL | **−0.115** | 0.500 | 0.000 | 0.500 | 0.000 | 0.000 | | | | | | |
| FR | **−0.019** | 0.313 | 0.187 | 0.313 | 0.187 | 0.000 | | | | | | |
| IT | 0.05 | 0.322 | 0.000 | 0.322 | 0.000 | 0.356 | 0.605 | 0.040 | 0.087 | 0.070 | **−0.612** | NL (1) |
| BE | 0.024 | 0.286 | 0.152 | 0.286 | 0.152 | 0.123 | **−0.433** | 0.150 | **−0.012** | 0.643 | 0.250 | PL (1) |
| FI | 0.050 | 0.259 | 0.115 | 0.259 | 0.115 | 0.252 | **−0.314** | 0.090 | **−0.166** | 0.016 | 0.643 | HU (1) |
| ES | 0.111 | 0.295 | 0.205 | 0.295 | 0.205 | 0.000 | 0.000 | 0.007 | 0.059 | 0.453 | 0.226 | FR;PL (0.275;0.725) |
| DE | 0.125 | 0.259 | 0.115 | 0.259 | 0.115 | 0.252 | 0.000 | 0.191 | 0.038 | 0.430 | 0.172 | FR;PL (0.275;0.725) |
| CZ | 0.129 | 0.259 | 0.115 | 0.259 | 0.115 | 0.252 | 0.000 | 0.207 | 0.112 | 0.648 | 0.004 | FR;PL (0.275;0.725) |
| AT | 0.132 | 0.286 | 0.152 | 0.286 | 0.152 | 0.123 | 0.000 | 0.024 | 0.072 | 0.552 | 0.198 | FR;PL (0.275;0.725) |
| UK | 0.433 | 0.000 | 0.301 | 0.000 | 0.301 | 0.397 | 0.133 | 0.395 | 0.977 | 0.866 | 0.135 | PL (1) |

Bold denotes efficiency

Thus, the highest the value achieved, the better.[5]

The performances of each country being evaluated for the different criteria in each model are shown in the left-hand side of Tables 9.6 and 9.7 (see Sect. 9.4.2).

### 9.4.2 Elicitation of Value Functions

In Value-Based DEA method, the construction of value functions is made to reveal the preferences of the DM (in this setting an expert in fiscal policy). The elicitation protocol used to construct the value functions for most factors (criteria) consists of comparing the merit of increasing a factor to be maximized (or decreasing a factor to be minimized) from $a$ to $b$ versus increasing the same factor to be maximized (or decreasing the same factor to be minimized) from $a'$ to $b'$, all other performance levels being equal. The DM is asked to set one of these four values (knowing the other three values) such that the increase in merit would be approximately equal. Therefore, we elicit the difference in the DMU's merit that corresponds to decreases in factors to be minimized or increases in factors to be maximized, rather than the value of having these factors to be minimized available or factors to be maximized produced. This elicitation protocol has already been used by Almeida and Dias (2012) and Gouveia et al. (2015, 2016) in other application contexts. Note that this conversion is performed assuming the continuity of functions.

The answers about the differences of merit between the performance levels in each factor enable a piecewise linear approximation to represent the value functions for the factors to be minimized: $x_{\mathrm{CO_2Red\_M1}}$ and $x_{\mathrm{CO_2Red\_M2}}$ (see Table 9.6; Fig. 9.1). For the factors to be maximized $y_{\mathrm{PGR\_M1}}$ and $y_{\mathrm{PGR\_M2}}$ in Table 9.5, it was possible to adjust the DM's answers to non-linear value functions (see Fig. 9.1). The factor $y_{\mathrm{ATFT}}$ in Table 9.5 has a value function obtained by means of a linear transformation. Since $p_{ATFTj}$ is the performance of DMU $j$ (Country $j$) in factor $y_{\mathrm{ATFT}}$, the factor performances (column 6 of Table 9.5) are converted into values in a linear way using formulation (9.5) since this factor should be maximized. The two limits, $M_{ATFT}^{L}$ and $M_{ATFT}^{U}$, were defined such that $M_{ATFT}^{L} = \min\{p_{ATFTj}, j = 1, \ldots, 12\}$ and $M_{ATFT}^{U} = \max\{p_{ATFTj}, j = 1, \ldots, 12\}$.

$$v_{ATFT}(DMU_j) = \frac{p_{ATFTj} - M_{ATFT}^{L}}{M_{ATFT}^{U} - M_{ATFT}^{L}}, \quad j = 1, \ldots, 12 \tag{9.5}$$

The piecewise linear value functions and the non-linear value functions are displayed in Fig. 9.1.

[5]When examined across the expenditure distribution "progressive" means that households in lower expenditure deciles spend a lower share of expenditure on fuel taxes, "regressive" means that the share decreases as expenditure increases, and "proportional" indicates that the share does not depend on expenditure.

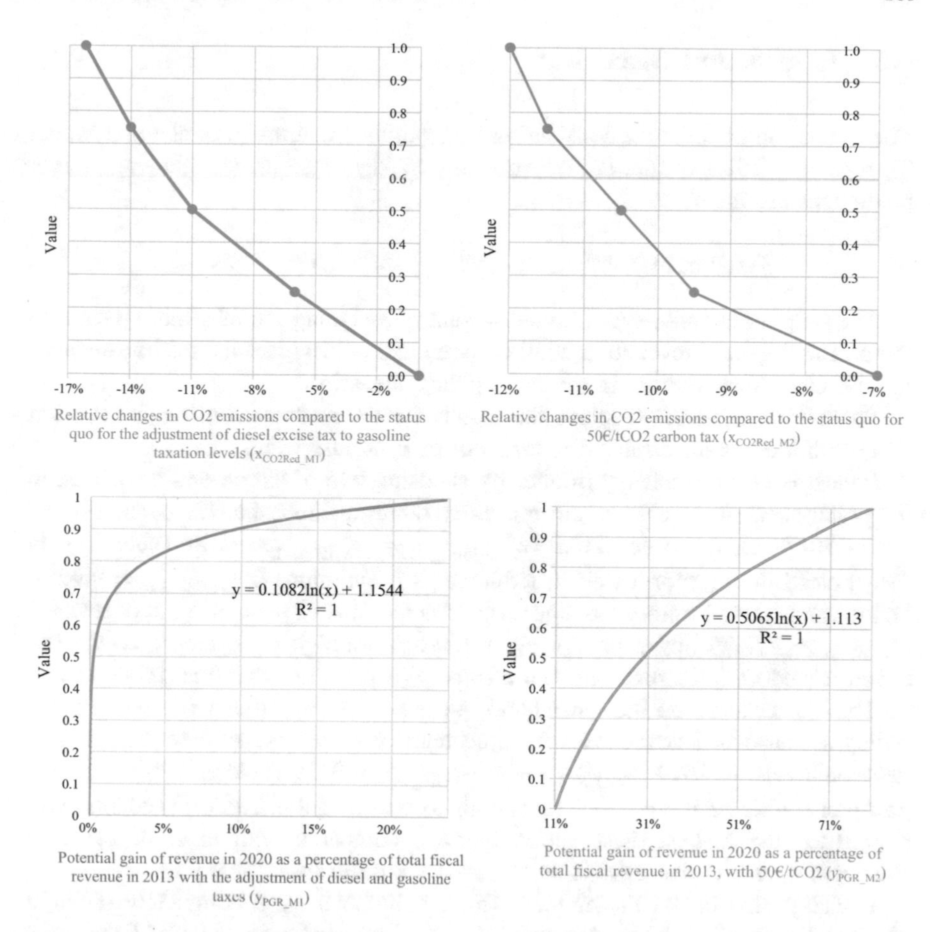


**Fig. 9.1** Value functions elicited for factors $x_{\mathrm{CO_2Red\_M1}}$, $x_{\mathrm{CO_2Red\_M2}}$, $y_{\mathrm{PGR\_M1}}$ and $y_{\mathrm{PGR\_M2}}$ associated with the two fiscal policies (Model 1 and Model 2)

For example, for $x_{\mathrm{CO_2Red\_M1}}$ : $v_{\mathrm{CO_2Red\_M1}}(-16.2\%) - v_{\mathrm{CO_2Red\_M1}}(-14\%) = v_{\mathrm{CO_2Red\_M1}}(-14\%) - v_{\mathrm{CO_2Red\_M1}}(-11\%) = v_{\mathrm{CO_2Red\_M1}}(-11\%) - v_{\mathrm{CO_2Red\_M1}}(-6\%) = v_{\mathrm{CO_2Red\_M1}}(-6\%) - v_{\mathrm{CO_2Red\_M1}}(0\%)$, all other performance levels being equal. The value function for factor $x_{\mathrm{CO_2Red\_M2}}$ is similar to the value function of $x_{\mathrm{CO_2Red\_M1}}$. For factors $y_{\mathrm{PGR\_M1}}$ and $y_{\mathrm{PGR\_M2}}$, the value functions were obtained by making the corresponding adjustment of a known function to the preferences bared by the DM.

The elicited ranges were chosen to include the observed performance ranges plus or minus the highest tolerance value considered (in this case $\delta = 10\%$).

Table 9.6 displays the data in the original scales and the data converted into the value scale for all the factors in evaluation.

## 9.5 Results and Discussion

The evaluation of countries considering both policy scenarios (Model 1 and Model 2), regarding 2020, of fuel tax reforms only makes sense if the following weight restrictions are imposed:

$$w_{\mathrm{CO_2Red\_M1}} = w_{\mathrm{PGR\_M1}} \quad \text{and} \quad w_{\mathrm{CO_2Red\_M2}} = w_{\mathrm{PGR\_M2}} \tag{9.6}$$

The weight restrictions (9.6) avoid a country assigning the maximum weight to the potential gain of revenue in 2020 criterion and to disregard the relative changes in $CO_2$ emissions criterion in the same policy scenario.

Phase 1 in Table 9.7 displays the results for the evaluation using formulation (9.1) with the countries ranked in terms of optimal loss value $d^*$.

Poland is at the efficiency frontier by choosing two of the criteria associated to the policy scenario of fuel tax reforms with the introduction of a $CO_2$ content-based tax of 50€/t$CO_2$, regarding 2020 ($w^*_{\mathrm{CO_2Red\_M1}} = w^*_{\mathrm{PGR\_M1}} = 0$, in Table 9.7). In fact, Poland does perform well in reducing $CO_2$ emissions ($x_{\mathrm{CO_2Red\_M2}} = 9.90\%$) and it stands out a lot from the other countries in terms of potential gain of revenue ($y_{\mathrm{PGR\_M2}} = 72.47\%$) even though Poland has also a high performance on the criterion related with the distributional effects ($y_{\mathrm{ATFT}} = 139.29\%$) (see Table 9.6).

The Netherlands, on the other hand, picks two of the criteria related with the policy scenario that advocates the adjustment of diesel excise taxes to gasoline taxation levels, in 2020 ($w^*_{\mathrm{CO_2Red\_M2}} = w^*_{\mathrm{PGR\_M2}} = 0$, in Table 9.7) to stay in the group of efficient countries. Due to its high tax differential in 2013, this country will experience the highest diesel price increase compared with most of the East European countries that exhibit much lower tax differ (see Table 9.6).

Hungary also lies at the efficient frontier, but making a completely different choice. Disregarding the criteria related to the expected reductions of $CO_2$ emissions and the potential gain of revenue in 2020, Hungary considers only the criteria that translates the distributional effects ($w^*_{\mathrm{ATFT}} = 1$) to be positioned as best as possible in relation to the other countries. The choice falls on this the factor since Hungary is the country with the best performance in this criterion (see Table 9.6).

To position itself as best as possible vis-à-vis the other countries in the sample, being part of the efficiency frontier, France chooses to consider the criteria related to the reduction of $CO_2$ emissions and the potential gain of revenue in 2020, associated with both policy scenarios. There is no policy scenario that favours France more, in detriment of the other; however, there is more weight on the criteria that are allied with the adjustment of diesel excise taxes to gasoline taxation levels, in 2020.

With the exception of Hungary, the other countries that form the efficiency frontier ignore the criterion related to the distributional effects ($w^*_{ATFT} = 0$).

Although classified as inefficient, Italy chose the criteria associated with the policy scenario of regarding the effects in 2020 of the adjustment of diesel excise taxes to gasoline taxation levels hypothetically implemented in 2013, in order to position itself as best as possible 2020 ($w^*_{CO_2Red\_M2} = w^*_{PGR\_M2} = 0$, in Table 9.7).

The United Kingdom occupies the last position despite of choosing the policy scenario of regarding the introduction, in 2013, of a carbon content-based tax of 50 € per ton of $CO_2$ on both diesel and gasoline, with effects assessed in 2020, which benefits it the most. This may be due to the fact that in 2013 this was the only country already taxing equally diesel and gasoline; moreover, no $CO_2$ emissions reductions are expected in 2020 ($x_{CO_2Red\_M1} = 0\%$, in Table 9.6).

The solution obtained from formulation (9.2) of the Value-Based DEA method is a proposal of an efficiency target (projection) for each inefficient country. To attain an efficiency status these inefficient countries must change their value in each factor by the amount indicated by $s^*$. Note that not all the proposed changes correspond to improvements since some of the inefficient countries have negative slacks. So, for an inefficient country to equalize its peers on the efficient frontier, if it has a negative slack corresponding to a factor to be minimized that should be increased and if it has a negative slack corresponding to a factor to be maximized that should be reduced. For instance, Belgium and Finland in order to become efficient as their peers (Poland and Hungary, respectively) should change its performances in each factor by the amount indicated by $s^*$ variables (Phase 2 in Table 9.7). Both countries need to reduce the potential gain of revenue ($s^*_{PGR\_M1} < 0$), but they are allowed to increase the $CO_2$ expected emissions in 2020 w.r.t. 2013 ($s^*_{CO_2Red\_M1} < 0$) on the policy scenario of adjusting diesel excise tax and improve all the other factors related with the policy scenario of introducing a $CO_2$–content based tax as well as the distributional effects factor.

Italy needs to reduce the $CO_2$ emissions and increase the potential gain of revenue in 2020, associated with both policy scenarios, but it could worsen the distributional effects factor ($s^*_{ATFT} < 0$) to emulate Netherlands.

The United Kingdom could be efficient in 2020 if the performances on all factors are improved to be as the same as those in Poland (the benchmark at the efficiency frontier).

## 9.6 Robustness Analysis and Stability Intervals

In this section, and according to Sect. 9.3, the robustness of the status (efficient or not) was assessed, for each DMU (Gouveia et al. 2013). The lower limit comes from the solution to the linear problem (9.3) and the upper limit is the solution to linear problem (9.4). The results considering a tolerance $\delta$ equal to 5% and 10% for the factors $x_{CO_2Red\_M1}, x_{CO_2Red\_M2}, y_{PGR\_M1}$ and $y_{PGR\_M2}$ are displayed in Table 9.8. The DM considered more realistic not to apply any tolerance to the factor related to the distributional effects.

It is also possible to apply different tolerance values to different criteria in the same analysis to see which are the criteria (Model 1) to which the country under

**Table 9.8** Lower and upper limits for the value loss $\left[d^*_{opt}, d^*_{pes}\right]$, for each country, considering δ = 5% for the factors $x_{CO_2Red\_M1}$, $x_{CO_2Red\_M2}$, $y_{PGR\_M1}$ and $y_{PGR\_M2}$ and δ = 10% for the factors $x_{CO_2Red\_M1}$, $x_{CO_2Red\_M2}$, $y_{PGR\_M1}$ and $y_{PGR\_M2}$

| DMUs | $p_{cj}(1-5\%) \le p_{cj} \le p_{cj}(1+5\%)$, $c = CO_2Red_{M1}, PGR_{M1}, CO_2Red\_M2, PGR\_M2$ | $p_{cj}(1-10\%) \le p_{cj} \le p_{cj}(1+10\%)$, $c = CO_2Red\_M1, PGR\_M1, CO_2Red\_M2, PGR\_M2$ |
|---|---|---|
| Poland | [**−0.466**; **−0.178**] | [**−0.609**; **−0.055**] |
| Hungary | [**−0.249**; **−0.232**] | [**−0.294**; **−0.232**] |
| Netherlands | [**−0.186**; **−0.050**] | [**−0.252**; **−0.007**] |
| France | [**−0.100**; 0.064] | [**−0.175**; 0.136] |
| Italy | [**−0.011**; 0.019] | [**−0.025**; 0.033] |
| Belgium | [**−0.061**; 0.092] | [**−0.145**; 0.142] |
| Finland | [0.012; 0.090] | [**−0.024**; 0.129] |
| Spain | [0.035; 0.175] | [**−0.032**; 0.225] |
| Germany | [0.074; 0.167] | [0.018; 0.209] |
| Czech Republic | [0.088; 0.164] | [0.050; 0.200] |
| Austria | [0.068; 0.181] | [0.001; 0.222] |
| United Kingdom | [0.379; 0.492] | [0.323; 0.519] |

Bold denotes efficiency

**Table 9.9** Lower and upper limits for the value loss $\left[d^*_{opt}, d^*_{pes}\right]$, for each country, considering $\delta = 5\%$ for factors $x_{CO_2Red\_M1}$ and $y_{PGR\_M1}$ and $\delta = 10\%$, for factors $x_{CO_2Red\_M1}$ and $y_{PGR\_M2}$ and vice versa

| DMUs | $p_{cj}(1-5\%) \le p_{cj} \le p_{cj}(1+5\%)$, c = $CO_2$Red_M1, PGR_M1 and $p_{cj}(1-10\%) \le p_{cj} \le p_{cj}(1+10\%)$, c = $CO_2$Red_M2, PGR_M2 | $p_{cj}(1-10\%) \le p_{cj} \le p_{cj}(1+10\%)$, c = $CO_2$Red_M1, PGR_M1 and $p_{cj}(1-5\%) \le p_{cj} \le p_{cj}(1+5\%)$, c = $CO_2$Red_M2, PGR_M2$n$ |
|---|---|---|
| Poland | [**−0.609**; **−0.055**] | [**−0.466**; **−0.178**] |
| Hungary | [**−0.294**; **−0.232**] | [**−0.249**; **−0.232**] |
| Netherlands | [**−0.186**; **−0.050**] | [**−0.252**; **−0.007**] |
| France | [**−0.148**; 0.107] | [**−0.131**; 0.096] |
| Italy | [**−0.011**; 0.019] | [**−0.025**; 0.033] |
| Belgium | [**−0.116**; 0.113] | [**−0.095**; 0.121] |
| Finland | [**−0.006**; 0.102] | [**−0.007**; 0.110] |
| Spain | [**−0.020**; 0.206] | [0.020; 0.193] |
| Germany | [0.040; 0.184] | [0.050; 0.189] |
| Czech Republic | [0.059; 0.184] | [0.076; 0.177] |
| Austria | [0.014; 0.204] | [0.053; 0.198] |
| United Kingdom | [0.323; 0.515] | [0.379; 0.493] |

Bold denotes efficiency

evaluation is most sensitive. So, a combination of the previous tolerances was also considered, such as $\delta = 5\%$, for factors $x_{CO_2Red\_M1}$ and $y_{PGR\_M1}$ (criteria of Model 1) and $\delta = 10\%$ for factors $x_{CO_2Red\_M2}$ and $y_{PGR\_M2}$ (criteria of Model 2). Then we interchange the tolerances of factors, applying a tolerance $\delta = 10\%$ for factors $x_{CO_2Red\_M1}$ and $y_{PGR\_M1}$, and $\delta = 5\%$, for factors $x_{CO_2Red\_M2}$ and $y_{PGR\_M2}$ (see Table 9.9).

Table 9.8 shows the results for each country considering the optimistic (for lower limits of the ranges) and pessimistic (for upper limits of the ranges) perspectives, using formulations (9.3) and (9.4) with the tolerance values of 5% (left side), 10% (right side) for the factors $x_{CO_2Red\_M1}$, $x_{CO_2Red\_M2}$, $y_{PGR\_M1}$ and $y_{PGR\_M2}$. Poland, Hungary and Netherlands are efficient and remain in this state when the tolerance of 5% is considered. Although France is classified as an efficient country in the ranking obtained (see Table 9.7), it does not maintain efficiency for this tolerance value (see Table 9.8). Poland, Hungary and Netherlands are surely efficient and France, Italy and Belgium are potentially efficient; Finland, Spain, Germany, Czech Republic, Austria and United Kingdom are surely inefficient for $\delta = 5\%$. If the uncertainty intervals are set to a tolerance of 10%, the surely efficient units (Poland, Hungary and Netherlands) remain and Finland and Spain become potentially efficient, instead of surely inefficient (for $\delta = 10\%$).

In Table 9.9 we have the results for both perspectives (optimistic and pessimistic) with the combination of tolerance values $\delta = 5\%$, for factors $x_{CO_2Red\_M1}$ and $y_{PGR\_M1}$ and $\delta = 10\%$, for factors $x_{CO_2Red\_M2}$ and $y_{PGR\_M2}$ (left side) and vice versa (right side). For the selected tolerance values applied to the criteria, Poland,

Hungary and Netherlands are still surely efficient, France, Italy, Belgium, Finland are potentially efficient and Spain is only potentially efficient if the tolerance $\delta = 5\%$ is considered for factors $x_{CO_2Red\_M1}$ and $y_{PGR\_M1}$ and $\delta = 10\%$. Germany, Czech Republic, Austria and United Kingdom are surely inefficient.

With this type of robustness analysis it is possible to conclude that in the group of the surely efficient countries The Netherlands would be the most fragile when viewed under a pessimistic light, in particular when the 10% tolerance is considered for $x_{CO_2Red\_M1}$ and $y_{PGR\_M1}$, and $\delta = 10\%$ for $x_{CO_2Red\_M2}$ and $y_{PGR\_M2}$ (see column 2 of Tables 9.8 and 9.9). With the exchange of the tolerance values The Netherlands is less sensitive to variations in factors. Comparing the second column of Table 9.8 with the fist column of Table 9.9 we may notice that the potentially efficient countries are the same and limits of the ranges of Poland and Hungary do not vary.

## 9.7 Concluding Remarks

Given its share in total GHG emissions, the transport sector plays a critical role in EU climate and energy policies that aim to curb greenhouse gas emissions. Fuel taxation can be used to internalize a wide range of externalities, including those directly linked to fuel use, such as GHG emissions and local air pollution. Although tax road fuels had primarily aimed to raise fiscal revenue, recent events such as the Paris Agreement on Climate Change have motivated a new interest in fuel pricing policies as an important instrument in the mitigation efforts of transition to a low-carbon economy.

Inspired on the economic principles underlying efficient and fair tax design, we developed a multi-criteria evaluation framework based on the Value-Based DEA method to assess the impact of two fuel pricing policy scenarios to reduce GHG emissions: abandoning of the diesel tax advantage and introducing a $CO_2$ content-based tax. The application of this approach was illustrated using data of 12 European countries.

This framework of analysis enabled taking into account multiple, conflicting and incommensurate criteria to assess the performance of countries to develop insights that may reveal useful to shape fuel taxation policies. A relevant result is the identification of the efficient countries that can be used as reference (benchmark) for inefficient countries to improve, thus unveiling best practices. Considering that each country has different economic and political agendas, which constrain their individual fiscal policies, the classical trade-off between equity and efficiency or the political pressure of different priorities can be (partially) resolved in the framework of the proposed approach. This is facilitated by offering the decision makers the possibility to incorporate their priorities in the design and evaluation of fuel tax policies.

The need to obtain robust conclusions in face of several sources of uncertainty is taken into account. For this purpose, we consider that perturbations in the coefficients in each factor (to be minimized or to be maximized) are captured through interval coefficients and then converted into value scales. Using two efficiency measures (an optimistic efficiency measure and a pessimistic efficiency measure) we can classify each DMU as surely efficient, potentially efficient, or surely inefficient for a given tolerance (level of uncertainty) value.

The results obtained in this illustrative case study indicate that the two policy options, the fiscal reform scenarios under analysis, will not be sufficient to place the majority of the countries on the efficient frontier. Additionally, with this type of robustness analysis, the results indicate that in the group of the surely efficient countries some countries will be more sensitive than others when considering different tolerance values. This kind of analysis is especially important when we consider a period of economic disturbance, such as the one after the 2008 crisis. The countries were differently vulnerable to economic changes and had different degrees of freedom in conducting their fiscal policies in order to overcome public finance constraints.

**Acknowledgements** This work has been supported by FCT—the Portuguese Foundation for Science and Technology under project grant UID/MULTI/00308/2013.

## References

Ajanovic, A., & Haas, R. (2012). The role of efficiency improvements vs. price effects for modelling passenger car transport demand and energy demand—Lessons from European countries. *Energy Policy, 41,* 36–46.

Ali, A. I., Lerme, C. S., & Seiford, L. (1995). Components of efficiency evaluation in data envelopment analysis. *European Journal of Operational Research, 80*(3), 462 473.

Almeida, P. N., & Dias, L. C. (2012). Value-based DEA models: application-driven developments. *Journal of the Operational Research Society, 63*(1), 16–27.

Andersen, P., & Petersen, N. C. (1993). A procedure for ranking efficient units in data envelopment analysis. *Management Science, 39*(10), 1261–1264.

Arbolino, R., & Romano, O. (2014). A methodological approach for assessing policies: The case of the environmental tax reform at European level. *Procedia Economics and Finance., 17,* 202–210.

Bonilla, D. (2009). Fuel demand on UK roads and dieselisation of fuel economy. *Energy Policy, 37*(10), 3769–3778.

Carreno, M., Ge, Y. E., & Borthwick, S. (2014). Could green taxation measures help incentivise future Chinese car drivers to purchase low emission vehicles? *Transport, 29*(3), 260–268.

Charnes, A., Cooper, W. W., Golany, B., Seiford, L., & Stutz, J. (1985). Foundations of data envelopment analysis for Pareto-Koopmans efficient empirical production functions. *Journal of Econometrics, 30,* 91–107.

Charnes, A., Cooper, W. W., & Rhodes, E. (1978). Measuring the efficiency of decision making units. *European Journal of Operational Research, 2,* 429–444.

Cook, W. D., Tone, K., & Zhu, J. (2014). Data envelopment analysis: Prior to choosing a model. *Omega, 44,* 1–4.

Coria, J. (2012). Fuel taxation in Europe. In *Cars and carbon* (pp. 201–222). Netherlands: Springer.

European Commission. (2003). Restructuring the community framework for the taxation of energy products and electricity. Council Directive 2003/96/EC.

European Commission. (2007). Communication from the Commission to the Council and the European Parliament Results of the review of the Community Strategy to reduce $CO_2$ emissions from passenger cars and light-commercial vehicles. COM/2007/0019 final. Brussels.

European Commission. (2011). Road map to a single European Transport Area—towards a competitive and resource efficient transport system. White Paper.

Filipović, S., & Golušin, M. (2015). Environmental taxation policy in the EU—New methodology approach. *Journal of Cleaner Production, 88,* 308–317.

Flues, F., & Thomas, A. (2015). *The distributional effect of energy taxes*. OECD. http://www.oecd-ilibrary.org/taxation/the-distributional-effects-of-energy-taxes_5js1qwkqqrbv-en.

Gago, A., Labandeira, X., & López-Otero, X. (2014). A panorama on energy taxes and green tax reforms. *Hacienda Pública Española, 208*(1), 145–190.

González, R. M., & Marrero, G. A. (2012). The effect of dieselization in passenger cars emissions for Spanish regions: 1998–2006. *Energy Policy, 51,* 213–222.

Gouveia, M. C., Dias, L. C., & Antunes, C. H. (2008). Additive DEA based on MCDA with imprecise information. *Journal of the Operational Research Society, 59*(1), 54–63.

Gouveia, M. C., Dias, L. C., & Antunes, C. H. (2013). Super-efficiency and stability intervals in additive DEA. *Journal of the Operational Research Society, 64*(1), 86–96.

Gouveia, M. C., Dias, L. C., Antunes, C. H., Boucinha, J., & Inácio, C. F. (2015). Benchmarking of maintenance and outage repair in an electricity distribution company using the Value-Based DEA method. *Omega, 53,* 104–114.

Gouveia, M. C., & Dias, L. C., Antunes, Mota, M. A., Duarte, E. M., & Tenreiro, E. M. (2016). An application of an additive DEA model to identify best practices in primary health care. *OR Spectrum, 38*(3), 743–767.

Harding, M. (2014). The diesel differential—Differences in the tax treatment of gasoline and diesel for road use. OECD Taxation paper no. 21.

Kloess, M., & Müller, A. (2011). Simulating the impact of policy. energy prices and technological progress on the passenger car fleet in Austria—A model based analysis 2010–2050. *Energy Policy, 39*(9), 5045–5062.

Mardani, A., Zavadskas, E. K., Streimikiene, D., Jusoh, A., & Khoshnoudi, M. (2017). A comprehensive review of data envelopment analysis (DEA) approach in energy efficiency. *Renewable and Sustainable Energy Reviews, 70,* 1298–1322.

Newbery, D. M. (1992). Should carbon taxes be additional to other fuel transport taxes? *The Energy Journal, 13*(2), 49–60.

Newbery, D. M. (2001). Harmonizing energy taxes in the EU. In *Tax Policy in the European Union Conference*, Erasmus University, October 17–19, 2001.

Newbery, D. M. (2005). Why tax energy? Towards a more rational energy policy. *The Energy Journal, 26*(3), 1–39.

Parry, I. W. H. (2007). Are the costs of reducing greenhouse gases from passenger vehicles negative? *Journal of Urban Economics., 62*(2), 273–293.

Parry, I. W. H., & Small, K. A. (2005). Does Britain or the United States have the right gasoline tax? *American Economic Review, 95*(4), 1276–1289.

Rodríguez-López, J., Marrero, G. A., & González-Marrero, R. M. (2015). Dieselization, $CO_2$ emissions and fuel taxes in Europe. In *Working Papers*, November 15, 2015.

Ryan, L., Ferreira, S., & Convery, F. (2009). The impact of fiscal and other measures on new passenger car sales and $CO_2$ emissions intensity: evidence from Europe. *Energy Economics, 31,* 365–374.

Santos, G. (2017). Road fuel taxes in Europe: Do they internalize road transport externalities? *Transport Policy, 53,* 120–134.

Schipper, L., & Fulton, L. (2013). Dazzled by diesel? The impact on carbon dioxide emissions of the shift to diesels in Europe through 2009. *Energy Policy, 54,* 3–10.

Seiford, L. M., & Zhu, J. (1998a). Stability regions for maintaining efficiency in data envelopment analysis. *European Journal of Operational Research, 108,* 127–139.

Seiford, L. M., & Zhu, J. (1998b). Sensitivity analysis of DEA models for simultaneous changes in all the data. *Journal of the Operational Research Society, 49,* 1060–1071.

Sterner, T. (2007). Fuel taxes: An important instrument for climate policy. *Energy Policy, 35*(6), 3194–3202.

Sterner, T., & Köhlin, G. (2015). Pricing carbon: The challenges. In S. Barrett, C. Carraro, & J. de Melo (Eds.), *Towards a workable and effective climate regime* (p. 251).

United States Environmental Protection Agency. (2011). *Greenhouse gas emissions from a typical passenger vehicle.* www.epa.gov/oms/climate/documents/420f11041.pdf.

von Winterfeldt, D., & Edwards, W. (1986). *Decision analysis behavioral research.* New York: Cambridge University Press.

Zhu, J. (1996). Robustness of the efficient DMUs in data envelopment analysis. *European Journal of Operational Research, 90,* 451–460.

Zhu, J. (2001). Super-efficiency and DEA sensitivity analysis. *European Journal of Operational Research, 129,* 443–455.

Zhu, J. (2003). Imprecise data envelopment analysis (IDEA): A review and improvement with an application. *European Journal of Operational Research, 144,* 513–529.

Zimmer, A., & Koch, N. (2016). *Fuel consumption dynamics in Europe—Implications of fuel tax reforms for air pollution and carbon emissions from road transport.* https://ssrn.com/abstract=2813534. Available at August 28, 2016.

# Part IV
# Strategic Analysis

# Chapter 10
# Strategic Analysis of Solar Energy Pricing Process with Hesitant Fuzzy Cognitive Map

**Veysel Çoban and Sezi Çevik Onar**

**Abstract** Sun is the leading renewable energy source for satisfying energy demand. Solar energy systems, which have direct and indirect energy generation technologies, require high initial costs and low operation costs. The right determination of solar energy price has an important role on efficient solar energy investment decisions. In this study, the critical factors for the solar energy price are defined and the causal relationships among them are represented with a Hesitant Fuzzy Cognitive Map (HFCM) model. The causal relations among the factors and the initial state values of the factors are defined with the linguistic evaluations of the experts by using Hesitant Fuzzy Linguistic Term Sets (HFLTSs). The linguistic expressions are converted into Trapezoid Fuzzy Membership Functions (TFMFs). The obtained HFCM model is used for simulating various scenarios, and the equilibrium state values of the factors are obtained. The results indicate that the factors affecting solar energy systems have an important effect in determining the solar energy price. The solar energy price adapts to the general energy price market in the long term.

## 10.1 Introduction

Energy which allows people to live a more productive life is basically provided with six power sources and they can be transformed from one form into another as mechanical, chemical, thermal, radiant, nuclear, and electric. The primary energy sources commonly used in the world (85.52% of total energy consumption) are fossil based (coal, gasoline, natural gas) (BP 2017). The increase in fossil fuel consumption leads to the increased atmospheric release of greenhouse gases, especially $CO_2$. Global warming and climate changes caused by greenhouse gases are an essential part of the economic, social and environmental problems. Therefore,

V. Çoban (✉) · S. Çevik Onar
Industrial Engineering Department, Istanbul Technical University, Maçka, İstanbul, Turkey
e-mail: cobanv@itu.edu.tr

© Springer International Publishing AG, part of Springer Nature 2018

C. Kahraman and G. Kayakutlu (eds.), *Energy Management—Collective and Computational Intelligence with Theory and Applications*, Studies in Systems, Decision and Control 149, https://doi.org/10.1007/978-3-319-75690-5_10

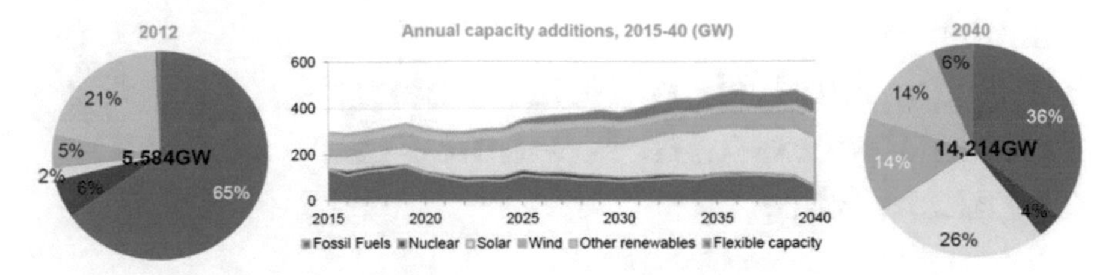


**Fig. 10.1** Annual capacity additions and expectations until 2040 (Bloomberg New Energy Finance 2015)

the use of fossil-based fuels in energy-intensive conditions is the most precise indication of the human impact on climate change (Stern 2015). Greenhouse effect, global warming, and climate change have led governments to turn to renewable energy sources (sun, wind, hydroelectric, biomass) as an alternative to fossil energy sources. The National Science Academies of the G8 countries reported that a joint action against climate change should be undertaken and urged governments to reduce $CO_2$ emissions by 50% below 1990 levels by 2050 (Academies 2009).

In the long run, production of electricity from coal, oil and natural gas is expected to be replaced by entirely renewable energy sources (Fig. 10.1). However, the significant disadvantages of renewable energy production against traditional energy sources are that they are more expensive and less reliable (Conkling 2011). The difficulties in using renewable resources are also as follows: political uncertainty, the tendency of countries to move away from FITs and green certificates, changes in subsidies and the need to integrate renewable-based systems with existing power plants.

In order to promote the use of solar energy, which is the most important renewable energy source, it must be able to compete with other renewable and traditional energy types. Regulatory policies, fiscal incentives, and public financing bases shape the countries' support for developing solar energy capacity. For example, Turkey uses feed-in tariff/premium payment, biofuels obligation/mandate, capital subsidy, grant, or rebate, and public investment, loans, or grants methods as promotion policy (Crawley 2016).

Price, which is the most important competitive factor in energy and all other markets, is an important measure for the adoption and diffusion of solar energy technologies. Therefore, the fundamental factors that are effective in solar energy pricing are defined and the relationship between them is modelled in this study. The national and international factors that influence solar energy pricing in the energy market include uncertainty and hesitancy. Hence, the HFCM model is utilized for developing the causal relationships of solar energy pricing. The causal relationships among the active factors in the solar energy HFCM pricing model are defined, and their effects on the solar energy pricing are reflected in the equilibrium state.

The organization of the chapter is as follows: Sect. 10.2 explains the pricing in solar energy, the effective factors in the solar energy system The HFCM model and its preliminaries the FCM, the hesitant fuzzy set and the HFLTS subjects are mentioned in the Sect. 10.3. The processing process and calculation methods of the

HFCM model are mentioned in Chap. 4. Chapter 5 gives the results of simulation evaluations of the solar energy price based HFCM model developed with two different scenarios. The study is completed in the conclusion section with the general results and future studies.

## 10.2 Solar Energy Pricing

Governments should access sustainable, quality and cheap energy sources to support and sustain their economic and social development. Increasing population leads to further increase in demand, hence, new energy generation methods are developed to meet this increasing demand. However, the use of fossil-based energy sources to meet rising energy demand creates environmental and economic problems (Thomas et al. 2011). Therefore, the countries have turned to renewable energy sources, especially solar with the support of national and international decisions and agreements. India, for example, has set a goal of increasing solar energy capacity from 5.2 GW in 2016 to 100 GW by 2022 (Council 2016). Similarly, Turkey has set a target to increase the solar energy capacity of 2 GW at the end of 2017 to 5 GW in 2023 (PV-Magazine 2017).

The price of energy is the most critical determining factor for the acceptance of renewable energies by the society and investors. Correct pricing is advantageous for energy providers to optimize capacity planning and for consumers to minimize energy costs. Energy pricing and forecasting of energy needs allow appropriate energy capacity planning, financing technologies and investments in energy diversity, and enabling investors and governments to develop stable policies (Mir-Artigues and Del Río 2016). In particular, the economic depression and poverty caused by the rise in energy prices in the 1970s and 1980s led to the development of new policies and models based on energy availability and cost (Timilsina et al. 2012). Knowing the factors that affect energy prices and understanding their impact on the energy market is the starting point for solar energy pricing.

The energy price (EP) is determined by the installed capacity, not by the actual energy production (Zatzman 2012). The total energy price is calculated taking into account factors that cause economic effects and components based on performance. Factors defined in the price calculation include uncertainty, which may vary locally and temporally.

***Factors affecting solar energy pricing***

In this section, factors affecting solar energy pricing are defined, and causal relationships among factors are explained in the model. Causal relationships are evaluated and how the factors affect each other in the long run are observed under the HFCM model. Thus, factors that determine solar energy prices and the causal relationships between the factors shown and the decision making processes of the government and investors in the long term accurately directed.

The main factor driving a country to use renewable energy from fossil energy consumption is the conscious governments know that fossil fuels are behind their country's environmental, economic and social problems (Environmental Effects, EE and Eco-social Effects, ESE). The anticipation of the deterioration of agricultural production and living conditions caused by the change of climate change and vegetation cover is at the basis of environmental concerns. Governments develop national and international directive laws and regulations (Global treaties, GT such as Kyoto Protocol) to manage the use and widespread of renewable energies in the community. Laws and regulations differ among countries according to their renewable energy potentials (Çoban and Onar 2017). National regulations are severely affected by international agreements and supportive policies.

The trend towards renewable energies revealed the technical and technological infrastructure problems (IP). Having different characteristics of the environmental conditions of the energy plants reveals the infrastructure requirements of the plants and affects the initial costs and solar pricing. Therefore, the use of renewable energy has a significant price disadvantage against the use of fossil based energy. In contrast, governments' policies to support renewable energies provide price competition against fossil fuels.

The supportive laws and legislations (SLLs), which aim to generate electricity from solar energy source, specify the procedures and principles for the realization of electricity generation in the country. Incentives (Fig. 10.2), which are applied in many countries around the world, are aimed at eliminating energy dependency by supporting renewable energy sources. Supporting, encouraging and inhibiting the solar energy investments can be covered by the cost of energy companies that do not produce renewable energy (Ministry 2017). This situation, which causes fluctuations in the general energy prices, causes solar energy prices to fluctuate in an indefinite range.

The most common support scheme for the development of solar energy systems is feed-in tariff (FIT). FIT for entire production (FITEP) guarantees that the electricity generated from a solar energy system and transmitted to the grid is purchased

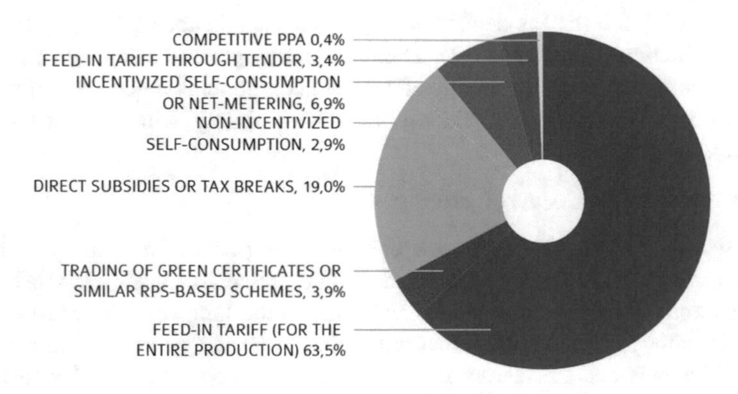


**Fig. 10.2** Historical market incentives and enablers (IEA 2016)

at a predefined price for a specified period (Crawley 2016). For example, the purchase of electricity generated by facilities that produce electricity using solar energy sources is committed by the state at a fixed price (0.133 $/kWh) for ten years in Turkey (Ministry 2017). The support provided by the FIT policy is financed by tax revenues or by taxation on companies generating energy without renewable energy.

The accurate determination of support price and duration values in FIT plans has a critical precaution for solar energy and general energy pricing. High-priced and long-term support leads to a decline in energy prices and a deterioration of the market balance (Spain 2008, Czech Republic 2010, Italy 2011) (Mir-Artigues and Del Río 2016). In addition, if domestic producers supply mechanical and/or electromechanical components used in grid-connected solar energy generation facilities, these facilities benefit from price supports. For example, if the solar energy plant established in Turkey supplies its equipment and materials from domestic manufacturers, it wins an additional domestic contribution to FIT for five years (Ministry 2017). If the PV modules in the installation of the PV solar energy plant are produced in Turkey, 1.3 US cents/kWh domestic production contribution is given for five years.

FIT with tender (FITT) is an alternative method of providing FIT support to reduce the cost of PV electricity. Competition with the tender procedure enables to draw the solar energy price to the lowest possible level and to reduce margins. This support method reflects how low the bids can be under competitive bidding conditions. Low bids can only be realized if the market has low capital costs, low component costs, and a low risk (IEA 2016).

The direct capital subsidy (DS) is the most straightforward way for governments to promote solar energy installations. The system investment cost is made attractive with this single-process subsidy method (CEDE 2014). Direct capital subsidies are the financial support through taxation (tax breaks, TB) for upfront investments in solar energy systems according to their off-grid and on-grid connections (Crawley 2016).

The Renewable Portfolio Standard (RPS) is a market mechanism based on a plan to gradually increase electricity generated from renewable energy sources (wind, biomass, geothermal and solar). Competition between renewable energies is achieved through this market-based approach. Thus, the use of renewable energy sources is continuously promoted and the cleanest energy is achieved at the lowest price (Scientists 2015). RPS and related approaches determine a share of electricity that must be generated from a particular renewable energy source. This incentive plan allows renewable electricity producers to charge a market-based fee for the electricity they give to the grid (Crawley 2016).

Self-consumption is the independent supply of individual energy needs from the small scale solar energy systems established on the residential roof. Although self-consumption systems have high costs compared to utility-scale systems, their price advantages provided by individual investors (non-incentivized self-consumption, SCNI), in the long run, can make their use even more widespread. Sustainable building regulations have increased the interest in using solar tools (photovoltaic, solar water heater and passive solar energy) as a source of heating and electricity. The use of solar energy technologies in construction can be supported by environmental regulations and legislation to reduce the energy footprint of buildings (incentivized self-consumption, SCI). Net metering (NM) is a system that helps to arrange energy billing by following the consumed electricity by the structures and the generated electricity by the solar energy system. Electricity generated by the net-metered solar power system of the residence meets the residential energy demand primarily, and the increasing electricity is supplied to the grid. If the residential solar power system generates more electricity than needed during the billing period, net metering customers receive bill credits. The investment of solar energy is depreciated in a shorter period, and solar energy prices are affected positively with this system (IEA 2016).

Power Purchasing Agreements (PPAs) are a standard business model for meeting near-energy demands with electricity generated from grid-connected medium-sized solar power systems inbuilt. The system owner sells the generated electricity through a direct connection to the nearby consumers. Thus, consumers' demand for mains electricity is reduced, and a lower selling price for electricity generated from solar energy arises. The system based on profitability is affected by the electricity cost of the grid that is shaped by the electricity supply-demand (Crawley 2016; IEA 2016).

Technical and political support increases the installation of solar energy facilities and solar energy generation capacity. The increase in the generation capacity of solar energy contributes to the stabilization of total energy demand and energy supply (ES) and determines the price of solar energy in the energy market. Improvements in the solar energy technology (SETI) increase solar energy production potential by reducing the costs of solar energy installation and by developing the infrastructure requirements for installation. Economic components (EC) for solar energy pricing can be shaped with capital cost, return on equity, interest on loan, depreciation, operation and maintenance expense, insurance, taxes, service charges, and cost escalation factor (Thomas et al. 2011). In addition to these critical economic factors, installed capacity, capacity utilization and penalty factors have an important influence on pricing.

The ever-increasing world population brings with it the increase in production and consumption. The total amount demanded of energy, which is the basic element of production and consumption, is called energy demand. In order to meet the rising energy demand, it is necessary to use the sun and other energy resources together to generate energy supply. Energy prices (as a combination of renewable and non-renewable energy prices, REP/NREP) are an important factor in achieving a balance between energy demand and supply. The installation of new solar power plants and the increase in the total installed solar capacity contribute to energy supply and indirectly affect energy prices.

The support and encouragement for the dissemination of solar energy reduce the availability of existing and widespread fossil-based energy production systems and equipment. In this case, the perceived damage of fossil-based investments, existing industrial system, and production technologies lead to the emergence of anti-solar energy policies. However, the ease of transportation and storage of fossil-based energies causes serious cost disadvantages to solar-based energy systems.

## 10.3 Fuzzy Cognitive Maps

FCMs, which are an extension of cognitive maps with fuzzy logic, enable recognizing causal relationships among components in complex systems. Fuzzy Cognitive Maps (FCM), fuzzy graphical structures representing causal reasoning, are defined by Kosko (1986). An FCM is represented with signed and directed graphs showing concepts and causal relationships among concepts. In graphical notation, concepts are represented by nodes and denoted as $C_i$, and causal relations among nodes are represented by weighted edges and denoted as $w_{ij}$. The weight value of the edges expresses the fuzzy strength of causal relations and is represented by fuzzy numbers. Graphical representation of complex systems with FCM allows visual representation of concepts and directional relationships among them.

Figure 10.3 represents a simple FCM model with five members $(C_i)$ and six weighted edges $(w_{ij})$ between these members. Weights expressed regarding the causal relationship between concepts are expressed as the values between $[-1, 1]$. The positive and negative sign of relationship weight expresses the direction of the relationship between concepts. The absence of a causal relationship between concepts indicates that the weight value is zero. The change in any concept in the FCM model, which has the fuzzy feedback loop feature, causes to change the current state of the other concepts in the system. If all the factors in the model reach an equilibrium state, the feedback loop process is terminated (Kosko 1997). Since the factors in the FCM are not self-feedback (i.e., no self-causal relationship), the diagonal value of the weighted relationship matrix is zero. Subjective information based on expert knowledge and experience or objective information obtained through methods such as literature review is used to identify the concepts and fuzzy causal relationships between concepts in the FCM (Çoban and Onar 2017).

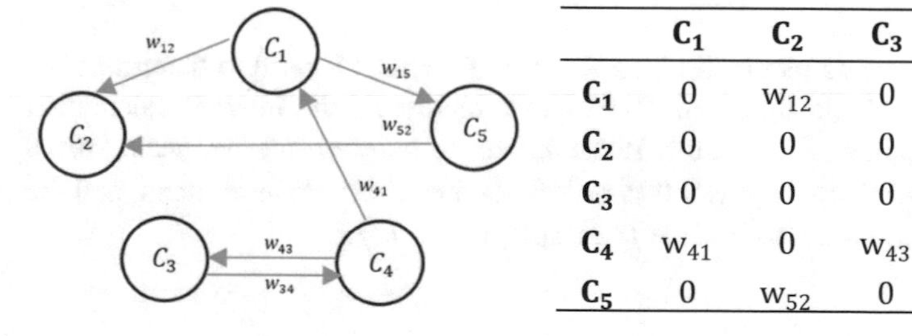


| | $C_1$ | $C_2$ | $C_3$ | $C_4$ | $C_5$ |
|---|---|---|---|---|---|
| $C_1$ | 0 | $w_{12}$ | 0 | 0 | $w_{15}$ |
| $C_2$ | 0 | 0 | 0 | 0 | 0 |
| $C_3$ | 0 | 0 | 0 | $w_{34}$ | 0 |
| $C_4$ | $w_{41}$ | 0 | $w_{43}$ | 0 | 0 |
| $C_5$ | 0 | $w_{52}$ | 0 | 0 | 0 |

**Fig. 10.3** A sample FCM and relation matrix

Some structural criteria reflect the model factors and general model characteristics. Transmitter refers to the factor affecting other factors but not being affected by other factors. Receiver refers to the factor affected by other factors but not affecting other factors. Ordinary refers to the factor affecting other factors and influenced by other factors. Centrality represents the sum of the influence values of the factor (Papageorgiou 2013). The total value of the relationships that are directed to a factor is defined as "in-degree." The sum of relations from one factor to the other is called "out-degree."

The weight values of causal relations in the FCM can be determined using triangular, trapezoidal, sigmoid, Gaussian functions or fuzzy linguistic terms. The FCM model operates using fuzzy arithmetic operators, and defuzzification methods (weight centers, center area, and weighted average method) are used to transform the fuzzy values reached in the steady state to crisp values in the range [−1, 1].

$A_i^t$ denotes the state value of the concept $C_i$ in time t, and the general state values for all concepts in FCM can be shown in the form $A^t = [A_1^t, A_2^t, \ldots, A_n^t]$. The next state value of concept $i$ ($C_i$) reaches after each iteration is defined as:

$$A_i^{t+1} = f\left(\sum_{j=1}^{n} A_j^t w_{ij} + A_i^t\right) \tag{10.1}$$

where f(.) is the threshold function that is used to transform the sum of the previous state value ($A_i^t$) and the total causal effects. The most commonly used transformation (threshold) functions are hyperbolic tangent and sigmoid functions that get values in the range [0,1] and [−1,1] respectively.

$$f(t) = \frac{1}{1 + e^{-\lambda t}} \tag{10.2}$$

$$f(t) = tanh(\lambda t) = \frac{e^{\lambda t} - e^{-\lambda t}}{e^{\lambda t} + e^{-\lambda t}} \tag{10.3}$$

The optional lambda parameter ($\lambda > 0$) in the functions is used to determine the appropriate slope of the function. The value x represents the internal calculation performed on the new state vector. If the difference between the two state values ($A_i^{t+1} - A_i^t$) for each concept is 0.001 or less, the iterations are terminated, and the final state is called as a steady state (Papageorgiou 2013).

## *10.3.1 Preliminaries*

### 10.3.1.1 Hesitant Fuzzy Sets

Fuzzy set theory was developed by Zadeh (1996) to model and calculate uncertainty and vagueness using mathematical methods. The fuzzy set theory, which is oriented towards solving complex everyday life problems, has been applied to a wide range of scientific fields such as decision theory, energy management, and artificial intelligence methods (Papageorgiou 2013; Michael 2010). New extensions of fuzzy sets are developed to produce more accurate approaches and solutions to the complex and ambiguous problems encountered in everyday life (Mizumoto and Tanaka 1976; Atanassov 1986; Torra 2010). The Hesitant Fuzzy Sets (HFSs), developed by Torra (2010), are aimed at dealing with the situations where more than one value of a membership of the fuzzy clusters may be possible. In HFS, a function is defined that returns a set of member values for each element in the domain (Torra 2010).

HFS, defined on the reference set (*X*), is expressed as a function (*h*) that returns a subset in the range [0,1]. The mathematical representation of the expression is as follows:

$$h : X \rightarrow \{[0, 1]\} \tag{10.4}$$

The association of HFSs for a set of N membership functions is represented as $\mathrm{M} = \{\mu_1, \mu_2, \ldots, \mu_N\}$ and shown as:

$$h_M : M \rightarrow \{[0, 1]\} \text{ and } h_M(x) = \cup_{\mu \epsilon M}\{\mu_x\} \tag{10.5}$$

The upper and lower bound of the hesitant fuzzy set h is given as Torra (2010):

$$h^-(x) = \min h(x) \text{ and } h^+(x) = \max h(x) \tag{10.6}$$

Some basic operations (complement, union, and intersection) of the HFSs can be defined as follows (Torra 2010):

$$h^c = \cup_{\gamma \in h(x)}\{1 - \gamma\} \tag{10.7}$$

$$(h_1 \cup h_2)(x) = \{h \in (h_1(x) \cup h_2(x)) | h \geq \max\{h_1^-, h_2^-\}\} \tag{10.8}$$

$$(h_1 \cap h_2)(x) = \{h \in (h_1(x) \cup h_2(x)) | h \leq \min\{h_1^+, h_2^+\}\} \tag{10.9}$$

where h represents the hesitant fuzzy set.

#### 10.3.1.2 Hesitant Fuzzy Linguistic Term Sets

Linguistic knowledge using words or phrases is applied to solve daily life problems which cannot be expressed by numerical values. The linguistic expressions used to identify and solve problems are a tool that best reflects people's perceptions and knowledge (Zadeh 1975). The fuzzy set theory is dependent on linguistic variables which are fuzzy variables. The fuzzy linguistic approach, which uses a single language term, is insufficient to express and evaluate language variants involving hesitation. HFLTSs have been proposed as a solution to these common problems by Rodriguez et al. (2012).

An ordered finite subset of consecutive linguistic terms of linguistic term set $S = \{s_0, s_1, \ldots, s_g\}$ is represented with $\mathrm{H_s}$ (HFLTS). For example, a sample HFLTS can be defined as $\mathrm{H_s} = \{\mathrm{s_2, s_3, s_4}\}$ where linguistic term set S is determined as $\mathrm{S} = \{\mathrm{s_0}: \text{nothing}, \mathrm{s_1}: \text{very low}, \mathrm{s_2}: \text{low}, \mathrm{s_3}: \text{medium}, \mathrm{s_4}: \text{high}, \mathrm{s_5}: \text{very high}, \mathrm{s_6}: \text{perfect}\}$. The upper/lower bounds $(\mathrm{H_{S^+}}, \mathrm{H_{S^-}})$, complement $(\mathrm{H_s^c})$ and basic operations of the HFLTSs $(\mathrm{H_s}, \mathrm{H_s^1}, \mathrm{H_s^2})$ are shown as:

$$\begin{aligned} \mathrm{H_{S^+}} &= \max(\mathrm{s_i}) = \mathrm{s_j}, \mathrm{s_i} \in \mathrm{H_s} \text{ and } \mathrm{s_i} \le \mathrm{s_j} \forall_\mathrm{i} \text{ and} \\ H_{S^-} &= \min(s_i) = s_j, s_i \in H_s \text{ and } s_i \ge s_j \forall_i \end{aligned} \tag{10.10}$$

$$H_s^c = S - H_s = \{s_i | s_i \in S \, and \, s_i \, not \in H_s\} \text{ and } \left(H_s^c\right)^c = H_s \tag{10.11}$$

$$(h_1 \cap h_2)(x) = \{h \in (h_1(x) \cup h_2(x)) | h \le \min\{h_1^+, h_2^+\}\} \tag{10.12}$$

Generated new values for these operations also will be an HFLTS.

#### 10.3.1.3 OWA Operators

Collecting a set of information to obtain a new information is called aggregation and the operators used for this purpose are called the aggregation operator (mean, arithmetic mean, weighted arithmetic mean) (MDAI 2014). The ordered weighted averaging (OWA) aggregation operator is applied to aggregate the HFLTSs and obtain a universal HFLTS.

$$OWA(x_1, x_2, \ldots, x_k) = \sum_{i=1}^{k} l_i w_i \tag{10.13}$$

where $l_i$ is the $i$. largest member of the aggregated elements $\mathrm{x_1, x_2, \ldots, x_k}$. $w_i$ is a weight of the ordered $i$. data in [0,1] interval and is defined the weighting vector $W$, $W = (w_1, w_2, \ldots, w_k)^T$. The sum of the weights defined in W equals one as $\sum_{i=1}^{k} w_i = 1$ (Yager 1988). The methods (maximum, minimum, average) applied to determine the weighting values enable differentiation of OWA operators. The OWA collection operator, introduced by Yager, had the opportunity to practice in different branches of science (Yager 1988). The ability of the OWA operator to collect and model linguistic expressions allows it to be used extensively in computational intelligence and fuzzy logic-based calculations as an aggregation operator.

The orness method can represent the degree of optimism and pessimism of the OWA operator (Liu and Rodríguez 2014). Because of this feature, orness method which is widely used in researches is also used in this study. The mathematical representation of the orness method is as follows.

$$orness(W) = \frac{1}{k-1}\sum_{i=1}^{k} w_i(k-i) \tag{10.14}$$

where $0 \leq \text{orness}(W) \leq 1$. $orness \geq 0.5$ condition points to optimistic OWA operators and $orness < 0.5$ state points to pessimistic OWA operators (Yager 1993).

## 10.4 Hesitant Fuzzy Cognitive Maps

FCM is a dynamic modelling tool that reflects the concepts and causal relationships between concepts in complex and uncertain systems. Hesitant fuzzy sets (HFS) provide ease of assessment by allowing more than one value to identify membership in a situation (Kahraman et al. 2016). HFCM is a fuzzy method that models the causal relationships of linguistic evaluations defined by HFLTS. Hesitant linguistic expressions that are natural translations of experts' cognitive assessments with words or phrases are used to define concepts and their initial states. The process flow of HFCM is as follows:

**Stage 1. Development of relationship model**

Factors and the relationships between the factors of the HFCM model are determined by the common opinions of experts' knowledge and experiences. In the model, the system members are represented by nodes $(C_i)$, and the causal relationships between the members are indicated by directed linguistic edges. A simple HFCM in Fig. 10.4 is represented with five concepts $(C_1, C_2, C_3, C_4, C_5)$ and six

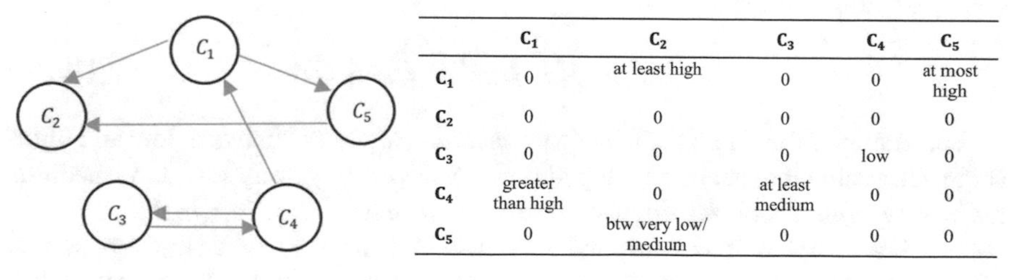


| | $C_1$ | $C_2$ | $C_3$ | $C_4$ | $C_5$ |
|---|---|---|---|---|---|
| $C_1$ | 0 | at least high | 0 | 0 | at most high |
| $C_2$ | 0 | 0 | 0 | 0 | 0 |
| $C_3$ | 0 | 0 | 0 | low | 0 |
| $C_4$ | greater than high | 0 | at least medium | 0 | 0 |
| $C_5$ | 0 | btw very low/ medium | 0 | 0 | 0 |

**Fig. 10.4** A simple HFCMs and HFLTS matrix

directed linguistic edges. Since the HFCM does not contain any self-loop concept, their values in the weight matrix are defined as zero ($w_{ii} = 0$).

**Step 2. Collection of experts' information using HFLTS**

Uncertain and dynamic system conditions cause experts to ambiguously identify the concepts and relationships between concepts in the cognitive map. Hence, experts use hesitant linguistic terms to convey ideas more naturally. Natural hesitant linguistic expressions of experts are defined by using context-free grammar, $G_H$ that is generated with 4-tuple $(V_N, V_T, I, P)$ (Rodriguez et al. 2012; Bordogna and Pasi 1993).

Hesitant linguistic expressions are defined by using a linguistic term set where $S = \{s_0 : nothing, s_1 : very\,low, s_2 : low, s_3 : medium,$ $s_4 : high, s_5 : very\,high, s_6 : absolute\}$ and context-free grammar. The sample hesitant linguistic statements are as follows: at most high, smaller than low, and between medium and high. Hesitant linguistic expressions provide flexibility to define and evaluate the hesitant concept and causal relationships among them.

The linguistic expressions obtained by expert evaluations must be converted to HFLTS for use in HFCM model calculations (Rodriguez et al. 2012). The transformation function, $E_{G_H}$ developed by Rodriguez et al. (2012) is used in the conversion process. The methods applied according to the linguistic term set, $S$ in the conversion process are as follows.

$$E_{G_H}(s_i) = \{s_i | s_i \in S\}$$
$$E_{G_H}(at\,least\,s_i) = \{s_j | s_j \in S\,and\,s_j \geq s_i\}, E_{G_H}(at\,most\,s_i) = \{s_j | s_j \in S\,and\,s_j \leq s_i\}$$
$$E_{G_H}(lower\,than\,s_i) = \{s_j | s_j \in S\,and\,s_j < s_i\}, E_{G_H}(greater\,than\,s_i) = \{s_j | s_j \in S\,and\,s_j > s_i\}$$
$$E_{G_H}(between\,s_i\,and\,s_j) = \{s_k | s_k \in S\,and\,s_i \leq s_k \leq s_j\}$$

For example, $\{medium, high, very\,high\}$ is a sample HFLTS that is transformed form of the "*between medium and* very *high*" linguistic expression; $E_{G_H}(between\,low\,and\,high) = \{low, medium, high\}$.

**Step 3. Fuzzy envelope of HFLTS**

The enveloping method is used to compare the HFLTS converted from the linguistic expressions of the experts and to start the calculation processes in the HFCM model. Envelopment of an HFLTS,$env(H_S)$ is indicated by upper $(H_{S^+})$ and lower $(H_{S^-})$ bounds as follows:

$$env(H_S) = [H_{S^-}, H_{S^+}], \quad H_{S^-} \leq H_{S^+} \tag{10.15}$$

For example, the HFLTS, $H_s = \{low, medium, high\}$ of "between low and high" linguistic evaluation can be enveloped under S = {nothing, very low, low, medium, high, very high, absolute} linguistic terms set as $env(H_S) = [low, high]$.

The OWA operator is contacted to obtain the fuzzy membership function of HFLTS and bring these membership functions together (Liu and Rodríguez 2014).

To reflect the linguistic uncertainties expressed by HFLTS, it is appropriate to use the trapezoidal membership function, $\tilde{A} = (\mathrm{a, b, c, d})$ in the OWA operator procedure (Delgado et al. 1998). The process stages for calculating the coefficients expressing the trapezoidal membership function are as follows (Liu and Rodríguez 2014):

**Stage 1. Defining the aggregation elements**

Linguistic terms are applied to calculate the parameters of the trapezoidal fuzzy membership function, $\tilde{A} = (\mathrm{a, b, c, d})$, as $A^k = T\{a_l^k, a_m^k, a_m^k, a_r^k\}, k = 0, 1, \ldots, g$. The set of aggregation elements of the linguistic terms in the HFLTS $H_s = \{s_i, s_{i+1}, \ldots, s_j\}$ are shown as;$T = \left\{a_L^i, a_M^i, a_L^{i+1}, a_R^i, a_M^{i+1}, a_L^{i+2}, a_R^{i+1}, \ldots, a_L^j, a_R^{j-1}, a_M^j, a_R^j\right\}$.

The set of aggregation elements can be simplified with fuzzy partition under $a_R^{k-1} = a_M^k = a_L^{k+1}, k = 1, 2, \ldots, g-1$ acceptance and defined as Ruspini (1969) $T = \{a_L^i, a_M^i, a_M^{i+1}, \ldots, a_M^j, a_R^j\}$.

**Stage 2. Calculation of the TFMF's parameters**

Parameters of the TFMF, $\tilde{A} = (\mathrm{a, b, c, d})$ that defines the fuzzy envelope, $env_F(H_S)$ of the HFLTS, $H_S$ are determined using the set of aggregation elements, $T = \{a_L^i, a_M^i, a_M^{i+1}, \ldots, a_M^j, a_R^j\}$. Limit values, $a$ and $d$, are defined by the linguistic limits as $s_i = \min H_s$ and $s_j = \max H_s$.

$$\begin{aligned} a &= \min\{a_L^i, a_M^i, a_M^{i+1}, \ldots, a_M^j, a_R^j\} = a_L^i \text{ and} \\ d &= \max\{a_L^i, a_M^i, a_M^{i+1}, \ldots, a_M^j, a_R^j\} = a_R^i \end{aligned} \tag{10.16}$$

The intermediate parameters, $b$ and $d$, of the TFMF are calculated using OWA aggregation operator.

$$b = OWA_{W^s}\left(a_M^i, a_M^{i+1}, \ldots, a_M^j\right) \text{ and } c = OWA_{W^t}\left(a_M^i, a_M^{i+1}, \ldots, a_M^j\right) \tag{10.17}$$

where $s, t = 1, 2; s \neq t$ or $s = t$. Filev and Yager's methods is used calculate the weighting vectors, $W^s$ and $W^t$, in the OWA aggregation operations (Filev and Yager 1998).

The first type of OWA weights $W^1 = \left(w_1^1, w_2^1, \ldots, w_n^1\right)^T, 0 \leq \alpha \leq 1$.

$$env(H_S) = [H_{S^-}, H_{S^+}], \quad H_{S^-} \leq H_{S^+} \tag{10.18}$$

The second type of OWA weights $W^2 = \left(w_1^2, w_2^2, \ldots, w_n^2\right)^T, 0 \leq \alpha \leq 1$.

$$\begin{aligned} ew_1^2 = \alpha^{n-1}, w_2^2 = (1-\alpha)\alpha^{n-2}, w_3^2 = (1-\alpha)\alpha^{n-3}, \ldots, w_{n-1}^2 = (1-\alpha)\alpha, w_n^2 \\ = (1-\alpha) \end{aligned} \tag{10.19}$$

The orness measures, $orness(W^1)$ and $orness(W^2)$, are calculated with the weighting vectors as follow:

$$\begin{aligned} orness\left(W^1\right) &= \sum_{i=1}^{n} w_i^1 \left(\frac{n-i}{n-1}\right) = \frac{n-1}{n-1}\alpha + \frac{n-2}{n-1}\alpha(1-\alpha) \\ &+ \frac{n-3}{n-1}\alpha(1-\alpha)^2 + \cdots + \frac{1}{n-1}\alpha(1-\alpha)^{n-2} + \frac{0}{n-1}(1-\alpha)^{n-1} \\ &= \frac{n}{n-1} - \frac{1-(1-\alpha)^n}{(n-1)\alpha} \end{aligned} \tag{10.20}$$

$$orness\left(W^2\right) = \frac{\alpha - \alpha^n}{(n-1)(1-\alpha)} \tag{10.21}$$

The orness value whose OWA operator is described in the [0,1] interval is used to measure the importance of the HFLTS.

**Stage 3. Sample fuzzy envelope**

In this section, the transformation of a sample linguistic expression into a TFMF form is illustrated to clarify the fuzzy envelope. The linguistic term set, $S = \{s_0 = nothing, s_1 = very\,low, s_2 = low, s_3 = medium,$ $s_4 = high, s_5 = very\,high, s_6 = absolute\}$ is used in the sample application and its graphical representations is as follows (Fig. 10.5):

The following process steps are as follows:

a. The comparative linguistic evaluation is defined by the context-free grammar form: *between low and high.*
b. Linguistic evaluation is converted into HFLTS as $E_{G_H}(between\,low\,and\,high) = \{s_2, s_3, s_4\}$.
c. The set of aggregation elements of the HFLTS is defined: $T = \{a_L^2, a_R^1, a_M^2, a_L^3, a_R^2, a_M^3, a_L^4, a_R^3, a_M^4, a_R^4\}$.
   where $a_R^1 = a_M^2 = a_L^3$, $a_R^2 = a_M^3 = a_L^4$, and $a_R^3 = a_M^4$, so set $T$ can be simplified as $T = \{a_L^2, a_M^2, a_M^3, a_M^4, a_R^4\}$.

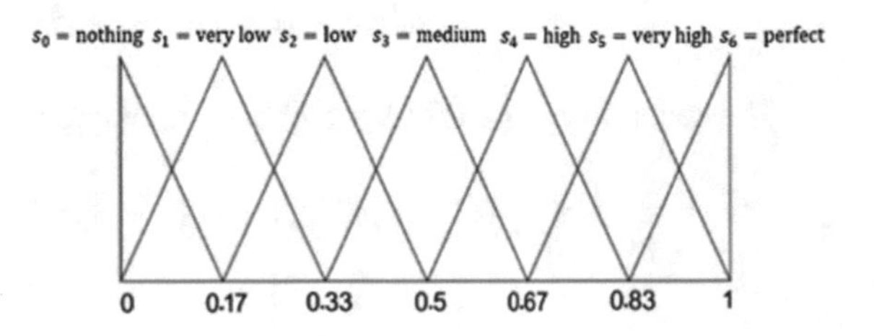


**Fig. 10.5** Graphical representation of a sample linguistic term set $S$

d. The parameters of the TFMF, $env_F(H_{s_4}) = T(a_4, b_4, c_4, d_4)$, are calculated as:

$$a_4 = \min\{a_L^2, a_M^2, a_M^3, a_M^4, a_R^4\} = a_L^2 = 0.17 \text{ and}$$
$$d_4 = \max\{a_L^2, a_M^2, a_M^3, a_M^4, a_R^4\} = a_R^4 = 0.83$$
$$b_4 = OWA_{W^2}(a_M^2, a_M^3) \text{ and } c_4 = OWA_{W^1}(a_M^3, a_M^4)$$

while $i = 2$ and $g = 6$, $\alpha$ is calculated as $\alpha = (g - (j - i))/(g - 1) = 0.8$ and OWA weights are defined as:

$$W^2 = (w_1^1, w_2^1)^T = (0.8, 0.2)^T \text{ and } W^1 = (w_1^1, w_2^1)^T = (0.2, 0.8)^T$$
$$b_4 = a_M^2 * 0.2 + a_M^3 * 0.8 = 0.466 \text{ and } c_4 = a_M^2 * 0.2 + a_M^3 * 0.8 = 0.636$$

e. TFMF of the fuzzy envelope of $H_{s_4}$, $env_F(H_{s_4})$ is defined: $T = (0.17, 0.466, 0.636, 0.83)$.

**Step 4. Operation of HFCM**

Linguistic evaluations of experts define the causal relationship between concepts in the HFCM model. The linguistic expressions are transformed into the crisp values in [−1, 1] interval using defuzzification methods. Thus, the causal relationships between the concepts in the dynamic HFCM can be calculated, and the stable states of the concepts to be reached in the long term can be determined. The crisp values obtained by the defuzzification method express the causal relationship strength between the concepts $(C_i, C_j)$ and are called directed weight $(w_{ij})$. The sign of the directed weight represents the directly related or inverse relationship among concepts. The weights of all causal relationships in HFCM are defined in the weight matrix (*W*) whose diagonal elements, $w_{ii}$ are equal to zero because of absence of self-loop in model. Experts' linguistic expressions can define the initial state of the concepts. New state value of the concept $C_i$ at the time $t$ time iteration is represented as $A_i^t$ in the interval [−1, 1]. $A^t = [A_1^t, A_2^t, \ldots, A_n^t]$ representation is also shows general state values for all concepts. The new state vector of a concept, $C_i$ in the next iteration is measured as follows:

$$A_i^{t+1} = f\left(\sum_{j=1}^{n} w_{ij} A_j^t + A_j^t\right) \quad (10.22)$$

Threshold function operation is an essential step in the new state calculation process. The hyperbolic tangent function is chosen for this study among the most commonly used threshold functions (sign, trivalent, sigmoid, and hyperbolic tangent function). The hyperbolic tangent function is chosen because the [−1,1] values obtained after the threshold calculation are compatible with real-life problems.

$$f(x) = tanh(\lambda x) = \frac{e^{\lambda x} - e^{-\lambda x}}{e^{\lambda x} + e^{-\lambda x}} \quad (10.23)$$

The value of the lambda ($\lambda > 0$) constant, which shapes the slope of the hyperbolic tangent function, is predefined by researchers according to their research characteristics (Bueno and Salmeron 2009). The calculation of the causal relationship in the HFCM model is terminated when the difference between the two consecutive iteration values of all concept relationship values is less than 0.0001 (i.e., $A^{t+1} - A^{t} \leq 0.0001$), and the last state reached is defined as the "steady state".

## 10.5 Application: Solar Energy Price Modelling with Hesitant Fuzzy Cognitive Map

The increasing importance of energy in life affects and is influenced by economic, social and environmental factors. The pricing of solar energy that is a developing member of the energy sector is shaped under similar factorial circumstances. The definition of causal relations among the main factors that determine solar energy prices is defined by the opinions and evaluations of experts under uncertainty and unpredictability conditions. Experts of this study is both the academic and energy-business community. Experts selected from the academic community are preferred because of their studies on economic analysis and decision making of renewable energies. The experts selected from the energy sector consist of analysts and specialists who make installation assessments of large-scale renewable and solar energy systems. Since the renewable energy pricing mechanisms are similar each other, the experts in renewable sector are consulted in assessing the factors that determine the solar energy price. Under these conditions, the HFCM model is used to describe the causal relationships among factors and the initial states of the factors around the solar energy price. The initial states of the factors are randomly developed, and different scenarios are defined. The scenarios are operated on the HFCM model, and the solar energy price and other factors' reactions are observed during the model.

### *10.5.1 Determining of Weight Matrix and Initial State Vector*

Firstly, the causal relationship between the factors and the powers of the causal relations among them must be defined for the operating the HFCM model. The "solar energy price" -based model is defined by twenty different models and the direction and sign of the causal relationship between the factors is determined by the collective opinion of experts (Fig. 10.6). The orange causality represents the inverse proportion (that is, the value of a factor increases while the value of the other factor that is related decreases) and the blue causality represents the direct proportion (that is, the value of a factor increases while the value of the other factor that is related increases).

The HFCM consists of twenty-four factors, three of which are transmitters (GT, EE, ESE), eighteen of which are ordinary, and no receiver factor (Fig. 10.6). The highest in-degree factor is the SEP (Solar Energy Price) that is also the center of the model structure, and the highest out-degree factor is SLL (Supportive Law and Legislations). The highest centrality factor is SLL with twelve value and followed by SEP and EP (Energy Price).

The causal relationship between the factors of the model that is shaped around the solar energy price factor is determined by the academic and sectoral specialists in the field of solar energy and solar economics. Experts make evaluations based on hesitant linguistic terms to express causal relationships among factors more realistically and explicitly. Linguistic term set $S = \{s_0 : nothing, s_1 : very\, low, s_2 : low, s_3 : medium, s_4 : high, s_5 : very\, high, s_6 : absolute\}$ is used to generate linguistic assessments based on the context free grammar. The relationship matrix, which is defined linguistically by the common view of experts, is shown in Table 10.1. Expressions of linguistic evaluation are shown in abbreviation on Table 10.1. The " + " sign indicates a positive relationship, and the "–" sign indicates a negative relationship. Explanations of other abbreviations are as follows; btw: between, atl: at least, gth: greater than, lth: lower than, atm: at most, n: neighter, vl: very low, low:l, m: medium, h: high, vh: very high, and a:absolute. Linguistic expressions that describe the causal relationship between the factors are transformed into HFLTS using conversion functions (Table 10.2).

HFLTSs transformed from linguistic evaluations are converted into a trapezoidal fuzzy membership function $A^{\sim} = (a, b, c, d)$ (Table 10.3). The intermediate b and c parameters of the trapezoidal fuzzy membership function are calculated using the OWA aggregation operator at this stage. The linguistic expressions converted to numerical values by the trapezoidal fuzzy membership function are transformed

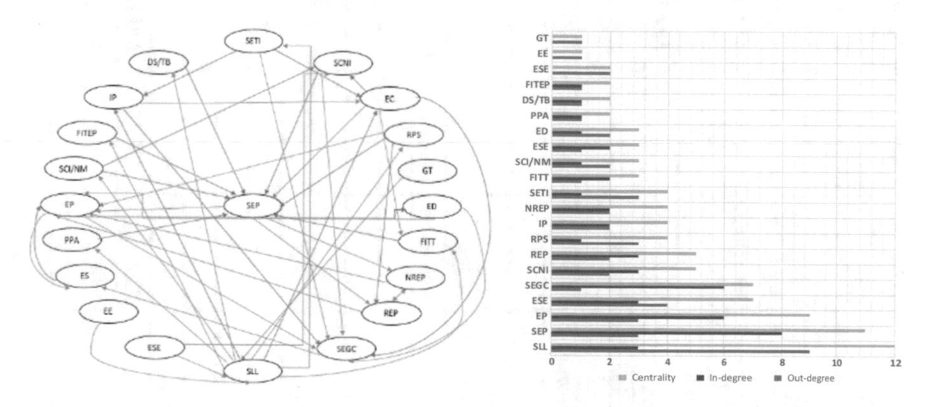


**Fig. 10.6** Solar energy price centered HFCM and the values of the model structure metrics

**Table 10.1** Experts' shared views on the causal relationships among the factors

| | RPS | PPA | FITT | SCNI | DS/TB | ESE | FITEP | IP | SCI/NM | EP |
|---|---|---|---|---|---|---|---|---|---|---|
| RPS | | | | | | | | | | +atm_l |
| PPA | | | | | | | | | | |
| FITT | | | | | | | | | | |
| SCNI | | | | | | | | | | |
| DS/TB | | | | | | | | | | |
| ESE | | | +btw_lvh | −atm_m | | | | | | |
| FITEP | | | | | | | | | | |
| IP | | | | | | +atm_h | | | | |
| SCI/NM | | | | −btw_vlm | | | | | | |
| EP | | | | | | | | | | |
| REP | | | | | | | | | | +gth_m |
| NREP | | | | | | | | | | +gth_l |
| SLL | +atl_m | +gth_m | +lth_h | | +atm_m | −btw_lh | +lth_h | −btw_vlm | +lth_h | |
| EE | | | | | | | | | | |
| ESE | | | | +lth_m | | | | | | |
| SEP | | | | | | | | | | +gth_m |
| ESE | | | | | | | | | | −atl_h |
| ED | | | | | | | | | | −btw_ma |
| SEGC | | | | | | | | | | |
| GT | | | | | | | | | | |
| SETI | | | | | | −atm_h | | −gth_m | | |

| | REP | NREP | SLL | EE | ESE | SEP | ESE | ED | SEGC | GT | SETI |
|---|---|---|---|---|---|---|---|---|---|---|---|
| RPS | −atm_m | | | | | −atm_m | | | | | |
| PPA | | | | | | −btw_vlm | | | | | |

(continued)

**Table 10.1** (continued)

| | REP | NREP | SLL | EE | ESE | SEP | ESE | ED | SEGC | GT | SETI |
|---|---|---|---|---|---|---|---|---|---|---|---|
| FITT | | | | | | −ism | | | | | |
| SCNI | | | | | | −gth_vl | | | +lth_h | | |
| DS/TB | | | | | | −btw vlh | | | | | |
| ESE | | | | | | +gth_m | | | −ish | | |
| FITEP | | | | | | −atl_m | | | | | |
| IP | | | | | | | | | −atm_m | | |
| SCI/NM | | | | | | −lth_l | | | | | |
| EP | | | | | | | +lth_h | −gth_h | +btw_vlm | | |
| REP | | +gth_l | | | | | | | | | |
| NREP | +lth_h | | | | | | | | | | |
| SLL | | | | | | | | | | | +lth_m |
| EE | | | +atl_h | | | | | | | | |
| ESE | | | +gth_vh | | | | | | | | |
| SEP | +btw_lm | +atm_m | | | | | | | | | |
| ESE | | | | | | | | | | | |
| ED | | | | | | | | | +gth_vl | | |
| SEGC | | | | | | | +btw_lm | | | | |
| GT | | | +atm_h | | | | | | | | |
| SETI | | | | | | | | | +lth_l | | |

**Table 10.2** HFLTS transformation of experts' linguistic evaluations

| | RPS | PPA | FITT | SCNI | DS/TB | ESE | FITEP | IP | SCI/NM | EP |
|---|---|---|---|---|---|---|---|---|---|---|
| RPS | | | | | | | | | | {+}{n,vl,l} |
| PPA | | | | | | | | | | |
| FITT | | | | | | | | | | |
| SCNI | | | | | | | | | | |
| DS/TB | | | | | | | | | | |
| ESE | | | {+}{l,m,h,vh} | {−}{n,vl,l,m} | | | | | | |
| FITEP | | | | | | | | | | |
| IP | | | | | | {+}{n,vl,l,m,h} | | | | |
| SCI/NM | | | | {+}{vl,l,m} | | | | | | |
| EP | | | | | | | | | | |
| REP | | | | | | | | | | {+}{h,vh,a} |
| NREP | | | | | | | | | | {+}{m,h,vh,a} |
| SLL | {+}{m,h,vh,a} | {+}{h,vh,a} | {+}{n,vl,l,m} | | {+}{n,vl,l,m} | {−}{l,m,h} | {+}{n,vl,l,m} | {−}{vl,l,m} | {+}{h,vh,a} | |
| EE | | | | | | | | | | |
| ESE | | | | {+}{n,vl,l} | | | | | | |
| SEP | | | | | | | | | | {+}{h,vh,a} |

(continued)

**Table 10.2** (continued)

| | RPS | PPA | FITT | SCNI | DS/TB | ESE | FITEP | IP | SCI/NM | EP |
|---|---|---|---|---|---|---|---|---|---|---|
| ESE | | | | | | | | | | {−}{h,vh, a} |
| ED | | | | | | | | | | {−}{m,h, vh,a} |
| SEGC | | | | | | | | | | |
| GT | | | | | | | | | | |
| SETI | | | | | | {−}{n,vl,l, m,h} | | {−}{h, vh} | | |

| REP | NREP | SLL | EE | ESE | SEP | ESE | ED | SEGC | GT | SETI |
|---|---|---|---|---|---|---|---|---|---|---|
| {−}{n,vl,l, m} | | | | | {−}{n,vl,l,m} | | | | | |
| | | | | | {−}{vl,l,m} | | | | | |
| | | | | | {−}{m} | | | | | |
| | | | | | {−}{l,m,h,vh, a} | | | {+}{n,vl,l,m} | | |
| | | | | | {−}{h,vh,a} | | | | | |
| | | | | | {+}{h,vh,a} | | | {−}{h} | | |
| | | | | | {−}{m,h,vh, a} | | | | | |
| | | | | | | | | {−}{n,vl,l,m} | | |
| | | | | | {−}{n,vl} | | | | | |
| | | | | | | {−}{n,vl,l, m} | {−}{vh, a} | {+}{vl,l,m} | | |
| | {+}{m,h,vh, a} | | | | | | | | | |

(continued)

**Table 10.2** (continued)

| REP | NREP | SLL | EE | ESE | SEP | ESE | ED | SEGC | GT | SETI |
|---|---|---|---|---|---|---|---|---|---|---|
| {+}{n,vl,l, m} | | | | | | | | | | |
| | | | | | | | | | | {+}{n,vl, l} |
| | | {+}{h,vh,a} | | | | | | | | |
| | | {+}{vh,a} | | | | | | | | |
| {+}{l,m} | {+}{n,vl,l, m} | | | | | | | | | |
| | | | | | | | | {+}{l,m,h,vh, a} | | |
| | | | | | | {+}{l,m} | | | | |
| | | {+}{n,vl,l,m, h} | | | | | | | | |
| | | | | | | | | {+}{n,vl} | | |

**Table 10.3** Trapezoidal representation of HFLTS

| | RPS | PPA | FITT | SCNI | DS/TB | ESE | FITEP | IP | SCI/NM | EP |
|---|---|---|---|---|---|---|---|---|---|---|
| RPS | (0,0,0,0) | (0,0,0,0) | (0,0,0,0) | (0,0,0,0) | (0,0,0,0) | (0,0,0,0) | (0,0,0,0) | (0,0,0,0) | (0,0,0,0) | (0,0,0.15,0.5) |
| PPA | (0,0,0,0) | (0,0,0,0) | (0,0,0,0) | (0,0,0,0) | (0,0,0,0) | (0,0,0,0) | (0,0,0,0) | (0,0,0,0) | (0,0,0,0) | (0,0,0,0) |
| FITT | (0,0,0,0) | (0,0,0,0) | (0,0,0,0) | (0,0,0,0) | (0,0,0,0) | (0,0,0,0) | (0,0,0,0) | (0,0,0,0) | (0,0,0,0) | (0,0,0,0) |
| SCNI | (0,0,0,0) | (0,0,0,0) | (0,0,0,0) | (0,0,0,0) | (0,0,0,0) | (0,0,0,0) | (0,0,0,0) | (0,0,0,0) | (0,0,0,0) | (0,0,0,0) |
| DS/TB | (0,0,0,0) | (0,0,0,0) | (0,0,0,0) | (0,0,0,0) | (0,0,0,0) | (0,0,0,0) | (0,0,0,0) | (0,0,0,0) | (0,0,0,0) | (0,0,0,0) |
| ESE | (0,0,0,0) | (0,0,0,0) | (0.17,0.44,0.77,1) | (0,0,−0.36,−0.67) | (0,0,0,0) | (0,0,0,0) | (0,0,0,0) | (0,0,0,0) | (0,0,0,0) | (0,0,0,0) |
| FITEP | (0,0,0,0) | (0,0,0,0) | (0,0,0,0) | (0,0,0,0) | (0,0,0,0) | (0,0,0,0) | (0,0,0,0) | (0,0,0,0) | (0,0,0,0) | (0,0,0,0) |
| IP | (0,0,0,0) | (0,0,0,0) | (0,0,0,0) | (0,0,0,0) | (0,0,0,0) | (0,0,0.59,0.83) | (0,0,0,0) | (0,0,0,0) | (0,0,0,0) | (0,0,0,0) |
| SCI/ NM | (0,0,0,0) | (0,0,0,0) | (0,0,0,0) | (0,−0.3,−0.47, −0.67) | (0,0,0,0) | (0,0,0,0) | (0,0,0,0) | (0,0,0,0) | (0,0,0,0) | (0,0,0,0) |
| EP | (0,0,0,0) | (0,0,0,0) | (0,0,0,0) | (0,0,0,0) | (0,0,0,0) | (0,0,0,0) | (0,0,0,0) | (0,0,0,0) | (0,0,0,0) | (0,0,0,0) |
| REP | (0,0,0,0) | (0,0,0,0) | (0,0,0,0) | (0,0,0,0) | (0,0,0,0) | (0,0,0,0) | (0,0,0,0) | (0,0,0,0) | (0,0,0,0) | (0.5,0.86,1,1) |
| NREP | (0,0,0,0) | (0,0,0,0) | (0,0,0,0) | (0,0,0,0) | (0,0,0,0) | (0,0,0,0) | (0,0,0,0) | (0,0,0,0) | (0,0,0,0) | (0.33,0.65,1,1) |
| SLL | (0.33,0.65,1,1) | (0.5,0.86,1,1) | (0,0,0.36,0.67) | (0,0,0,0) | (0,0,0.36,0.67) | (−0.17,−0.47,−0.64, −0.83) | (0,0,0.36,0.67) | (0,−0.3,−0.47, −0.67) | (0,0,0.36,0.67) | (0,0,0,0) |
| EE | (0,0,0,0) | (0,0,0,0) | (0,0,0,0) | (0,0,0,0) | (0,0,0,0) | (0,0,0,0) | (0,0,0,0) | (0,0,0,0) | (0,0,0,0) | (0,0,0,0) |
| ESE | (0,0,0,0) | (0,0,0,0) | (0,0,0,0) | (0,0,0.15,0.5) | (0,0,0,0) | (0,0,0,0) | (0,0,0,0) | (0,0,0,0) | (0,0,0,0) | (0,0,0,0) |
| SEP | (0,0,0,0) | (0,0,0,0) | (0,0,0,0) | (0,0,0,0) | (0,0,0,0) | (0,0,0,0) | (0,0,0,0) | (0,0,0,0) | (0,0,0,0) | (0.5,0.86,1,1) |
| ESE | (0,0,0,0) | (0,0,0,0) | (0,0,0,0) | (0,0,0,0) | (0,0,0,0) | (0,0,0,0) | (0,0,0,0) | (0,0,0,0) | (0,0,0,0) | (−0.5,−0.86,−1,−1) |
| ED | (0,0,0,0) | (0,0,0,0) | (0,0,0,0) | (0,0,0,0) | (0,0,0,0) | (0,0,0,0) | (0,0,0,0) | (0,0,0,0) | (0,0,0,0) | (−0.33,−0.61,−0.94, −1) |
| SEGC | (0,0,0,0) | (0,0,0,0) | (0,0,0,0) | (0,0,0,0) | (0,0,0,0) | (0,0,0,0) | (0,0,0,0) | (0,0,0,0) | (0,0,0,0) | (0,0,0,0) |
| GT | (0,0,0,0) | (0,0,0,0) | (0,0,0,0) | (0,0,0,0) | (0,0,0,0) | (0,0,0,0) | (0,0,0,0) | (0,0,0,0) | (0,0,0,0) | (0,0,0,0) |
| SETI | (0,0,0,0) | (0,0,0,0) | (0,0,0,0) | (0,0,0,0) | (0,0,0,0) | (0,0,−0.59,−0.83) | (0,0,0,0) | (−0.5,−0.86,−1, −1) | (0,0,0,0) | (0,0,0,0) |

**Table 10.3** (continued)

| | REP | NREP | SLL | EE | ESE | SEP | ESE | ED | SEGC | GT | SETI |
|---|---|---|---|---|---|---|---|---|---|---|---|
| RPS | (0,0,−0.36,−0.67) | (0,0,0,0) | (0,0,0,0) | (0,0,0,0) | (0,0,0,0) | (0,0,−0.36,−0.67) | (0,0,0,0) | (0,0,0,0) | (0,0,0,0) | (0,0,0,0) | (0,0,0,0) |
| PPA | (0,0,0,0) | (0,0,0,0) | (0,0,0,0) | (0,0,0,0) | (0,0,0,0) | (0,−0.3,−0.47,−0.67) | (0,0,0,0) | (0,0,0,0) | (0,0,0,0) | (0,0,0,0) | (0,0,0,0) |
| FITT | (0,0,0,0) | (0,0,0,0) | (0,0,0,0) | (0,0,0,0) | (0,0,0,0) | (−0.33,−0.5,−0.5, −0.67) | (0,0,0,0) | (0,0,0,0) | (0,0,0,0) | (0,0,0,0) | (0,0,0,0) |
| SCNI | (0,0,0,0) | (0,0,0,0) | (0,0,0,0) | (0,0,0,0) | (0,0,0,0) | (−0.17,−0.42,−1,−1) | (0,0,0,0) | (0,0,0,0) | (0,0,0.36,0.67) | (0,0,0,0) | (0,0,0,0) |
| DS/TB | (0,0,0,0) | (0,0,0,0) | (0,0,0,0) | (0,0,0,0) | (0,0,0,0) | (0,−0.27,−0.61, −0.83) | (0,0,0,0) | (0,0,0,0) | (0,0,0,0) | (0,0,0,0) | (0,0,0,0) |
| ESE | (0,0,0,0) | (0,0,0,0) | (0,0,0,0) | (0,0,0,0) | (0,0,0,0) | (0.5,0.86,1,1) | (0,0,0,0) | (0,0,0,0) | (−0.5,−0.67, −0.67,0.83) | (0,0,0,0) | (0,0,0,0) |
| FITEP | (0,0,0,0) | (0,0,0,0) | (0,0,0,0) | (0,0,0,0) | (0,0,0,0) | (−0.33,−0.65,−1,−1) | (0,0,0,0) | (0,0,0,0) | (0,0,0,0) | (0,0,0,0) | (0,0,0,0) |
| IP | (0,0,0,0) | (0,0,0,0) | (0,0,0,0) | (0,0,0,0) | (0,0,0,0) | (0,0,0,0) | (0,0,0,0) | (0,0,0,0) | (0,0,−0.6,−0.67) | (0,0,0,0) | (0,0,0,0) |
| SCI/ NM | (0,0,0,0) | (0,0,0,0) | (0,0,0,0) | (0,0,0,0) | (0,0,0,0) | (0,0,−0.03,−0.33) | (0,0,0,0) | (0,0,0,0) | (0,0,0,0) | (0,0,0,0) | (0,0,0,0) |
| EP | (0,0,0,0) | (0,0,0,0) | (0,0,0,0) | (0,0,0,0) | (0,0,0,0) | (0,0,0,0) | (0,0,0.36,0.67) | (−0.67,−0.98,−1, −1) | (0,0.3,0.47,0.67) | (0,0,0,0) | (0,0,0,0) |
| REP | (0,0,0,0) | (0.33,0.65,1,1) | (0,0,0,0) | (0,0,0,0) | (0,0,0,0) | (0,0,0,0) | (0,0,0,0) | (0,0,0,0) | (0,0,0,0) | (0,0,0,0) | (0,0,0,0) |
| NREP | (0,0,0.36,0.67) | (0,0,0,0) | (0,0,0,0) | (0,0,0,0) | (0,0,0,0) | (0,0,0,0) | (0,0,0,0) | (0,0,0,0) | (0,0,0,0) | (0,0,0,0) | (0,0,0,0) |
| SLL | (0,0,0,0) | (0,0,0,0) | (0,0,0,0) | (0,0,0,0) | (0,0,0,0) | (0,0,0,0) | (0,0,0,0) | (0,0,0,0) | (0,0,0,0) | (0,0,0,0) | (0,0,0.15,0.5) |
| EE | (0,0,0,0) | (0,0,0,0) | (0.5,0.86,1,1) | (0,0,0,0) | (0,0,0,0) | (0,0,0,0) | (0,0,0,0) | (0,0,0,0) | (0,0,0,0) | (0,0,0,0) | (0,0,0,0) |
| ESE | (0,0,0,0) | (0,0,0,0) | (0.83,1,1,1) | (0,0,0,0) | (0,0,0,0) | (0,0,0,0) | (0,0,0,0) | (0,0,0,0) | (0,0,0,0) | (0,0,0,0) | (0,0,0,0) |
| SEP | (0.17,0.33,0.5,0.67) | (0,0,0.36,0.67) | (0,0,0,0) | (0,0,0,0) | (0,0,0,0) | (0,0,0,0) | (0,0,0,0) | (0,0,0,0) | (0,0,0,0) | (0,0,0,0) | (0,0,0,0) |
| ESE | (0,0,0,0) | (0,0,0,0) | (0,0,0,0) | (0,0,0,0) | (0,0,0,0) | (0,0,0,0) | (0,0,0,0) | (0,0,0,0) | (0,0,0,0) | (0,0,0,0) | (0,0,0,0) |
| ED | (0,0,0,0) | (0,0,0,0) | (0,0,0,0) | (0,0,0,0) | (0,0,0,0) | (0,0,0,0) | (0,0,0,0) | (0,0,0,0) | (0.17,0.42,1,1) | (0,0,0,0) | (0,0,0,0) |
| SEGC | (0,0,0,0) | (0,0,0,0) | (0,0,0,0) | (0,0,0,0) | (0,0,0,0) | (0,0,0,0) | (0.17,0.33,0.5,0.67) | (0,0,0,0) | (0,0,0,0) | (0,0,0,0) | (0,0,0,0) |
| GT | (0,0,0,0) | (0,0,0,0) | (0,0,0.59,0.83) | (0,0,0,0) | (0,0,0,0) | (0,0,0,0) | (0,0,0,0) | (0,0,0,0) | (0,0,0,0) | (0,0,0,0) | (0,0,0,0) |
| SETI | (0,0,0,0) | (0,0,0,0) | (0,0,0,0) | (0,0,0,0) | (0,0,0,0) | (0,0,0,0) | (0,0,0,0) | (0,0,0,0) | (0,0,0.03,0.33) | (0,0,0,0) | (0,0,0,0) |

into crisp values using the weighted average method to obtain computation values. The causal relations between the factors in the HFCM model are expressed by a single numerical value in the range [−1,1] by defuzzification of the trapezoidal representations (Table 10.4). This obtained table is defined as the weight matrix of HFCM and expresses the causal relationship between the factors.

Since the initial states of the factors cannot be expressed with definite values within the dynamic energy system, the initial states of the factors are linguistically defined by the experts. The combination of different initial states of the factors reveals different solar energy price centered scenarios. The process steps followed in obtaining the weight matrix are also followed when the initial state table is obtained. Defuzzified trapezoidal fuzzy membership functions give crisp valued at the initial state scenarios. Applications are made on two scenarios (Case1, Case2) selected from ten cases developed by experts.

Equation (10.22) based on the initial states of the factors $(A^0)$ and the weight matrix of HFCM ($W$) is run to obtain the next state vector $A^1$. The equation that calculates the next state vector $A^{t+1}$ from the previous state vector $A^t$ is repeated until the state vector reaches the steady state, $A^l$. It is a common practice to terminate the iteration if the difference is less than 0.001 for each factor in the two consecutive state vectors $(A^{t+1} - A^t \leq 0.001)$. In the application section, the hyperbolic tangent function (Eq. 10.23), which derives values in the range [−1,1], is used as a threshold function and the value of lambda is assumed to be 0.7 $(\lambda = 0.7)$.

### *10.5.2 Case Studies Base on Solar Energy Price*

The two initial state vectors selected from the scenarios identified by the experts' joint evaluations are evaluated in this section. The values in the initial state are defined in the range [−1,1], and the value of the corresponding factor reflects the current state of the solar energy price in the HFCM model. The positive (negative) value means that the factor has increased (decreased), while the zero value means that the factor has not changed. The changes in the initial state values of the factors over time and the convergence values of the factors are graphically displayed throughout the iterations. The number of iterations and the steady-state values of the factors change for each scenario. The order of the factors in the initial state vector is: Renewable Portfolio Standard (RPS), Power Purchasing Agreements (PPA), FIT with Tender (FITT), Self-consumption (non-incentivized) (SCNI), Direct Subsidies or Tax Breaks (DS/TB), Economic Components (EC), FIT for Entire Production (FITEP), Infrastructure Problems (IP), Self-consumption or Net-metering (incentivized) (SCI/NM), Energy Price (EP), Renewable Energy Price (REP), Non-renewable Energy Price (NREP), Supportive Law and Legislations (SLL), Environmental Effects (EE), Eco-social Effects (ESE), Solar Energy Price (SEP),

**Table 10.4** Weight matrix of solar energy price HFCM model

| | RPS | PPA | FITT | SCNI | DS/TB | ESE | FITEP | IP | SCI/NM | EP |
|---|---|---|---|---|---|---|---|---|---|---|
| RPS | 0 | 0 | 0 | 0 | 0 | 0 | 0 | 0 | 0 | 0.132593 |
| PPA | 0 | 0 | 0 | 0 | 0 | 0 | 0 | 0 | 0 | 0 |
| FITT | 0 | 0 | 0 | 0 | 0 | 0 | 0 | 0 | 0 | 0 |
| SCNI | 0 | 0 | 0 | 0 | 0 | 0 | 0 | 0 | 0 | 0 |
| DS/TB | 0 | 0 | 0 | 0 | 0 | 0 | 0 | 0 | 0 | 0 |
| ESE | 0 | 0 | 0.594 | −0.23 | 0 | 0 | 0 | 0 | 0 | 0 |
| FITEP | 0 | 0 | 0 | 0 | 0 | 0 | 0 | 0 | 0 | 0 |
| IP | 0 | 0 | 0 | 0 | 0 | 0.334 | 0 | 0 | 0 | 0 |
| SCI/NM | 0 | 0 | 0 | −0.366 | 0 | 0 | 0 | 0 | 0 | 0 |
| EP | 0 | 0 | 0 | 0 | 0 | 0 | 0 | 0 | 0 | 0 |
| REP | 0 | 0 | 0 | 0 | 0 | 0 | 0 | 0 | 0 | 0.867 |
| NREP | 0 | 0 | 0 | 0 | 0 | 0 | 0 | 0 | 0 | 0.77 |
| SLL | 0.770 | 0.867 | 0.23 | 0 | 0.23 | −0.534 | 0.23 | −0.366 | 0.23 | 0 |
| EE | 0 | 0 | 0 | 0 | 0 | 0 | 0 | 0 | 0 | 0 |
| ESE | 0 | 0 | 0 | 0.133 | 0 | 0 | 0 | 0 | 0 | 0 |
| SEP | 0 | 0 | 0 | 0 | 0 | 0 | 0 | 0 | 0 | 0.867 |
| ESE | 0 | 0 | 0 | 0 | 0 | 0 | 0 | 0 | 0 | −0.867 |
| ED | 0 | 0 | 0 | 0 | 0 | 0 | 0 | 0 | 0 | −0.733 |
| SEGC | 0 | 0 | 0 | 0 | 0 | 0 | 0 | 0 | 0 | 0 |
| GT | 0 | 0 | 0 | 0 | 0 | 0 | 0 | 0 | 0 | 0 |
| SETI | 0 | 0 | 0 | 0 | 0 | −0.334 | 0 | −0.867 | 0 | 0 |

| REP | NREP | SLL | EE | ESE | SEP | ESE | ED | SEGC | GT | SETI |
|---|---|---|---|---|---|---|---|---|---|---|
| −0.22958 | 0 | 0 | 0 | 0 | −0.23 | 0 | 0 | 0 | 0 | 0 |
| 0 | 0 | 0 | 0 | 0 | −0.366 | 0 | 0 | 0 | 0 | 0 |

(continued)

**Table 10.4** (continued)

| REP | NREP | SLL | EE | ESE | SEP | ESE | ED | SEGC | GT | SETI |
|---|---|---|---|---|---|---|---|---|---|---|
| 0 | 0 | 0 | 0 | 0 | −0.5 | 0 | 0 | 0 | 0 | 0 |
| 0 | 0 | 0 | 0 | 0 | −0.666 | 0 | 0 | 0.23 | 0 | 0 |
| 0 | 0 | 0 | 0 | 0 | −0.428 | 0 | 0 | 0 | 0 | 0 |
| 0 | 0 | 0 | 0 | 0 | 0.867 | 0 | 0 | −0.668 | 0 | 0 |
| 0 | 0 | 0 | 0 | 0 | −0.77 | 0 | 0 | 0 | 0 | 0 |
| 0 | 0 | 0 | 0 | 0 | 0 | 0 | 0 | −0.23 | 0 | 0 |
| 0 | 0 | 0 | 0 | 0 | −0.064 | 0 | 0 | 0 | 0 | 0 |
| 0 | 0 | 0 | 0 | 0 | 0 | 0.23 | −0.936 | 0.366 | 0 | 0 |
| 0 | 0.77 | 0 | 0 | 0 | 0 | 0 | 0 | 0 | 0 | 0 |
| 0.23 | 0 | 0 | 0 | 0 | 0 | 0 | 0 | 0 | 0 | 0 |
| 0 | 0 | 0 | 0 | 0 | 0 | 0 | 0 | 0 | 0 | 0.133 |
| 0 | 0 | 0.867 | 0 | 0 | 0 | 0 | 0 | 0 | 0 | 0 |
| 0 | 0 | 0.972 | 0 | 0 | 0 | 0 | 0 | 0 | 0 | 0 |
| 0.417 | 0.23 | 0 | 0 | 0 | 0 | 0 | 0 | 0 | 0 | 0 |
| 0 | 0 | 0 | 0 | 0 | 0 | 0 | 0 | 0 | 0 | 0 |
| 0 | 0 | 0 | 0 | 0 | 0 | 0 | 0 | 0.666 | 0 | 0 |
| 0 | 0 | 0 | 0 | 0 | 0 | 0.417 | 0 | 0 | 0 | 0 |
| 0 | 0 | 0.334 | 0 | 0 | 0 | 0 | 0 | 0 | 0 | 0 |
| 0 | 0 | 0 | 0 | 0 | 0 | 0 | 0 | 0.064 | 0 | 0 |

Energy supply (ES), Energy Demand (ED), Solar Energy Generation Capacity (SEGC), Global Treaties (GT), Solar Energy Technological Improvement (SETI).

**Case1**:

In this scenario, different initial states of the factors that affect the solar energy price are examined. The scenario analyzes the long-term effects of increasing and decreasing states of factors on solar energy price and other factors. Initial state vector is defined as $A_1^0$ = [0.332 −0.972 0.332 0.168 −0.668 0.972 0.028 −0.5 0.028–0.168 0.832 0.972 0.972 0.668 0.168 0.028 −0.168 0.168 0.028 0.5 0.028]. According to the scenario, the EC, NREL, SLL and REP factors appear with the highest increase of 0.972 in the initial state of the system. The increase in investment costs leads to an increase in renewable and non-renewable energy costs and energy prices. Laws and regulations that increase solar energy production capacity arise in order to balance high energy production prices. The scenario also includes high environmental sensitivity and high international agreement impacts with the other high positive values. The factors that decrease in the initial state are EP, ES, IP, DS / TB, and PPA factors and the greatest reduction is seen in DS / TB and PPA factors supporting solar energy production capacity. Other remaining factors have a weak impact on the system with their low growth rates.

The simulated solar energy price-centered HFCM model simulated under the defined initial state reaches equilibrium state after 45 iterations. In the equilibrium model (Fig. 10.7), the final state vector of the factors is as follows: $A_1^l$ =[0 0–0.003 0.001 0–0.001 0 0 0–0.862 −0.193 −0.337 0 0 0 0–0.050 0.813 0.429 0 0]. The steady-state values obtained by simulating the HFCM model are interpreted as follows:

- The system shows complex fluctuations in the first twelve iterations. After twelfth iteration, the tendencies of the factors begin to appear more clearly. Factors which direct to the converged values after forty-first iteration reaches to the final state values after forty-fifth iteration.
- The factors of RPS, PPA, DS/TB, FITEP, IP, SCI/NM, SLL, EE, ESE, SEP, GT, and SETI converge to zero in the balanced system. This state means that there is no enhancing or reducing effects on these factors. The solar energy price among these factors is also a constant value; there is no tendency to increase or decrease. The state of the solar energy price causes the general system factors to remain unchanged.
- In this scenario, ED, SEGC, and SCNI factors show an increasing tendency with 0.813, 0.429, and 0.001 values. The main reason for the high increase in energy demand is the decrease in renewable energy, non-renewable energy, and general energy prices. Increased energy demand is balanced by increasing solar energy production capacity. The installation of individual solar energy systems without incentive contributes to increasing solar energy capacity with a low increase.
- While the highest reductions are seen in EP, NREP, and REP factors (−0.862, −0.337, −0.193 respectively); ES, FITT, and EC factors affect the system with low reductions (−0.020, −0.003, −0.001 respectively). The inverse relationship

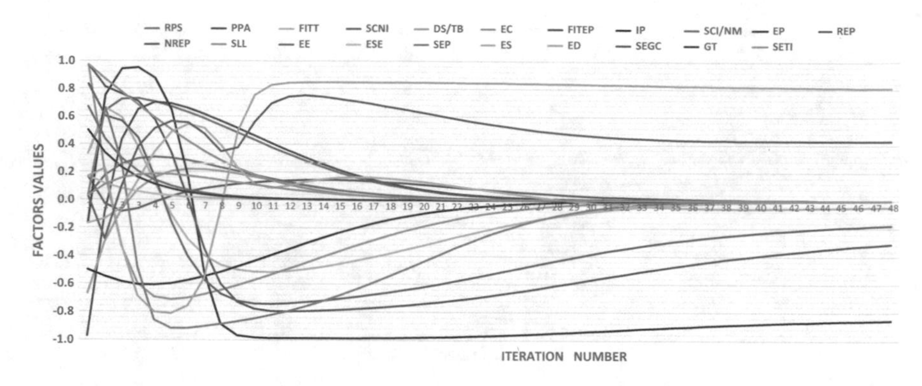


**Fig. 10.7** HFCM simulation of Case1

between energy demand and energy prices is the leading cause of the decline in energy prices. The decline in energy prices is also triggered by reductions in renewable energy (excluding solar energy) and non-renewable energy prices. Reductions in support and incentives for solar energy generation may have led to a lower reduction in overall energy supply. The scale economy created by the developing solar energy systems technology leads to a reduction in installation costs, though there are no incentives for solar energy.

- **Case2**:

This scenario is designed to examine the effects of the supports and programs for improving solar energy systems on solar energy price and general system factors. Therefore, the system consists of supportive factors with increasing GT, FITT, PPA, SLL, DS/TB and RPS (0.972, 0.972, 0.972, 0.832, 0.668, 0.500, 0.168) and decreasing EC and IP (−0.028, −0.500). Nevertheless, there are some mitigating factors to increase the solar energy capacity and reduce the solar energy price in the system such as FITEP, ESE, SETI, SCNI, and SCI/NM (−0.168, −0.668, −0.668, −0.832, −0.832). The solar energy price, solar energy generation capacity, demand for energy, energy supply, and other supportive factors tend to increase in this scenario. Initial state vector is defined as $A_2^0$ =[−0.972 0.5 0.028 0.332 −0.668 0.168 −0.832 0.332 −0.332 −0.028 0.972 0.168 0.832 0.832 0.168 −0.028 −0.668 −0.168 −0.332 0.5 0.332]. The HFCM simulation model reaches to the equilibrium state at the 37th iteration and obtained steady state vector is as follows: $A_2^l$ =[ 0 0 0.004–0.001 0 0.001 0 0 0–0.724 0.007 0.012 0 0 0 0.001 0.047 0.766 0.451 0 0]. The general behavior of the factors according to the graphical representation (Fig. 10.8) and the obtained values are summarized as follows.

- The highest tendency to increase is seen in the energy demand (0.766) that is resulted from the high reduction in energy price (−0.724). The high decline in energy prices in the initial state and the increase in solar energy production support will lead to an increase in energy supply and cause energy prices to continue at low levels over the long run. The second highest increase is seen in

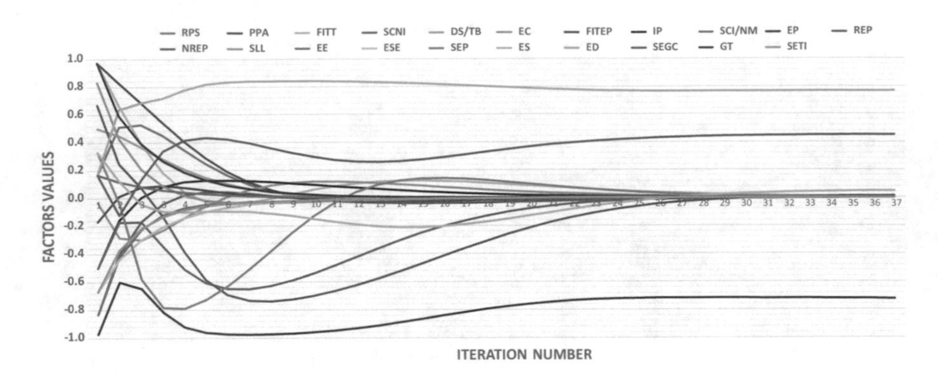


**Fig. 10.8** HFCM simulation of Case2

solar energy generation capacity (0.451) which is the result of the focus of this scenario that includes the high increase in solar energy support. The increase in solar energy capacity provides a low level of energy supply (0.047) and causes the installation costs to increase (EC, 0.001). The increase in the installation costs of solar energy systems can lead to an increase in the bid prices for the FIT (0.004) and solar energy prices. The high increase in energy demand also leads to renewable (including solar energy, 0.007) and non-renewable energy prices (0.012).

- When the system reaches equilibrium, the factors of IP, GT, EE, ESE, SLL, SETI, DS / TB and FITEP go into inertia position, and the steady-state values of these factors are indicated by zero. The inertia of most of the policies that support the development of solar energy systems proves that the system can stand independently as unsupported and uncontrolled. The inertia results show that the widespread of solar energy use has removed environmental, economic and social problems, has stopped national and international concerns, and has led to the more balanced distribution of incentives.
- The establishment of individual solar energy systems (−0.001) and energy price (−0.724) factors show a decline in an equilibrium state. Solar energy capacity reaches satisfying with high support and encouragement, so support for the establishment of new systems is either eliminated or reduced. Increasing energy demand with energy price can be balanced by increasing solar energy generation capacity or raising the renewable or non-renewable energy prices.

The solar energy price HFCM model is evaluated through two different scenarios that are developed according to the initial state values of the factors. Scenario models reach equilibrium state with different iteration numbers, and model factors take different equilibrium state values according to their initial state values. The solar energy price reaches a constant equilibrium value at the end of calculations made with different initial state vectors. The inertia state of the solar energy price causes the factors supporting and disturbing the solar energy system to go into inertia state. The continuation of the equilibrium energy model by energy supply, energy demand, and energy price factors shows that solar energy price is evaluated within renewable and non-renewable energy prices. The solar energy generation capacity increase occurs depending on energy demand in market conditions without any incentive mechanism.

## 10.6 Conclusion

In this chapter, the causal relationships among the solar-energy price factors are identified. The economic, environmental and social conditions and the uncertainties are considered in the model. This HFCM model is used for identifying and assessing relationships among factors accurately. The causal relationships among the factors are determined by linguistic evaluation depending on the experts' knowledge, skills, and experience. Since the initial states of the factors cannot be measured, the initial state values of factors are determined by the knowledge of the experts. The linguistic expressions used in determining the causal relationships between the factors and the initial state value of the factors are transformed into the HFLTS and the trapezoidal fuzzy membership function. The weight matrix and the initial state vector are obtained by defuzzification of the TFMF and they are used to simulate the HFCM model. The simulated HFCM models result in different equilibrium state values for the factors. For example, the energy price reaches −0.862 in the first scenario and solar energy generation capacity converges 0.451 in the second scenario.

Similar results are obtained in the equilibrium states of the HFCM models that are developed at the solar energy price cantered. Although the initial state vectors are different, the solar energy price does not tend to increase or decrease at the end of the simulations. The inertia state of the solar energy price causes the factors directly affecting the solar energy system to go into an inertia state for each scenario. The energy model, which continues its existence by energy supply, energy demand, and energy price factors, accommodates solar energy price in renewable and non-renewable energy prices. Also, equilibrium situations show that the development of new solar power generation systems takes place in market conditions without any incentive mechanism.

In the future studies, the applied HFCM model-based pricing studies can be extended for other renewable energy sources since the renewable energy sector has similar pricing structure to solar energy. Thus, renewable energy price prediction models can be improved by considering the factors affecting renewable energy price and their impact levels on pricing. As a result, more accurate energy price estimation can be made with a more realistic price estimates of renewable energies, which will have significant shares in meeting future energy demand.

## References

Atanassov, K. T. (1986). Intuitionistic fuzzy sets. *Fuzzy Sets and Systems, 20*(1), 87–96.
BP. (2017). Statistical Review of World Energy June 2017; 66 [cited 2017 10 December]. Available from: https://www.bp.com/content/dam/bp/en/corporate/pdf/energy-economics/statistical-review-2017/bp-statistical-review-of-world-energy-2017-full-report.pdf.
Bloomberg New Energy Finance. (2015). New Energy Outlook [cited 2017 10 December]. Available from: https://data.bloomberglp.com/bnef/sites/4/2015/06/BNEF-NEO2015_Executive-summary.pdf.
Bordogna, G., & Pasi, G. (1993). A fuzzy linguistic approach generalizing boolean information retrieval: A model and its evaluation. *Journal of the American Society for Information Science, 44*(2), 70.
Bueno, S., & Salmeron, J. L. (2009). Benchmarking main activation functions in fuzzy cognitive maps. *Expert Systems with Applications, 36*(3), 5221–5229.
Çoban, V., & Onar, S. Ç. (2017). Modelling solar energy usage with fuzzy cognitive maps. In *Intelligence Systems in Environmental Management: Theory and Applications* (pp. 159–187). Springer.
Conkling, R. L. (2011). *Energy pricing: Economics and principles*. Berlin: Springer Science & Business Media.
Crawley, G. M. (2016). *Solar energy*. Hackensack, NJ: World Scientific Publishing Co., Pte. Ltd.
Delgado, M., et al. (1998). Combining numerical and linguistic information in group decision making. *Information Sciences, 107*(1–4), 177–194.
Filev, D., & Yager, R. R. (1998). On the issue of obtaining OWA operator weights. *Fuzzy Sets and Systems, 94*(2), 157–169.
International Energy Agency (IEA). (2016). Trends 2016 in Photovoltaic Applications [cited 2017 22 December]. Available from: http://iea-pvps.org/fileadmin/dam/public/report/national/Trends_2016_-_mr.pdf.
Jaimol, T., Ashok, S., & Jose, T. (2011). *Hybrid pricing strategy for solar energy. Proc. Sustainable Energy and Intelligent Systems (SEISCON)*, Chennai, July 2011, pp. 121–125.
Kahraman, C., Kaymak, U., & Yazici, A. (2016). Fuzzy logic in its 50th year: New developments, directions and challenges (Vol. 341). Berlin: Springer.
Kosko, B. (1986). Fuzzy cognitive maps. *International Journal of Man-Machine Studies, 24*(1), 65–75.
Kosko, B. (1997). *Fuzzy engineering*. New Jersey: Prentice Hall.
Liu, H., & Rodríguez, R. M. (2014). A fuzzy envelope for hesitant fuzzy linguistic term set and its application to multicriteria decision making. *Information Sciences, 258*, 220–238.
MDAI. (2014). M.D.f.A.I. A few aggregation operators 2014. Available from: http://www.mdai.cat/ifao/operadors/index.php?llengua=en.
Michael, G. (2010). *Fuzzy cognitive maps: Advances in theory, methodologies, tools and applications. Studies in fuzzines and soft computing* (Vol. 10, pp. 978–3). Berlin: Springer.
Ministry. (2017). E.a.N.R Renewable Energy Resources Support Mechanism (YEKDEM) [cited 2017 20 November]. Available from: http://www.eie.gov.tr/yenilenebilir/YEKDEM.aspx.
Mir-Artigues, P., & Del Río, P. (2016). *The Economics and Policy of Solar Photovoltaic Generation*. Cham: Springer.
Mizumoto, M., & Tanaka, K. (1976). Some properties of fuzzy sets of type 2. *Information and Control, 31*(4), 312–340.
Papageorgiou, E. I. (2013). *Fuzzy cognitive maps for applied sciences and engineering: from fundamentals to extensions and learning algorithms* (Vol. 54). Heidelberg: Springer Science & Business Media.
PV-Magazine. (2017). Turkey adds 553 MW of solar in H1 2017. [cited 2017 10 November]. Available from: https://www.pv-magazine.com/2017/09/11/turkey-adds-553-mw-of-solar-in-h1-2017/.

Rodriguez, R. M., Martinez, L., & Herrera, F. (2012). Hesitant fuzzy linguistic term sets for decision making. *IEEE Transactions on Fuzzy Systems, 20*(1), 109–119.

Ruspini, E. H. (1969). A new approach to clustering. *Information and Control, 15*(1), 22–32.

Scientists. (2015). U.o.C Increase Renewable Energy [cited 2017 10 November]. Available from: http://www.ucsusa.org/clean_energy/smart-energy-solutions/increase-renewables/real-energy-solutions-the.html#.WhualTeZk2x.

Stern, N. (2015). *Why are we waiting? The logic, urgency, and promise of tackling climate change*. Cambridge, MA: Mit Press.

The National Academies of Sciences, E., and Medicine. (2009). G8+5 Academies' joint statement: Climate change and the transformation of energy technologies for a low carbon future [cited 2017 12 December]. Available from: http://www.nationalacademies.org/includes/G8+5energy-climate09.pdf..

Timilsina, G. R., Kurdgelashvili, L., & Narbel, P. A. (2012). Solar energy: Markets, economics and policies. *Renewable and Sustainable Energy Reviews, 16*(1), 449–465.

Torra, V. (2010). Hesitant fuzzy sets. *International Journal of Intelligent Systems, 25*(6), 529–539.

World Energy Council. (2016) World Energy Resources [cited 2017 10 December]. Available from: https://www.worldenergy.org/publications/2016/world-energy-resources-2016/.

Yager, R. R. (1988). On ordered weighted averaging aggregation operators in multicriteria decisionmaking. *IEEE Transactions on systems, Man, and Cybernetics, 18*(1), 183–190.

Yager, R. R. (1993). Aggregating fuzzy sets represented by belief structures. *Journal of Intelligent & Fuzzy Systems, 1*(3), 215–224.

Zadeh, L. A. (1975). The concept of a linguistic variable and its application to approximate reasoning—I. *Information Sciences, 8*(3), 199–249.

Zadeh, L. A. (1996). *Fuzzy sets, in fuzzy sets, fuzzy logic, and fuzzy systems: Selected papers by Lotfi A Zadeh* (pp. 394–432). River Edge, NJ: World Scientific.

Zatzman, G. M. (2012). *Sustainable energy pricing: Nature, sustainable engineering, and the science of energy pricing*. Hoboken, NJ: Wiley.

# Chapter 11
# Strategic Renewable Energy Source Selection for Turkey with Hesitant Fuzzy MCDM Method

**Gülçin Büyüközkan, Yağmur Karabulut and Merve Güler**

**Abstract** Renewable energy sources (RES) strengthen their hold on emerging economies. Record numbers of newly installed RES capacity are being observed in recent years. In 2016, the addition of renewable resources were more than 60% of new capacity investments globally, surpassing fossil fuel-based investments. The majority of these additions take place in developing countries, indicating the vital importance of selecting the best RES technologies for Turkey, an emerging economy. RES is not only becoming less expensive, they also contribute to employment and environmental protection. Selecting the most appropriate RES strategy among alternatives involves many criteria. This chapter introduces a novel RES evaluation model that can guide investors in identifying the most suitable RES strategy from a sustainability perspective. Complex socio-economic decision problems often make it more difficult for Decision Makers to consider different aspects, and to provide exact numerical values. Considering many, usually conflicting sustainability factors that affect this selection process, the chapter proposes a Multi-Criteria Decision-Making (MCDM) model by implementing hesitant fuzzy linguistic term sets (HFLTS) for an effective RES strategy evaluation problem. Group Decision Making (GDM) is also integrated to the method, as it is capable to offset individual DMs' bias and partiality. HFLTS enables DMs to accurately provide their linguistic expressions. An integrated HFL SAW method (Simple Additive Weighting) and HFL TOPSIS method (Technique for Order Performance by Similarity to Ideal Solution) are employed for this purpose. The criteria priorities are determined with the HFL SAW method and the final RES strategy ranking results are determined with HFL TOPSIS method. The plausibility of the proposed framework is tested in a case study. This combination of MCDM techniques is applied for the first time in the literature for dealing with this problem setting.

---

G. Büyüközkan (✉) · M. Güler
Department of Industrial Engineering, Galatasaray University,
Çırağan Caddesi No: 36, Ortaköy Istanbul, 34357 Istanbul, Turkey
e-mail: gulcin.buyukozkan@gmail.com

Y. Karabulut
Mavi Consultants, Altunizade Mah. Kisikli Caddesi
No: 28 Avrupa Is Merkezi K.1/2, 34662 Uskudar/Istanbul, Turkey

© Springer International Publishing AG, part of Springer Nature 2018

C. Kahraman and G. Kayakutlu (eds.), *Energy Management—Collective and Computational Intelligence with Theory and Applications*, Studies in Systems, Decision and Control 149, https://doi.org/10.1007/978-3-319-75690-5_11

## 11.1 Introduction

Today, around 1 billion people globally have no access to electricity. Providing these people, and the other parts of the world, with clean, affordable and sustainable electricity still remains a challenge today. Despite many challenges, Renewable Energy Sources (RER) have become the strategic first choice of investors in recent years. In 2016 alone, renewables accounted for more than 60% of new capacity additions globally. Most of this addition came from solar PV for the first time, which accounts for about 47% of new renewable power capacity additions in 2016, while wind and hydropower contributed 34 and 15.5%, respectively (REN 21 2017). This sustained growth and geographical expansion can be mostly attributed to the continued decline of installation costs, particularly for wind and solar PV, as wells as continually increasing power demand in developing countries and governmental support mechanisms. Innovations in solar PV manufacturing and installation techniques, as well as cell and module efficiency and performance, are major causes for this wide adoption. Similarly, recent improvements in wind turbine materials, design, operation, and maintenance lead to lower operational costs and higher energy generation for the same wind turbine capacity. New advances in power grids are able to support more RES plants. Improvements in the production of advanced biofuels are also observed.

Today, the world is adding renewable power capacity at unprecedented rates, it even surpasses all fossil fuels combined (International Energy Agency 2015). Some mature RES options, such as hydropower and geothermal energy, are already competitive in terms of costs with thermal power plants run with fossil fuels. Solar PV and wind power are converging to these well-known and established power sources due to recent technological developments. Moreover, the flexibility of capacity, ease of deployment in remote areas and low maintenance requirements increasingly favor such newer RES technologies. Distributed, off-grid RES projects in rural areas present strategic sustainable alternatives over conventional power plants not only thanks to their competitive capital investment and low maintenance costs but also their environmental benefits and new job creation opportunities locally. The development of these community RES investments continued in 2016. Moreover, these emerging RES alternatives bring about significant employment opportunities, technology transfer, local economic activity, lower greenhouse gas emissions, less environmental footprint and many other co-benefits.

This trend is especially true for developing countries. In 2016, most of new RES capacity installations took place in developing countries. For the first time, developing economies overtook the level of RES investment of developed countries in 2015. Although developed countries took the lead back the next year, developing countries are becoming a significant market for the RES industry due to natural potential and willingness of investors. This is also true for Turkey, an emerging economy. Turkey takes steps to minimize its dependency on unsustainable fossil fuels and to reduce pollution caused by power generation (Büyüközkan and Güleryüz 2017). Together with Indonesia, Turkey is leading the world in new geothermal power installations by adding 10 new geothermal plants in 2016 to its existing cohort of 10 plants. For the wind industry, Turkey had a record year in 2016 as well. It added ca. 1.4 GW new wind power capacity and ranked among the top 10 countries globally (REN 21 2017). This also reflected in employment numbers in this industry. As of 2016, more than 1 million people around the world are employed in businesses related to wind power. More than half a million of this employment takes place in China, followed by Germany, the United States, India, and Turkey. In the face of high energy prices, global warming, lack of decent employment opportunities, ecologic deterioration and development priorities, the selection of the most sustainable RES strategy is becoming a key decision problem in Turkey that can ensure environmental protection, lower pollution, and new jobs. These developments lead to higher interest by investors, who seek to strategically balance profits, good governance, community dialogue, environmental integrity and compliance with national policies at the same time.

Low-cost and environmentally friendly energy supply is a pre-requisite for a sustainable power supply. There exist many RES strategies, each having their advantages and disadvantages. Investors, as Decision Makers (DMs) of RES projects, are therefore faced with a multitude of factors that shall be considered to come to a thorough decision. As the number of RES options expand, this decision process also becomes more complex for DMs. The long-term value of this RES strategy selection problem necessitates powerful decision support systems to aid DMs in determining which RES is the best by considering qualitative and quantitative sustainability aspects.

Decision-making activities aim to select the best from two or more of alternatives. Deciding on a suitable RES strategy is a complex process, and can be overwhelming for DMs in the presence of many decision factors, if not treated with proper methods. Traditional single-criterion decision-making approaches are unable to cope with such complex systems, as the problem involves the assessment of many criteria which shall be assisted by DMs (Taha and Daim 2013; Ishizaka and Nemery 2013; Kahraman et al. 2015). To address this need, the literature offers to treat it as a Multi-Criteria Decision-Making (MCDM) problem (Iskin et al. 2012; Kabak and Dağdeviren 2014; Pak et al. 2015; Şengül et al. 2015; Ishizaka et al. 2016). MCDM methods can solve various energy management and planning problems, especially complex issues that feature low certainty, conflicting goals, multiple interests and differing points of view. They provide researchers with many effective tools that can be used individually or in combination for reaching the intended results. In MCDM, criteria and alternatives should be determined at the beginning and evaluated one by one by DMs in a particular way.

MCDM processes can be improved with Group Decision Making (GDM) approaches by involving several DMs at once that possess different notions and ideas. Each DM can approach the decision problem from different angles, and their collective assessments can be integrated into the procedure. Furthermore, there are many MCDM techniques offered in the literature. While DMs evaluate the alternatives, they might be guided by their personal feelings, uncertainty, and hesitancy in their opinions. To add to these challenges, DMs can have difficulty in expressing their assessment numerically, especially for qualitative criteria.

This chapter presents an integrated MCDM model that addresses these complications. This approach consists of HFL SAW (Simple Additive Weighting) method and HFL TOPSIS method (Technique for Order Performance by Similarity to Ideal Solution). HFL SAW method is applied for determining the criteria weights, while HFL TOPSIS is employed for obtaining the final RES strategy rankings. The alternatives are ranked according to their proximity to the positive ideal solution and negative ideal solution (Chen et al. 1992). This approach can sort and select the best RES from a number of alternatives by comparing their sustainability performance. This chapter discusses this new approach, which integrates SAW and TOPSIS under a hesitant fuzzy environment with GDM. It differentiates from the literature by using HFL SAW and HFL TOPSIS with GDM approach for the RES strategy selection problem with technical, social, environmental and economic aspects in a developing country setting.

The chapter continues with Sect. 11.2 to give a snapshot of the state of the art. Then, Sect. 11.3 will follow, where the methods are described in detail. Section 11.4 demonstrates the proposed method's application on a case study from Turkey, while Sect. 11.5 summarizes the results and concludes this chapter.

## 11.2 Literature Review

There is extensive research in the literature that deploy MCDM tools, e.g. AHP, TOPSIS, DEMATEL, ELECTRE, PROMETHEE, and VIKOR, as well as fuzzy logic and GDM applications, to deal with RES strategy selection problems. In this field, Kumar et al. (2017) recently reviewed the literature on MCDM applications for sustainable RES strategy selection and provided a good overview of the state of the art. Suganthi et al. (2015) reviewed the literature on the fuzzy logic application in RES problems and found that fuzzy-based MCDM methods are applied for RES site assessment, strategy selection, and optimization of conflicting criteria, among others. Considering the multitude of research, readers are kindly referred to these articles.

Among the publications that use MCDM methods for selecting RES strategies with a specific focus on Turkey, Önüt et al. (2008) applied ANP to assess RES strategies for the Turkish manufacturing industry. Kahraman et al. (2009) deployed a fuzzy AHP approach for selecting the most suitable renewable energy strategy for Turkey and came to the conclusion that wind energy generates the best effects. In a study by Kaya and Kahraman (2011), a new fuzzy TOPSIS technique is presented for energy planning. Kabak and Dağdeviren (2014) employed a hybrid ANP model to consider the benefits, opportunities, costs, and risks of RES strategies in Turkey. Büyüközkan and Güleryüz (2014) constructed an evaluation method to rank alternative strategies for RES. In another paper, Erdogan and Kaya (2015) first deployed fuzzy AHP using interval type-2 fuzzy sets to calculate the priorities of the evaluation criteria. Then, they fuzzified the TOPSIS method by interval type-2 fuzzy sets to put strategic alternatives into order. Şengül et al. (2015) utilized fuzzy TOPSIS technique to rank RES strategies for Turkey. A similar goal was pursued by Büyüközkan and Güleryüz (2016), who combined DEMATEL with ANP for identifying the best RES option in Turkey from an investor point of view. Recently, Büyüközkan and Karabulut (2017) came up with an evaluation method fusing AHP with VIKOR for selecting energy projects from a sustainability point of view, and Büyüközkan and Güleryüz (2017) applied linguistic interval fuzzy preferences with DEMATEL, ANP, and TOPSIS to pinpoint the most appropriate energy strategy for Turkey. The same objective was explored by Çolak and Kaya (2017), who merged AHP based on interval type-2 fuzzy sets with hesitant fuzzy TOPSIS methods, as well as Balin and Baraçli (2017), who integrated fuzzy AHP-based type-2 fuzzy sets with interval type-2 TOPSIS method.

Publications that use the techniques proposed in this chapter are also reviewed. In literature, those papers that integrate HFLTS and MCDM are dispersed to a few fields, such as finance, technology, and management. The integrated use of HFLTS and MCDM tools began in 2013 with the studies of Zhang and Beg. Zhang and Wei (2013) developed the HFL VIKOR technique, an effective MCDM method for determining the best compromise solution by collecting linguistic expressions. They also compared this method to HFL TOPSIS. In another article, Beg and Rashid (2013) proposed Hesitant Fuzzy Linguistic TOPSIS for aggregating the opinions of experts and DMs on various criteria by GDM. Senvar et al. (2016) applied Hesitant Fuzzy TOPSIS to pinpoint to the best hospital site. Zhang et al. (2015) applied Hesitant Fuzzy TOPSIS and linear programming for selecting the best supplier. Onar et al. (2016) employed Hesitant Fuzzy Linguistic AHP, Hesitant Fuzzy Linguistic TOPSIS, and QFD methods and explored the applicability and effectiveness of their approach by a case study. Zhou et al. (2016) proposed Hesitant TOPSIS and Hesitant TODIM and combined it with linguistic hesitant fuzzy sets (LHFS) with the evidential reasoning (ER) approach. Since it is a very new combined method, HFL TOPSIS's applications are limited. One example is by Cevik Onar et al. (2014), who developed a Hesitant Fuzzy TOPSIS model that considers the complexity and imprecision of strategic decisions and presented a case study for an electronics company. Büyüközkan and Güler (2017) integrated HFLTS, OWA operator and TOPSIS method for evaluating smart glasses alternatives.

Chou et al. (2008) used SAW method in fuzzy environment. However, the integrated use of HFLTS and SAW method is a research gap in the literature. Therefore, this is the first publication in the literature that integrates HFLTS, SAW and TOPSIS methods in the field of energy in general, and for RES strategy selection more specifically. Furthermore, HFLTS, SAW and TOPSIS methods are not operated before with GDM in any publication, marking another scientific contribution of this chapter.

## 11.3 Proposed RES Strategy Selection Model

RES can be defined as energy sources that are continually replenished by nature, such as the solar radiation, wind, water and geothermal heat. These resources do not originate from fossil fuels, have lower emissions, are renewed in continuous cycles and are available in nature to utilize (Şengül et al. 2015). The most important RES strategies for Turkey are wind, solar PV, biogas, geothermal and hydro energy (Büyüközkan and Güleryüz 2017).

The RES model introduced in this chapter is based on a set of evaluation criteria and an integrated MCDM method for processing the criteria evaluations of the DMs. MCDM allows DMs to have a systematic overview of the decision problem so that the problem can be investigated and scaled according to specific needs (Işıklar and Büyüközkan 2007). The proposed RES strategy selection model is based on the criteria introduced in Sect. 11.3.1.1, and on a combination of MCDM techniques. In this approach, MCDM methods will be deployed in a certain order and pre-defined setting. This approach applies HFL SAW and HFL TOPSIS techniques in a GDM environment. HFL SAW is put into use for finding the weights of the evaluation criteria, and HFL TOPSIS is used for ranking the energy strategy alternatives in an optimal manner. This algorithm can be described with the following phases:

1. **Problem definition**: Initially, the goal of the decision problem is determined. Then, the DMs, who will be involved in the process, are chosen. Next, available RES strategies to be considered are defined. At the final stage of this first phase, the evaluation criteria are established.
2. **Criteria weights**: In this second phase, HFL SAW will be applied. First, linguistic opinions of DMs are gathered for each criterion about their perceived impact on the group decision. Based on these data, the criteria decision matrix is constructed. Then, linguistic data are transformed into HFLTS, which are then converted to trapezoidal fuzzy numbers (TFNs). The aggregated fuzzy weights are calculated, and criteria weights are eventually found by de-fuzzifying and normalizing them.
3. **Ranking of alternatives**: In this third phase, HFL TOPSIS will be deployed. First, linguistic opinions of DMs are gathered for each alternative, according to each criterion about how well the alternatives fare. Once the evaluation alternative matrix is constructed, the linguistic judgment matrix is converted into the HFLTS judgment matrix. The standardized decision matrix and weighted standardized decision matrix are constructed. Then, the positive and negative ideal solutions are determined and the distance between alternatives are computed using a special distance measure named as Hamming distance. The proximity coefficients of the alternatives are calculated, which are then ranked according to their proximity coefficients.

The research methodology of the study is provided in Fig. 11.1.

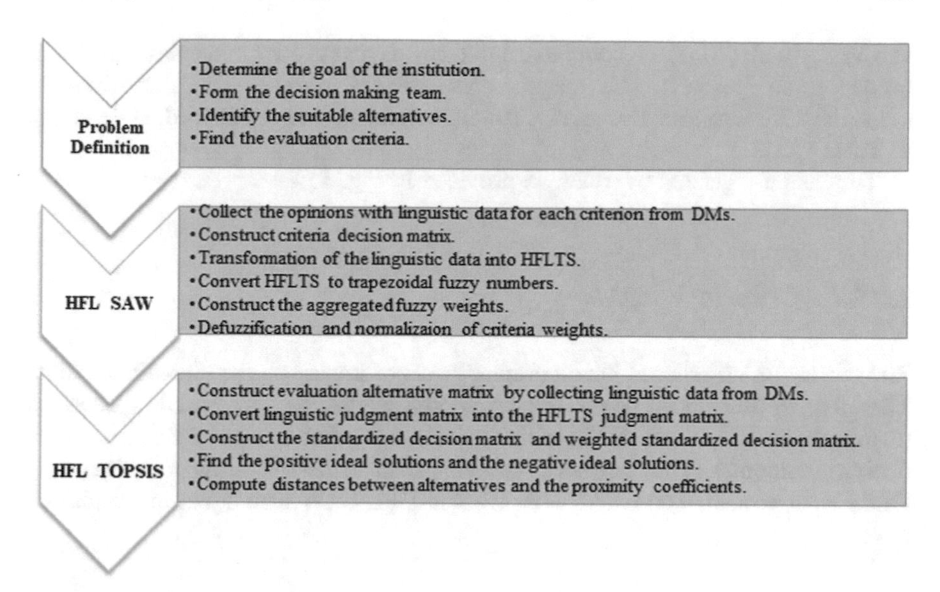


**Fig. 11.1** Research methodology of the proposed model

## *11.3.1 Problem Definition*

In this first phase of the model, the goal is determined as selecting the most appropriate RES strategy by taking various aspects, including sustainability-related factors, into account from an investor point of view. This decision will be taken with the support of industry experts. While RES strategies can be expanded according to local circumstances and availability of natural resources, usually these options include wind, solar PV, biogas, hydropower and geothermal alternatives.

### 11.3.1.1 Evaluation Criteria

Evaluation criteria are identified based on a detailed literature survey of existing models and consultations with three experts from the energy industry.

Compared to conventional energy strategies, RES offers many economic, social and environmental benefits. Each type of RES has its own attributes, as benefits or harms, that make it uniquely suitable for the specific use (Kabak and Dağdeviren 2014). Certainly, the identification of suitable criteria is one of the most important prerequisites for DMs (Pak et al. 2015). For this model, evaluation criteria from the literature, mostly from Ishizaka et al. (2016), Taha and Daim (2013), Kahraman et al. (2015), and Wang et al. (2009) are compiled, and then adapted to RES strategies with DMs' guidance and feedback. This chapter thus provides a novel criteria structure for assessing RES strategies. Eventually, 10 selection criteria are determined, as described as Table 11.1.

The model's general overview is provided in Fig. 11.2.

## *11.3.2 Criteria Weights*

The evaluation criteria can have different levels of impact on the ultimate decision. Therefore, in this second phase of the proposed model, the selected DMs are asked to provide their opinions about which criteria has what level of influence on the decision outcomes. This process is accomplished with HFL SAW technique in a GDM environment. The GDM, HFLTS, and HFL SAW techniques are explained next.

### 11.3.2.1 Group Decision Making

RES strategies are inherently subject to different opinions and views. This subjectivity embedded in human judgments can lead to biased perception in individual decisions, even for experts, as an expert might not always have the necessary knowledge about the problem. Different DMs can provide different points of view (Pohekar and Ramachandran 2004). Depending on a single DM, therefore, poses subjectivity risks due to limited experiences and personal preferences. These risks can effectively be reduced by including more than one DM in the process. A GDM process involves two or more industry specialists, who understand the common problem and have a common interest in reaching a collective decision (Herrera et al. 1995). Therefore, GDM is often superior for evading the prejudice and subjectivity of individual DMs.

### 11.3.2.2 Hesitant Fuzzy Linguistic Term Sets

In decision-making processes, experts are usually inclined to express their judgments with words, which correspond to imprecise, and unquantifiable ratings, since it might be difficult for DMs to precisely estimate their preference degrees numerically. Values of linguistic information can include words, phrases or sentences instead of numbers (Tapia Garcı´a et al. 2012). Linguistic assessment tend to be more flexible, practical and suitable for the real world (Rodríguez et al. 2013). These linguistic judgments can be taken into account with the fuzzy set theory, developed by Zadeh (1965), to deal with the uncertainty and vagueness.

**Table 11.1** Evaluation criteria of the proposed model

| Criterion | | Explanation |
|---|---|---|
| $C_1$ | Investment and O&M Costs | Investment costs represent those expenditures that occur at the beginning for establishing the energy strategy alternative. Operation and maintenance (O&M) costs refer to production costs that are associated with running a power plant |
| $C_2$ | Price tariff and incentives | RES strategies are often supported with attractive legal and financial mechanisms to stabilize cash inflows for investors and reduce various costs and red tape. This criterion affects the return on investment and the economic success of the strategy |
| $C_3$ | Maturity and serviceability | Maturity is related to technological penetration, availability of and maintenance knowhow and services, familiarity of investors and suppliers and technical development for reliable operation. Serviceability becomes especially important for remote RES installations, where a breakdown may not be fixed locally |
| $C_4$ | Grid connectivity | RES strategies are often halted due to unavailable capacity at local power grids. Many RES strategies are not able to carry the base load in a grid, therefore additional transformer capacity can be needed to connect renewables. The lack of such capacity can delay, or prevent, the realization of RES strategies |
| $C_5$ | Greenhouse gas emissions | RES strategies reduce greenhouse gas emissions indirectly by substituting electricity in the grid generated with fossil fuels. However, the manufacturing of RES equipment (steel, silicon wafers, concrete etc.) has a carbon footprint, as does the operation (e.g. hydropower plants with shallow reservoirs, or geothermal power plants). Therefore, the emission reductions shall incorporate a Life Cycle Analysis (LCA) approach |
| $C_6$ | Land use and ecologic footprint | Most RES strategies are bound to specific geographies and locations, so that their impact on their immediate surroundings can vary according to the regional ecologic sensitivity. This impact is amplified, as the physical size of the RES facility increases (such as solar PV covering large areas, or high wind turbines in bird migration routes) |
| $C_7$ | Job creation | Creation of decent, full-time, diverse, and permanent employment opportunities for local communities is a central priority for sustainable development |
| $C_8$ | Social acceptability | Many energy facilities are subject to opposition by residents for a new development because it is close to them, which can be due to environmental pollution, poor air quality, increased traffic, visual beauty, sharing of limited local resources etc. These challenges shall be overcome by enhanced dialogue, voluntary actions and social responsibility |
| $C_9$ | Supply security | Energy supply in a grid is expected to be resilient to international political developments, price volatility of fuels and market shortages. The supply of natural resources is prone to such shocks, but can be sensitive to ecologic and climatic variations |
| $C_{10}$ | Policy compatibility | The RER strategy shall be in line with national energy policies, compatible with regional priorities and relevant legislation. These policies can aim to improve international competitiveness, technology transfer, trade balance, job creation and environmental protection, among others |

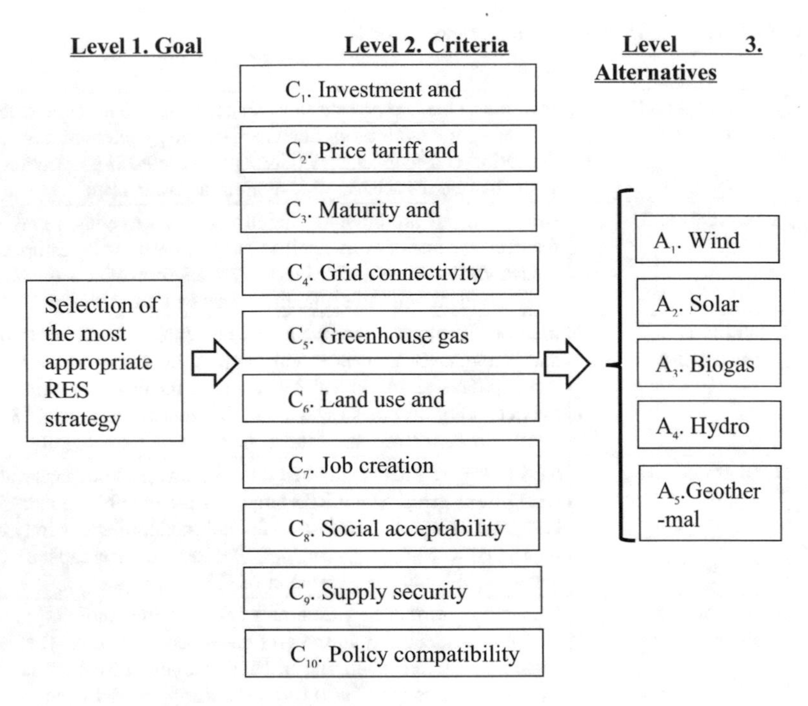


**Fig. 11.2** The overall structure of the proposed model

However, DMs also can find it difficult to identify the best fitting linguistic term for voicing their opinions in. Hesitant Fuzzy Sets (HFS), which constitute the extension of classical fuzzy sets, prove helpful in such settings.

Extending the classical fuzzy set theory to the HFS method is first developed by Torra and Narukawa (2009). It defines the degree of adhesion of an element with a set of possible values between 0 and 1. This method is useful when DMs hesitate in expressing a certain evaluation. It is based on the following definitions:

**Definition 1** Let X be a universal set. HFS over X is defined as ssa function that will render a subset of [0, 1] when applied to X, which is defined as the following (Torra 2010):

$$E = \{\langle x, h_E(x) \rangle x \in X \tag{11.1}$$

In this definition, $H$ is the set of all Hesitant Fuzzy Element (HFE), with HFEs $h_E(x)$ between [0, 1]. Possible degrees of adhesion of the element $x \in X$ to the set $E$ are specified.

**Definition 2** X is defined as a reference set. HFS over *X* is a function *h* which assigns values between [0, 1]:

$$h\colon X \to \{[0,1]\} \tag{11.2}$$

Then, an HFS is represented with the union of their membership functions.

**Definition 3** $M = \{\mu 1, \mu 2, \ldots, \mu n\}$ is defined as a set of membership functions n. HFS is linked to *M*. Here,, $h_M$ gives values between 0 and 1:

$$h_M : M \to \{[0,1]\} \tag{11.3}$$

$$h_M(x) = U_{\mu \in M}\{\mu(x)\} \tag{11.4}$$

**Definition 4** The lower and upper boundaries of *h*, an HFS, are defined as (Torra 2010):

$$h^-(x) = \min\ h(x) \tag{11.5}$$

$$h^+(x) = \max\ h(x) \tag{11.6}$$

**Definition 5** When *h* is defined as an HFS, its envelope $A_{env}(h)$ is defined as:

$$A_{env(h)} = \{x,\ \mu_A(x), v_A(x)\} \tag{11.7}$$

Where $A_{env(h)}$ is an intuitionistic fuzzy set of *h*. Accordingly, $\mu$ and $v$ are represented as:

$$\mu_A(x) = h^-(x) \tag{11.8}$$

$$v_A(x) = 1 - h^+(x) \tag{11.9}$$

Rodriguez et al. (2012) developed an MCDM method, where DMs voice their evaluations with linguistic expressions as HFLTS.

**Definition 6** $S = \{s_0, \ldots, s_g\}$ is defined as a set of linguistic expressions. An HFLTS, $H_s$, is an ordered finite subset of the consecutive linguistic elements of *S*, which can also be shown as a subscript-symmetric linguistic term set as $S = \{s_i | i = -\tau, \ldots, -1, 0, 1, \ldots, \tau\}$.

**Definition 7:** HFLTS's upper and lower bounds, $H_s$, $H_{s+}$ and $H_{s-}$ respectively, are formulated as:

$$H_{s+} = \max\ (s_i) = s_j, s_i \in H_S \text{ et } s_i \le s_j \,\forall\, i \tag{11.10}$$

$$H_{s-} = \min(s_i) = s_j, s_i \in H_S \text{ et } s_i \le s_j \,\forall\, i \tag{11.11}$$

**Definition 8** $E_{GH}$ is defined as a function which transforms linguistic expressions into HFLTS, $H_S$. Then, $G_H$ is defined as an out-of-context grammar that utilizes the linguistic term set in *S*. $S_{ll}$ is defined as the expression domain generated by $G_H$. This mapping can be represented as:

$$E_{GH}: S_{ll} \rightarrow H_s \tag{11.12}$$

Comparative linguistic expressions are converted into HFLTS with the following formulae;

$$E_{GH}(s_i) = \{s_i | s_i \in S\} \tag{11.13}$$

$$E_{GH}(\text{at most } s_i) = \{s_j | s_j \in S \text{ et } s_j \leq s_i\} \tag{11.14}$$

$$E_{GH}(\text{lower than } s_i) = \{s_j | s_j \in S \text{ et } s_j < s_i\} \tag{11.15}$$

$$E_{GH}(\text{at least } s_i) = \{s_j | s_j \in S \text{ et } s_j \geq s_i\} \tag{11.16}$$

$$E_{GH}(\text{greater than } s_i) = \{s_j | s_j \in S \text{ et } s_j > s_i\} \tag{11.17}$$

$$E_{GH}\left(\text{between } s_i \text{ and } s_j\right) = \{s_k | s_k \in S \text{ et } s_i \leq s_k \leq s_j\} \tag{11.18}$$

**Definition 9** When $H_s$ is defined as an HFLTS, based on $H_{s+}$ and $H_{s-}$ as introduced in *Definition* 7, its envelope *env*($H_s$) is shown as:

$$\text{env}(H_S) = [H_{s-}, H_{s+}], \quad H_{s-} \leq H_{s+} \tag{11.19}$$

#### 11.3.2.3 HFL SAW Method

Simple Additive Weighting (SAW) method is developed by Hwang and Yoon (1981). Still, it is counted among the most popular MCDM techniques thanks to its simplicity. It is based on a simple aggregation concept that is useful for positive values only. This makes it mandatory to transform negative criteria into positive values first with a normalization process. As an extension of SAW, Chou et al. (2008) introduced the combined Fuzzy Simple Additive Weighting (FSAW) technique as a way to approach decision problems with fuzzy aspects. Similarly, the SAW method is combined with HFLTS in this chapter, the steps of which are explained next in consecutive steps (Chou et al. 2008).

**Step 1**. DMs voice their opinions, in words, about the importance of the evaluation criteria. These opinions are expressed with a context-free grammar, as shown in *Definition* 6 and Table 11.2.

**Table 11.2** Linguistic terms for HFL SAW (Chou et al. 2008)

| Linguistic term | $S_i$ | Abb. | Fuzzy numbers |
|---|---|---|---|
| Very low | $s_{-2}$ | VL | (0, 0, 0, 3) |
| Low | $s_{-1}$ | L | (0, 3, 3, 5) |
| Medium | $s_0$ | M | (2, 5, 5, 8) |
| High | $s_1$ | H | (5, 7, 7, 10) |
| Very high | $s_2$ | VH | (7, 10, 10, 10) |

**Step 2**. Linguistic judgment matrix is transformed into HFLTS judgment matrix on the basis of the scale provided in Table 11.2 by using the conversion function $E_{GH}$, as in *Definition* 8.

**Step 3**. The alternatives are formulated as $A_i = \{a_1, a_2, \ldots, a_I\}$ with $I$ members. The evaluation criteria are represented as $C_j = \{c_1, c_2, \ldots, c_J\}$ with J members. The decision committee is formulated as $D_t = \{d_1, d_2, \ldots, d_k\}$ with $k$ DMs. The DMs do not necessary possess equal say on the decision, and $I_t$ delineates the degree of importance of each DM, with $0 \leq I_t \leq 1$, $t = 1, 2, \ldots, k$, and $\sum_{t=1}^{k} I_t = 1$, $\tilde{\omega}_t$ being the fuzzy weight of the DMs. $I_t$ is found as:

$$I_t = \frac{d(\tilde{w}_t)}{\sum_{t=1}^{k} d(\tilde{w}_t)}, \quad t = 1, 2, \ldots, k \tag{11.20}$$

Here, $d(\tilde{w}_t)$ stands for the de-fuzzified value of the fuzzy weight according to its signed distance.

**Step 4**. Aggregated fuzzy weights of the evaluation criteria $C_j, \tilde{w}_j = (a_j, b_j, c_j, d_j)$, are computed:

$$\widetilde{W}_j = (I_1 \otimes \widetilde{W}_{j1}) \oplus (I_2 \otimes \widetilde{W}_{j2} \oplus \ldots \oplus (I_k \otimes \widetilde{W}_{k1}) \tag{11.21}$$

Here, $a_j = \sum_{t=1}^{k} I_t a_{jt}$, $b_j = \sum_{t=1}^{k} I_t b_{jt}$, $c_j = \sum_{t=1}^{k} I_t c_{jt}$, $d_j = \sum_{t=1}^{k} I_t d_{jt}$.

**Step 5**. Criteria's fuzzy weights are de-fuzzified. The de-fuzzified $\widetilde{W_j}$, shown as $d(\widetilde{W_j})$, is calculated as:

$$d\left(\widetilde{W_j}\right) = \frac{1}{4}(a_j + b_j + c_j + d_j), \quad \text{where } j = 1, 2, \ldots, n \tag{11.22}$$

**Step 6**. Normalized weight of the criteria $C_j$, shown as $W_j$, is calculated as:

$$W_j = \frac{d(\tilde{w}_j)}{\sum_{j=1}^{n} d(\tilde{w}_j)}, \quad j = 1, 2, \ldots, n \tag{11.23}$$

Here, the normalized weights add to 1, i.e. $\sum_{j=1}^{n} W_j = 1$. Eventually, the weight vector $W = (W_1, W_2, \ldots, W_n)$ is established.

## 11.3.3 Ranking of Alternatives

After criteria weights are known, DMs are asked to rate the RES strategy alternatives according to the evaluation criteria, one by one. This 3rd phase is guided by HFL TOPSIS technique, again in a GDM environment with consensus process. HFLTS and GDM approach are explained in Sect. 11.3.2. Therefore, the algorithmic steps of HFL TOPSIS are described next.

### 11.3.3.1 HFL TOPSIS Method

Technique for Order Preference by Similarity to an Ideal Solution (TOPSIS) method is presented by Chen and Hwang (1992). It is based on the concept that the chosen alternative should have the smallest geometric distance from the positive ideal solution (PIS) and the largest geometric distance from the negative ideal solution (NIS).

Cevik Onar et al. (2014) came up with a Hesitant Fuzzy TOPSIS model that considers the complexity and imprecision of strategic decisions and presented a case study for an electronics company.

The steps of HFL TOPSIS method are:

**Step 1**. DMs express their opinions by using linguistic expressions about criteria. The linguistic expression is voiced by the DM based on a context-free grammar, as shown in *Definition 6*.

**Step 2**. The linguistic judgment matrix is converted to the HFLTS judgment matrix with the help of the transformation function $E_{GH}$ as given in *Definition 8*. Table 11.3 shows the scale used in HFL TOPSIS method.

**Table 11.3** Linguistic terms for HFL TOPSIS (Beg and Rashid 2013)

| Linguistic term | $S_i$ | Abb. | Fuzzy numbers |
|---|---|---|---|
| None | $s_{-3}$ | N | (0, 0, 0.17) |
| Very bad | $s_{-2}$ | VB | (0, 0.17, 0.33) |
| Bad | $s_{-1}$ | B | (0.17, 0.33, 0.5) |
| Medium | $s_0$ | M | (0.33,0.5,0.67) |
| Good | $s_1$ | G | (0.5, 0.67, 0.83) |
| Very good | $s_2$ | VG | (0.67, 0.83, 1) |
| Perfect | $s_3$ | P | (0.83, 1, 1) |

**Step 3**. The positive and negative ideal solutions are determined as:

$$A^* = \{h_1^*, h_2^*, \ldots, h_n^*\} \tag{11.24}$$

where $h_j^* = \cup_{i=1}^m h_{ij} = \cup_{\gamma_{1j} \in h_{1j}, \ldots, \gamma_{mj} \in h_{mj}} \max\{\gamma_{1j}, \ldots, \gamma_{mj}\} \quad j = 1, 2, \ldots, n$

$$A^- = \{h_1^-, h_2^-, \ldots, h_n^-\} \tag{11.25}$$

where $h_j^* = \cap_{i=1}^m h_{ij} = \cap_{\gamma_{1j} \in h_{1j}, \ldots, \gamma_{mj} \in h_{mj}} \min\{\gamma_{1j}, \ldots, \gamma_{mj}\} \quad j = 1, 2, \ldots, n$

**Step 4**. Separation measures of each alternative from the ideal solution are calculated. As the separation measure, the weighted hesitant normalized Hamming distance is applied. The proximity of an alternative to the positive ideal is calculated as:

$$D_i^+ = \sum_{j=1}^{n} w_j \left\| h_{ij} - h_j^* \right\| \tag{11.26}$$

where $w_j$ is the weight of the jth criterion determined by hesitant AHP. The distance from the negative ideal solution is given as:

$$D_i^- = \sum_{j=1}^{n} w_j \left\| h_{ij} - h_j^- \right\| \tag{11.27}$$

The distance between two hesitant fuzzy numbers is found as:

$$\|h_1 - h_2\| = \frac{1}{l} \sum_{j=1}^{l} w_j \left| h_{1\sigma(j)} - h_{2\sigma(j)} \right| \tag{11.28}$$

**Step 5**. The relative proximity to the ideal solution is found as:

$$C_i = \frac{D_i^-}{D_i^+ + D_i^-} \tag{11.29}$$

**Step 6**. The alternatives are ranked in increasing order, based on their relative closeness index. The alternative that has the highest value is determined to be the best alternative.

With this step, the ranking of RES strategies is accomplished.

## 11.4 Case Study

The proposed model is applied on a case study, in which a number of RES strategies from Turkey are assessed and then ranked. The most important alternative strategies for Turkey, i.e. wind, solar PV, biogas, hydro, and geothermal, are chosen for this comparison.

Energy demand in Turkey, electricity consumption, in particular, grows at high rates since decades, requiring continuous new capacity additions. The rapidly increasing electricity need is covered by installing large fossil fuel-powered power plants, mostly coal and natural gas. Due to their environmental impacts, such as greenhouse gas emissions and pollution, as well as social impacts, such as local acceptability, renewables remain top on the energy agenda of Turkey, which has abundant natural resources and willingness of investors. While RES strategies, in general, are considered to be a priority, investors find it difficult to select which RES strategy to prioritize in their investment decisions.

The integrated MCDM model presented previously is applied for finding the most suitable RES strategy by first forming a decision committee with 3 industry experts. These experts support the process of defining the criteria set, weighing these criteria, and rating the alternatives. All three DMs have sufficient knowledge about energy strategies and are adequately qualified for this evaluation.

## *11.4.1 Application of the Proposed Model*

The criteria are introduced next. C1 is Investment and O&M cost, C2 is Price tariff and incentives, C3 is Maturity and serviceability, C4 is Grid connectivity, C5 is LCA greenhouse gas emissions, C6 is Land use and ecologic footprint, C7 is Job creation, C8 is Social acceptability, C9 is Supply security and C10 is Policy compatibility.

There are five possible alternatives: A1 is Wind, A2 is Solar, A3 is Biogas, A4 is Hydro and A5 is Geothermal.

### 11.4.1.1 Criteria Weight Calculation with HFL SAW Method

In the first stage, DMs evaluated the criteria by using linguistic term sets given in Table 11.2. Table 11.4 shows the assessments of DMs.

**Table 11.4** DMs evaluation about criteria

| Criteria | DM1 | DM2 | DM3 |
|---|---|---|---|
| C1 | Between H and VH | At least VH | At least H |
| C2 | Between H and VH | Between L and H | Between L and H |
| C3 | At most VL | At most VL | Between L and H |
| C4 | Between L and H | Between L and H | At most VL |
| C5 | At most VL | between VL and L | between L and H |
| C6 | At most VL | Between VL and L | Between L and H |
| C7 | Between VL and L | between VL and L | between VL and L |
| C8 | Between VL and L | Between VL and L | Between VL and L |
| C9 | At most VL | Between VL and L | Between L and H |
| C10 | Between VL and L | Between VL and L | Between VL and L |

Table 11.5 Criteria weights

| Criteria | Defuzzified value | Normalized value | Ranking |
|---|---|---|---|
| C1 | 7.917 | 0.215 | 1 |
| C2 | 5.750 | 0.156 | 2 |
| C3 | 2.833 | 0.077 | 4 |
| C4 | 3.583 | 0.097 | 3 |
| C5 | 2.833 | 0.077 | 4 |
| C6 | 2.833 | 0.077 | 4 |
| C7 | 2.750 | 0.075 | 8 |
| C8 | 2.750 | 0.075 | 8 |
| C9 | 2.833 | 0.077 | 4 |
| C10 | 2.750 | 0.075 | 8 |
| Total | 36.833 | | |

Based on these assessments in Table 11.4, the linguistic expressions are converted into HFLTS by using (11.13)–(11.18). The HFLTS are converted into fuzzy numbers by using the scale given in Table 11.1. Based on these numbers, the fuzzy weights of individual criteria are calculated by (11.21). The de-fuzzified values of the aggregated fuzzy weights are computed using (11.22) and the normalized weights of criteria are found using (11.23). Table 11.5 depicts the criteria weights.

The most important criterion is Investment, O&M cost (C1), and the second important criterion is Price tariff and incentives (C2).

#### 11.4.1.2 Ranking E-Health Technology Alternatives with HFL TOPSIS Method

Initially, the DMs evaluated the alternatives with regard to criteria via comparative linguistic expressions and the linguistic scale given in Table 11.3.

In the initial phase, the DMs reached consensus by using Delphi Method and a series of questionnaires (Hsu and Sandford 2007; Marchais-Roubelat and Roubelat 2011). The consensus evaluation with linguistic expressions is listed in Table 11.6.

Linguistic expressions are converted into HFLTS by using Eqs. (11.13)–(11.18). The positive ideal and the negative ideal solution are found with Eqs. (11.24) and (11.25). The Hamming distances are calculated by using Eqs. (11.26) and (11.27). The distance between two hesitant fuzzy numbers is found with Eq. (11.28). Finally, the proximity to the ideal solution is found with Eq. (11.29). Table 11.7 shows the results of HFL TOPSIS methodology and ranking of alternatives.

The results about alternatives give an idea to find the best alternative. As a result, Hydro (A4) is the most desirable energy alternative through these alternatives, with the nearest competitor Geothermal (A5). Solar (A2) has become the third, and the fourth one is Wind (A1), as depicted in Table 11.6.

**Table 11.6** DMs evaluation about alternatives

| Ai | C1 | C2 | C3 | C4 | C5 |
|---|---|---|---|---|---|
| A1 | Between VB and M | Between VB and M | Between B and G | At most VB | Between B and G |
| A2 | Between VB and M | Between M and VG | Between VB and M | Between VB and M | Between B and G |
| A3 | Between VB and M | Between M and VG | Between VB and M | Between B and G | At least VG |
| A4 | At least VG | At most VB | Between M and VG | Between M and VG | Between B and G |
| A5 | Between B and G | At least VG | Between B and G | Between M and VG | At most N |
| Ai | C6 | C7 | C8 | C9 | C10 |
| A1 | Between B and G | Between B and G | At most VB | Between VB and M | Between VB and M |
| A2 | At most N | At most VB | At least VG | At least VG | Between VB and M |
| A3 | At least VG | Between M and VG | Between M and VG | Between VB and M | Between M and VG |
| A4 | At most VB | Between B and G | At most N | Between VB and M | Between B and G |
| A5 | At least VG | Between B and G | At least VG | Between B and G | At least VG |

**Table 11.7** Ranking of alternatives

| Ai | Di+ | Di− | Ci | Ranking |
|---|---|---|---|---|
| A1 | 0.403 | 0.440 | 0.522 | 4 |
| A2 | 0.354 | 0.489 | 0.580 | 3 |
| A3 | 0.412 | 0.431 | 0.511 | 5 |
| A4 | 0.312 | 0.540 | 0.634 | 1 |
| A5 | 0.360 | 0.535 | 0.598 | 2 |

The main ranking list of alternatives is:
A4 ≻ A5 ≻ A2 ≻ A1 ≻ A3

## 11.5 Conclusion

The main objective of this chapter is to identify the most applicable RES strategy with a sustainability point of view and developing country perspective. This decision-making process is governed by a set of evaluation factors that are assessed by a decision committee. In such complex problems with conflicting criteria, uncertainty, and vagueness MCDM methods can prove very useful. For this reason, this decision-making problem is approached by proposing a new set of criteria and integrating it with MCDM methods in a GDM setting. The proposed model is based on 10 criteria, the weights of which are determined with HFL SAW method. The results are then fed into the HFL TOPSIS to find the ranking of selected RES strategies. The combined method offers superior solutions, as it is able to successfully capture DMs' opinions.

The plausibility and practical usefulness of the proposed model are shown in a case study from Turkey. The case study revealed Hydro to be the best RES strategy for Turkey, followed by Geothermal and Solar. These findings can be associated with legal difficulties for getting permits for wind farms in Turkey in recent years, as well as the economic performance of hydro energy plants. Investors can benefit from these results by applying similar practices in comparing different RES strategies available to them.

Individually, HFL SAW and HFL TOPSIS techniques are recent and novel methods. In the literature, publications applying these methods are very few. Using these methods together with GDM, therefore, presents a scientific contribution. Therefore, this model is unique in its application of HFL SAW and HFL TOPSIS in combination in a GDM setting for the RES strategy selection problem. It not only contributes to the RES strategy evaluation literature by developing a new evaluation model, it also provides a case study to illustrate how the proposed method can be utilized to solve real problems. The introduction of a new criteria set, adapted to developing economies, adds to its research value. The proposed model can be applied in other developing countries as well by re-weighing the criteria and assessing different alternative RES strategies with other experts.

The proposed model also has some limitations. One of these limitations is its focus on developing countries when it comes to selecting evaluation criteria, which can show differences from a developed country perspective. Future research therefore can consider the adaptation of these criteria to other circumstances and geographies. Moreover, the criteria set consists of one level, with no hierarchical structure. In the future, the criteria set can be extended. In terms of MCDM methods, future research can also use other similar techniques, such as HFL VIKOR, instead of HFL TOPSIS, and compare the findings.

**Acknowledgements** The authors express their sincere thanks and gratitude to the industry experts for their invaluable feedback and support in the evaluations. This research was supported by Galatasaray University Research Fund (Projects number: 17.402.004 and 17.402.009).

## References

Balin, A., & Baraçli, H. (2017). A fuzzy multi-criteria decision making methodology based upon the interval Type-2 fuzzy sets for evaluating renewable energy alternatives in Turkey. *Technological and Economic Development of Economy, 23,* 742–763. https://doi.org/10.3846/20294913.2015.1056276.

Beg, I., & Rashid, T. (2013). TOPSIS for hesitant fuzzy linguistic term sets. *International Journal of Intelligent Systems, 28,* 1162–1171. https://doi.org/10.1002/int.21623.

Büyüközkan, G., & Güler, M. (2017). Hesitant fuzzy linguistic VIKOR method for e-health technology selection (in Press).
Büyüközkan, G., & Güleryüz, S. (2017). Evaluation of Renewable Energy Resources in Turkey using an integrated MCDM approach with linguistic interval fuzzy preference relations. *Energy, 123,* 149–163. https://doi.org/10.1016/j.energy.2017.01.137.
Büyüközkan, G., & Güleryüz, S. (2014). A new GDM based AHP framework with linguistic interval fuzzy preference relations for renewable energy planning. *Journal of Intelligent & Fuzzy Systems, 27,* 3181–3195.
Büyüközkan, G., & Güleryüz, S. (2016). An integrated DEMATEL-ANP approach for renewable energy resources selection in Turkey. *International Journal of Production Economics, 182,* 435–448. https://doi.org/10.1016/j.ijpe.2016.09.015.
Büyüközkan, G., & Karabulut, Y. (2017). Energy project performance evaluation with sustainability perspective. *Energy, 119,* 549–560. https://doi.org/10.1016/j.energy.2016.12.087.
Cevik Onar, S., Oztaysi, B., & Kahraman, C. (2014). Strategic decision selection using hesitant fuzzy TOPSIS and interval type-2 fuzzy AHP: A case study. *International Journal of Computational intelligence systems, 7,* 1002–1021.
Chen, S.-J. J., Hwang, C.-L., Beckmann, M. J., & Krelle, W. (1992). *Fuzzy multiple attribute decision making: Methods and applications*. New York Inc: Springer.
Chou, S.-Y., Chang, Y.-H., & Shen, C.-Y. (2008). A fuzzy simple additive weighting system under group decision-making for facility location selection with objective/subjective attributes. *European Journal of Operational Research, 189,* 132–145. https://doi.org/10.1016/j.ejor.2007.05.006.
Çolak, M., & Kaya, İ. (2017). Prioritization of renewable energy alternatives by using an integrated fuzzy MCDM model: A real case application for Turkey. *Renewable and Sustainable Energy Reviews, 80,* 840–853. https://doi.org/10.1016/j.rser.2017.05.194.
Erdogan, M., & Kaya, I. (2015). An integrated multi-criteria decision-making methodology based on type-2 fuzzy sets for selection among energy alternatives in Turkey. *Iranian Journal of Fuzzy Systems, 12,* 1–25.
Herrera, F., Herrera-Viedma, E., & Verdegay, J. L. (1995). A sequential selection process in group decision making with a linguistic assessment approach. *Information Sciences, 85,* 223–239.
Hsu, C.-C., & Sandford, B. A. (2007). The Delphi technique: Making sense of consensus. *Practical assessment, research & evaluation, 12,* 1–8.
Hwang, C., & Yoon, K. (1981). Multiple attribute decision making. *Lecture notes in Economics and Mathematical Systems*. http://www.doi.org/10.1007/978-3-642-48318-9.
International Energy Agency (Ed.). (2015). *World energy outlook 2016*. Paris: OECD.
Ishizaka, A., Nemery, P. (2013). *Multi-criteria decision analysis: Methods and software*. Wiley.
Ishizaka, A., Siraj, S., & Nemery, P. (2016). Which energy mix for the UK (United Kingdom)? An evolutive descriptive mapping with the integrated GAIA (graphical analysis for interactive aid)–AHP (analytic hierarchy process) visualization tool. *Energy, 95,* 602–611. https://doi.org/10.1016/j.energy.2015.12.009.
Işıklar, G., & Büyüközkan, G. (2007). Using a multi-criteria decision making approach to evaluate mobile phone alternatives. *Computer Standards & Interfaces, 29,* 265–274. https://doi.org/10.1016/j.csi.2006.05.002.
Iskin, I., Daim, T., Kayakutlu, G., & Altuntas, M. (2012). Exploring renewable energy pricing with analytic network process—Comparing a developed and a developing economy. *Energy Economics, 34,* 882–891. https://doi.org/10.1016/j.eneco.2012.04.005.
Kabak, M., & Dağdeviren, M. (2014). Prioritization of renewable energy sources for Turkey by using a hybrid MCDM methodology. *Energy Conversion and Management, 79,* 25–33. https://doi.org/10.1016/j.enconman.2013.11.036.
Kahraman, C., Kaya, İ., & Cebi, S. (2009). A comparative analysis for multiattribute selection among renewable energy alternatives using fuzzy axiomatic design and fuzzy analytic hierarchy process. *Energy, 34,* 1603–1616. https://doi.org/10.1016/j.energy.2009.07.008.

Kahraman, C., Onar, S. C., & Oztaysi, B. (2015). Fuzzy multicriteria decision-making: A literature review. *International Journal of Computational Intelligence Systems, 8,* 637–666. https://doi.org/10.1080/18756891.2015.1046325.

Kaya, T., & Kahraman, C. (2011). Multicriteria decision making in energy planning using a modified fuzzy TOPSIS methodology. *Expert Systems with Applications, 38,* 6577–6585. https://doi.org/10.1016/j.eswa.2010.11.081.

Kumar, A., Sah, B., Singh, A. R., et al. (2017). A review of multi criteria decision making (MCDM) towards sustainable renewable energy development. *Renewable and Sustainable Energy Reviews, 69,* 596–609. https://doi.org/10.1016/j.rser.2016.11.191.

Marchais-Roubelat, A., & Roubelat, F. (2011). The Delphi method as a ritual: Inquiring the Delphic Oracle. *Technological Forecasting and Social Change, 78,* 1491–1499.

Onar, S. Ç., Büyüközkan, G., Öztayşi, B., & Kahraman, C. (2016). A new hesitant fuzzy QFD approach: An application to computer workstation selection. *Applied Soft Computing, 46,* 1–16.

Önüt, S., Tuzkaya, U. R., & Saadet, N. (2008). Multiple criteria evaluation of current energy resources for Turkish manufacturing industry. *Energy Conversion and Management, 49,* 1480–1492. https://doi.org/10.1016/j.enconman.2007.12.026.

Pak, B. K., Albayrak, Y. E., & Erensal, Y. C. (2015). Renewable energy perspective for Turkey using sustainability indicators. *International Journal of Computational Intelligence Systems, 8,* 187–197. https://doi.org/10.1080/18756891.2014.963987.

Pohekar, S. D., & Ramachandran, M. (2004). Application of multi-criteria decision making to sustainable energy planning—A review. *Renewable and Sustainable Energy Reviews, 8,* 365–381. https://doi.org/10.1016/j.rser.2003.12.007.

REN 21. (2017). Renewables 2017 Global Status Report. REN 21, Paris, France.

Rodriguez, R. M., Martinez, L., & Herrera, F. (2012). Hesitant fuzzy linguistic term sets for decision making. *IEEE Transactions on Fuzzy Systems, 20,* 109–119. https://doi.org/10.1109/tfuzz.2011.2170076.

Rodríguez, R. M., Martínez, L., & Herrera, F. (2013). A group decision making model dealing with comparative linguistic expressions based on hesitant fuzzy linguistic term sets. *Information Sciences, 241*, 28–42. https://doi.org/10.1016/j.ins.2013.04.006.

Şengül, Ü., Eren, M., Eslamian Shiraz, S., et al. (2015). Fuzzy TOPSIS method for ranking renewable energy supply systems in Turkey. *Renewable Energy, 75,* 617–625. https://doi.org/10.1016/j.renene.2014.10.045.

Senvar, O., Otay, I., & Bolturk, E. (2016). Hospital site selection via hesitant fuzzy TOPSIS. *IFAC-PapersOnLine, 49,* 1140–1145. https://doi.org/10.1016/j.ifacol.2016.07.656.

Suganthi, L., Iniyan, S., & Samuel, A. A. (2015). Applications of fuzzy logic in renewable energy systems—A review. *Renewable and Sustainable Energy Reviews, 48,* 585–607. https://doi.org/10.1016/j.rser.2015.04.037.

Taha, R. A., & Daim, T. (2013). Multi-criteria applications in renewable energy analysis, a literature review. In T. Daim, T. Oliver & J. Kim (Eds.), *Research and technology management in the electricity industry* (pp. 17–30). London: Springer.

Tapia Garcı´a, J. M., del Moral, M. J., Martínez, M. A., Herrera-Viedma, E. (2012). A consensus model for group decision making problems with linguistic interval fuzzy preference relations. *Expert Systems with Applications, 39*, 10022–10030. https://doi.org/10.1016/j.eswa.2012.02.008.

Torra, V. (2010). Hesitant fuzzy sets. *International Journal of Intelligent Systems, 25,* 529–539. https://doi.org/10.1002/int.20418.

Torra, V., Narukawa, Y. (2009). On hesitant fuzzy sets and decision. In: *2009 IEEE International Conference on Fuzzy Systems*, pp. 1378–1382.

Wang, J.-J., Jing, Y.-Y., Zhang, C.-F., & Zhao, J.-H. (2009). Review on multi-criteria decision analysis aid in sustainable energy decision-making. *Renewable and Sustainable Energy Reviews, 13,* 2263–2278. https://doi.org/10.1016/j.rser.2009.06.021.

Zadeh, L. A. (1965). Fuzzy sets. *Information and Control, 8,* 338–353.

Zhang, N., & Wei, G. (2013). Extension of VIKOR method for decision making problem based on hesitant fuzzy set. *Applied Mathematical Modelling, 37,* 4938–4947. https://doi.org/10.1016/j.apm.2012.10.002.

Zhang, Y., Xie, A., & Wu, Y. (2015). A hesitant fuzzy multiple attribute decision making method based on linear programming and TOPSIS**This work was supported by the Specialized Research Fund for the Doctoral Program of Higher Education under Project No. 20130009120040. *IFAC-PapersOnLine, 48,* 427–431. https://doi.org/10.1016/j.ifacol.2015.12.165.

Zhou, H., Wang, J., Zhang, H., & Chen, X. (2016). Linguistic hesitant fuzzy multi-criteria decision-making method based on evidential reasoning. *International Journal of Systems Science, 47,* 314–327.

# Chapter 12
# Modeling and Solution Approaches for Crude Oil Scheduling in a Refinery

**Antonios Fragkogios and Georgios K. D. Saharidis**

**Abstract** One of the most critical activities in a refinery is the scheduling of loading and unloading of crude oil. Better analysis of this activity gives rise to better use of a system's resources, decrease losses, increase security as well as control of the entire supply chain. It is important that the crude oil is loaded and unloaded contiguously in storage tanks, primarily for security reasons (e.g. possibility of system failures) but also to reduce the setup costs incurred when flow between a dock/ports and a tank and/or between a tank and a crude distillation unit is reinitialized. The aim of this book chapter is to present a review on modeling and solution approaches in refinery industry. Mathematical programming modeling approaches are presented as well as exact, heuristic and hybrid solution approaches, widely applicable to most refineries where several modes of blending and several recipe preparation alternatives are used.

## 12.1 Introduction

In the late 1800s and early 1900s, when the large oil deposits were discovered, the second Industrial Revolution took place. Since then, oil has been being used, in almost all human activities, from transportation (cars, airplanes, ships etc.) to heating and from road building to goods production. However, does anyone wonder what path does crude oil follow from the earth deposits to our car reservoir or to the street we step on? After its mining, crude oil is transferred to a refinery mainly by ship, train or pipeline. The refinery is the main "factory" that transforms crude oil to its various products. The procedure is called distillation.

---

A. Fragkogios · G. K. D. Saharidis (✉)
Department of Mechanical Engineering, Polytechnic School,
University of Thessaly, Volos, Greece
e-mail: saharidis@gmail.com

A. Fragkogios
e-mail: fragkogiosantonios@gmail.com

© Springer International Publishing AG, part of Springer Nature 2018 
C. Kahraman and G. Kayakutlu (eds.), *Energy Management—Collective and Computational Intelligence with Theory and Applications*, Studies in Systems, Decision and Control 149, https://doi.org/10.1007/978-3-319-75690-5_12

The loading and unloading of crude oil in a refinery is a very complicated problem with a lot of restrictions concerning the system safety and the final product quality. The pipeline network needs to transfer crude oil from the docks to the storage tanks and from the tanks to the crude distillation units (CDUs). The capacity of the storage tanks, the delivery rate of crude oil by the ships in the docks, the rate of distillation of crude oil in the CDUs, the capacity and complexity of the pipeline network are some of the parameters that make the problem of loading and unloading crude oil in a refinery very difficult to solve.

In this chapter, a review will be made as to present the various developed methods for modelling and solving the problem of crude oil transferring inside a refinery. The next sections are organized as follows: In Sect. 12.2, the problem of crude oil scheduling is described and in Sect. 12.3, a literature review is made, mentioning the most significant studies that have been published to deal with this problem. Finally, Sect. 12.4 contains concluding remarks and future challenges.

## 12.2 Problem Description

The structure of a refinery may not be exactly the same in the various refineries built all over the world. However, all refineries follow a general structure that consists of the basic major parts. These parts are docks, pipelines, a series of tanks to store the crude oil (and prepare the different blends), Crude Distillation Units (CDUs) and/or Vacuum Distillation Units (VDUs), production units (such as reforming, cracking, alkylating and hydrotreating), blenders and tanks to store the raw materials and the final products. These major parts of a refinery are depicted in Fig. 12.1.

As presented by Shah et al. (2009), in general there are two decision levels in refinery process operations-the planning and the scheduling level. The planning level determines the volume of raw materials needed for the upcoming months (typically 12 months), and the type of final products and the estimated quantities to be ordered,

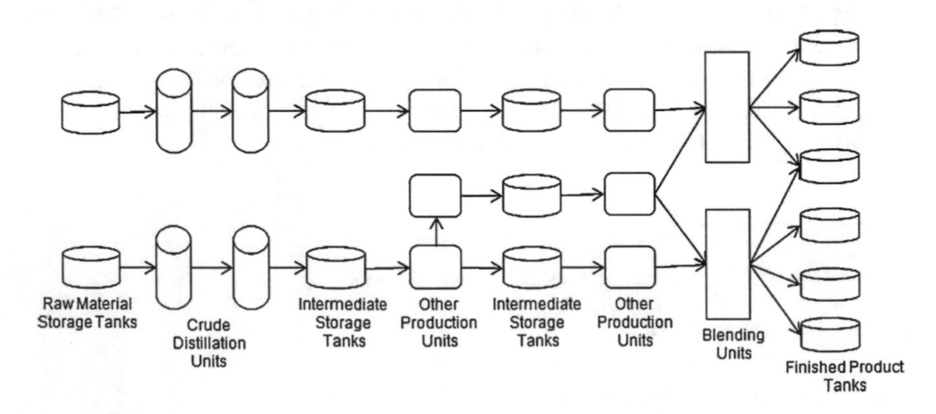


**Fig. 12.1** Graphic overview of the refinery production system (Shah et al. 2009)

depending on demand forecasts. After determining the yearly plan in the second level we have to determine the optimal production scheduling. The scheduling level determines the detailed schedule of each CDU and other production unit for a shorter period (typically 20 days) by taking into account the operational constraints of the system under study. Once the plan is known (the quantities and the types of final products ordered as well as the arrival of raw materials), managers must schedule the production of each unit based on the objective which usually is minimization of the overall makespan or maximization of the total profit. Each production unit is defined as a continuous processing element that transforms the input streams into several products according to the recipe.

The two decision levels exchange information and are strongly related with each other. However, they can be dealt with as two separate problems. One could argue that the first level of planning is the independent one, which basically takes information from the predictions of demand, and the second level of scheduling is the dependent one, which takes information from the first level of planning.

In most of the studies, the long term plan is assumed to be given and the objective is to define the optimal production scheduling. In such a case the key information available to the managers is the proportion of material produced or consumed at each production unit. Most of the times, these recipes are assumed fixed to maintain the model's linearity. The managers also know the minimum and maximum flow-rates for each production unit and the minimum and maximum inventory capacities for each storage tank. The different types of material that can be stored in each storage tank are known as well as the demand of final products at the end of time horizon.

Moreover, even the scheduling is divided into two main parts of the refinery. The first part is known as the front-end crude transfer, which contains the part of the refinery from the berth until the Crude Distillation Units. The front-end crude transfer part is depicted on Fig. 12.2. The second part of the refinery is the rest of the infrastructure after Crude Distillation Units until the finished product tanks. In the literature most of the studies deal with either with only the first or only the second part of refinery scheduling. However, some studies approach the refinery scheduling as a whole.

Focusing on the front-end crude transfer, once the quantities and the types of crude oil required are known, schedulers must schedule the loading and unloading of tanks (Fig. 12.2). The problem that arises then is how to schedule the transfer of crude oil from the docks to the tanks and from the tanks to the CDUs/VDUs, minimizing the setup cost of the system. More specifically, the scheduler should decide for every time period which tanks will be loaded with what type and quantity of crude oil from the dock(s) and which tanks will unload what type and quantity of the crude oil they have to which CDU/VDU in order to meet the demand of each CDU/VDU. There are several constraints that govern this procedure. Shah et al. (2009) introduce seven major groups of constraints which guarantee the operational conditions of the system. These are allocation constraints, production and storage capacity constraints, material balance constraints for units and storage tanks, demand constraints, and time sequence constraints.

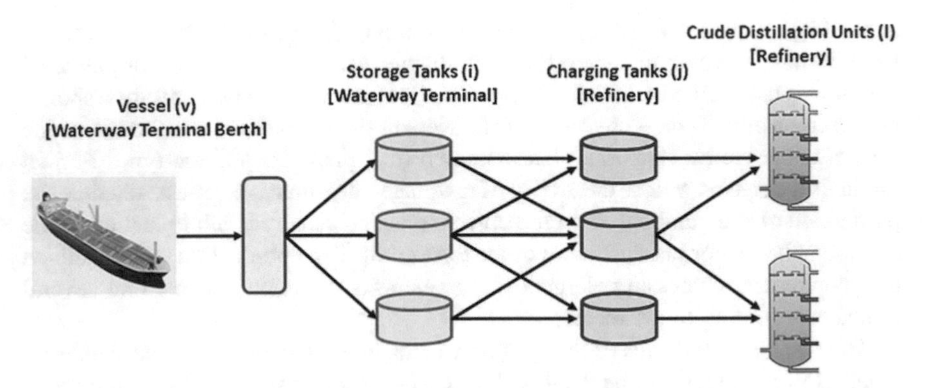


**Fig. 12.2** Typical configuration of the first major part of a refinery, where loading and unloading of crude oil storage and charging tanks takes place (Cortez and Pessoa 2016)

At this point, it should be noticed that there are different types of configurations that are associated with several modes of blending combined with the type of recipe preparation. As presented by Saharidis et al. (2009), in general, the different blends could be produced using the pipelines just before the CDU/VDU in a place called the manifold, where liquids circulate through a petroleum refinery and schedulers control the pumping systems, or in the tanks. For the first mode, the blending of different types of crude oil is made just before the distillation units through the use of pipelines, which meet in the manifold. It is a continuous process and in this case, only one type of crude oil can be stored in each tank at a time. The second mode is to prepare the blends required by the CDU/VDU in the tanks themselves. In this case, a quantity of a given type of crude oil is already loaded in a tank then stored and kept on standby until a quantity of another type of crude oil is unloaded into the same tank, in order to produce the required blend.

Moreover, there are two recipe preparation alternatives, the standard and the flexible recipe preparation. For the first alternative, the required blend must satisfy an exact composition of crude before being distilled in a CDU, regardless of blending mode, whereas in the second alternative of recipe preparation, the required blend must satisfy lower and upper bounds for each type of crude oil. The second alternative is a relaxation of the first one and is more commonly used in practice.

The goal of the above procedure is to result in a feasible schedule, taking into account all the constraints, which either minimizes the operational cost or maximizes the profit. The main cost of the process of loading and unloading the tanks is the setup cost incurred when flow between a dock and a tank or between a tank and a CDU/VDU is reconfigured. As presented by Saharidis and Ierapetritou (2009), the setup of tanks requires a series of operations, which are expensive for the refinery. The most critical and expensive operations associated with the tank's setup at each stage are as follows: before loading/unloading, (1) configuring the pipeline networks (e.g., opening of valves, configuration of pumps, etc.); (2) filling pipelines with crude oil; (3) sampling of crude oil for chemical analyses; (4) measuring of the crude oil stock in tank before loading/unloading; (5) starting the loading/unloading;

and (6) stopping the loading/unloading; and after loading/unloading, (1) configuring the pipeline networks (e.g., closing of valves, configuration and maintenance of pumps, etc.); (2) emptying pipelines; and (3) measuring the crude oil loaded/unloaded in the tanks.

Another important problem in many refineries is the storage capacity, which appears increasingly more often, and multiple uses of tanks become necessary. Consequently, reducing the number of tanks used in a scheduling period becomes critical. The reason is that the minimum number of setups is associated with the minimum number of tanks used for the scheduling, which is in turn associated with an increase in the number of tanks available for other uses (e.g., storage of the finished products or raw materials).

By minimizing the setups and the number of used tanks, the scheduler results in: (a) the optimal schedule and the extra production cost incurred in a case where fewer tanks are available to stock the crude oil and (b) the profit reduction or increase with the use of tanks in other operations. This information provides the flexibility to a system to change the use of tanks due to an unexpected event or changes in the market or the company's strategy.

Due to the above reasons, one could conclude that loading and unloading of crude oil into the tanks is one of the most critical activities in a refinery. Saharidis and Ierapetritou (2009) claim that better analysis of this activity gives rise to better use of a system's resources, as well as improved total visibility and control of production units and the entire supply chain. This is the reason why many researchers have focused their studies on this major part of a refinery. The need for the development of a systematic methodology and optimization tool for these activities is clearly justified. The potential financial and operational benefits associated with the development of an optimization tool are enormous, as an advanced optimization tool for scheduling could allow the refinery to minimize the flow problems and the loss of crude oil and to obtain the optimal periodic schedules of production.

The researchers have dealt with the problem with different ways of modelling and solving. Moreover, they study various configurations of the system by adding or removing several restrictions (blending inside the tanks or in a manifold tank) or by targeting to a different goal (minimizing cost, maximize profit etc.). In the next sections of the chapter, a literature review is made and various studies are presented with different approaches on the problem of crude oil scheduling.

## 12.3 Literature Review

Whereas the oldest continuously operated refinery in the United States of America has been in service for more than 130 years, it was not until the 1970s that the first studies appear for managing the crude oil pipelines. In 1974, Chaumeau and Vonner (1974) presented a heuristic arborescent procedure for the problems of automatic batch scheduling in a crude oil line and associated storage capacity control implemented at the particular case of the "Pipe-line de l'Ile-de-France". One year later, Speur et al. (1975) introduced a computerized advisory system for pipeline scheduling implemented on a crude-oil pipeline which stretches from the Netherlands to cities in Germany, which was a reliable and versatile tool, quicker than the manual scheduling and better than the latter in reducing pumping costs. However, these studies deal with the manipulation of crude oil pipelines and do not deal with crude oil scheduling in a refinery. It was not until the 1990s, when the first papers considering the refinery problem are published. Floudas and Lin (2004) present a very interesting and complete overview of the developments in the scheduling of multiproduct/multipurpose batch and continuous processes, whose subcategory is the crude oil scheduling. The authors discuss all continuous as well as discrete time formulations existing in the literature before 2004 and examine their strengths and limitations through computational studies. In the following section a detailed and updated overview of scheduling of crude oil is presented.

The developed methods for the resolution of this problem are classified into four general groups: the exact methods that use either discrete or continuous time representation, the heuristic methods and the hybrid methods. Figure 12.3 depicts the two different representations of time.

### *12.3.1 Methods Based on Discrete Time Representation*

Chronologically, the first approach for the problem of scheduling in a refinery introduces the use of Mathematical Programming for the modelling of the problem. One of the first published approaches, presented by Shah (1996), uses a discrete time formulation. The principle of discrete time representation is to split the scheduling time horizon into intervals of equal size and use binary variables to specify whether an action starts or finishes during an interval. The author presents a model based on discrete time representation for the Scheduling of Crude Oil (SCO) leading to the resolution of a mixed integer linear program (MILP). In this formulation, due to nonlinearity issues the problem was broken up into two sub-problems: the upstream, which considers the loading of crude oil from docks to

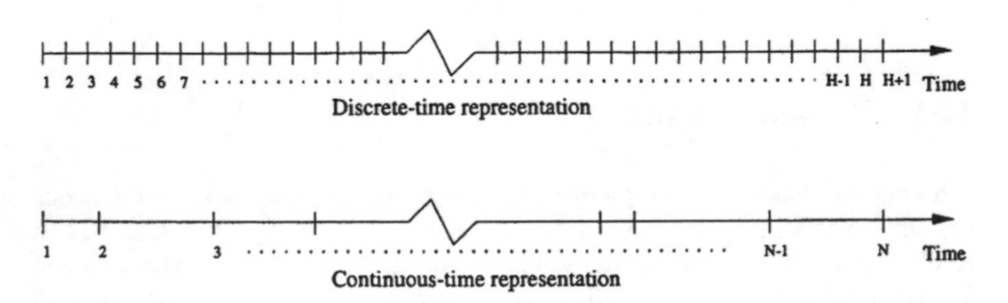


**Fig. 12.3** Discrete and continuous representations of time (Floudas and Lin 2004)

the storage tanks, and the downstream, which refers to the unloading from the tanks to the CDUs. This approach guarantees a feasible, but not optimal, solution for the system. Shah (1996) has considered a system where the available tanks can feed only one CDU at a time and where a CDU can be fed by only one tank at a time. Moreover, the author subdivides the scheduling time horizon into intervals of equal duration and each activity must start and finish within the boundaries of these intervals. The objective is to minimize the heel of crude oil left in a tank after its content has been transformed to CDU.

The problem of inventory management of a refinery that imports several types of crude oil which are delivered by different vessels is addressed by Lee et al. (1996). A mixed-integer optimization model is developed which relies on time discretization. The system studied is composed by two types of tanks (storage tanks and charging tanks) which are used to blend different types of crude oil. The obtained bi-linear term due to the mixing operations is replaced with individual component flows maintaining the linearity of the developed model. The objective is the minimization of the operation cost which includes unloading cost, cost for vessels waiting in the sea, inventory cost and change over cost.

Joly et al. (2002) and Pinto et al. (2000) dekveloped two models, the first for planning and the second for scheduling in a refinery. Concerning the latter, a MILP optimization model was developed for the loading and unloading of crude oil between the tankers and CDUs (the application considered concerns a refinery in Brazil), based on **both continuous and discrete time representations**. The presented model maximizes the operating profit by maximizing production while minimizing the number of tanks used for the loading and unloading of the crude oil, giving the optimal schedule for the following week. The schedule fulfills all the general operational rules but takes into account only one blending mode and one recipe preparation alternative. These modeling and solution strategies enable the scheduler to explore market opportunities, mainly in the short term or in any unexpected situation, but are susceptible to infeasible times to obtain global optimal solutions.

An extension of this model is presented in Neiro and Pinto (2004). The authors propose a complete model for the same refinery by taking into account all the different operational options possible in this refinery. Before the presentation of the case study, a (large scale) mixed integer nonlinear problem (MILNP) is proposed. The authors give a detailed description of the planning problem in this refinery. Several models were tested and presented before the development of the complete model for the entire system. The first model corresponds to the production units, the second corresponds to the operations taking place in tanks, and, finally, the third relates to crude oil, raw materials, and the final products transferred between tanks and production units. The complete model is presented after the illustration of these three models. In conclusion, the authors propose the linearization of nonlinear constraints by applying decomposition methods while at the same time testing several scenarios in parallel.

Furthermore, Persson and Gothe-Lundgren (2005) present a study, where shipment planning and production scheduling complement each other well. The shipment planning problem is the problem of simultaneously planning ship routes and planning the quantities to ship in order to satisfy the demand at the depots, i.e., the aggregated forecasted demand at the asphalt producers. At the same time, the shipment plan needs to be realizable in terms of production, and it is important to make a crosschecking between the shipment plan and the corresponding process schedules to make sure that the products can be produced in time. The time horizon used is equal to 1 month decomposed into intervals of equal duration (time discretization). The demand of final products is given and the objective is to minimize the total cost of production. The authors propose a solution strategy using Column Generation, Constraint Branching and Valid Inequalities. Moreover, the authors propose an extension to their model, in particular for the case where the arrival dates of tankers are not fixed but based on a prediction. The stochastic aspect is taken into account and the schedule is based on a time window in which the tankers expect to arrive at the refinery.

In Wu et al. (2008, 2009), the authors have introduced a two-layer hierarchical approach for short-term crude oil scheduling. At the upper level, a realizable refining schedule to optimize some objectives is reached and at the lower level, a detailed schedule is obtained to realize it. In these studies the lower level is solved from a control perspective and schedulability conditions are derived. Based on the schedulability conditions, given a realizable refining schedule, it is easy to find a detailed schedule at the lower level. With the schedulability conditions, Wu et al. (2012) solve the upper level problem of finding the optimal and realizable refining schedule using a novel method based on the results obtained at the lower level. Their method is a three-phase one. In phase 1, a linear problem is solved in order to determine the production rate to maximize the production. The amount of crude oil to be processed for each distiller during each bucket is, therefore, found. Then, in phase 2, a transportation problem is solved so as to assign the crude oil types and amount of crude oil to the distillers with a goal to minimize the crude oil type assignment cost. In phase 3, the result obtained in phase 2 is adjusted, and the crude oil parcels and their sequence are determined to minimize the changeover cost. By this three-phase approach, the authors decompose a hybrid optimization problem into sub-problems such that each sub-problem contains continuous or discrete event variables only but not both. In addition, multiple objectives are effectively handled in different phases. The computational results from the application of the approach on an industrial case study show that it is quite efficient. Extending their previous work, Wu et al. (2013) introduced a Hybrid Timed Petri Net Approach for modeling and scheduling of Crude Oil Operations in a refinery. The problem is studied in a control theory perspective and a two-level control architecture is presented. At the lower level, the model solves the schedulability and detailed scheduling problem in a hybrid control theory perspective. At the upper level, it solves a refining scheduling problem, a relative simple problem, with the schedulability conditions as constraints. Consequently, it results in a breakthrough solution such that the large practical application problem can be solved.

Another study applying discrete time formulation was proposed by Saharidis et al. (2009). The authors present a generic model for the scheduling of loading and unloading of tanks which is applicable for several modes of blending and for several alternatives of recipe preparation. A series of Valid Inequalities are introduced for each one of the different blending modes, which help to reduce the computational time needed for the solution of the model. The main contribution of the paper is a novel time formulation, called event-based time representation where the intervals are based on events instead of hours. Thus, instead of discretizing the time horizon into time intervals of the same duration, the time intervals are defined as a period when an event starts and finishes. The events which define the time intervals can be either, (1) boat arrivals, and/or (2) the change in blend constitution, as required by a CDU/VDU. With this discretization, the authors avoid the increased complexity of a larger number of variables and constraints, which is the case when the fixed time interval period is decreased, leading correspondingly to larger number of intervals in model formulation.

The same year, Saharidis and Ierapetritou (2009) published another exact solution approach based on a generic mixed integer model, which provides not only the optimal schedule of loading and unloading of crude oil minimizing the setup cost, but also the optimal type of mixture preparation. Comparison between an empirical approach applied in a real refinery and the developed exact approach is presented to justify the advantage of optimization tools. However, the high computational time needed to solve the proposed exact approach led the authors to the development of a series of valid inequalities based on data and operational rules in order to speed up the convergence of the resolution approach of the model. Concerning the mixture preparation, the two classical types are (a) the mixture preparation in the manifold just before the CDUs through the use of pipelines and use of mixers existing in the manifold and (b) the mixture preparation in the tanks through the use of the tanks mixers. The authors investigate the combination of these two types of mixture preparation as compared to each one separately. The comparative results depict the superiority of the optimal mixture preparation (combination of the two classical types of mixture) in terms of solution quality as compared to the two other approaches separately. Although the CPU time is longer for the combined case, because the model is nonlinear and has to be linearized increasing the number of variables and constraints, as compared to the other approaches, it does not exceed the time limits defined by the managers of the refinery due to the use of valid inequalities. The potential advantage of applying the combined approach is fewer setups of the refinery process, which justifies spending more time applying the combined method because the potential gain is significant.

Moreover, Shah et al. (2009) propose a structural decomposition scheme that generates smaller sub-systems that can be solved to global optimality. The original problem of refinery scheduling is decomposed at intermediate storage tanks such that inlet and outlet streams of the tank belong to the different sub-systems. Following the decomposition, each decentralized problem is solved to optimality and the solution to the original problem is obtained by integrating the optimal schedule of each sub-system. The decentralized system model results in fewer constraints and fewer continuous and binary variables compared to the centralized system. The paper presents a problem where both optimization strategies result in the same optimal makespan but the computational time for decentralized system is reduced significantly compared to that of centralized system.

Next year, Robertson et al. (2010) presented an integrated optimization approach for both production and scheduling of crude oil in a refinery. The production plan is modeled as a Non Linear Problem and includes manipulation of unit operating conditions in order to optimize the energy integration of the fractionating section of the refinery as well as minimize environmental cost such as the burning of high sulfur fuels in furnaces. The schedule is modeled as a Mixed Integer Linear Problem, based on the model introduced by Saharidis et al. (2009), and decides which tanks to store the incoming crude oil and which tanks should feed the refinery distillation units. The authors consider a time horizon of 15 days discretized into time intervals of 1 h. The nonlinear simulation model for the production process is used to derive individual crude costs using multiple linear regressions of the individual crude oil flow rates around the crude oil percentage range allowed by a specific production facility. These individual crude costs are then used to derive a linear cost function that is optimized in the MILP scheduling model, along with logistics costs. By applying this approach, Robertson et al. (2010) concluded, with just the minimization of setup costs, which happens when only scheduling is considered, the operational cost could be minimized further. Thus, their proposed method of minimizing the sum of both the refining (in production) and setup costs (in scheduling) gave the minimal total cost. The authors present their methodology more analytically in Robertson et al. (2011).

Saharidis et al. (2011) applied Benders decomposition on the SCO problem. More specifically, introduced a series of generic valid inequalities for the initialization of the Benders Master Problem for the fixed-charge network problem. As such a network, a case study of refinery system is addressed. The problem of loading and unloading the crude oil tanks is modeled as a Mixed Integer Linear Program, based on Saharidis et al. (2009) and is decomposed into a Master and a Slave model, based on Benders decomposition (Benders 1962). The methodology is quite efficient in solving the scheduling problem, while the added valid inequalities in the Master model result in significant reduce of computational time.

Zhang et al. (2012) proposed a general MINLP (mixed-integer nonlinear programming) model in order to address the profit optimization of crude-oil blending and purchase planning in refinery plants. The authors consider simultaneously both the short-time crude-oil blending and the long-time crude-oil purchase planning. Taking into account delivery delay uncertainties greatly increases the potential profitability and production flexibility of refineries. For this purpose, an inventory-related time flexibility index is created so as to characterize the ability of a refinery for handling the uncertainty of crude-oil delivery delays. The profit maximization and the production flexibility maximization are generally two contradictory aspects that should be well balanced. To quantify their relationship under crude-oil delivery uncertainty, a systematical study has been conducted. Meanwhile, the in-depth relations between the production flexibility and the plant profit are also disclosed. The applicability of the developed methodology is demonstrated by industrial case studies.

A study that was not directly applied on the SCO problem, but addresses the general chemical production scheduling was published by Velez and Maravelias (2013). The authors propose a discrete modeling of time by using a common uniform grid in the formulation of mixed-integer programming (MIP) models for scheduling, production planning, and operational supply chain planning problems. First, it is described that multiple grids can actually be employed in discrete-time models. Second, it is shown that not only unit-specific but also task-specific and material-specific grids can be generated. Third, the paper presents methods to systematically formulate discrete-time multi-grid models that allow different tasks, units, or materials to have their own time grid. Two different algorithms to find the grid are introduced. The first algorithm determines the largest grid spacing that will not eliminate the optimal solution. The second algorithm allows the user to adjust the level of approximation; more approximate grids may have worse solutions, but many fewer binary variables. Due to the fact that the proposed models have exactly the same types of constraints (unit-task assignment, material balance, batch capacity and storage capacity constraints) as models relying on a single uniform grid, the proposed models are proven to be tight and known solution methods can be employed. The proposed methods lead to substantial reductions in the size of the formulations and thus the computational requirements. The way to select the different time grids and state the formulation is described in the paper and computational results show that better solutions can be yield than formulations that use approximations.

Zimberg et al. (2015) proposed a MILP model for the problem of planning daily operations in a crude oil terminal, taking into the costs associated with deviations of total volume and quality with respect to required quantities, unfilled tanks during cargo unloading, quality adjustments, mixture of qualities and requirements to keep tanks empty or full during certain periods for maintenance reasons. Thus, constraints are imposed to ensure material balance and operating rules, inventory levels in the tanks are discretized and linear constraints with an adjustment term for composition discrepancies are formulated to force the concentration in a tank to be equal to the concentration of the outlet volume. The model consists of a discrete-time formulation where each period corresponds to 1 day of operations at the terminal. In order to achieve good results in an affordable time, the rolling horizon strategy (RHS) is applied to determine the optimal schedule of crude oil operations over a time horizon. For a real-world problem, a solution was reached by the RHS in less than 5 min. An extension of this model was made by de Assis et al. (2017). The authors propose an iterative two-step MILP-NLP algorithm based on piecewise McCormick relaxation and a domain-reduction strategy for handling bilinear terms. More specifically, the first step of the proposed algorithm is the construction of a MILP relaxation by applying piecewise McCormick envelopes for relaxing the bilinear terms, providing a lower bound on the MINLP. Following, the solution of the MILP is used as an initial point and its logistics decisions (binary variables) are fixed into the MINLP, resulting in a non-linear programming (NLP) problem. Finally, after solving the NLP and obtaining an upper bound, the domain of each variable involved in the bilinear terms is tightened for the next iteration. The algorithm stops when the difference between the upper and lower bounds is within the tolerance or a maximum solution time is achieved. For small instances for which an optimal solution is known, the proposed strategy consistently finds optimal or near-optimal solutions. It also solves larger instances which are, in some cases, intractable by a global optimization solver.

A more recent work integrating supply uncertainty in a refinery terminal was made by Oliveira et al. (2016). The proposed framework comprises a stochastic optimization model based on Mixed-Integer Linear Programming (MILP) for scheduling a crude oil pipeline connecting a marine terminal to an oil refinery and a method for representing oil supply uncertainty. Uncertainties are inherent in the process of supplying crude oil from a marine terminal to a refinery, because of the dependence between the oil supply and maritime conditions. Maritime transportation and docking activities are subject to weather and ocean stream conditions, which directly impact the arrival time of the vessels and/or the start of unloading activities. This uncertainty concerning when the oil supply will be available in the terminal jeopardizes the possibility of anticipating decisions concerning pumping activities from the terminal to the refinery, which is crucial to achieve efficiency in this context. Other uncertainty sources, such as refinery demand or product prices, are relatively stable in the short-time planning and thus can be considered deterministic in the scheduling of pipeline pumping and vessel berthing activities. The scenario generation method aims at generating a minimal number of scenarios while preserving as much as possible of the uncertainty characteristics. More specifically, the author's methodology combines a two-stage stochastic MILP model with a problem-driven scenario generation methodology. The objective is to obtain an adherent representation of the uncertainty regarding vessel arrival so that the scheduled activities can be implemented in practice. The proposed framework was evaluated considering real-world data. The computational results showed the importance of considering uncertainty within decision support tool systems. The authors conclude that feasibility becomes the main issue when a manager focuses on short-term planning such as scheduling so it is imperative to be able to represent this uncertainty within the optimization model. It gives a wider reach to the decision-making process in terms of foreseeing possible out- comes and making resilient decisions.

Another recent work, made by Cortez and Pessoa (2016), focuses on the non-linearities and concavities in mathematical programming models, which are due to blending and splitting operations. The authors show that the existence of splitting operations can lead to inconsistencies in the solutions obtained by the previous MILP models from the literature, which use simplifying assumptions to keep the formulation computationally tractable. More specifically, they doubt that the nonlinear mixing equations can be reformulated into linear ones since the scheduling system involves only mixing operations without splitting operations. The authors claim that this statement leads to an imprecision in the mathematical model and propose a more consistent way to handle with flow splitting issues without using non-linear inequalities. It considers an aggregated inventory capacity for the storage tanks combined to a disaggregation algorithm for the flows among stages. Furthermore, a mathematical reformulation that improves the solving efficiency of the method is developed. Computational results show that the reformulated MILP model presents significant gains concerning linear relaxation gaps and runtimes, and the disaggregation algorithm leads to feasible solutions for all the tested instances.

The same year, Castro (2016) introduced a source-based mixed-integer nonlinear programming for discrete and continuous representations of time. The sources are the crudes initially present in the system or due to arrival through marine vessels, with the properties of a crude mixture being computed using the properties of the individual crudes, assuming linear blending rules. This is a problem with important logistic constraints that were modeled through Generalized Disjunctive Programming, using logic propositions involving binary variables of conflicting connections between units, while a specialized algorithm featuring relaxations from multi-parametric disaggregation handles the bilinear terms. Results over a set of test problems from the literature show that the discrete-time approach finds better solutions when minimizing cost (avoids source of bilinear terms). In contrast, solution quality is slightly better for the continuous-time formulation when maximizing gross margin. The results also show that the specialized global optimization algorithm can lead to lower optimality gaps, but overall the performance of commercial solvers is better.

Xu et al. (2017) deal with the simultaneous scheduling of front-end crude-oil transfer and refinery operations, which results in a large scale and complex optimization problem. The authors propose a systematic methodology, which provides a large-scale continuous-time based scheduling model for crude unloading, transferring, and processing (CUTP) to simulate and optimize the front-end and refinery crude-oil operations simultaneously. The CUTP model consists of a newly developed Refinery Processing Status Transition (RPST) sub-model, a crude processing status transition sub-model, and a borrowed front-end crude transferring sub-model. The objective is to maximize the total operational profit while satisfying various constraints such as operation and production specifications, inventory limits, and production demands. The authors' methodology is depicted on Fig. 12.4. Apart from the fact that the simultaneous scheduling of front-end crude transfer and refinery processing has been achieved, another contribution of the paper is that RPST has been considered in the CUTP model for seamlessly connecting both front-end crude transfer and refinery processing models. The efficiency of the proposed scheduling model has been demonstrated by an industrial-scale case study.

### *12.3.2 Methods Based on Continuous Time Representation*

Being considered as a generic model for the continuous time representation approaches, Ierapetritou and Floudas (1998a, b) present a formulation for the scheduling of operations in a refinery based on continuous time reformulation.

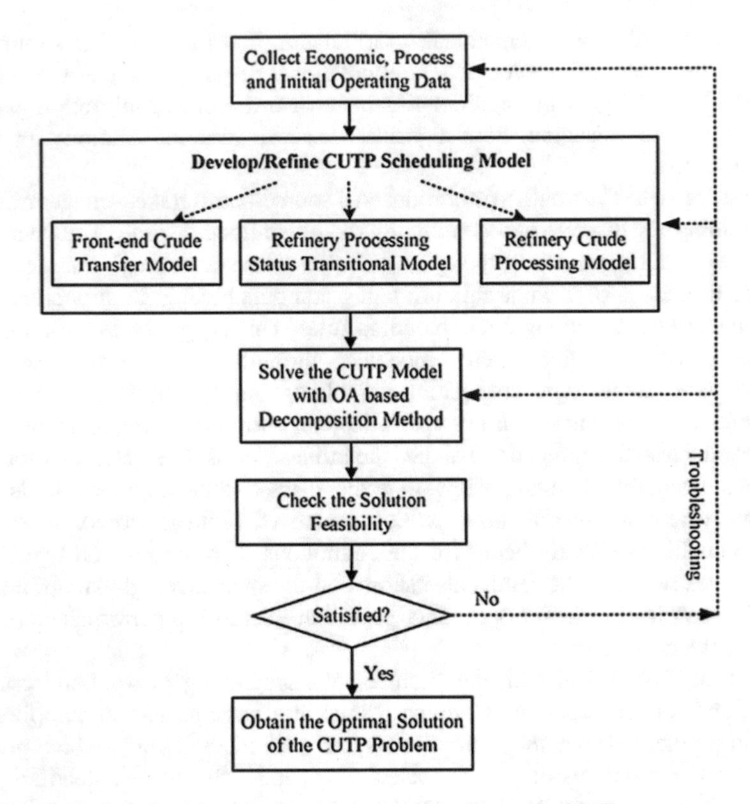


**Fig. 12.4** General methodology framework proposed by Xu et al. (2017)

A general refinery system divided into three sub-systems is addressed. The first subsystem is related to crude oil operations, the second to the processes of refining and intermediate tanks, and the third to the ends of operation processes and the stock of final products. The article treats the first sub-problem using the material balance constraints introduced by Lee et al. (1996). The setup cost of tanks is not taken into account in this model. Also the proposed model does not allow certain types of system configurations, such as the feeding of a CDU by several tanks or a single tank feeding several CDUs, because the industrial application did not render it necessary. Moreover, the two articles introduce a new idea for modeling. The authors develop a decomposition approach for the refinery systems giving rise to several sub-problems, the objective being to solve them in a reasonable time frame and find a feasible solution for the entire system. Similar approaches are made in Jia et al. (2003), Jia and Ierapetritou (2004), where again continuous time formulations are used for short-term crude oil scheduling.

Another interesting model is presented by Furman et al. (2007) where a generalized model is proposed for the continuous time scheduling problem of fluid transfer in tanks. This model generally and more robustly handles the synchronization of time events with material balances than previously proposed models in the literature without approximations and addresses all the identified drawbacks. A novel method for representing the flow to and from a tank is developed with the potential for significant reduction in the number of necessary time events required for continuous time scheduling formulations. The authors make the assumption of no simultaneous input and output flow to a tank for fluid streams comprised of multiple components. Thus, modeling the input and output within the same time event can potentially reduce the number of binary variables by a factor of 2 for such problems and can result in a significant reduction of combinatorial complexity. The model is formulated as a nonconvex mixed integer nonlinear programming (MINLP) problem, where the nonlinearity arises in the form of bilinear terms used in the calculation of component fractions for blending and pooling. Their method can be applied to general scheduling problems of transfer in tanks of a multi-component fluid and the authors provide its implementation on four case studies of refinery scheduling.

A year later, Karuppiah et al. (2008) introduced a non-convex mixed integer nonlinear model (MINLP) for the scheduling of crude oil movement at the front-end of a refinery, i.e. the supply stream, the tanks and the CDUs. The model applies continuous time representation making use of transfer events. The novelty of the proposed formulation is that the number of transfer events needed to characterize the time horizon for each stream is not known as in other continuous time models, and is chosen arbitrarily before the optimization, significantly decreasing the size of the model. Moreover, as in Furman et al. (2007), the authors allow inputs and outputs in the same transfer event, thus postulating fewer transfer events for a given problem and reducing the number of binary variables. In order to obtain the global optimum solution, the authors propose a specialized outer approximation algorithm. The latter focuses on effectively solving a mixed-integer linear programming (MILP) relaxation of the non-convex MINLP to obtain a rigorous lower bound (LB) on the global optimum. Cutting planes derived by spatially decomposing the network are added to the MILP relaxation of the original non-convex MINLP in order to reduce the solution time for the MILP relaxation. The solution of this relaxation is used in order to obtain a feasible solution to the MINLP which serves as an upper bound (UB). The lower and upper bounds are made to converge to within a specified tolerance in the proposed outer-approximation algorithm. On applying the proposed technique to test examples, significant savings are realized in the computational effort required to obtain provably global optimal solutions.

In 2010, Mouret et al. (2010) proposed four different time representations with corresponding strengthened formulations that rely on exploiting the non-overlapping graph structure of process operation scheduling problems through maximum cliques and bicliques. The representations are: (a) Multi-operation sequencing (MOS), (b) Multi-operation sequencing with synchronized start times (MOS-SST), (c) Multi-operation sequencing with fixed start times (MOS-FST) and (d) Single operation sequencing (SOS). Using the common concept of priority-slot, the authors show that it is possible to derive relationship results between these time representations. These formulations are compared, and applied to single-stage and multi-stage batch scheduling problems, as well as crude-oil operations scheduling problems. Computational results show that the Multi-Operation Sequencing time representation is superior to the others as it allows efficient symmetry-breaking and requires fewer priority-slots, thus leading to smaller model sizes.

Another comprehensive integrated optimization model based on continuous-time formulation for the scheduling problem of production units and end-product blending problem is presented by Shah and Ierapetritou (2011). The addressed problem is to determine the detailed schedule of each production unit and each demand order unloading for a short time period (typically 10 days–1 month) by taking into account the operational constraints of the plant. The schedule defines which products should be produced and which materials should be consumed to meet the market needs satisfying the demand and product specifications. The model, which is based on Ierapetritou and Floudas (1998b), incorporates quantity, quality, and logistics decisions related to real-life refinery operations. These feature start-up, minimum run length, fill-draw-delay, one-flow out of blender, sequence-dependent changeovers, maximum heel quantity, and downgrading of product. The objective function maximizes both the performance (by minimizing the use of units and tanks, all the connection between production units and tanks, start up setups, changeovers, and production downgrading) and the profit of total production. The authors introduce valid inequalities in their model, apply their method in two case studies of real-life large-scale problems (Honeywell refinery) and present their results. Thanks to the valid inequalities the computational time needed to reach optimal solution is significantly reduced. However, it is still high and the authors conclude that decomposition approaches or heuristics are required for future work in order to achieve even more reduced computational time.

In 2012, a comparative study was published considering three continuous-time models for scheduling of crude oil operations a refinery Chen et al. (2012). The authors compare the event-based model, proposed by Jia et al. (2003), the unit slot model, proposed by Hu and Zhu (2007), and the multi-operations sequence (MOS) model, proposed by Mouret et al. (2010). Pros and cons of different models are highlighted based on modeling and computational experiments. The models are reviewed, analyzed, modified and implemented. Experimental results show that the MOS model needs less computational time than the other two models, but fails to find the best solution of some problems.

The same year, Yadav and Shaik (2012) proposed a simplified State-Task-Network (STN) based formulation to address the problem of short-term scheduling of crude oil operations using unit-specific event-based continuous-time representation by incorporating explicit storage tasks for handling material transfer from storage and charging tanks. In a traditional STN representation tasks are normally defined based on the different processing operations occurring in a unit. In the proposed model, in order to effectively handle the unloading and loading operations the transfer of oil from one storage unit to another unit is treated as a task. Unloading of crude-oil from vessels to storage tank is termed as unloading task, transfer of crude-oil from storage tank to charging tank is termed as transfer task, and charging of CDUs from crude-mix stored in charging tank as charging task. The corresponding units suitable for these tasks would be the respective pipelines connected between the upstream and downstream units. Because of the STN representation, which naturally treats all tasks and states in a unified way, there is no need to consider explicit variables and constraints based on each unit (vessels, storage tanks, charging tanks, etc.). And hence, there is no need to write separate allocation constraints, material balances, and duration constraints for each unit (vessels, storage tanks, charging tanks, etc.) in a repeated manner. Constraints are developed for different cases corresponding to material flow in a tank based on (i) whether bypassing is allowed or not, (ii) whether crude mixing is allowed or not, and based on (iii) whether simultaneous input and output is allowed or not. The original model developed by the authors corresponds to a mixed integer nonlinear programming problem (MINLP) which is difficult to solve and requires high computational effort. To reduce the computational effort, they initially relax our MINLP model by dropping nonlinear constraints and they deal with the composition discrepancies, which the MILP may have, by solving the original MINLP model, as well, in order to rectify composition discrepancies. A few benchmark examples from the literature are solved using the proposed model giving promising results with improved objective functions compared with the results reported in previous works.

A couple of years later, Zhang and Xu (2014) introduced a reactive scheduling methodology for short-term crude oil operations so as to manage crude movements from ship unloading to distillation processing under various uncertainties. It contains a two-stage solving procedure for handling uncertainties such as shipping delay, crude mixture demand change, and tank unavailability. On the first stage, a deterministic schedule is initially obtained based on a continuous-time global event model using nominal data provided/projected at the beginning of each scheduling time horizon. On the second stage, indicator binary variables are implemented respectively under any occurrence of different uncertainties to collect information from the current deterministic schedule through defining "executed" tasks. On the basis of the information, rescheduling models are configured correspondingly under diverse uncertainty scenarios by combining the first-stage scheduling model and amendment constraints associated with those "executed" tasks. In their implementation of their method, the authors have taken into consideration four cases of uncertainties: (a) shipping delay, (b) crude mixture demand increase, (c) storage tank malfunction and (d) multiple uncertainties. The raised case studies have demonstrated that the developed reactive scheduling methodology can generate optimal rescheduling solutions seamlessly connected to the current completed scheduling operations and thus ensure the optimal operational continuity under impacts from various uncertainties. In addition, the methodology can systematically explore possible rescheduling actions for continuous crude operations without any presumed heuristics.

Another study focusing on scheduling of refinery operations from crude oil processing to the blending and dispatch of finished products was published by Shah and Ierapetritou (2015). Based on their previous work Shah and Ierapetritou (2011), the authors introdced a Lagrangian decomposition (LD) algorithm to solve the integrated scheduling problem using an iterative procedure and applied it to realistic large scale refinery scheduling problem to evaluate its efficiency. The LD algorithm involves relaxing complicating constraints to the objective function by introducing Lagrange multipliers to form a relaxed version of a primal problem. In LD algorithm, one can obtain lower bound and upper bound of the optimal value of the initial problem at each iteration. A novel strategy was presented to formulate restricted relaxed sub-problems based on the solution of the Lagrangian relaxed sub-problems that take into consideration the continuous process characteristic of the refinery. More specifically, the integrated full-scale scheduling problem is decomposed into two independent sub-problems: (a) the production unit scheduling problem (PSP) and (b) the blend scheduling problem (BSP) using spatial decomposition. In the proposed algorithm, referred to as restricted Lagrangian decomposition algorithm, the best lower bound is obtained amongst the restricted-relaxed sub-problems and relaxed sub-problems in each iteration. To improve the performance of the algorithm, a preprocessing step, constraints for decomposed sub-problems, and inclusion of the best upper bound's solution in lower bounding problems are proposed. The proposed restricted relaxed sub-problems produce better lower bounds and better upper bounds. Computational results of a real case study show that the proposed algorithm is very effective and provide better solutions in reasonable times.

### *12.3.3 Heuristic*

As previously analyzed, the problem of crude oil scheduling in a refinery is a very complicated problem. Solving it by modelling it as a Mixed Integer Program can be very demanding in terms of CPU time. This is due to the fact that the optimization tools based on branch-and-bound or branch-and-cut methods can be difficult and sometimes inadequate due to the associated complexity (i.e. number of variables and nonlinear constraints). The impractical amount of time needed to reach the first integer/feasible solution has led several researchers to the development of heuristic algorithms. The advantage of such algorithms is their ability to solve large-scale problems. However, they do not guarantee either global optimality or feasibility.

One of the first heuristics applied on the problem of Scheduling of Crude Oil (SCO) was proposed by Kelly (2002). The heuristic, known as the Chronological Decomposition Heuristic (CDH), is a time-based divide-and-conquer strategy intended to efficiently find integer-feasible solutions to practical scale production scheduling optimization problems, such as the SCO problem. The CDH is not an exact algorithm in that it will not find the global optimum, although it does use either branch-and-bound or branch-and-cut. The CDH is specifically designed for production scheduling optimization problems, which are formulated by discrete time representation using a pre-specified time grid with fixed time period spacing. However, the approach can easily be tailored to continuous time formulations. The basic idea of the CDH is to chop the scheduling time horizon into aggregate time intervals (time-chunks), which are a multiple of the base time period. Each time-chunk is solved using mixed-integer linear programming (MILP) techniques, beginning from the first time-chunk and moving forward in time using the technique of chronological backtracking if required. The efficiency of the heuristic is that it decomposes the temporal dimension into smaller-sized time-chunks which are solved in succession instead of solving one large problem over the entire scheduling horizon. The CDH is also able to handle other more complicated logic-type constraints such as mixing delays (a time lag between flows in and flows out of a tank) and production run-lengths with straightforward modifications. As the author mentions, the CDH should be considered a step in the direction of aiding the scheduling user in finding integer-feasible solutions of reasonable quality quickly.

One year later, the same author published another heuristic for the problem of Scheduling of Crude Oil (SCO) (Kelly 2003). The author presents an effective primal heuristic to encourage a significant reduction of binary variables. This heuristic can be applied before an implicit enumerative type search to find integer-feasible solutions for the SCO problem which is formulated as a discrete time, mixed integer linear programming problem (MILP). The basis of the technique is to employ four different well-known smoothing functions into the framework of a Smooth-and-Dive Accelerator (SDA) algorithm. The first smoothing function is the quadratic smoothing function, proposed in Raghavachari (1969). The second smoothing function, the sigmoidal smoothing function introduced by Chen and Mangasarian (1996), also known as the neural network smoothing function, is formulated to smooth the well-known sum of the integer-infeasibility metric. The third smoothing function is the interior-point smoothing function, also known as the Chen–Harker–Kanzow–Smale function. Finally, the fourth smoothing function is known as the Fischer–Burmeister smoothing function. The focus of the SDA is to decrease the time required to find locally optimized solutions using branch-and-bound or branch-and-cut. The basic algorithm of SDA is to successively solve the linear relaxation of the initial MILP with the smoothing functions added to the existing problem's objective function and to use, if required, a sequence of binary variable fixing known as diving. If the smoothing function term is not driven to zero as part of the recursion then a branch-and-bound or branch-and-cut search heuristic is used to close the procedure, finding at least integer-feasible primal infeasible solutions. The heuristic's effectiveness is illustrated by its application to an oil refinery's crude oil scheduling problem. The main benefit of the SDA over the standalone branch-and-cut implementation is the reduction in the number of 0–1 variables from the root LP relaxation to the invocation of the branch-and-cut.

A more recent study based on a heuristic was published by An et al. (2017), focusing on the case when the number of charging tanks is not sufficient. This happens when some charging tanks have to be maintained or more distillers are installed without enough charging tanks being built. In this case, in order to make a refinery able to operate, a charging tank has to be in simultaneous charging and feeding to a distiller for some time, called simultaneously-charging-and-feeding (SCF) mode, leading to disturbance to the oil distillation in distillers. A hybrid Petri net model is developed to describe the behavior of the system. Then, a scheduling method is proposed to find a schedule such that the SCF mode is minimally used. It is computationally efficient. An industrial case study is given to demonstrate the obtained results.

### 12.3.4 Hybrid Methods

Apart from the methods that use solely Mathematical Formulations and the methods that apply solely heuristics, there have been studies in the literature combining both approaches. Due to this combination, they are considered as hybrid methods. The hybrid methods generally try to exploit the benefits of both the heuristics and the Mathematical Programming. In fact, they solve a problem nearly as fast as a corresponding heuristic, but with reduced optimality gap, meaning that they are closer to the global optimal solution derived by Mathematical Programming.

Such a hybrid algorithm was proposed by Wenkai et al. (2002) in order to solve the mixed-integer nonlinear programming (MINLP) model of scheduling of crude oil unloading, storage, and processing. The algorithm combines two mixed-integer linear programming (MIP) models and a nonlinear programming (NLP) model, in order to solve the cases in which the required blend properties are not obtained. The number of binary variables is reduced by incorporating the tri-indexed binary variables in bi-indexed variables. The authors also incorporated multiple oil types, multiple docks, and multiple processing units, but maintained the constraint specifying that a CDU can be loaded only by two different tanks. For handling large-scale problems, heuristics are included in the formulations to further reduce the solution time. The proposed algorithm requires an iterative solution of an integer NLP problem but does not guarantee a feasible solution.

A couple of years later, Reddy et al. (2004a) present a mixed-integer nonlinear programming (MINLP) formulation and a mixed-integer linear programming (MILP)–based solution approach for optimizing crude oil unloading, storage, and processing operations in a multi-CDU (crude distillation unit) refinery receiving crude from multiparcel VLCCs (very large crude carriers) through a high-volume, single-buoy mooring (SBM) pipeline and/or single-parcel tankers through multiple jetties. The authors implement a hybrid time representation, where the MINLP is a discrete-time model that allows multiple sequential crude transfers to begin even at intermediate points in a period, thus mimicking a continuous-time formulation. Their solution approach could be considered as a hybrid one, since they impose a heuristic iterative algorithm and do not solve a simple Mathematical Program, thus leading to quick, but sub-optimal solutions. More specifically, to avoid MINLP/NLP solutions, the MINLP is relaxed by dropping the non-linear constraints associated the parcel-to-tank and tank-to-CDU allocations. The resulting relaxation is a MILP, and will inevitably suffer from the composition discrepancy, because the optimizer can push arbitrary amounts of individual crudes rather than the correct mixture to CDU. However, their solution approach corrects composition discrepancy without solving a single NLP, although the problem is inherently nonlinear, by fixing parcel-to-tank and tank-to-CDU allocations based on corrected compositions. Thus, their method cannot produce infeasible results. This is achieved by identifying a part of the horizon, for which MINLP's linear relaxation is exact, and then solving this MILP repeatedly with progressively shorter horizons. With this procedure, a MILP is iteratively solved, avoiding the additional resolution of a NLP, as was the case in Wenkai et al. (2002). Moreover, Reddy et al. (2004a) include in their models several real features such as multiple tanks feeding one CDU, one tank feeding multiple CDUs, SBM pipeline, brine settling, tank-to-tank transfers, and so forth. The scheduling objective is the maximization of total gross profit instead of the minimization of operating cost because the former includes the effect of crude compositions and crude margins, whereas the latter does not. The authors define gross profit as the sum of crude margins (netbacks) minus the operating costs related to logistics, where crude margin is the total value of cuts from a crude oil minus the costs of purchasing, transporting, and processing the crude.

Extending their previously referred work, Reddy et al. (2004b) presented a continuous-time mixed integer linear programming (MILP) formulation for the short-term scheduling of operations in a refinery that receives crude from very large crude carriers via a high-volume single buoy mooring pipeline. The solution approach is again a hybrid one, since it incorporates a heuristic iterative algorithm in order to eliminate the crude composition discrepancy and it does not guarantee global optimality. A direct head-to-head comparison between the discrete-time and continuous-time formulations reveals that the proposed continuous-time approach outperformed the discrete-time one for problems with longer horizons, although the latter appeared to be better for smaller and more complex problems.

A study focusing on the in-line diesel blending and distribution subsystem of an oil refinery was made by Neiro et al. (2014). The proposed formulation is based on a hybrid time representation in which time horizon is partitioned into periods of equal length representing days at which points demand may be incurred. Within each day a pre-specified number of time slots is postulated, whose duration is to be determined by the optimization problem. The set of slots is subdivided in mutually exclusive subsets and assigned to each of the days. Consequently, slots used to represent events in the first day are not the same as those used in the second day and so forth (Fig. 12.5a). Moreover, throughout the time horizon slots are placed in monotonically increasing order. Resource-centric representations are considered for the continuous timing decisions (Fig. 12.5b). Overlapping operations must be assigned to the same time slot in each day, however, neither durations nor start/finish times need to coincide. The hybrid time representation takes advantage of the flexibility of the continuous time representation and enables handling of intermediate due dates with the use of fixed time points. Time variables are defined in terms of resources instead of transfer operations leading to a smaller size formulation. Results from a real-world case are used to validate the proposed formulation that takes into consideration capacities and operating rules while minimizing costs. Solution time is dramatically decreased by adding valid inequalities for symmetry breaking. Two scenarios are analyzed to illustrate the importance of the short-run optimized schedule as opposed to the one-day planning.

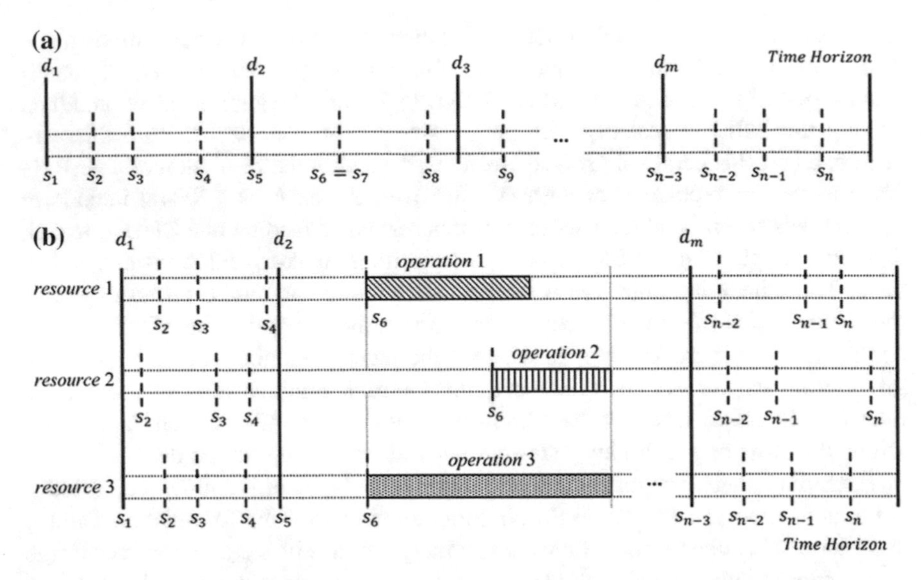


**Fig. 12.5** Time Representation: **a** hybrid time representation framework, **b** resource-centric representation for continuous timing decisions (Neiro et al. 2014)

## 12.4 Conclusion

Having reviewed the literature on the Scheduling of Crude Oil (SCO), one definitely would conclude that it is a very complex and demanding problem to be solved. The researchers have approached the problem from various aspects, coming up with mathematical models with either discrete or continuous time representation, heuristic and hybrid methods. The progress made over the last two decades for the solution of the SCO problem is huge, thanks to the novel introduced methodologies.

Many of the challenges of the SCO problem have been approached, but yet future research will have to deal with them in a more detailed and holistic way. One main aspect of the problem, which has concerned some researchers, but will still concern future researchers, is the uncertainty of the refinery system. Shipping delay/early arrival, crude mixture demand variation, tanks malfunction are some of the stochastic events that influence the schedule of a refinery. A second aspect of the problem to be dealt with in future research is the globality and feasibility/optimality of the overall schedule in a refinery. Instead of focusing only on the refinery front-end or focusing only on the refinery processing, resulting in a global optimal solution for the schedule of the whole system is a challenge for future work. Finally, despite the numerous studies, dealing with the Non-Linearity of the SCO problem will still be subject of future papers. If one would consider all these three aspects (stochasticity, globality, non-linearity) together, he/she would reach the conclusion that a lot remains to be done on the way solving the complicated problem of Crude Oil Scheduling.

## References


An, Y., Wu, N., Hon, C. T. & Li, Z., (2017). Scheduling of crude oil operations in refinery without sufficient charging tanks using petri nets. *Applied Sciences, 7.*

Benders, J. F. (1962). Partitioning procedures for solving mixed-variables programming problems. *Numerische Mathematik, 4,* 238–252.

Castro, P. M. (2016). Source-based discrete and continuous-time formulations for the crude oil pooling problem. *Computers & Chemical Engineering, 93,* 382–401.

Chaumeau, J., & Vonner, R. (1974). Management of a crude oil line and associated storage. *Rev Fr Autom Inf Rech Oper, 8,* 51–61.

Chen, C., & Mangasarian, O. (1996). A class of smoothing functions for nonlinear and mixed complementarity problems. *Computational Optimization and Applications, 5,* 97–138.

Chen, X., Grossmann, I., & Zheng, L. (2012). A comparative study of continuous-time models for scheduling of crude oil operations in inland refineries. *Computers & Chemical Engineering, 44,* 141–167.

Cortez, L. C. S., & Pessoa, A. A. (2016). A new model and a reformulation for the crude distillation unit charging problem with oil blends and sequence-dependent changeover costs. *Computers and Chemical Engineering*, 49–62.

de Assis, L. S., et al. (2017). A piecewise McCormick relaxation-based strategy for scheduling operations in a crude oil terminal. *Computers & Chemical Engineering, 106,* 309–321.

Floudas, C. A., & Lin, X. (2004). Continuous-time versus discrete-time approaches for scheduling of chemical processes: A review. *Computers & Chemical Engineering, 28,* 2109–2129.

Furman, K. C., Jia, Z., & Ierapetritou, M. G. (2007). A robust event-based continuous time formulation for tank transfer scheduling. *Industrial and Engineering Chemistry Research, 46,* 9126–9136.

Hu, Y., & Zhu, Y. (2007). An asynchronous time slot-based continuous time formulation approach for crude oil scheduling. *Computers and Applied Chemistry, 24,* 713–719.

Ierapetritou, M. G., & Floudas, C. A. (1998a). Effective continuous-time formulation for short-term scheduling. 1. multipurpose batch processes. *Industrial and Engineering Chemistry Research, 37,* 4341–4359.

Ierapetritou, M. G., & Floudas, C. A. (1998b). Effective continuous-time formulation for short-term scheduling. 2. continuous and semicontinuous processes. *Industrial and Engineering Chemistry Research, 37,* 4360–4374.

Jia, Z., & Ierapetritou, M. (2004). Efficient short-term scheduling of refinery operations based on a continuous time formulation. *Computers & Chemical Engineering, 28,* 1001–1019.

Jia, Z., Ierapetritou, M., & Kelly, J. D. (2003). Refinery short-term scheduling using continuous time formulation: Crude-oil operations. *Industrial and Engineering Chemistry Research, 42,* 3085–3097.

Joly, M., Moro, L., & Pinto, J. (2002). Planning and scheduling for petroleum refineries using mathematical programming. *Brazilian Journal of Chemical Engineering, 19,* 207–228.

Karuppiah, R., Furman, K. C., & Grossmann, I. E. (2008). Global optimization for scheduling refinery crude oil operations. *Computers & Chemical Engineering, 32,* 2745–2766.

Kelly, J. D. (2002). Chronological decomposition heuristic for scheduling: Divide and conquer method. *AIChE Journal, 48.*

Kelly, J. D. (2003). Smooth-and-dive accelerator, a pre-MILP primal heuristic applied to scheduling. *Computers and Chemical Engineering, 27.*

Lee, H., Pinto, J. M., Grossman, I. E., & Park, S. (1996). Mixed-integer linear programming model for refinery short-term scheduling of crude oil unloading with inventory management. *Industrial & Engineering Chemical Research, 35,* 1630–1641.

Mouret, S., Grossman, I. E., & Pestiaux, P. (2010). Time representations and mathematical models for process scheduling problems. *Computers & Chemical Engineering, 35,* 1038–1063.

Neiro, S. M., & Pinto, J. M. (2004). A general modeling framework for the operational planning of petroleum supply chains. *Computers & Chemical Engineering, 28,* 871–896.

Neiro, S. M. S., Murata, V. V., & Pinto, J. M. (2014). Hybrid time formulation for diesel blending and distribution scheduling. *Industrial and Engineering Chemistry Research, 53,* 17124–17134.

Oliveira, F., Nunes, P. M., Blajberg, R., & Hamacher, S. (2016). A framework for crude oil scheduling in an integrated terminal-refinery system under supply uncertainty. *European Journal of Operational Research, 252,* 635–645.

Persson, J. A., & Gothe-Lundgren, M. (2005). Shipment planning at oil refineries using column generation and valid inequalities. *European Journal of Operational Research, 163,* 631–652.

Pinto, J., Joly, M., & Moro, L. (2000). Planning and scheduling models for refinery operations. *Computers & Chemical Engineering, 24,* 2259–2276.

Raghavachari, M. (1969). On connections between zero-one integer programming and concave programming under linear constraints. *Operations Research, 17,* 680–684.

Reddy, P. C. P., Karimi, I. A., & Srinivasan, R. (2004a). Novel solution approach for optimizing crude oil operations. *AIChE Journal, 50,* 1177–1197.

Reddy, P. C. P., Karimi, I. A., & Srinivasan, R. (2004b). A new continuous-time formulation for scheduling crude oil operations. *Chemical Engineering Science, 59,* 1325–1341.

Robertson, G., Palazoglu, A., & Romagnoli, J. (2010). Refining scheduling of crude oil unloading, storing, and processing considering production level cost. *Computer Aided Chemical Engineering, 28,* 1159–1164.

Robertson, G., Palazoglu, A., & Romagnoli, J. (2011). A multi-level simulation approach for the crude oil loading/unloading scheduling problem. *Computers & Chemical Engineering, 35,* 817–827.

Saharidis, K. D. G., & Ierapetritou, G. M. (2009). Scheduling of loading and unloading of crude oil in a refinery with optimal mixture preparation. *Industrial & Engineering Chemistry Research, 48*, 2624–2633.

Saharidis, G. K., Boile, M., & Theofanis, S. (2011). Initialization of the Benders master problem using valid inequalities applied to fixed-charge network problems. *Expert Systems with Applications, 38,* 6627–6636.

Saharidis, G. K., Minoux, M., & Dallery, Y. (2009). Scheduling of loading and unloading of crude oil in a refinery using event-based discrete time formulation. *Computers & Chemical Engineering, 33*(8), 1413–1426.

Shah, N. (1996). Mathematical programming techniques for crude oil scheduling. *Computers & Chemical Engineering, 20,* S1227–S1232.

Shah, N. K., & Ierapetritou, M. G. (2011). Short-term scheduling of a large-scale oil-refinery operations: Incorporating logistics details. *Process Systems Engineering, 57,* 1570–1584.

Shah, N. K., & Ierapetritou, M. G. (2015). Lagrangian decomposition approach to scheduling large-scale refinery operations. *Computers & Chemical Engineering, 79,* 1–29.

Shah, N., Saharidis, K. D. G., Jia, Z., & Ierapetritou, G. M. (2009). Centralized–decentralized optimization for refinery scheduling. *Computers and Chemical Engineering, 33*, 2091–2105.

Speur, A., Markusse, A., van Aarle, L., & Vonk, F. (1975). Computer schedules Rotterdam-Rhine line. *Oil & Gas Journal, 73,* 93–95.

Velez, S., & Maravelias, C. T. (2013). Multiple and nonuniform time grids in discrete-time MIP models for chemical production scheduling. *Computers & Chemical Engineering, 53,* 70–85.

Wenkai, L., Hui, C.-W., Hua, B., & Tong, Z. (2002). Scheduling crude oil unloading, storage, and processing. *Industrial & Engineering Chemical Research, 41,* 6723–6734.

Wu, N., et al. (2012). A novel approach to optimization of refining schedules for crude oil operations in refinery. *IEEE Transactions on Systems, Man, and Cybernetics Part C (Applications and Reviews), 42,* 1042–1053.

Wu, N., Chu, F., Chu, C., & Zhou, M. (2008). Short-term schedulability analysis of crude oil operations in refinery with oil residency time constraint using petri nets. *IEEE Transactions on Systems, Man, and Cybernetics Part C (Applications and Reviews), 38,* 765–778.

Wu, N., Chu, F., Chu, C., & Zhou, M. (2009). Short-term schedulability analysis of multiple distiller crude oil operations in refinery with oil residency time constraint. *IEEE Transactions on Systems, Man, and Cybernetics Part C (Applications and Reviews), 39,* 1–16.

Wu, N., Zhou, M., Chu, F., & Mammar, S. (2013). Modeling and scheduling of crude oil operations in refinery: A hybrid timed petri net approach. In: I. Global (Ed.) *Embedded computing systems: applications, optimization, and advanced design*. s.l.:s.n., p. 49.

Xu, J., et al. (2017). Simultaneous scheduling of front-end crude transfer and refinery processing. *Computers & Chemical Engineering, 96,* 212–236.

Yadav, S., & Shaik, M. A. (2012). Short-term scheduling of refinery crude oil operations. *Industrial and Engineering Chemistry Research, 51,* 9287–9299.

Zhang, J., Wen, Y., & Xu, Q. (2012). Simultaneous optimization of crude oil blending and purchase planning with delivery uncertainty consideration. *Industrial & Chemistry Research, 51,* 8453–8464.

Zhang, S., & Xu, Q. (2014). Reactive scheduling of short-term crude oil operations under uncertaintes. *Industrial and Engineering Chemistry Research, 53,* 12502–12518.

Zimberg, B., Camponogara, E., & Ferreira, E. (2015). Reception, mixture, and transfer in a crude oil terminal. *Computers & Chemical Engineering, 82,* 293–302.

# Chapter 13
# Fuzzy MCDM Methods in Sustainable and Renewable Energy Alternative Selection: Fuzzy VIKOR and Fuzzy TODIM

**Zeynep Kezban Turgut and Abdullah Çağrı Tolga**

**Abstract** In recent years, there has been a remarkable trend toward sustainable and renewable energy sources due to environmental problems and depletion of fossil energy sources. Authorities encourage energy investors to tend this field and consequently many investors develop an energy planning for next decades based on renewable energy sources. At this stage, it is a crucial step to choose energy type and develop a strategic plan based on it. In this study, we aimed to find out the best performing sustainable and/or renewable energy alternative and thus guide decision makers on energy investments. We evaluated four energy power plant types, which are solar, wind, hydroelectric and landfilled gas (LFG). Multi-criteria decision making (MCDM) methods are very appropriate for the evaluation of the renewable energy alternatives, with many factors to consider. We conducted a real life case study with two different MCDM methods; VIKOR and TODIM, and compared their results. In order to cope with vagueness and uncertainty in this evaluation process, we integrated fuzzy sets into both methods. Finally, we presented a sensitivity analysis to see how robust decisions we obtained.

## 13.1 Introduction

There are different kinds of energy sources in nature; among these sources fossil fuels have major use rate to meet energy need (Tasri and Susiwalati 2014). This high-level consumption rate has caused a rapid reduction of reserves and has been creating serious environmental problems. Fossil fuel utilization is a primary source of $CO_2$ emission, which causes heating of Earth's surface and thus results in dangerous climate change. Critique temperature for Earth's surface, 2 °C, will be exceeded by 2100 according to forecasts of leading institutions such as International Energy Agency, European Commission (Myers 2015). Besides, fossil

Z. K. Turgut · A. Ç. Tolga (✉)
Galatasaray University, Istanbul, Turkey
e-mail: ctolga@gsu.edu.tr

© Springer International Publishing AG, part of Springer Nature 2018

C. Kahraman and G. Kayakutlu (eds.), *Energy Management—Collective and Computational Intelligence with Theory and Applications*, Studies in Systems, Decision and Control 149, https://doi.org/10.1007/978-3-319-75690-5_13

fuel reduction causes energy shortage in the next decades and therefore in both energy supply and environmental pollution side, unconscious consumption of fossil fuels should be lowered to an acceptable level. As a consequence of these issues, governments have put emphasis on and legislative regulations such as Kyoto Protocol and Paris Agreement have emerged as a concrete step for taking precautions against global warming and its impacts. In relation to that, authorities have been seeking for a solution to overcome these problems. As a result of this, an orientation has been occurred towards renewable energy sources (RES) since it was a solution to both supplying energy need and reducing carbon emissions. Therefore, it became a trend followed by governments, companies, and researchers to utilize clean energy sources to meet energy demand and to solve environmental problems.

Through legislative regulations, many governments have started to prepare a strategic energy plan based on RES and set a target to reduce greenhouse gas (GHG) emissions by certain levels. In order to achieve these goals, policy makers have been trying to replace conventional energy sources with sustainable and renewable energy sources to meet energy requirements. Accordingly, new energy production projects have been created and investors have been encouraged to invest in this field. For example; purchasing guarantee per kilowatt-hour generated electricity during specified years is provided under the name of government support policy.

As well as legislative regulations, other factors such as advancing energy technologies, carbon regulation and trading, trend towards clean energy, rising decision maker's expectations boost energy sector and create competitive energy market. After these developments in energy field, sustainable and renewable energy has become a growing sector and it has a room for improvements. This case triggers firms situated in energy sector and they question about what sustainable and/or renewable energy source they should invest in for next decades. This issue requires evaluating technical, economical, environmental and social aspects. As a consequence, the firms need to develop a strategic energy plan. It is a road map of the firms in their sector and describes organizational goals and how to attain them over the next years. A strategic energy plan becomes more important to be successful in relatively new renewable energy sector. At this stage it is a crucial step to choose energy type and develop a strategic plan based on it. Managers may have difficulty making decision and they can use special tools to assist them in determining sustainable and renewable energy source to produce energy. One of the tools is multi-criteria decision making (MCDM) methodology, which assesses alternatives and develops a ranking system. MCDM approach is one of the most adopted technique in energy planning studies in the literature (Sellak et al. 2017).

This study suggests two different MCDM methods, namely VIKOR and TODIM, in order to choose the best sustainable and renewable energy alternatives. The first method is VIKOR technique with several additional benefits, which enable maximum group utility of the majority with minimum individual regret of the opponent (Opricovic and Tzeng 2004). Many researchers apply MCDM methods by integrating one method with another to reach better results. It is also the case with VIKOR studies. The most preferred integration with VIKOR is fuzzy set approaches (Yazdani and Graeml 2014; Mardani et al. 2016). The fuzzy set theory was developed by Zadeh in 1965 to cope with vagueness and impreciseness in real world problems. In this regard, when we analyze the decisions with energy power plants, we face fuzziness in data. For instance, annual electricity production from solar energy, wind energy or hydropower energy heavily depend on seasonal conditions. The annual production amount may not be regular. Through the fuzzy set theory, we are able to define an accurate interval rather than assigning an exact value for the annual production estimation. Therefore we integrated VIKOR method with fuzzy sets in order to improve the quality of results in our study.

In real life problems, risk always exists and it is an important factor in decision-making process. However, most of the MCDM techniques are not able to cope with risk or do not consider risk factor in their methodologies. As the second MCDM technique in our study, we chose TODIM (an acronym in Portuguese of Interactive and Multicriteria Decision Making) method so that we can add risk factor to our decision-making problem. Renewable energy power plants include a lot of risk from many different aspects. Especially solar, wind and hydraulic power plants are dependent on season conditions. For example, rain level is a risk factor for hydraulic energy power plants. If the geographical areas of hydraulic power plants have low rain rate in any year, it affects the energy production amount negatively. Therefore adding risk factor to energy power plant evaluation problems is a critical issue to receive consistent and reliable results.

TODIM is a discrete MCDM technique based on prospect theory and deals with risk in decision-making process. Prospect theory is developed by Kahneman and Tversky (1979) and it is proposed to be a descriptive model, alternative to utility theory for decision making under the condition of risk. The theory reveals that people rely on the potential value of gains and losses rather than the final outcome when they make a decision. This feature of the theory contradicts with utility theory because utility theory assumes that people make rational decisions based on final outcome. The prospect theory has a value function and it is defined on deviations from a reference point (Kahnemann and Tversky 1979). The value function has an S-shaped curve and shows gains and losses on it. The function generally shows a concave characteristic above the reference point, meaning risk aversion in case of gains; and commonly convex characteristic below the reference point, which represents propensity to risk in case of losses (Rangel et al. 2011). Risk aversion in the case of gain means that people prefer certain or high probable gains even they have a chance to earn far more than that gain. Risk propensity in the case of losses refers that when people are faced with loss, they are willing to take a risk if there is a chance to earn. After that, it is understood that equal amount of gain and loss does not have equal importance for people, fear of loss outweighs gain. This finding of Kahneman brings him Nobel economy prize in 2002.

In TODIM method gains and losses of each alternative over another are calculated for each criterion. Pairwise comparison of alternatives leads us to find the best option among the alternatives. As it is in the prospect theory, TODIM has a value function as well and shape of the function of TODIM is the same as the value function of prospect theory. TODIM is a discrete method and is not able to cope with uncertain conditions. Therefore, we need to integrate fuzzy sets into TODIM methodology to increase the quality of results as we did in fuzzy VIKOR. Integrated fuzzy TODIM method is not one of the widespread studies in literature and early studies can be found at the beginning of the 2000s (Nobre et al. 1999). In the last decade, there have been limited numbers of fuzzy TODIM applications such as studies of Tosun and Akyüz (2015), Krohling and Souza (2012), Hanine et al. (2016). Besides, Gomes and Rangel conducted the early studies of discrete TODIM method in 1992.

In this chapter, a real life study in energy field is presented. The evaluation criteria constitute the most significant aspects of a power plant. To sum up, in this study, fuzzy VIKOR and fuzzy TODIM methods have been used to find best sustainable and renewable energy power plant option among the alternatives. The study was formatted in the following way: After this introduction section, fuzzy MCDM models are analyzed. In the next part, application of VIKOR and TODIM methods are conducted and finally ended up with conclusion section.

## 13.2 Literature Review

In this section, we analyze the renewable energy (RE) studies over the past two decades. Mirasgedis and Diakoulaki (1997) performed a cost analysis of electricity production systems including REs. They used MCDM methods for identifying their environmental impacts. Iniyan and Sumathy (1998) presented a study to find an optimal RE model reducing cost-efficiency ratio and they also presented best utilization fields of REs. Beccali et al. (2003) prepared an action plan to spread RE technologies and used ELECTRE method to find the best technology. Afgan and Carvalho (2001) made an assessment study to specify RE power plant evaluation criteria in sustainability frame. They created sustainability index of the alternatives and accordingly made some comparisons.

Kaya and Kahraman (2010a) applied AHP and VIKOR techniques to obtain the best renewable energy option and the plant site for Istanbul under fuzzy environment. They used AHP method to obtain criteria weights and utilized VIKOR for the remaining part. The same topic with different techniques and criteria was investigated to reach best energy policy and technology. In this regard, Kaya and Kahraman (2010b) preferred fuzzy AHP technique while Kahraman and Kaya (2011) applied modified fuzzy TOPSIS. Zerpa and Yusta (2015) applied an integrated AHP-VIKOR method in their study for energy planning for Istanbul. They asked four groups of experts' opinions in different sectors such as academia, private companies and determined the criteria weights. The authors highlighted that for the remote-rural area electricity production projects; there is a conflict between technical, economical criteria and social, environmental criteria. Finally, hybrid renewable technology systems were found as the best solution for their problem.

Şengül et al. (2015) analyzed RE resources for Turkey with fuzzy TOPSIS and applied Shannon's entropy methodology. According to their criteria, the best option was hydropower for Turkey.

Tasri and Susilawati (2014) conducted a study for Indonesia and aimed to find the best RE alternative in terms of generating electricity. They evaluated RE resources with fuzzy AHP technique and found that hydropower is the most appropriate alternative for Indonesia. Streimikiene et al. (2012) had same research with MULTIMOORA and TOPSIS to find best sustainable electricity generation technologies. The authors suggest water and solar thermal resources in this regard.

Zhang et al. (2015) emphasized the conflicting criteria when RE alternatives are evaluated and stated that traditional MCDM methods are inadequate to overcome this matter. They proposed an improved model integrated Choquet Integral with fuzzy sets.

Qin et al. (2017) extended classical TODIM method in their studies and proposed a fuzzy TODIM technique to solve multi-criteria group decision-making (MCGDM) problems under fuzzy environment. They tried different values of attenuation factor, however the best alternative stayed the same as hydropower.

Almost every country goes through choosing an appropriate electricity production system. All over the world, there are many researchers who perform MCDM selection process for their countries energy planning. For example, San Cristobal (2011) worked on renewable energy project alternatives provided by Spanish Government within its energy policy. He utilized AHP method for weighting process and performed VIKOR method for selection among alternatives. Abdullah and Najib (2014) realized a similar study for Malaysia with a different technique. In order to cope with uncertainty, the researchers applied intuitionistic fuzzy analytic hierarchy process (IF- AHP). They used a different scale to convert linguistic variables to numbers. Streimikiene et al. (2016) carried out a study for Lithuania to choose the best electricity production system. They applied AHP methodology and selected the biomass energy as the best option. Also, a sensitivity analysis was realized by ARAS (Additive Ratio Assessment method). Zhao and Guo (2015) performed a study for Chinese government to constitute right energy policies. They implemented a hybrid MCDM method in two phases: the superiority linguistic ratings and entropy weighting method for index weight determination and the fuzzy grey relation analysis for ranking alternatives. Their results showed that solar energy type is the best option followed by wind and biomass power. Al Garni et al. (2016) conducted a study using AHP method for Saudi Arabia to evaluate renewable power generation sources and obtained solar photovoltaic as the most favorable technology. Greece and Iran have more specific studies of wind energy in renewable energy alternatives. Shirgholami et al. (2016) applied selection process for wind turbine technologies by AHP method in Iran. Vagiona and Karanikolas (2012) used AHP to find out the most efficient area in electricity production to construct offshore wind farms in Greece.

## 13.3 Fuzzy Multi Criteria Decision Making Models

Fuzzy Multi-Criteria Decision Making (FMCDM) models are decision-making methods integrated with fuzzy approaches. Fuzzy techniques are the most preferred combination with MCDM methods in the literature (Asemi et al. 2014). It copes with uncertain, vague and ambiguous situations of real life problems. There are several types of fuzzy sets used in the literature, i.e. type-2 fuzzy sets, intuitionistic fuzzy sets, and hesitant fuzzy sets. Applying FMCDM methods increases the quality of decision-making process. The most used application areas of fuzzy MCDM methods are computer science, engineering, mathematics, decision sciences, business and management, and environmental sciences (Kahraman et al. 2015).

### *13.3.1 Fuzzy Set Theory*

Zadeh (1965) specified a fuzzy set such that it is a class of objects with a continuum of grades of membership and this set allows its members to have different grade of membership from 0 to 1. The definition of a fuzzy set is as follows (Zimmermann 2010):

If $X$ is a collection of objects denoted generically by $x$, then a fuzzy set $\widetilde{A}$ in $X$ is a set of ordered pairs:

$$\widetilde{A} = \left\{ \left( x, \mu_{\widetilde{A}}(x) | x \in X \right) \right\} \tag{13.1}$$

$\mu_{\widetilde{A}}(x)$ is called the membership function (generalized characteristic function) which maps $X$ to the membership space $M$. Its range is the subset of nonnegative real numbers whose supremum is finite. If a fuzzy set is convex and normalized, and its membership function is defined in $\mathbb{R}$ and piecewise continuous, it is called as fuzzy number. Normalization of a fuzzy set means that maximum degree of membership function is 1 (Gao et al. 2009).

There are different types of fuzzy numbers defined such as triangular, bell shaped, trapezoidal. We have preferred triangular fuzzy numbers (TFN) to implement fuzzy MCDM techniques in this study. It provides more accuracy in results and ease of computation (Tsai and Chou 2011). A TFN is defined as follows (Chen et al. 1992):

Let $x, l, m, u \in \mathbb{R}$ and $\mu_{\widetilde{A}}(x)$ is a membership function of $x$ in $\widetilde{A}$. A triangular fuzzy number $\widetilde{A} = (l, m, u)$ is defined as in (13.2):

**Table 13.1** TFN values for the determination of criteria weights

| Linguistic terms | Corresponding TFNs |
|---|---|
| Very Low (VL) | (0.0, 0.1, 0.2) |
| Low (L) | (0.1, 0.2, 0.3) |
| Medium Low (ML) | (0.2, 0.35, 0.5) |
| Medium (M) | (0.4, 0.5, 0.6) |
| Medium High (MH) | (0.5, 0.65, 0.8) |
| High (H) | (0.7, 0.8, 0.9) |
| Very High (VH) | (0.8, 0.9, 1.0) |

$$\mu_{\widetilde{A}}(x) = \begin{cases} 0, & x \leq l, \\ \frac{(x-l)}{(m-l)}, & l < x \leq m, \\ \frac{(u-x)}{(u-m)}, & m < x \leq u, \\ 0, & x > u. \end{cases} \quad (13.2)$$

where $l$ is the lower bound and $u$ is the upper bound and $m$ is the most probable value of fuzzy number $\widetilde{A}$.

We utilized linguistic variables in this study to estimate importance weight of the evaluation criteria and to assess performance of the alternatives according to qualitative criteria. Linguistic variables were expressed by triangular fuzzy numbers. Tables 13.1 and 13.2 show the corresponding fuzzy numbers of the linguistic terms (Chang 2014).

In this chapter, we implemented fuzzy VIKOR and fuzzy TODIM methods in our problem. VIKOR is a distance-based technique while TODIM achieves results through pairwise comparison.

### *13.3.2 Fuzzy VIKOR*

VIKOR method was developed by Opricovic in 1990 for multicriteria optimization of complex systems. It solves MCDM problems containing conflicting and non-commensurable (different unit) criteria (Opricovic 2011). In case of having conflicting criteria in the problem, the method provides a compromise ranking list and a solution set. The compromise ranking is obtained by measuring distance of the alternatives to the ideal. In Fig. 13.1, it is illustrated that the compromise solution $F^c$ is the closest point to the ideal $F$. The strength of the method is to provide a maximum "group utility" for the "majority" and a minimum of an individual regret for the "opponent" (Opricovic and Tzeng 2004).

**Table 13.2** TFN values for the performance evaluation

| Linguistic variables | Corresponding TFNs |
|---|---|
| Very Poor (VP) | (0, 1, 2) |
| Poor (P) | (1, 2, 3) |
| Medium Poor (MP) | (2, 3.5, 5) |
| Fair (F) | (4, 5, 6) |
| Medium Good (MG) | (5, 6.5, 8) |
| Good (G) | (7, 8, 9) |
| Very Good (VG) | (8, 9, 10) |

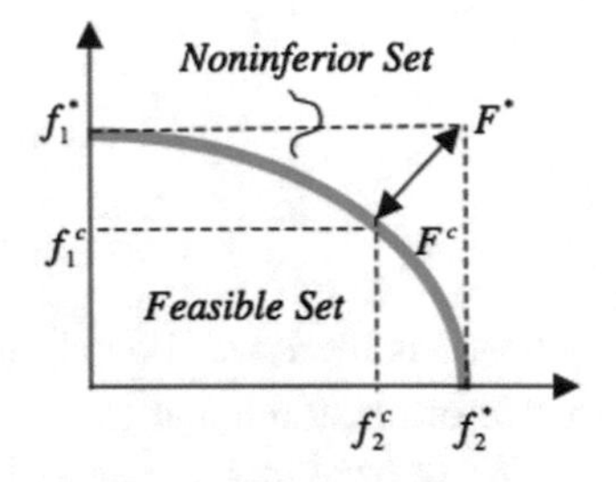


**Fig. 13.1** Ideal and compromise solutions

Development of the VIKOR technique is based on the following Lp metric form:

$$L_{pj} = \left\{ \sum_{i=1}^{n} \left[ w_i \left( f_i^* - f_{ij} \right) / \left( f_i^* - f_i^- \right) \right]^p \right\}^{1/p}, \quad 1 \ll p \ll \infty; \; j = 1, 2, \ldots J. \tag{13.3}$$

$L_{1j}$ produces $S_j$ in (13.10) and $L_{\infty j}$ produces $R_j$ in (13.11). The solution obtained by *minj* $S_j$ represents a maximum group utility and by *minj* $R_j$ represents minimum individual regret of the opponent (Opricovic and Tzeng 2007).

In a similar manner, the fuzzy VIKOR method has been developed to achieve a compromise solution in a multicriteria decision making problem under fuzzy environment where both criteria and weights could be fuzzy sets. In this fuzzy MCDM problem, there are $m$ number of alternatives $j = 1, 2, \ldots m$ and $n$ number of criteria $i = 1, 2, \ldots n$. $A_j$ indicates the $j$th alternative, $C_i$ indicates the $i$th criterion. $\widetilde{f}_{ij}$ is a triangular fuzzy number which is performance rating of $j$th alternative by $i$th criterion such that $\widetilde{f}_{ij} = (l_{ij}, m_{ij}, r_{ij})$, $l_{ij}$ and $r_{ij}$ are the lower and upper bounds respectively, $m_{ij}$ is most likely value of $\widetilde{f}_j$. $I^b$ denotes the set of benefit criteria and $I^c$ denotes cost criteria. To construct framework of the problem, we note that there are $m$ alternatives, $n$ evaluation criteria, and $k$ decision makers. This system can be expressed in a matrix format such that:

$$\widetilde{D} = \begin{array}{c} \\ C_1 \\ C_2 \\ \vdots \\ C_n \end{array} \begin{array}{c} \begin{array}{cccc} A_1 & A_2 & \cdots & A_m \end{array} \\ \begin{bmatrix} \widetilde{f}_{11} & \widetilde{f}_{21} & \cdots & \widetilde{f}_{1m} \\ \widetilde{f}_{21} & \widetilde{f}_{22} & \cdots & \widetilde{f}_{2m} \\ \vdots & \vdots & \vdots & \vdots \\ \widetilde{f}_{n1} & \widetilde{f}_{n2} & \cdots & \widetilde{f}_{nm} \end{bmatrix}_{j=1,2,\ldots,m,=1,2,\ldots,n} \end{array} \tag{13.4}$$

$\widetilde{D}$ is a performance matrix with $n \times m$ size, where $\widetilde{f}_{ij}$ is the performance rating of alternative $A_j$ evaluated by criterion $C_i$ It is formed as:

$$\widetilde{f}_{ij} = \frac{1}{k}\left[\widetilde{f}^1_{ij} \oplus \widetilde{f}^2_{ij} \oplus \cdots \oplus \widetilde{f}^k_{ij}\right] \tag{13.5}$$

where $\widetilde{f}^k_{ij}$ is the performance rating determined by $k$th decision maker of alternative $j$ evaluated by $i$th criterion.

The fuzzy VIKOR method is described in the following steps (Opricovic 2011).

Step 1: Determination of fuzzy best $\widetilde{f}^*_i = (l^*_i, m^*_i, r^*_i)$ and fuzzy worst $\widetilde{f}^\circ_i = (l^\circ_i, m^\circ_i, r^\circ_i)$ values of all criteria

$$\widetilde{f}^*_i = MAX_j\, \widetilde{f}_{ij}, \quad \widetilde{f}^\circ_i = MIN_j \; for\, i \in I^b; \tag{13.6}$$

$$\widetilde{f}^*_i = MIN_j\, \widetilde{f}_{ij}, \quad \widetilde{f}^\circ_i = MAX_j \widetilde{f}_{ij} \; for\, i \in I^c. \tag{13.7}$$

Step 2: Computation of normalized fuzzy difference $\widetilde{d}_{ij}$

$$\widetilde{d}_{ij} = \left(\widetilde{f}^*_i \ominus \widetilde{f}_{ij}\right)/(r^*_i - l^\circ_i) \quad for\, i \in I^b; \tag{13.8}$$

$$\widetilde{d}_{ij} = \left(\widetilde{f}_{ij} \ominus \widetilde{f}^*_i\right)/(r^\circ_i - l^*_i) \quad for\, i \in I^c. \tag{13.9}$$

Step 3: Computation of $\widetilde{S}_j = \left(S^l_j, S^m_j, S^r_j\right)$ and $\widetilde{R}_j = \left(R^l_j, R^m_j, R^r_j\right)$. $\widetilde{S}_j$ refers to distance of alternative $j$ from the fuzzy best value, similarly $\widetilde{R}_j$ is the distance from the fuzzy worst value.

$$\widetilde{S}_j = \sum_{i=1}^{n} \oplus\left(\widetilde{w}_i \otimes \widetilde{d}_{ij}\right) \tag{13.10}$$

$$\widetilde{R}_j = MAX_i\left(\widetilde{w}_i \otimes \widetilde{d}_{ij}\right) \tag{13.11}$$

$$\widetilde{w}_i = \frac{1}{k}\left[\widetilde{w}_i^1 \oplus \widetilde{w}_i^2 \oplus \cdots \oplus \widetilde{w}_i^k\right] \quad (13.12)$$

where $\widetilde{w}_i$ is the fuzzy importance weight of $i$th criterion, which is determined by decision makers. $\widetilde{w}_i^k$ shows the fuzzy importance weight of $i$th criterion and determined by $k$th decision maker.

Step 4: Computation of the values $\widetilde{Q}_j = \left(Q_j^l, Q_j^m, Q_j^r\right)$ by the formula

$$\widetilde{Q}_j = v\left(\widetilde{S}_j \ominus \widetilde{S}^*\right)/\left(S^{\circ r} - S^{*l}\right) \oplus (1-v)\left(\widetilde{R}_j \ominus \widetilde{R}^*\right)/\left(R^{\circ r} - R^{*l}\right) \quad (13.13)$$

where $\widetilde{S}^* = MIN_j\widetilde{S}_j$, $S^{\circ r} = MAX_jS_j^r$, $\widetilde{R}^* = MIN_j\,\widetilde{R}_j$, $R^{\circ r} = MAX_jR_j^r$ and while $v$ is a weight to represent the maximum group utility, $1 - v$ indicates the weight of the individual regret. $v$ value can be estimated by $v = (n+1)/2n$ or could be 0.5 to compromise both side.

Step 5: Defuzzification of $\widetilde{S}_j, \widetilde{R}_j\, and \widetilde{Q}_j$. There are various ways of defuzzification operation applied in different studies. In this study, we prefer to use the equation that Opricovic (2011) used in his study to convert fuzzy numbers into crisp scores. It is given by (13.14):

$$Crisp\left(\widetilde{N}\right) = (2m+l+r)/4 \quad (13.14)$$

Step 6: Ranking the alternatives by crisp value of $S$,$R$ and $Q$ in ascending order. There are three ranking lists $\{A\}_S, \{A\}_R, \{A\}_Q$.

Step 7: Reaching the compromise solution

The alternative having the smallest $Q$ value indicates the best option among the alternatives if the following conditions are satisfied.

$C_1$. Acceptable Advantage

$$Q(A^{(2)}) - Q\left(A^{(1)}\right) \geq DQ \quad (13.15)$$

where $A^{(1)}$ and $A^{(2)}$ are first and second best alternative respectively in the $Q$ ranking list. The threshold $DQ = 1/(J-1)$

$C_2$. Acceptable stability in decision making

The best alternative $A^{(1)}$ must also be the best ranked by $S$ or/and $R$. If one of the conditions is not satisfied, then a set of compromise solutions is proposed, which consists of:

Alternatives $A^{(1)}$ and $A^{(2)}$ if only the condition $C_2$ is not satisfied, or

Alternatives $A^{(1)}$, $A^{(2)}$, …, $A^{(M)}$ if the condition $C_1$ is not satisfied; $A^{(M)}$ is determined by the relation $Q(A^{(M)}) - Q\left(A^{(1)}\right) < DQ$ for maximum $M$ (the positions of these alternatives are "in closeness").

### 13.3.3 Fuzzy TODIM

Prospect theory creates the infrastructure of TODIM method. The theory and method presented individually in the following sections.

#### 13.3.3.1 Preliminaries on Prospect Theory

TODIM method (an acronym in Portuguese for iterative multicriteria decision making) is an MCDM method based on prospect theory and it was proposed by Gomes and Lima (1992). Prospect theory was developed by Kahneman and Tversky (1979) and it is a proposed descriptive model for decision making under condition of risk. Prospect theory has a value function indicating risk aversion and risk propensity and it is described in the following expression:

$$v(x) = \begin{cases} x^{\propto} & if\, x \geq 0 \\ -\theta(-x)^{\beta} & if\, x < 0 \end{cases} \tag{13.16}$$

where $\propto$ and $\beta$ are parameters related to gains and losses, respectively. Parameter $\theta$ represents a characteristic of being steeper for losses than for gains. In case of risk aversion, $\theta > 1$. Kahneman and Tversky (1979) experimentally determined the values of $\propto = \beta = 0.88$, and $\theta = 2.25$. Further, they suggest that the value of $\theta$ is between 2.0 and 2.5 (Krohling and Souza 2012). This function is S-shaped as shown in Fig. 13.2.

Concave curve represents the gains and convex curve represents the losses. As it is in the prospect theory, TODIM has a value function as well and its shape is the same as the value function of prospect theory as shown in Fig. 13.2.

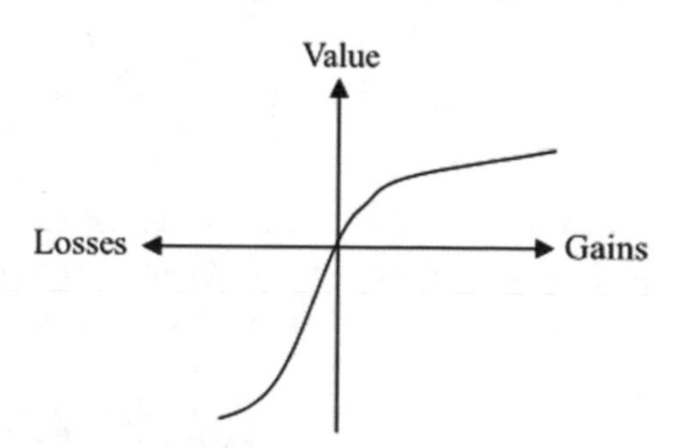


**Fig. 13.2** The value function of prospect theory

#### 13.3.3.2 Fuzzy TODIM Method

Fuzzy TODIM is an integrated model of fuzzy sets with traditional TODIM. The method makes pairwise comparison between alternatives with regard to each criterion and gains and losses of each alternative over the others are obtained. The sum of gains and losses of each alternative gives dominance degree of that alternative. In the final step, alternatives are ranked by these dominance degrees.

Let there are $m$ number of alternatives $i = 1, 2 \ldots m$ and $n$ number of evaluation criteria $j = 1, 2 \ldots n$. $A_i$ denotes the $i$th alternative, $C_j$ denotes the $j$th criterion. Each criterion has different importance degree and $w = (w_1, w_2 \ldots w_n)^T$ is a weight vector, where $w_j$ denotes the importance weight of criterion $C_j$, such that $\sum_{j=1}^{n} w_j = 1$ and $0 \leq w_j \leq 1$. Alternatives have a performance value for each criterion. $\widetilde{x}_{ij}$ is a performance value of $i$th alternative with respect to $j$th criterion. Note that $w_j$ is a discrete number and $\widetilde{x}_{ij}$ is a triangular fuzzy number.

The steps of the fuzzy TODIM method are organized by using studies of Tosun and Akyüz (2015), Xiao and Zhi-ping (2011), Sen et al. (2016).

Step 1: Determination of criteria weight and performance values of alternatives.

For the performance evaluation of alternatives according to qualitative criteria and determination of criteria weights, triangular fuzzy numbers are used in this fuzzy TODIM method. Alternatives have numerical values for quantitative criteria. Performance evaluation and weight determination processes are conducted by decision makers. The equations of TODIM are given in the following:

$$\widetilde{x}_{ij} = \frac{1}{k}\left[\sum_{e=1}^{k} \widetilde{x}_{ij}^{e}\right] \quad i = 1, 2 \ldots m \tag{13.17}$$

where $\widetilde{x}_{ij}^{e}$ is the performance rating determined by $e$th decision maker of alternative $i$ evaluated by $j$th criterion. $k$ is the number of decision makers.

$$\widetilde{w}_j = \frac{1}{k}\left[\sum_{e=1}^{k} \widetilde{w}_j^{e}\right] \quad j = 1, 2 \ldots n \tag{13.18}$$

where $\widetilde{w}_j^{e}$ is the weight of $j$th criterion, determined by $e$th decision maker. If performance values are in different units, normalization of the values is necessary. The fuzzy normalized value of $\widetilde{x}_{ij} = (l_{ij}, m_{ij}, u_{ij})$ is $\widetilde{r}_{ij}$ and calculated as:

$$\widetilde{r}_{ij} = \left(\frac{l_{ij}}{u_j^*}, \frac{m_{ij}}{u_j^*}, \frac{u_{ij}}{u_j^*}\right), \quad j \in B \tag{13.19}$$

$$\widetilde{r}_{ij} = \left(\frac{l_j^-}{u_{ij}}, \frac{l_j^-}{m_{ij}}, \frac{l_j^-}{l_{ij}}\right), \quad j \in C \tag{13.20}$$

$B$ and $C$ are the sets of benefit and cost criteria respectively. $u_j^* = max_i u_{ij}$ if $j \in B$, $l_j^- = min_i l_{ij}$ if $j \in C$. This normalization method standardizes the fuzzy performance values and makes the value range between 0 and 1, i.e. [0,1].

Step 2: Defuzzification of fuzzy criteria weights.

Defuzzification method used in this study belongs to Abdel-Kader and Dugdale (2001). $\alpha$ is index of optimism. Bigger values of $\alpha$ represent an optimistic decision maker, whereas smaller values represent a pessimistic decision maker. $\alpha$ parameter reflects the decision maker's risk attitude. For example, a decision maker who avoids risk because of uncertain situations may prefer a low value of $\alpha$. Different index of optimism values can be used in researches for sensitivity analysis. In this study index of optimism ($\alpha$) is accepted as 0.5, which is a neutral point in order to balance between optimism and pessimism.

Let $\alpha \in [0, 1]$ be index of optimism. For a triangular fuzzy number $\widetilde{F}_j = (l_j, m_j, u_j)\, j = 1, 2\ldots n$; let $V(\widetilde{F}_j)$ be the value of $\widetilde{F}_j$ and ordering can be calculated as;

$$V(\widetilde{F}_j) = m_j \left\{ \alpha \left[ \frac{u_j - x_{min}}{x_{max} - x_{min} + u_j - m_j} \right] + (1 - \alpha) \left[ 1 - \frac{x_{max} - l_j}{x_{max} - x_{min} + m_j - l_j} \right] \right\} \tag{13.21}$$

where $x_{min} = \inf S$, $x_{max} = \sup S$

$$S = \bigcup_{j=1}^{n} S_j \tag{13.22}$$

and

$$S_j = (l_1, m_1, u_1, \ldots, l_n, m_n, u_n) \quad j = 1, 2\ldots n \tag{13.23}$$

Calculated weights with the ordering method are normalized by the following formula:

$$w_j = \frac{V(\widetilde{F}_j)}{\sum_{j=1}^{n} V(\widetilde{F}_j)} \tag{13.24}$$

Step 3: Calculation of Gains and Losses.

Gains and losses of an alternative over the other alternatives are estimated by pairwise comparison. Let $\widetilde{x}_{ij}$ and $\widetilde{x}_{kj}$ are performance values of alternative $A_i$ and $A_k$ respectively regarding to criterion $C_j$, $k = 1, 2\ldots m$. The performance values $\widetilde{x}_{ij}$ and

$\widetilde{x}_{kj}$ are represented by TFNs. The Euclidian distance between them are calculated by the following equation:

$$d\left(\widehat{x}_{ij}, \widehat{x}_{kj}\right) = \sqrt{\frac{1}{3}\left[\left(x_{ij}^{l} - x_{kj}^{l}\right)^{2} + \left(x_{ij}^{m} - x_{kj}^{m}\right)^{2} + \left(x_{ij}^{u} - x_{kj}^{u}\right)^{2}\right]} \tag{13.25}$$

Gains ($G_{ik}^{j}$) and losses ($L_{ik}^{j}$) of $A_i$ against $A_k$ regarding to criterion $C_j$ are given as:

For benefit criteria:

$$G_{ik}^{j} = \begin{cases} d\left(\widehat{x}_{ij}, \widehat{x}_{kj}\right), & \widehat{x}_{ij} \geq \widehat{x}_{kj} \\ 0, & \widehat{x}_{ij} < \widehat{x}_{kj} \end{cases} \tag{13.26}$$

$$L_{ik}^{j} = \begin{cases} 0, & \widehat{x}_{ij} \geq \widehat{x}_{kj} \\ -d\left(\widehat{x}_{ij}, \widehat{x}_{kj}\right), & \widehat{x}_{ij} < \widehat{x}_{kj} \end{cases} \tag{13.27}$$

For cost criteria:

$$G_{ik}^{j} = \begin{cases} 0, & \widehat{x}_{ij} \geq \widehat{x}_{kj} \\ d\left(\widehat{x}_{ij}, \widehat{x}_{kj}\right), & \widehat{x}_{ij} < \widehat{x}_{kj} \end{cases} \tag{13.28}$$

$$L_{ik}^{j} = \begin{cases} -d\left(\widehat{x}_{ij}, \widehat{x}_{kj}\right), & \widehat{x}_{ij} \geq \widehat{x}_{kj} \\ 0, & \widehat{x}_{ij} < \widehat{x}_{kj} \end{cases} \tag{13.29}$$

It is obvious that $G_{ik}^{j} + L_{ki}^{j} = 0$ and $G_{ii}^{j} = L_{ii}^{j} = 0$. Using the equations, gain matrix $G_j = \left[G_{ik}^{j}\right]_{m \times m}$ and loss matrix $L_j = \left[L_{ik}^{j}\right]_{m \times m}$ are constructed for each criterion.

Step 4: Calculation of criteria's relative weights $w_{jr}$.

Relative weights of criteria are estimated based on a reference criterion. It is the criterion with highest weight. Let $C_r$ be the reference criterion, the relative weight $w_{jr}$ of criterion $C_j$ to the reference criterion $C_r$ is found as follows:

$$w_{jr} = w_j / w_r \tag{13.30}$$

where $w_j$ is the weight of criterion $C_j$ and $w_r$ is the weight of the reference criterion $C_r$.

Step 5: Construction of dominance degree matrix.

$\phi_{ik}^{j(+)}$ denotes the dominance degree of gain and $\phi_{ik}^{j(-)}$ denotes dominance degree of loss. To construct the matrix, dominance degree of alternative $A_i$ over $A_k$ for criterion $C_j$ is calculated with the following equations.

$$\phi_{ik}^{j(+)} = \sqrt{G_{ik}^{j} w_{jr} / \left( \sum_{j=1}^{n} w_{jr} \right)} \tag{13.31}$$

$$\phi_{ik}^{j(-)} = -\frac{1}{\theta} \sqrt{-L_{ik}^{j} \left( \sum_{j=1}^{n} w_{jr} \right) / (w_{jr})} \tag{13.32}$$

where $\theta$ is attenuation factor of the loss. Overall dominance degree $\phi_{ik}^{j}$ is found as follows:

$$\phi_{ik}^{j} = \phi_{ik}^{j(+)} + \phi_{ik}^{j(-)} \tag{13.33}$$

after that dominance degree matrix $\phi_j = [\phi_{ik}^{j}]_{m \times m}$ for criterion $C_j$ can be constructed.

Step 6: Construction of overall dominance degree matrix.

Overall dominance degree of alternative $i$ on alternative $k$ is calculated by:

$$\delta_{ik} = \sum_{j=1}^{n} \phi_{ik}^{j} \tag{13.34}$$

It creates an $m \times m$ size dominance degree matrix $\Delta$ and $\Delta = [\delta_{ik}]_{m \times m}$.

Step 7: Calculation of overall value of each alternative and ranking the alternatives.

Based on matrix $\Delta$, the overall value of alternative $A_i$ can be calculated as follows:

$$\xi(A_i) = \frac{\sum_{k=1}^{m} \delta_{ik} - min_{i \in M} \{ \sum_{k=1}^{m} \delta_{ik} \}}{max_{i \in M} \{ \sum_{k=1}^{m} \delta_{ik} \} - min_{i \in M} \{ \sum_{k=1}^{m} \delta_{ik} \}} \tag{13.35}$$

$0 \leq \xi(A_i) \leq 1$ and greater $\xi(A_i)$ indicates better alternative. Therefore the alternatives are ranked according to descending order of overall value $\xi(A_i)$.

## 13.4 An Application: A Strategic Selection of a Firm in Sustainable and Renewable Energy Sector

In this section we present a real life case from Turkey. We aim to find out the best performing sustainable and renewable energy alternative and by means of this to lead the energy investors. We conducted this study based on four most common sustainable and renewable energy power plant types, which are solar energy (SE), wind energy (WE), hydraulic energy (HE) and specifically land filled gas energy (LFG-E) consisting of solid waste under the category of biomass.

In this study, we worked with experts in their field and powerful companies in energy sector. There are four decision makers, two of them are academicians whose area of expertise is renewable energy and the others are expert engineers in the field of energy trading and investment. An assistant professor from energy institute of İstanbul Technical University helped us for technical aspects of the power plants and made significant review for the criteria determination and performance evaluation. Another assistant professor who studies on renewable energy area from İstanbul University contributed our study for criteria and performance evaluation. Besides we have received data of LFG, solar and wind power plants from a company working internationally and an important actor of energy sector of Turkey with 30 years of experience. We also received help from the manager of this company's energy trading investments department. Our forth decision maker is a general manager of a consultancy company which work with renewable energy companies. They analyzed our criteria and alternatives, reflected their field viewpoint and knowledge on our study. Lastly hydraulic energy data was taken from another company, which develops and invests in power and water infrastructure and it is qualified by World Bank.

## *13.4.1 Application of Fuzzy VIKOR Technique*

### 13.4.1.1 Determination of Evaluation Criteria

One of the most important aspects of the multicriteria problem is to determine evaluation criteria properly. In this study, firstly we utilized the literature to choose energy evaluation criteria afterwards revised with the decision makers. As the most frequently adopted criteria in the energy evaluation studies are used, there are some rarely used criteria such as government support rate and cost increasing rate (Büyüközkan and Güleryüz 2017). After the carefully investigation of the energy production subject, we determined the criteria list that needs to be considered to evaluate sustainable energy power plants. In the following table, the criteria list is presented and necessary information related to them is given (Şengül et al. 2015; Cavallaro and Ciraolo 2005) (Table 13.3).

We can categorize the criteria as technical from $C_1$ to $C_9$, economical from $C_{10}$ to $C_{16}$, and economical from $C_{17}$ to $C_{22}$.

To estimate importance weight, a questionnaire was prepared and sent to the decision makers. They evaluated all the criteria individually by referring the linguistic variables in Table 13.1. Table 13.4 shows the decision makers' opinions on how important the mentioned criteria are. By using (13.12), we synthesized four different opinions on one criterion by averaging corresponding TFN values given by the decision makers. Calculation steps were explicitly provided in the following:

Fuzzy importance weight and crisp score of criterion $C_1$:

**Table 13.3** Evaluation criteria for power plants

| Criterion | Description | Units |
|---|---|---|
| $C_1$: Technical efficiency | It is the amount of useful energy that we can gain from an energy source | ratio |
| $C_2$: Technical risk | The probability of loss resulted from process of a power plant and effects of environmental conditions on the plants i.e. rain, icing | – |
| $C_3$: Maturity | It measures the availability of technology and its reliability | – |
| $C_4$: Net annually electricity production | It refers to net amount of energy generated from an energy source at the end of the year | MWh/year |
| $C_5$: Construction time | It is the length of construction period for the RE plants | months |
| $C_6$: Land use | It represents annual net electricity per $m^2$ | $kWh/m^2$ |
| $C_7$: Per unit installed power | It is the installed power of the plant per $km^2$ | $MW/km^2$ |
| $C_8$: Plant lifetime | It is the service life of the plants | year |
| $C_9$: Reserve potential | It states Turkey's RE energy potential | MW |
| $C_{10}$: Annual income | It is the annual income obtained from power plants' operations | cent/kWh |
| $C_{11}$: Investment cost | It contains all type of costs related to equipment, installation, construction and engineering services | cent/kWh |
| $C_{12}$: Total operating cost | It refers to costs due to energy plants' s operation, repair and maintenance activities including personal and service facility costs | cent/kWh |
| $C_{13}$: Payback period | It is the time of repay period of investments | year |
| $C_{14}$: Government support rate | It refers to rate of guarantee of electricity purchase by the government | cent/kWh |
| $C_{15}$: Operation and maintenance cost increasing rate | It refers to increasing cost rate over the years related to RE plant's operations and maintenance activities | % (percentage) |
| $C_{16}$: Employment | It refers job creation in the RE plants | number |
| $C_{17}$: Lifecycle GHG emissions | Generation of greenhouse gas emissions due to plant operations. These gasses are hazardous and cause global warming. $CO_2$, $CH_4$, $N_2O$ etc | ton/year |
| $C_{18}$: GHG emissions avoided | When we produce an amount of electricity from clean energy sources, conventional energy systems don't have to be used produce that amount of electricity. In this case the RE system prevents $CO_2$ emissions generating from conventional plants | $CO_2$-eq kg/kWh |
| $C_{19}$: Impact on ecosystem | It refers to potential risk to ecosystem that may be caused by RE plants, including liquid and solid disposals and costs caused by them, magnetic hazard, changing of microclimate and causing bad smell | – |
| $C_{20}$: Social acceptability | It refers to public opinion about RE plants | – |
| $C_{21}$: Noise | It measures the noise level caused by RE plants | – |
| $C_{22}$: Visual impact | It evaluates visual pollution caused by RE plants | – |

*Note* [–] denotes not having a unit because they are qualitative criteria

**Table 13.4** Decision makers' opinions on criteria importance

| Criteria | $D_1$ | $D_2$ | $D_3$ | $D_4$ |
|---|---|---|---|---|
| $C_1$ | VH | H | VH | VH |
| $C_2$ | H | VH | MH | VH |
| $C_3$ | H | H | VH | H |
| $C_4$ | VH | M | VH | VH |
| $C_5$ | MH | L | MH | MH |
| $C_6$ | L | H | M | MH |
| $C_7$ | L | H | M | MH |
| $C_8$ | H | MH | VH | M |
| $C_9$ | VH | H | H | H |
| $C_{10}$ | VH | M | VH | VH |
| $C_{11}$ | VH | MH | VH | VH |
| $C_{12}$ | H | H | VH | H |
| $C_{13}$ | H | MH | H | VH |
| $C_{14}$ | VH | H | VH | H |
| $C_{15}$ | MH | H | VH | H |
| $C_{16}$ | MH | L | M | L |
| $C_{17}$ | H | VH | MH | H |
| $C_{18}$ | H | VH | VH | H |
| $C_{19}$ | H | VH | VH | H |
| $C_{20}$ | M | H | VH | H |
| $C_{21}$ | L | H | MH | H |
| $C_{22}$ | L | VH | MH | L |

$$
\begin{aligned}
\widetilde{w}_1 &= \frac{1}{4}[(0.8, 0.9, 1.0)\oplus(0.7, 0.8, 0.9)\oplus(0.8, 0.9, 1.0)\oplus(0.8, 0.9, 1.0)] \\
&= (0.775, 0.875, 0.975) \\
Crisp(\widetilde{w}_1) &= \frac{2(0.875) + 0.775 + 0.975}{4} = 0.875
\end{aligned}
$$

The other criteria weights and crisp scores are calculated in the same way. Consequently, the criteria and their fuzzy weights are shown in Table 13.5. The crisp score column of it indicates the order of importance of each evaluation criterion. According to Table 13.5, first three important criteria are technical efficiency, government support rate, GHG emission avoided, and impact on ecosystem and investment cost. It proves that technical, economical and environmental aspects should be analyzed together for a power plant evaluation.

#### 13.4.1.2 Creating of the Performance Matrix

Before creating the decision matrix, we need to specify the criteria by their features. We have 5 qualitative criteria such as visual impact, maturity and 17 quantitative

**Table 13.5** Fuzzy importance weights of the criteria

| Criteria | Fuzzy importance weight | Crisp score |
|---|---|---|
| $C_1$ | (0.775, 0.875, 0.975) | 0.875 [1] |
| $C_2$ | (0.7, 0.813, 0.925) | 0.813 [6] |
| $C_3$ | (0.725, 0.825, 0.925) | 0.825 [5] |
| $C_4$ | (0.7, 0.8, 0.9) | 0.8 [7] |
| $C_5$ | (0.175, 0.313, 0.45) | 0.313 [15] |
| $C_6$ | (0.433, 0.55, 0.667) | 0.55 [12] |
| $C_7$ | (0.433, 0.55, 0.667) | 0.55 [12] |
| $C_8$ | (0.6, 0.713, 0.825) | 0.713 [10] |
| $C_9$ | (0.733, 0.833, 0.933) | 0.833 [4] |
| $C_{10}$ | (0.7, 0.8, 0.9) | 0.8 [7] |
| $C_{11}$ | (0.725, 0.838, 0.95) | 0.838 [3] |
| $C_{12}$ | (0.725, 0.825, 0.925) | 0.825 [5] |
| $C_{13}$ | (0.675, 0.788, 0.9) | 0.788 [8] |
| $C_{14}$ | (0.75, 0.85, 0.95) | 0.85 [2] |
| $C_{15}$ | (0.675, 0.788, 0.9) | 0.788 [8] |
| $C_{16}$ | (0.275, 0.388, 0.5) | 0.389 [14] |
| $C_{17}$ | (0.675, 0.788, 0.9) | 0.788 [8] |
| $C_{18}$ | (0.75, 0.85, 0.95) | 0.85 [2] |
| $C_{19}$ | (0.75, 0.85, 0.95) | 0.85 [2] |
| $C_{20}$ | (0.65, 0.75, 0.85) | 0.75 [9] |
| $C_{21}$ | (0.5, 0.613, 0.725) | 0.613 [11] |
| $C_{22}$ | (0.375, 0.488, 0.6) | 0.488 [13] |

Note: [] denotes importance order of the criteria

criteria such as electricity production amount, payback period. $C_2$, $C_5$, $C_{11}$, $C_{12}$, $C_{13}$, $C_{15}$, $C_{17}$, $C_{19}$, $C_{22}$ are defined as cost criteria stating drawback and the rest are defined as benefit criteria stating advantage. For the qualitative criteria, the decision makers rated the alternatives by referring Table 13.2. In order not to cause confusion for the decision makers, we wanted they to assume all qualitative criteria as benefit. For example; when solar energy is rated in terms of noise, if a decision maker evaluates it as very good (8, 9, 10), it does not mean that solar energy is very noisy it states solar energy is in a very good condition in terms of noise, doesn't cause high undesirable noise level. Table 13.6 shows the decision makers' evaluation rates for the alternatives with respect to the qualitative criteria.

As in (13.5), we estimated the performance rating of the alternatives by averaging corresponding TFN values given by the decision makers. Performance value of Alternative 1, which is solar energy for the second criterion is presented as an example:

$$\tilde{f}_{21} = \frac{1}{4}[(4,5,6) \oplus (7,8,9) \oplus (7,8,9) \oplus (8,9,10)] = (6.5, 7.5, 8.5)$$

**Table 13.6** Decision makers' opinions on performance ratings of the alternatives

| | | $C_2$ | $C_3$ | $C_{19}$ | $C_{20}$ | $C_{21}$ | $C_{22}$ |
|---|---|---|---|---|---|---|---|
| **$D_1$** | | | | | | | |
| | SE | F | F | G | VG | G | G |
| | WE | F | G | G | VG | F | G |
| | HE | G | G | P | P | P | P |
| | LFG-E | P | G | VG | VG | F | F |
| **$D_2$** | | | | | | | |
| | SE | G | VG | VG | MG | VG | P |
| | WE | G | MG | VG | F | VP | P |
| | HE | G | MP | P | P | P | MP |
| | LFG-E | F | G | MP | VP | P | VP |
| **$D_3$** | | | | | | | |
| | SE | G | VG | G | VG | VG | G |
| | WE | G | VG | F | MG | F | MG |
| | HE | G | VG | MP | MG | F | F |
| | LFG-E | F | MG | VG | G | G | F |
| **$D_4$** | | | | | | | |
| | SE | VG | VG | VG | VG | VG | VG |
| | WE | VG | VG | VG | VG | F | G |
| | HE | VG | VG | G | VG | P | VG |
| | LFG-E | G | G | VP | MG | F | F |

On the other hand for the quantitative criteria, we did not consult the judgments of the decision makers on the power plants because we have given numeric data for each alternative. Fuzzy performance values of the alternatives regarding to qualitative and quantitative criteria were gathered and Table 13.7 shows the fuzzy performance ratings of all the alternatives.

#### 13.4.1.3 Calculation of Normalized Fuzzy Differences

After we obtained the performance matrix, Eqs. 13.6 and 13.7 were used to specify fuzzy best and worst values (Table 13.8). Here it is important to note that we assumed all qualitative criteria as benefit to make evaluation process convenient for the decision makers. In this case, cost criteria are $C_5$, $C_{11}$, $C_{12}$, $C_{13}$, $C_{15}$ and $C_{17}$ in our problem. They are construction time, investment cost, total operating cost, payback period, operation and maintenance cost increasing rate, and lifecycle GHG emissions respectively.

In the next step, Eqs. 13.8 and 13.9 were applied to calculate the normalized fuzzy difference. In the following, an example of normalization in terms of benefit and cost criterion is shown:

Normalization with benefit criterion:

**Table 13.7** Performance matrix of the alternatives

| | SE | WE | HE | LFG-E |
|---|---|---|---|---|
| $C_1$ | (0.15, 0.187, 0.22) | (0.25, 0.29, 0.4) | (0.3, 0.364, 0.5) | (0.8, 0.913, 0.95) |
| $C_2$ | (6.5, 7.5, 8.5) | (6.5, 7.5, 8.5) | (7.25, 8.25, 9.25) | (3.25, 4.25, 5.25) |
| $C_3$ | (7, 8, 9) | (7, 8.125, 9.25) | (6.25, 7.375, 8.5) | (6.5, 7.625, 8.75) |
| $C_4$ | (44, 580.375, 44, 625, 44, 669.625) | (79, 929.99, 80, 010, 80, 090.01) | (55, 069.875, 55, 125, 55, 180.125) | (293, 706, 294, 000, 294, 294) |
| $C_5$ | (11.538, 11.55, 11.562) | (14.685, 14.7, 14.715) | (25.175, 25.2, 25.225) | (12.587, 12.6, 12.613) |
| $C_6$ | (127.373, 127.5127.628) | (7.993, 8.001, 8.009) | (0.209, 0.21, 0.21) | (587.412, 588, 588.588) |
| $C_7$ | (77.922, 78, 78.078) | (3.147, 3.15, 3.153) | (0.066, 0.066, 0.066) | (73.427, 73.5, 73.574) |
| $C_8$ | (31.469, 31.5, 31.532) | (26.224, 26.25, 26.276) | (51.399, 51.45, 51.501) | (36.713, 36.75, 36.787) |
| $C_9$ | (58, 741.2, 58, 800, 58, 858.8) | (50, 349.6, 50, 400, 50, 450.4) | (49, 821.978, 49, 871.85, 49, 921.722) | (3978.667, 3982.65, 3986.633) |
| $C_{10}$ | (14.809, 14.824, 14.838) | (7.343, 7.35, 7.357) | (27.939, 27.967, 27.995) | (13.951, 13.965, 13.979) |
| $C_{11}$ | (71.082, 71.153, 71.224) | (64.699, 64.764, 64.829) | (57.143, 57.2, 57.257) | (17.607, 17.625, 17.643) |
| $C_{12}$ | (0.864, 0.865, 0.866) | (0.688, 0.689, 0.69) | (2.27, 2.272, 2.275) | (1.124, 1.125, 1.126) |
| $C_{13}$ | (7.343, 7.35, 7.357) | (10.49, 10.5, 10.511) | (10.49, 10.5, 10.511) | (5.245, 5.25, 5.255) |
| $C_{14}$ | (13.636, 13.65, 13.664) | (6.818, 6.825, 6.832) | (5.769, 5.775, 5.781) | (14.685, 14.7, 14.715) |
| $C_{15}$ | (3.409, 3.413, 3.416) | (7.552, 7.56, 7.568) | (13.636, 13.65, 13.664) | (5.245, 5.25, 5.255) |
| $C_{16}$ | (10.49, 10.5, 10.511) | (7.343, 7.35, 7.357) | (25.175, 25.2, 25.225) | (52.448, 52.5, 52.553) |
| $C_{17}$ | (13, 85, 731) | (6, 26, 124) | (2, 26, 237) | (10, 45, 101) |
| $C_{18}$ | (0.895, 0.896, 0.897) | (0.895, 0.896, 0.897) | (0.895, 0.896, 0.897) | (7.84, 7.848, 7.856) |
| $C_{19}$ | (7.5, 8.5, 9.5) | (6.75, 7.75, 8.75) | (2.75, 3.875, 5) | (4.5, 5.625, 6.75) |
| $C_{20}$ | (7.25, 8.375, 9.5) | (6.25, 7.375, 8.5) | (3.75, 4.875, 6) | (5, 6.125, 7.25) |
| $C_{21}$ | (7.75, 8.75, 9.75) | (3, 4, 5) | (1.75, 2.75, 3.75) | (4, 5, 6) |
| $C_{22}$ | (5.75, 6.75, 7.75) | (5, 6.125, 7.25) | (3.75, 4.875, 6) | (3, 4, 5) |

**Table 13.8** Fuzzy best and worst values of the alternatives

| | Fuzzy best value | | | Fuzzy worst value | | |
|---|---|---|---|---|---|---|
| | L | m | r | l | m | r |
| $C_1$ | 0.8 | 0.913 | 0.95 | 0.15 | 0.1866 | 0.22 |
| $C_2$ | 7.25 | 8.25 | 9.25 | 3.25 | 4.25 | 5.25 |
| $C_3$ | 7 | 8.125 | 9.25 | 6.25 | 7.375 | 8.5 |
| $C_4$ | 293,706 | 294,000 | 294,294 | 44,580.375 | 44,625 | 44,669.625 |
| $C_5$ | 11.538 | 11.55 | 11.561 | 25.174 | 25.2 | 25.225 |
| $C_6$ | 587.412 | 588 | 588.588 | 0.209 | 0.21 | 0.21 |
| $C_7$ | 77.922 | 78 | 78.078 | 0.066 | 0.066 | 0.066 |
| $C_8$ | 51.398 | 51.45 | 51.501 | 26.224 | 26.25 | 26.276 |
| $C_9$ | 58,741.2 | 58,800 | 58,858.8 | 3978.667 | 3982.65 | 3986.633 |
| $C_{10}$ | 27.939 | 27.967 | 27.995 | 7.343 | 7.35 | 7.357 |
| $C_{11}$ | 17.607 | 17.625 | 17.643 | 71.082 | 71.153 | 71.224 |
| $C_{12}$ | 0.688 | 0.689 | 0.69 | 2.270 | 2.272 | 2.275 |
| $C_{13}$ | 5.245 | 5.25 | 5.255 | 10.49 | 10.5 | 10.511 |
| $C_{14}$ | 14.685 | 14.7 | 14.715 | 5.769 | 5.775 | 5.781 |
| $C_{15}$ | 3.409 | 3.413 | 3.416 | 13.636 | 13.65 | 13.664 |
| $C_{16}$ | 52.448 | 52.5 | 52.553 | 7.343 | 7.35 | 7.357 |
| $C_{17}$ | 6 | 26 | 124 | 13 | 85 | 731 |
| $C_{18}$ | 7.84 | 7.848 | 7.856 | 0.895 | 0.896 | 0.897 |
| $C_{19}$ | 7.5 | 8.5 | 9.5 | 2.75 | 3.875 | 5 |
| $C_{20}$ | 7.25 | 8.375 | 9.5 | 3.75 | 4.875 | 6 |
| $C_{21}$ | 7.75 | 8.75 | 9.75 | 1.75 | 2.75 | 3.75 |
| $C_{22}$ | 5.75 | 6.75 | 7.75 | 3 | 4 | 5 |

$$\begin{aligned}\tilde{d}_{11} &= \frac{(0.8, 0.913, 0.95) \ominus (0.15, 0.187, 0.22)}{(0.95 - 0.15)} \\ &= \frac{(0.8 - 0.22), (0.913 - 0.187), (0,95 - 0.15)}{(0.95 - 0.15)} \\ &= (0.725, 0.908, 1)\end{aligned}$$

Normalization with cost criterion:

**Table 13.9** Normalized fuzzy difference values of alternatives

| Criteria | SE | WE | HE | LFG-E |
|---|---|---|---|---|
| $C_1$ | (0.725, 0.908, 1) | (0.5, 0.779, 0.875) | (0.375, 0.687, 0.813) | (−0.188, 0, 0.188) |
| $C_2$ | (−0.208, 0.125, 0.458) | (−0.208, 0.125, 0.458) | (−0.333, 0, 0.333) | (0.333, 0.667, 1) |
| $C_3$ | (−0.667, 0.042, 0.75) | (−0.75, 0, 0.75) | (−0.5, 0.25, 1) | (−0.583, 0.167, 0.917) |
| $C_4$ | (0.997, 0.999, 1) | (0.855, 0.857, 0.858) | (0.955, 0.957, 0.958) | (−0.002, 0, 0.002) |
| $C_5$ | (−0.002, 0, 0.002) | (0.228, 0.23, 0.232) | (0.995, 0.997, 1) | (0.075, 0.077, 0.078) |
| $C_6$ | (0.781, 0.783, 0.784) | (0.985, 0.986, 0.987) | (0.998, 0.999, 1) | (−0.002, 0, 0.002) |
| $C_7$ | (−0.002, 0, 0.002) | (0.958, 0.959, 0.961) | (0.998, 0.999, 1) | (0.056, 0.058, 0.06) |
| $C_8$ | (0.786, 0.789, 0.793) | (0.994, 0.997, 1) | (−0.004, 0, 0.004) | (0.578, 0.582, 0.585) |
| $C_9$ | (−0.002, 0, 0.002) | (0.151, 0.153, 0.155) | (0.161, 0.163, 0.165) | (0.998, 0.999, 1) |
| $C_{10}$ | (0.634, 0.636, 0.638) | (0.997, 0.998, 1) | (−0.003, 0, 0.003) | (0.676, 0.678, 0.68) |
| $C_{11}$ | (0.997, 0.998, 1) | (0.878, 0.879, 0.881) | (0.737, 0.738, 0.74) | (−0.001, 0, 0.001) |
| $C_{12}$ | (0.11, 0.111, 0.112) | (−0.001, 0, 0.001) | (0.996, 0.998, 1) | (0.274, 0.275, 0.276) |
| $C_{13}$ | (0.396, 0.399, 0.401) | (0.994, 0.997, 1) | (0.994, 0.997, 1) | (−0.002, 0, 0.002) |
| $C_{14}$ | (0.114, 0.117, 0.121) | (0.878, 0.88, 0.883) | (0.995, 0.998, 1) | (−0.003, 0, 0.003) |
| $C_{15}$ | (−0.001, 0, 0.001) | (0.403, 0.404, 0.406) | (0.997, 0.998, 1) | (0.178, 0.179, 0.18) |
| $C_{16}$ | (0.928, 0.929, 0.93) | (0.997, 0.999, 1) | (0.602, 0.604, 0.606) | (−0.002, 0, 0.002) |
| $C_{17}$ | (−0.153, 0.081, 1) | (−0.163, 0, 0.163) | (−0.168, 0, 0.319) | (−0.157, 0.026, 0.131) |
| $C_{18}$ | (0.997, 0.999, 1) | (0.997, 0.999, 1) | (0.997, 0.999, 1) | (−0.002, 0, 0.002) |
| $C_{19}$ | (−0.296, 0, 0.296) | (−0.185, 0.111, 0.407) | (0.37, 0.685, 1) | (0.111, 0.426, 0.741) |
| $C_{20}$ | (−0.391, 0, 0.391) | (−0.217, 0.174, 0.565) | (0.217, 0.609, 1) | (0, 0.391, 0.783) |
| $C_{21}$ | (−0.25, 0, 0.25) | (0.344, 0.594, 0.844) | (0.5, 0.75, 1) | (0.219, 0.469, 0.719) |
| $C_{22}$ | (−0.421, 0, 0.421) | (−0.316, 0.132, 0.579) | (−0.053, 0.395, 0.842) | (0.158, 0.579, 1) |

$$\widetilde{d}_{51} = \frac{(11.538, 11.55, 11.562) \ominus (11.538, 11.55, 11.562)}{(25.225 - 11.538)}$$
$$= \frac{(11.538 - 11.562), (11.55 - 11.55), (11.562 - 11.538)}{(25.225 - 11.538)}$$
$$= (-0.002, 0, 0.002)$$

All the other normalization calculations were done in the same way and the results are presented in Table 13.9.

#### 13.4.1.4 Calculation of $\widetilde{S}_j$, $\widetilde{R}_j$ and $\widetilde{Q}_j$ Values

$\widetilde{S}_j$ and $\widetilde{R}_j$ values computed were using Eqs. 13.10 and 13.11 with the data listed in Table 13.9. For $\widetilde{Q}_j$ value, 13 was used and $v$ value was estimated as 0.52 utilizing the formula in Step 4. Examples of calculation method were presented for fuzzy $S$, $R$ and $Q$ values. All the results of the computations are placed in Table 13.10.

$$\widetilde{S}_1 = [(0.775, 0.875, 0.975) \otimes (0.725, 0.908, 1)] \oplus \cdots \oplus [(0.375, 0.488, 0.6) \otimes (-0.421, 0, 0.421)] = (0.562, 0.795, 0.975) \oplus \cdots \oplus (-0.158, 0, 0.253) = (3.178, 5.846, 9.641)$$

$$\widetilde{R}_1 = MAX_1([(0.775, 0.875, 0.975) \otimes (0.725, 0.908, 1)], \ldots, [(0.375, 0.488, 0.6) \otimes (-0.421, 0, 0.421)] = (0.562, 0.795, 0.975), \ldots, (-0.158, 0, 0.253) = (0.748, 0.849, 0.9)$$

$$v = (22 + 1)/44 = 0.52$$
$$\widetilde{Q}_1 = 0.52 \frac{[(3.178, 5.846, 9.641) \ominus (1.710, 4.137, 7.096)]}{13.778 - 1.710} + \frac{(1 - 0.52)[(0.748, 0.849, 0.975) \ominus (0.732, 0.832, 0.933)]}{0.975 - 0.732}$$
$$= (0.534, 0.106, 0.822)$$

**Table 13.10** Fuzzy $S$, $R$ and $Q$ values of the alternatives

| | $\widetilde{S}_j$ | $\widetilde{R}_j$ | $\widetilde{Q}_j$ |
|---|---|---|---|
| SE | (3.178, 5.846, 9.641) | (0.748, 0.849, 0.975) | (−0.534, 0.106, 0.822) |
| WE | (5.507, 8.724, 12.38) | (0.748, 0.849, 0.95) | (−0.434, 0.266, 0.89) |
| HE | (6.312, 9.754, 13.778) | (0.748, 0.849, 0.95) | (−0.399, 0.275, 0.951) |
| LFG-E | (1.71, 4.137, 7.096) | (0.732, 0.832, 0.933) | (−0.63, 0, 0.63) |

**Table 13.11** *Q*, *S* and *R* ranking list of alternatives

| | Crisp scores | | | Ranking | | |
|---|---|---|---|---|---|---|
| Alternatives | $Q$ | $S$ | $R$ | $\{A\}_Q$ | $\{A\}_s$ | $\{A\}_R$ |
| SE | 0.125 | 6.128 | 0.855 | 2 | 2 | 3 |
| WE | 0.247 | 8.834 | 0.849 | 3 | 3 | 2 |
| HE | 0.275 | 9.899 | 0.849 | 4 | 4 | 2 |
| LFG-E | 0 | 4.27 | 0.832 | 1 | 1 | 1 |

#### 13.4.1.5 Defuzzification and Ranking the Alternatives

This study adopts the defuzzification method given as a formula in Step 5 to obtain crisp scores of fuzzy numbers. We obtained Table 13.11 showing the crisp score of $S$, $R$ and $Q$ values and ranked list of alternatives based on their crisp scores. Consequently, there are three ranking lists of alternatives and fourth alternative that is LFG power plants is in the first order in each ranking list.

According to the result of fuzzy VIKOR application, LFG is the best performing option among the alternatives. $Q$ value of first and second alternative is 0 and 0.125 respectively and our $DQ$ value is 0.33. According to the methodology the difference of $Q$ values of the first and second alternatives should be less than the threshold $DQ$. Therefore, the results do not satisfy Condition 13.1 in VIKOR methodology, which states there is a considerable difference "acceptable advantage" between the alternatives. It means that LFG is still our best compromise solution; on the other hand selection of LFG among the alternatives as a sustainable energy resource does not far outweigh the other alternatives. Rests of the alternatives too are in the set of compromise solutions and a decision maker may prefer one of them.

$$Crisp\left(\tilde{S}_1\right) = \frac{2(5.846) + 3.178 + 9.641}{4} = 6.128$$

$$Crisp\left(\tilde{R}_1\right) = \frac{2(0.849) + 0.748 + 0.975}{4} = 0.855$$

$$Crisp\left(\tilde{Q}_1\right) = \frac{2(0.106) - 0.534 + 0.822}{4} = 0.125$$

#### 13.4.1.6 Categorical Analysis of the Alternatives

We want to learn the performance of the alternatives separately by technical, economical and environmental aspects. Technical criteria are from $C_1$ to $C_9$, economical is from $C_{10}$ to $C_{16}$ and environmental $C_{17}$ to $C_{22}$. Same calculation steps of VIKOR technique were applied on corresponding criteria and ranks of the alternatives were obtained in three aspects.

**Table 13.12** Fuzzy and crisp *S*, *R* and *Q* values of the alternatives in technical aspect

| Technical | SE | WE | HE | LFG-E |
|---|---|---|---|---|
| $\widetilde{S}_j$ | (1.438, 2.722, 4.173) | (1.886, 3.448, 5.116) | (0.778, 1.982, 3.45) | (0.778, 1.982, 3.45) |
| $S_j$ | 2.764 | 3.475 | 3.146 | 2.048 |
| $\widetilde{R}_j$ | (0.698, 0.799, 0.975) | (0.599, 0.71, 0.853) | (0.669, 0.765, 0.925) | (0.732, 0.832, 0.933) |
| $R_j$ | 0.818 | 0.718 | 0.781 | 0.832 |
| $\widetilde{Q}_j$ | (−0.441, 0.2, 0.881) | (−0.503, 0.186, 0.854) | (−0.466, 0.21, 0.904) | (−0.484, 0.146,0.739) |
| $Q_j$ | 0.21 | 0.181 | 0.215 | 0.137 |
| Rank | 3 | 2 | 4 | 1 |

According to VIKOR technique application, LFG-E is the best alternative. Here we observe that LFG-E is in the first place in terms of three aspects. It shows the consistency of VIKOR application result. Similarly, HE always takes the last place in the technical, economical and environmental rank. Solar energy is the second

**Table 13.13** Fuzzy and crisp *S*, *R* and *Q* values of the alternatives in economical aspect

| Economical | SE | WE | HE | LFG-E |
|---|---|---|---|---|
| $\widetilde{S}_j$ | (1.854, 2.21, 2.569) | (3.209, 3.774, 4.341) | (3.51, 4.095, 4.683) | (0.787, 0.91, 1.036) |
| $S_j$ | 2.211 | 3.775 | 4.096 | 0.911 |
| $\widetilde{R}_j$ | (0.723, 0.836, 0.95) | (0.698, 0.799, 0.9) | (0.747, 0.848, 0.95) | (0.473, 0.542, 0.612) |
| $R_j$ | 0.836 | 0.799 | 0.848 | 0.542 |
| $\widetilde{Q}_j$ | (0.219, 0.455, 0.691) | (0.395, 0.65, 0.905) | (0.483, 0.742, 1) | (−0.162, 0, 0.162) |
| $Q_j$ | 0.455 | 0.65 | 0.742 | 0 |
| Rank | 2 | 3 | 4 | 1 |

**Table 13.14** Fuzzy and crisp *S*, *R* and *Q* values of the alternatives in environmental aspect

| Environmental | SE | WE | HE | LFG-E |
|---|---|---|---|---|
| $\widetilde{S}_j$ | (0.115, 0.913, 2.898) | (0.412, 1.502, 2.923) | (1.284, 2.54, 4.267) | (0.144, 1.245, 2.61) |
| $S_j$ | 1.152 | 1.584 | 2.658 | 1.311 |
| $\widetilde{R}_j$ | (0.748, 0.849, 0.95) | (0.748, 0.849, 0.95) | (0.748, 0.849, 0.95) | (0.109, 0.362, 0.704) |
| $R_j$ | 0.849 | 0.849 | 0.849 | 0.384 |
| $\widetilde{Q}_j$ | (−0.377, 0.243, 0.829) | (−0.307, 0.321, 0.822) | (−0.191, 0.458, 1) | (−0.661, 0.044, 0.658) |
| $Q_j$ | 0.235 | 0.289 | 0.431 | 0.021 |
| Rank | 2 | 3 | 4 | 1 |

best alternative. However it falls in the third order in technical category, wind energy is placed in the second order. A decision maker who puts great emphasize on technical performance may select wind energy type when compared to solar energy (Tables 13.12, 13.13 and 13.14).

## *13.4.2 Application of Fuzzy TODIM Technique*

### 13.4.2.1 Determination of Criteria Weight and Performance Values of the Alternatives

In this section we will apply a different technic TODIM to the same case. Criteria weighting and performance rating are mutual phases in both techniques. Therefore we can use the criteria weights and performance rating values calculated in VIKOR technic application.

Our data is in different units thus we need to normalize performance values by using Eqs. 13.19 and 13.20. An example of normalization in terms of benefit and cost criterion is shown:

For benefit criterion:

$$\tilde{r}_{11} = \left(\frac{0.15}{0.95}, \frac{0.187}{0.95}, \frac{0.22}{0.95}\right) = (0.158, 0.196, 0.232)$$

For cost criterion:

$$\tilde{r}_{15} = \left(\frac{11.538}{11.562}, \frac{11.538}{11.55}, \frac{11.538}{11.538}\right) = (0.998, 0.99, 1)$$

After this standardization process we constructed normalized performance matrix. It is shown in Table 13.15.

### 13.4.2.2 Defuzzification of Fuzzy Criteria Weights

Fuzzy criteria weights given in Table 13.5 were defuzzified by using Eqs. (13.21) and (13.22) and the obtained crisp weights were normalized by using Eq. (13.24) The results are presented in Table 13.16. In the following, examples of the calculations are given.

**Table 13.15** Normalized performance matrix of TODIM

| | SE | WE | HE | LFG-E |
|---|---|---|---|---|
| $C_1$ | (0.158, 0.196, 0.232) | (0.263, 0.305, 0.421) | (0.316, 0.383, 0.526) | (0.842, 0.961, 1) |
| $C_2$ | (0.703, 0.811, 0.919) | (0.703, 0.811, 0.919) | (0.784, 0.892, 1) | (0.351, 0.459, 0.568) |
| $C_3$ | (0.757, 0.865, 0.973) | (0.757, 0.878, 1) | (0.676, 0.797, 0.919) | (0.703, 0.824, 0.946) |
| $C_4$ | (0.151, 0.152, 0.152) | (0.272, 0.272, 0.272) | (0.187, 0.187, 0.188) | (0.998, 0.999, 1) |
| $C_5$ | (0.998, 0.999, 1) | (0.784, 0.785, 0.786) | (0.457, 0.458, 0.458) | (0.915, 0.916, 0.917) |
| $C_6$ | (0.216, 0.217, 0.217) | (0.014, 0.014, 0.014) | (0.0004, 0.0004, 0.0004) | (0.998, 0.999, 1) |
| $C_7$ | (0.998, 0.999, 1) | (0.04, 0.04, 0.04) | (0.001, 0.001, 0.001) | (0.94, 0.941, 0.942) |
| $C_8$ | (0.611, 0.612, 0.612) | (0.509, 0.51, 0.51) | (0.998, 0.999, 1) | (0.713, 0.714, 0.714) |
| $C_9$ | (0.998, 0.999, 1) | (0.855, 0.856, 0.857) | (0.846, 0.847, 0.848) | (0.068, 0.068, 0.068) |
| $C_{10}$ | (0.529, 0.53, 0.53) | (0.262, 0.263, 0.263) | (0.998, 0.999, 1) | (0.498, 0.499, 0.499) |
| $C_{11}$ | (0.247, 0.247, 0.248) | (0.272, 0.272, 0.272) | (0.308, 0.308, 0.308) | (0.998, 0.999, 1) |
| $C_{12}$ | (0.795, 0.796, 0.797) | (0.998, 0.999, 1) | (0.303, 0.303, 0.303) | (0.611, 0.612, 0.612) |
| $C_{13}$ | (0.713, 0.714, 0.714) | (0.499, 0.5, 0.5) | (0.499, 0.5, 0.5) | (0.998, 0.999, 1) |
| $C_{14}$ | (0.927, 0.928, 0.929) | (0.463, 0.464, 0.464) | (0.392, 0.392, 0.393) | (0.998, 0.999, 1) |
| $C_{15}$ | (0.998, 0.999, 1) | (0.45, 0.451, 0.451) | (0.25, 0.25, 0.25) | (0.649, 0.649, 0.65) |
| $C_{16}$ | (0.2, 0.2, 0.2) | (0.14, 0.14, 0.14) | (0.479, 0.48, 0.48) | (0.998, 0.999, 1) |
| $C_{17}$ | (0.003, 0.024, 0.154) | (0.016, 0.077, 0.333) | (0.008, 0.077, 1) | (0.02, 0.044, 0.2) |
| $C_{18}$ | (0.114, 0.114, 0.114) | (0.114, 0.114, 0.114) | (0.114, 0.114, 0.114) | (0.998, 0.999, 1) |
| $C_{19}$ | (0.789, 0.895, 1) | (0.711, 0.816, 0.921) | (0.289, 0.408, 0.526) | (0.474, 0.592, 0.711) |
| $C_{20}$ | (0.763, 0.882, 1) | (0.658, 0.776, 0.895) | (0.395, 0.513, 0.632) | (0.526, 0.645, 0.763) |
| $C_{21}$ | (0.795, 0.897, 1) | (0.308, 0.41, 0.513) | (0.179, 0.282, 0.385) | (0.41, 0.513, 0.615) |
| $C_{22}$ | (0.742, 0.871, 1) | (0.645, 0.79, 0.935) | (0.484, 0.629, 0.774) | (0.387, 0.516, 0.645) |

**Table 13.16** Defuzzificated, normalized and relative weights of the criteria

| | $V_j$ | $W_j$ | $W_{jr}$ |
|---|---|---|---|
| $C_1$ | 0.729 | 0.065 | 1 |
| $C_2$ | 0.618 | 0.055 | 0.847 |
| $C_3$ | 0.642 | 0.058 | 0.88 |
| $C_4$ | 0.6 | 0.054 | 0.823 |
| $C_5$ | 0.069 | 0.006 | 0.094 |
| $C_6$ | 0.26 | 0.023 | 0.357 |
| $C_7$ | 0.26 | 0.023 | 0.357 |
| $C_8$ | 0.464 | 0.042 | 0.636 |
| $C_9$ | 0.656 | 0.059 | 0.899 |
| $C_{10}$ | 0.6 | 0.054 | 0.823 |
| $C_{11}$ | 0.66 | 0.059 | 0.905 |
| $C_{12}$ | 0.642 | 0.058 | 0.88 |
| $C_{13}$ | 0.577 | 0.052 | 0.792 |
| $C_{14}$ | 0.685 | 0.061 | 0.939 |
| $C_{15}$ | 0.577 | 0.052 | 0.792 |
| $C_{16}$ | 0.114 | 0.01 | 0.157 |
| $C_{17}$ | 0.577 | 0.052 | 0.792 |
| $C_{18}$ | 0.685 | 0.061 | 0.939 |
| $C_{19}$ | 0.685 | 0.061 | 0.939 |
| $C_{20}$ | 0.521 | 0.047 | 0.714 |
| $C_{21}$ | 0.331 | 0.03 | 0.455 |
| $C_{22}$ | 0.197 | 0.018 | 0.27 |

$$V\left(\tilde{C}_1\right) = 0.875\left\{0.5\left[\frac{0.975-0.175}{0.975-0.175+0.975-0.875}\right] + (1-0.5)\left[1-\frac{0.975-0.775}{0.975-0.175+0.875-0.775}\right]\right\} = 0.729$$

$$\sum_{j=1}^{22} V\left(\tilde{C}_j\right) = 0.729 + 0.618 + \cdots + 0.197 = 11.147$$

$$w_1 = \frac{0.729}{11.147} = 0.065$$

#### 13.4.2.3 Calculation of Gains and Losses

After the standardization (normalization) process of the performance ratings, all criteria turned into benefit criteria features. For example before normalization, the biggest value is the best value for benefit criteria and the smallest value is the best value for cost criteria. After normalization, all criteria's best value is the biggest value of the performance ratings. Therefore gain and loss matrices were calculated according to the benefit criteria calculation method 26–27. An example of calculation are presented as follows:

$$\tilde{x}_{21} = (0.2631, 0.3052, 0.4211) > \tilde{x}_{12} = (0.1579, 0.1964, 0.2316)$$

$$d(\tilde{x}_{21}, \tilde{x}_{12}) = \sqrt{\frac{1}{3}\left[(0.2632 - 0.1579)^2 + (0.3052 - 0.1964)^2 + (0.4211 - 0.2316)^2\right]}$$
$$= 0.1400$$

$$G_{21}^{1} = 0.1400,\ L_{21}^{1} = 0\ and\ L_{12}^{1} = -0.1400$$

$$G_1 = \begin{bmatrix} 0 & 0 & 0 & 0 \\ 0.1400 & 0 & 0 & 0 \\ 0.2210 & 0.0814 & 0 & 0 \\ 0.7402 & 0.6058 & 0.5280 & 0 \end{bmatrix}$$

$$L_1 = \begin{bmatrix} 0 & -0.1400 & -0.2210 & -0.7402 \\ 0 & 0 & -0.0814 & -0.6058 \\ 0 & 0 & 0 & -0.5280 \\ 0 & 0 & 0 & 0 \end{bmatrix}$$

As it can be noticed, the loss matrix is transpose of the gain matrix with minus sign. These calculations were iterated for all the 22 criteria and at the end we have 44 matrices, half of them belong to gain matrix and the other half belong to loss matrix.

#### 13.4.2.4 Calculation of Criteria's Relative Weights $w_{jr}$

$w_{jr}$ values were calculated using Eq. 13.30. The reference criterion is $C_1$ with the highest importance weight. The results presented in Table 13.16 in the third column. An example is as follows:

$$w_{1r} = \frac{0.065}{0.065} = 1$$

#### 13.4.2.5 Construction of Dominance Degree Matrix

In this study $\theta$ values are specified as 1, 2.5, 3 and 4. An example of calculations regarding to first criterion and for $\theta$ equals 1 is presented in the following:

$$\phi_{12}^{1(+)} = \sqrt{(0)(1)/(15.287)} = 0$$

$$\phi_{12}^{1(-)} = -\frac{1}{1}\sqrt{(0.14)(15.287)/(1)} = -1.4630$$

$$\phi_{12}^{1} = -1.4630$$

$$\phi_1 = \begin{bmatrix} 0 & -1.4630 & -1.8379 & -3.3638 \\ 0.0957 & 0 & -1.1152 & -3.0431 \\ 0.1202 & 0.0729 & 0 & -2.8409 \\ 0.2200 & 0.1991 & 0.1858 & 0 \end{bmatrix}$$

Same calculations repeated separately for rest of the criteria when $\theta$ equals 1, 2.5, 3 and 4.

#### 13.4.2.6 Construction of Overall Dominance Degree Matrix

Overall dominance degree matrices were constructed for each $\theta$ value as in the following formula.

$$\delta_{12} = \phi_{12}^{1} + \phi_{12}^{2} + \cdots + \phi_{12}^{22} = -1.463 + 0 + \cdots + 0.038 = -6.1786$$

for $\theta = 1$.

$$\Delta_1 = \begin{bmatrix} 0 & -1.6934 & -6.0760 & -12.8445 \\ -15.5272 & 0 & -6.4835 & -20.9836 \\ -18.1072 & -10.7689 & 0 & -23.4762 \\ -10.7193 & -5.4432 & -4.8552 & 0 \end{bmatrix}$$

#### 13.4.2.7 Calculation of Overall Values and Ranking the Alternatives

In the following the calculations were made for $\theta = 1$. Results for the other $\theta$ values were given under the sensitivity analysis in Sect. 4.2.8. From first row of the $\Delta_1$ matrix:

$$\delta_{11} = 0,\ \delta_{12} = -6.1786,\ \delta_{13} = -17.5832 \text{ and } \delta_{14} = -33.8932$$

$$\sum_{k=1}^{4} \delta_{1k} = 0 - 6.1786 - 17.5832 - 33.8932 = -57.6549$$

$$\sum_{k=1}^{4} \delta_{2k} = -110.9697$$

$$\sum_{k=1}^{4} \delta_{3k} = -134.4190$$

$$\sum_{k=1}^{4} \delta_{4k} = -60.8709$$

As it is stated in Eq. 13.34:

$$\xi(\text{SE}) = [(-57.6549) - (-134.4190)]/[(-57.6549) - (-134.4190)] = 1$$

$\xi(\text{WE}) = 0.3055$, $\xi(\text{HE}) = 0$ and $\xi(\text{LFG} - \text{E}) = 0.9581$
According to the results rank of the alternatives for $\theta = 1$ were obtained as:

SE > LFG - E > WE > HE.

#### 13.4.2.8 Sensitivity Analysis for TODIM

We conducted a sensitivity analysis for TODIM to see the effect of different situations on the results. In this regard, in order to analyze the influence of the parameter $\theta$, we tried different $\theta$ values in the application. Table 13.17 shows overall value of the alternatives for each $\theta$ value and ranking lists are in Table 13.18. The order of the alternatives changed at $\theta = 3$ and LFG-E and SE interchanged. This order stayed the same when $\theta$ is equal to 4. The suggested $\theta$ interval is between 2 and 2.5 (Krohling and Souza 2012). In this case, energy investors may base on the ranking list of $\theta = 2.5$. On the other hand, an investor who concerns and wants to prevent risk may prefer the order for $\theta = 3\, or\, 4$.

The emergent shape in Fig. 13.3 has the same characteristic feature with S-shaped graph of prospect theory. In Fig. 13.3, for the gain part, $x$-axis represents real gain calculated by Eq. 13.26 and $y$-axis represents dominance degree for gain in Eq. 13.31. In a similar manner, for the loss part $x$-axis represents real loss calculated by Eq. 13.27 and $y$-axis represents dominance degree for lost in Eq. 13.32. Consequently in the graph of TODIM method $x$-axis is for the real gain and loss values, and $y$-axis reflects the effects of those gains and losses. In Fig. 13.3, we can observe that the effect of loss is more than the loss itself. This difference is decreasing as $\theta$ value increase and that is why bigger value of $\theta$ gives secure results in terms of risk.

In order to better visualize gain part of the Fig. 13.3, we drew it separately as Fig. 13.4 and concavity of the graph is apparent. Here the difference of gain and its effect is very close to each other. This situation results from prospect theory's characteristic features. For more explanation see the introduction section.

**Table 13.17** Overall value of the alternatives in different $\theta$ values

| | $\theta = 1$ | $\theta = 2.5$ | $\theta = 3$ | $\theta = 4$ |
|---|---|---|---|---|
| $\xi(SE)$ | 1 | 1 | 0.3068 | 0.9856 |
| $\xi(WE)$ | 0.3055 | 0.2948 | 0.0255 | 0.2809 |
| $\xi(HE)$ | 0 | 0 | 0 | 0 |
| $\xi(LFG - E)$ | 0.9581 | 0.9873 | 1 | 1 |

**Table 13.18** Rank of the alternatives by $\theta$ values

| $\theta$ values | Rank of the alternatives |
|---|---|
| $\theta = 1$ | $SE > LFG - E > WE > HE$ |
| $\theta = 2.5$ | $SE > LFG - E > WE > HE$ |
| $\theta = 3$ | $LFG - E > SE > WE > HE$ |
| $\theta = 4$ | $LFG - E > SE > WE > HE$ |

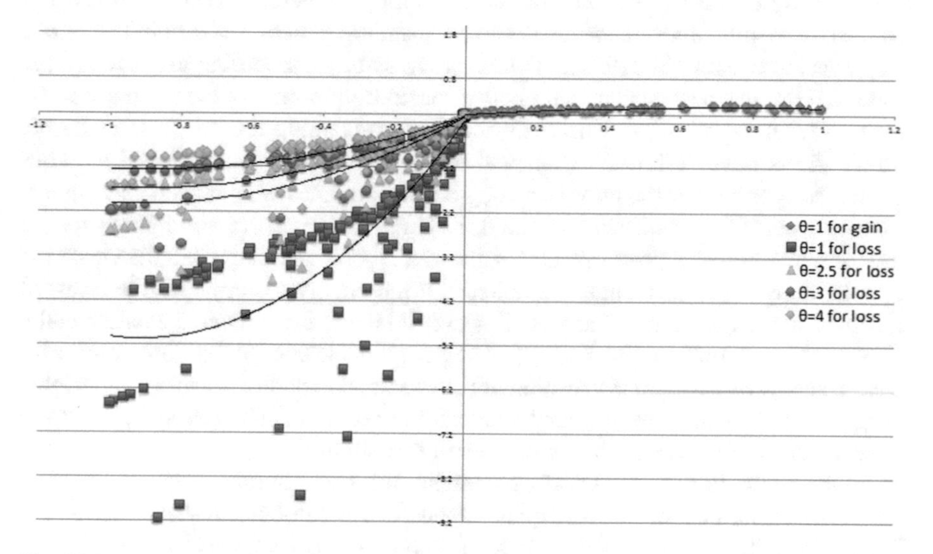


**Fig. 13.3** Value function of TODIM method application with different $\theta$ values

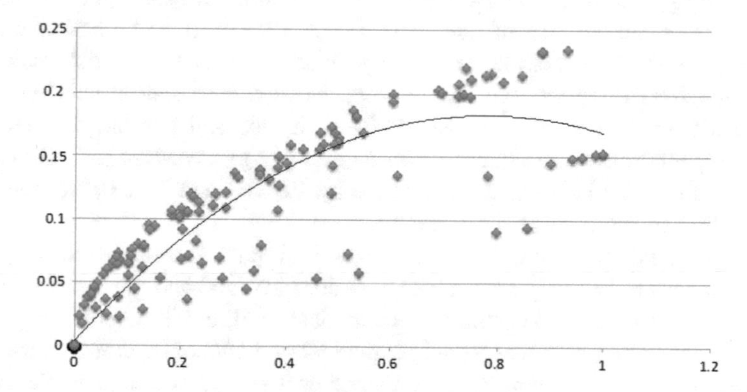


**Fig. 13.4** Gain function of TODIM application

## 13.5 Conclusion

In this chapter we have presented a solution regarding strategic energy source selection of a firm in energy sector. Sustainable and renewable energy alternatives have been evaluated by fuzzy VIKOR and fuzzy TODIM methods separately. VIKOR and TODIM are both multi-criteria decision-making techniques. Their basic principles are to rank the alternatives through specified criteria. Uncertainty analysis is a fundamental stage in energy planning process. We could integrate fuzzy sets into both VIKOR and TODIM successfully. According to results of the case, LFG is the best sustainable energy source option among the alternatives by fuzzy VIKOR application. The main reason for obtaining this result is that LFG power plants have performed very well in the technical, economical and environmental categories. For the criteria having high importance weight such as technical efficiency and GHG emissions avoided, LFG has best performance rating most of the time. Solar energy power plants are the second best alternative but the weakness of solar energy is in its technical efficiency. Wind energy performs well in terms of technical and environmental aspects however it is not very attractive economically. The worst alternative is hydraulic. Hydraulic power plants are not very environment-friendly energy production systems comparing to other renewable energy sources. Although they are technically and economically effective, hydraulic energy falls behind the other alternatives in this study.

In the evaluation with TODIM technique, the best alternative is solar energy under the normal conditions ($\theta$ equals 1 and 2.5). It can be said that VIKOR and TODIM results are consistent with each other. LFG-E is the best alternative but $Q$ values of the alternatives are close to each other, thus it doesn't have an acceptable advantage on the other alternatives. TODIM method results in LFG-E as the best energy alternative in the case of risk aversion when $\theta$ equals 3 and 4. In other words as the concern of risk increase, LFG becomes to be best alternative. According to TODIM method, an investor reluctant to deal with risk may prefer LFG-E and it satisfies the decision makers. However, to choose solar energy to invest can't be defined as a wrong decision, it also satisfies and is capable of meeting expectations. Consequently, we suggest energy investors LFG-E to invest, which reveals as the best energy alternative by VIKOR and TODIM technic and a secure option in the case of risk aversion.

In this evaluation system, first three most important criteria are technical efficiency, impact on the ecosystem, GHG emissions avoided and government support rate. As a consequence technical, economical and environmental aspects of renewable energies are almost equally important and cannot be thought separately. Analyzing a power plant considering only one or two aspects of renewable energies may mislead decision makers and the results may not be reliable. We conducted this study regarding all the important criteria within technical, economical and environmental scope. This makes our results more quality and improves the reliability.

Waste creates both economical and environmental problems in the cities and LFG power plants are a smart and efficient way of eliminating and utilizing of waste while producing energy. Therefore municipalities need a comprehensive waste management policy to use LFG opportunity and so to create a sustainable environment in the cities. LFG power plants are followed by solar, wind and hydraulic alternatives.

For the further studies, a research can be conducted locally in a specific region to find out best performing alternatives regarding that area to increase utilization of renewable energy. Different sophisticated economic applications like real options can be applied in lieu of net present value.

This is a comparative study of fuzzy VIKOR and fuzzy TODIM techniques. TODIM technic provides us a pairwise comparison between the alternatives and by this means we can check gains and losses of any two alternatives regarding any criteria. TODIM method differs from VIKOR by this feature. On the other hand, VIKOR is a distance based method and ranks the alternatives accordingly. Besides ranking the alternatives, VIKOR yields a solution set and an alternative can be preferred in that set. Finally, the main difference between the techniques is that TODIM adds risk factor in the system and it enriches the scope of evaluation process. In the literature, there are remarkable amount of VIKOR studies in different fields. VIKOR application papers in sustainable and renewable energy area are nearly at 8% (Mardani et al. 2016). VIKOR technic part of this comprehensive study enriches the literature. On the other hand, TODIM and fuzzy TODIM relatively new techniques; there are not sufficient numbers of studies yet. Especially in terms of fuzzy TODIM, there is very limited number of papers and this study is one of the first applications of evaluation of sustainable and renewable energy systems by fuzzy TODIM. Therefore this study may be a reference for next studies in the energy field.

## References

Abdel-Kader, M.G., & Dugdale, D. (2001). Evaluating investments in advanced manufacturing technology: a fuzzy set theory approach, *The British Accounting Review, 33*, 455–489.

Abdullah, L., & Najib, L. (2014). Sustainable energy planning decision using the intuitionistic fuzzy analytic hierarchy process: Choosing energy technology in Malaysia. *International Journal of Sustainable Energy, 35*(4), 360–377.

Afgan, N., & Carvalho, M. (2001). Multi-criteria assessment of new and renewable energy power plants. *Energy, 27,* 739–755.

Al Garni, H., Kassem, A., Awasthi, A., Komljenovic, D., & Al-Haddad, K. (2016). A MCDM approach for evaluating renewable power generation sources in Saudi Arabia. *Sustainable Energy Technologies and Assessments, 16,* 137–150.

Anadolu ajansı. (2017). Çöpten 400 bin konutun elektriği çıkıyor. URL: www.dunya.com.

Asemi, A., Baba, M., Asemi, A., Rukaini, B. H. A., & Idris, N. (2014). Fuzzy multi criteria decision making applications: A review study.

Beccali, M., Cellura, M., & Mistretta, M. (2003). Decision-making in energy planning. Application of the electre method at regional level for the diffusion of renewable energy technology. *Renewable Energy, 28*(13), 2063–2087.

Büyüközkan, G., & Güleryüz, S. (2017). Evaluation of renewable energy resources in Turkey using an integrated MCDM approach with linguistic interval fuzzy preference relations. *Energy, 123,* 149–163.

Cavallora, F., & Ciarolo, L. (2005). A multicritera approach to evaluate wind energy plants on an Italian island. *Energy Policy, 33,* 235–244.

Chang, T. H. (2014). Fuzzy VIKOR method: A case study of the hospital service evaluation in Taiwan. *Information Sciences, 271,* 2–196.

Chen, S. J., Hwang, C. L., & Hwang, F. P. (1992). *Fuzzy multiple attribute decision making methods and applications* (pp. 88–89). Berlin: Springer.

Dermencioğlu, M. (2017). Yerli güneş paneli hücresiyle maliyeti düşürecek. URL: http://aa.com.tr/tr/bilim-teknoloji/yerli-gunes-paneli-hucresiyle-maliyetidusurecek/770874.

DSİ. (2013). Türkiye'nin Hidrolik Potansiyeli, A., Baba, M., Asemi, A., Rukaini, B. H. A., Idris, N. *Fuzzy Multi Criteria Decision Making Applications: A Review Study* (2014) www.dsi.gov.tr.

Gao, S., Zhang, Z., & Cao, C. (2009). Multiplication operation on fuzzy numbers. *Journal Of Software, 4*(4), 331–338.

GAP Bölge Kalkınma İdaresi Başkanlığı. GAP Nedir? URL: www.gap.gov.tr.

Global Wind Energy Council. (2017a). URL: http://www.gwec.net/global-figures/wind-in-numbers/.

Global Wind Energy Council. (2017b). URL: http://www.gwec.net/global-figures/graphs/.

Gomes, L. F. A. M. & Lima, M. M. P. P. (1992). TODIM: basics and application to multicriteria ranking of projects with environmental impacts, *Foundations of Computing and Decision Sciences, 16*, 113–127.

Hanine, M., Boutkhoum, O., Tikniouine, A., & Agouti, T. (2016). Comparison of fuzzy AHP and fuzzy TODIM methods for land fill location selection. *SpringerPlus, 5,* 1–30. https://doi.org/10.1016/j.rser.2017.07.013.

Hwang, C. L., & Yoon, K. (1981). *Multiple attribute decision making: methods and applications*. New York: Springer.

Iniyan, S., & Sumathy, K. (1998). An optimal renewable energy model for various end-uses. *Energy, 25,* 563–575.

Kahneman, D., & Tversky, A. (1979). Prospect theory: An Analysis of decision under risk. *Econometrica, 47*(2), 263–291.

Kahraman, C., Çevik, S., & Öztaysi, B. (2015). Fuzzy Multicriteria Decision Making: Aliterature review. *International Journal of Computational Intelligence Systems, 8*(4), 637–666. https://doi.org/10.1080/18756891.2015.1046325.

Kaya, T., & Kahraman, C. (2010a). Multicriteria renewable energy planning using an integrated fuzzy VIKOR & AHP methodology: The case of Istanbul. *Energy, 35*(6), 2517.

Kaya, İ., & Kahraman, C. (2010b). A fuzzy multicriteria methodology for selection among energy alternatives. *Expert Systems with Applications, 37*(9), 6270–6281.

Kaya, T., & Kahraman, C. (2011). Multicriteria decision making in energy planning using a modified fuzzy TOPSIS methodology. *Expert Systems with Applications, 38*(6), 6577–6585.

Krohling, R. A., & Souza, T. T. M. (2012). Combining prospect theory and fuzzy numbers to multi-criteria decision making. *Expert Systems with Applications, 39,* 11487–11493.

Mardani, A., Zavadskas, E. K., Govindan, K., Senin, A. A., & Jusoh, A. (2016). VIKOR technique: A systematic review of the state of the art literature on methodologies and applications. *Sustainability, 8,* 1–38.

Mirasgedis, S., & Diakoulaki, D. (1997). Multicriteria analysis vs. externalities assessment for the comparative evaluation of electricity generation systems. *European Journal of Operational Research, 102*(2), 364–379.

Myers, Joe. (2015). 5 expert predictions on how warm the world will get. *World Economic Forum*. https://www.weforum.org/agenda/2015/12/5-expert-predictions-on-howwarm-the-world-will-get/.

Nobre, F. F., Trotta, L. T. F., & Gomes, L. F. A. M. (1999). Multi-criteria decision making—An approach to setting priorities in health care. *Statistics in Medicine, 18*(23), 3345–3354.

Opricovic, S. (2011). Fuzzy VIKOR with an application to water resources planning. *Expert System with Applications, 38*(10), 12983–12990.

Opricovic, S., & Tzeng, G. H. (2004). Compromise solution by MCDM methods: A comparative analysis of VIKOR and TOPSIS. *European Journal of Operational Research, 156*(2), 445–455.

Opricovic, S., & Tzeng, G. H. (2007). Extended VIKOR method in comparison with outranking methods. *European Journal of Operational Research, 178*(2), 514–529.

Paris Agreement. *European Commission,* URL: http://ec.europa.eu/clima/policies/international/negotiations/paris_en.

Qin, Q., Liang, F., Chen, Y. W., & Yu, G. F. (2017). A TODIM-based multi-criteria group decision making with triangular intuitionistic fuzzy numbers. *Applied Soft Computing, 55,* 93–107.

Rajaram, V., Siddiqui, F. Z., & Khan. M. E. (2012). *From Landfill Gas to Energy Technologies and Challenges.* URL: https://books.google.com.tr/books?id=finMBQAAQBAJ&printsec=frontcover&hl=tr#v=onepage&q&f=false.

Rangel, L., Gomes, L., & Cardoso, F. (2011). An Application of the TODIM method to the evaluation of broadband internet plans. *Pesquisa Operacional, 31*(2), 235–249.

Rüzgar Enerjisi Çalışmaları, Yenilenebilir Enerji Genel Müdürlüğü, URL: http://www.eie.gov.tr/eie-web/turkce/YEK/ruzgar/ruzgar_en_hak.html.

Renewable Energy Policy Network. (2016). Global status report. URL: http://www.ren21.net/wp-content/uploads/2016/06/GSR_2016_Full_Report.pdf.

San Cristobal, J. R. (2011). Multicriteria decision making in the selection of a renewable energy project in Spain: The Vikor method. *Renewable Energy, 36*(2), 498–502.

Sellak, H., Ouhbi, B., Frikh, B., & Palommares, I. (2017). Towards next-generation energy planning decision-making: An expert based framework for intelligent decision support. *Renewable and Sustainable Energy Reviews, 80,* 1544–1577. https://doi.org/10.1016/j.rser.2017.07.013.

Sen, D. K., Datta, S., Patel, S. K., & Mahapatra, S. S. (2016). Fuzzy-TODIM for industrial robot selection. In: *International Conference on Emerging Trends in Mechanical Engineering.* pp. 1–8.

Şengül, Ü., Eren, M., Shiraz, S. E., Gezder, V., & Şengül, A. B. (2015). Fuzzy TOPSIS method for ranking renewable energy supply systems in Turkey. *Renewable Energy, 75,* 617–625.

Shirgholami, Z., Zangeneh, A. N., & Bortolini, M. (2016). Decision system to support the practitioners in the wind farm design: A case study for Iran mainland. *Sustainable Energy Technologies and Assessments, 16,* 1–10.

Şimşek, C. (2014). *Avrupa'nın en büyük çöp biyogaz tesisi İstanbul'da.* URL: www.enerjienstitusu.com.

Şimşek, C. (2017). *Türkiye potansiyelinin 25'te birini kullanıyor.* URL: http://enerjienstitusu.com/2017/03/02/turkiye-ruzgar-enerji-potansiyelinde-25te-birini-kullaniyor/.

Streimikiene, D., Balezentis, T., Krisciukaitiene, I., & Balezentis, A. (2012). Prioritizing sustainable electricity production technologies: MCDM approach. *Renewable and Sustainable Energy Reviews, 16*(5), 3302–3311.

Streimikiene, D., Sliogeriene, J., & Turskis, J. (2016). Multi-criteria analysis of electricity generation technologies in Lithuania. *Renewable Energy, 85,* 148–156.

Tasri, A., & Susilawati, A. (2014). Selection among renewable energy alternatives based on a fuzzy analytic hierarchy process in Indonesia. *Sustainable Energy Technologies and Assessments, 7,* 34–44.

Tosun, Ö., & Akyüz, G. (2015). A fuzzy TODIM approach for the supplier selection problem. *International Journal of Computational Intelligent Systems, 8*(2), 317–329.

Tsai, K., & Chou, F. (2011). Developing a fuzzy multiattribute matching and negotiation mechanism for sealed—bid online reverse auctions. *Journal of Theoretical and Applied Electronic Commerce Research, 6*(3), 85–96.

Vagiona, D. G., & Karanikolas, N. M. (2012). A multicriteria approach to evaluate offshore wind farms siting In Greece. *Global NEST Journal, 14*(2), 235–243.

Xiao, Z. and Zhi-ping, F. (2011). A method for linguistic multiple attribute decision making based on TODIM. *International Conference on Management and Service Science*. (pp. 1–4).

Yazdani, M., & Graeml, F. (2014). VIKOR and its applications: A state of the art survey. *International Journal of Strategic Decision Sciences, 5*(2), 56–83.

YEGM. (2017). Güneş Enerjisi Potansiyeli Atlası. URL: http://www.eie.gov.tr/MyCalculator/Default.aspx.

YEGM. (2017). Türkiye' nin hidroelektrik potansiyeli. URL: http://www.eie.gov.tr/yenilenebilir/h_turkiye_potansiyel.aspx.

Yıldırım, G. (2015). Türkiye'nin ilk deniz üstü rüzgar gülü çiftliği projesine 2016 yılında başlanacak. URL: http://enerjienstitusu.com/2015/12/25/turkiyenin-ilk-deniz-ustu-ruzgar-gulu-ciftligi-projesine-2016-yilinda-baslanacak/.

Yıldırım, L., Ata, G., Taşkın, O., & Lale, T. (2014). Sera Gazı Salınımları ve Türkiye' de Yapılan Çalışmalar. *Su Dünyası,* Vol. 135 URL: http://www2.dsi.gov.tr/sudunyasi/135/index.htmlKahraman.

Zadeh, L. A. (1965). Fuzzy Sets. *Information and Control, 8,* 338–353.

Zerpa, J. C. R., & Yusta, J. M. (2015). Application of multicriteria decision methods for electric supply planning in rural and remote areas. *Renewable and Sustainable Energy Reviews, 52,* 557–571.

Zhang, L., Zhou, P., Newton, S., Fang, J., Zhou, D., & Zhang, L. P. (2015). Evaluating clean energy alternatives for Jiangsu, China: An improved multi-criteria decision making method. *Energy, 90,* 953–964.

Zhao, H., & Guo, S. (2015). External benefit evaluation of renewable energy power in China for sustainability. *Sustainability, 7*(5), 4783–4805.

Zimmerman, H. J. (2010). Fuzzy set theory, WIREs. *Computational Statistics, 2,* 317–332.

# Part V
# Operational Analysis

# Chapter 14
# Designing Distributed Real-Time Systems to Process Complex Control Workload in the Energy Industry

**Eduardo Valentin, Rosiane de Freitas and Raimundo Barreto**

**Abstract** The energy industry demands computing system technologies with advanced state-of-the-art techniques to achieve reliability and safety for monitoring and properly dealing with several complex constraints. These computing systems also require delivering correct data at the right time imposing hard real-time constraints, because there are lots of situations where missing critical data may be catastrophic. The challenges faced by computer engineers in the energy industry also include designing distributed real-time systems to process such complex control workload. Besides, the computing system may also demand high energy consumption on its own. In this chapter, we demonstrate how to construct a mathematical formulation applicable for these computing systems and how to solve it to distribute the hard real-time workload of the process control systems considering technological constraints and optimizing for low power consumption of such computing systems. We present two computational techniques of resolution: an exact algorithm based on Branch-and-Cut and a meta-heuristic based on Genetic Algorithm. While the exact algorithm combines a branch-and-cut strategy with response time based schedulability analysis, the genetic algorithm still considers the response time schedulability analysis but follows an evolutionary solving strategy. Both computational techniques deliver solutions for heterogeneous computing systems with a control application, considering precedence, preemption, mutual exclusion, timing, temperature, and capacity constraints. In computational experiments, we present the usage of such techniques in a case study based on a control system for a power plant monitoring application.

---

E. Valentin (✉) · R. de Freitas · R. Barreto
Instituto de Computação - Ufam, Federal University of Amazonas, Manaus, AM, Brazil
e-mail: eduardo.valentin@icomp.ufam.edu.br

R. de Freitas
e-mail: rosiane@icomp.ufam.edu.br

R. Barreto
e-mail: rbarreto@icomp.ufam.edu.br

© Springer International Publishing AG, part of Springer Nature 2018

C. Kahraman and G. Kayakutlu (eds.), *Energy Management—Collective and Computational Intelligence with Theory and Applications*, Studies in Systems, Decision and Control 149, https://doi.org/10.1007/978-3-319-75690-5_14

## 14.1 Introduction

Engineers working in the energy field have an increasing need for state-of-the-art computer system technology. Challenges involve reliability and safety in refinery and power distribution operations. Delivering correct data at the right time imposes hard real-time constraints in these systems, because in many situations missing critical data may be catastrophic. Nevertheless, the computing systems applied in the energy sector may also demand high energy consumption on its own due to the necessity to continuously deliver reliable and trustworthy results.

In large data centers, for example, power consumption is a major concern due to the increasing expense in room cooling systems and mainly due to the expensive power bills. For instance, according to Eric Schmidt, CEO of Google, "What matters most to the computer designers at Google is not speed but, power, low power, because data centers can consume as much electricity as a city" (Markoff and Lohr 2002). Also, according to the Department of Energy (DoE) of the United States, power is one of the major challenges to overcome to achieve the needed computing excellence required to advance in many applications of the energy industry (Department of Energy (DoE) 2014).

In this chapter, we demonstrate how to construct a mathematical formulation applicable for these computing systems and how to solve it to distribute the hard real-time workload of the process control systems considering technological constraints and optimizing for low power consumption of such computing systems.

The organization of this chapter is as follows. We present how a typical computing system architecture and how a hard real-time task model for a energy sector monitoring application look like in Sects. 14.2 and 14.3, respectively. We show how a mathematical formulation can be written associating combinatorial optimization with schedulability analysis in Sect. 14.4. We also explain two strategies to solve such formulation, evolutionary based and branch-and-cut based, in Sects. 14.5 and 14.6. We also exemplify how such advanced techniques can be applied in a case study for a monitoring control application of the energy sector in Sect. 14.7. We close this chapter with final comments in Sect. 14.8.

## 14.2 Typical Multi-processor Architecture

Practitioners execute applications with hard deadline restrictions on multiple heterogeneous processors due to the expected energy consumption reduction. Nevertheless, developing software with timing constraints for multiple heterogeneous processors is a complex task. Scheduling becomes especially hard to deal with, particularly under low power constraints.

Adopting multiple processing elements to enhance the computing capability and to reduce the power consumption is a common design strategy, especially for embedded systems. Therefore, the heterogeneous multicore platforms have become the de facto solution to cope with the rapid increase of system complexity, reliability, and energy consumption (He and Mueller 2012).

For this reason, a simple way to create a processing model for the energy industry applications is to use as reference a Multi-Processor System-On-Chip (MPSoC) architecture. We can state then that the system is composed by a set, H, of m processors, $H = \{H_1, H_2, \ldots, H_m\}$. Each core may operate on l different performance states, $1 \leq k \leq l$. The frequency of performance state k on the processor i is $F_{ik}$ and the power consumption is $P_{ik}$. The set of frequencies of one core is not necessarily the same of other cores. Also, a task may have different code size and execution time for different processors, due to instruction set and performance state differences. The idle power of processor i is $P_{idle,i}$.

## 14.3 Hard Real-Time Workload Model

A typical hard real-time workload can be represented by a task model of periodic tasks. A task model M is a set composed by n task $\tau_j$. A task $\tau_j \in M$, with $1 \leq j \leq n$, has the properties: worst-case execution cycle $WCEC_j$; worst-case execution time $C_j(f)$, which is a function of frequency f, thus $C_j(f) = \frac{WCEC_j}{f}$; period of execution $T_j$; deadline $D_j$. A task $\tau_j$ also has the following properties, specific to fixed priority policies: fixed priority $p_j$; set of high priority tasks $hp_j$ representing the tasks $\tau_j$ with a priority higher than the priority of $\tau_j$. The response time $R_j$ is dependent not only on task set characteristics, but also on the target platform, and on the task allocation and frequency distribution that have been selected for the workload. A task model can be locally processed in a single processor using a fixed priority based on-line scheduler, such as Deadline Monotonic.

Deadline Monotonic (DM) is a fixed priority based on-line scheduler in which task priorities decrease with larger deadlines. Audsley et al. (1993) extend the schedulability test proposed by Lehoczky et al. (1989) for DM, considering the release jitter $J_j$ and the local blocking delay $B_j$ due to semaphore usage. The delay $B_j$ caused by low priority tasks accessing shared resources in the same processors using Priority Ceiling Protocol can be estimated as $B_j = \max_{jk}\left\{D_{jk} | \left(p_j < p_i\right) \wedge (C(S_k) \geq p_i)\right\}$, where C $(S_k)$ is the ceiling priority of the shared resource $S_k$. The schedulability test proposed by Audsley is $R_j \leq D_j$, $\forall 1 \leq j \leq n$, where $R_j = I_j + J_j$.

The task influence $I_j$ in multiple processors may be calculated as $I_j^{n+1} = C_j + B_j^r + B_j + \sum_{p \in hp(j)} \frac{I_j^n + J_p + B_p^r}{T_p} \times C_p$. Precedence constraints can be represented by including the maximum response time of the predecessors tasks in the $J_j$ component of the task $\tau_j$. Also, when precedence constraints occur across different processors, this imposes an additional messaging cost that may be incorporated in the emitting task to perform inter-processor communication. When the Multiprocessor Priority Ceiling Protocol is in place to avoid priority inversion issues, the remote blocking delay $B_j^r$ is an upper bound for the blocking time suffered by task $\tau_j$ from other tasks in a different processor. Response time tests are computationally expensive but provide exact conditions, i.e., sufficient and necessary. The test uses task's WCEC, periods, and the concept of critical instant phasing (Lehoczky et al. 1989).

## 14.4 Mathematical Formulation

A classical mathematical model that resembles modern heterogeneous multicore platforms is the Multilevel Generalized Assignment Problem—MGAP (Glover et al. 1979), though it was originally conceived in the manufacturing context. The MGAP consists of minimizing the assignment cost of a set of jobs to machines, each having associated therewith a capacity constraint. Each machine can perform a job with different performance states that entail different costs and amount of resources required. The MGAP is originally in the context of large manufacturing systems as a more general variant of the well-known Generalized Assignment Problem (GAP). In this paper, we correlate MGAP model with the problem of assigning frequencies and distributing hard real-time tasks on heterogeneous processors, minimizing energy consumption.

Considering the schedulability test proposed by Audsley, we propose the MGAP formulation using tasks response times as seen in Eqs. 14.1a–14.1f, based on the formulation of Valentin et al. (2016b).

$$\text{Minimize}\,\Psi(x) \tag{14.1a}$$

$$\text{s.t.}: \sum_{i=1}^{m}\sum_{k=1}^{l} x_{ijk} = 1,\ j \in \{1,\ldots,n\} \tag{14.1b}$$

$$\sum_{j=1}^{n}\sum_{k=1}^{l}\left(\frac{WCEC_{ij}}{F_{ik}T_j}x_{ijk}\right) \leq 1,\quad i \in \{1,\ldots,m\} \tag{14.1c}$$

$$\Psi_i \leq \frac{\kappa_i^{max} - \kappa_{amb}}{\rho},\quad i \in \{1,\ldots,m\} \tag{14.1d}$$

$$R_j \leq D_j,\quad j \in \{1,\ldots,n\} \tag{14.1e}$$

$$x_{ijk} \in \{0,1\},\quad 1 \leq i \leq m,\ 1 \leq j \leq n,\ 1 \leq k \leq l \tag{14.1f}$$

where the tri-indexed decision variable $x_{ijk}$ represents the distribution and assignment, i.e. when $x_{ijk} = 1$ the task $\tau_j$ executes in the processor i at performance state k, or frequency $F_{ik}$, when $x_{ijk} = 0$, the task $\tau_j$ is distributed somewhere else.

A distribution is a partitioned approach in which each processor i executes a local scheduler responsible for a partition of the real-time task workload and migration is not allowed (see the set of constraints 14.1b). The set of constraints 14.1c represent the maximum system utilization capacity of each processor i. The set of constraints 14.1d represent the temperature limits by creating a linear relation, where $\kappa_{amb}$ is the ambient temperature, $\kappa_i^{max}$ is the maximum junction temperature of each processor i, $\rho$ thermal resistance constant, and $\psi_i$ is power consumption of each processor i. This formulation applies each task deadline as a constraint against their response time in the linear programming (see the set of constraints 14.1e). The matrix $R_j$ is the response time of tasks $\tau_j$ for a given allocation configuration. The response time of each task varies depending on the workload distribution and the frequency assignment of the configuration because a change in the value of $x_{ijk}$ may result in a different computation time ($C_i$). Equation 14.1 is applicable for DM scheduling policy ($D_j \leq T_j$).

We are using an objective function $\Psi(x)$ that minimizes energy consumption, accounting dynamic and idle energy, over the time window represented by the hyperperiod of the real-time tasks, i.e., the Least Common Multiple (LCM) of tasks periods. We extend the objective functions presented by Valentin et al. (2016b) by improving the idle energy estimation. Equations 14.2a–14.2c has the objective function.

$$\text{Minimize } \Psi(x) = \sum_{i=1}^{m} \left(E_{dyn,i}(x) + E_{idle,i}(x)\right) \tag{14.2a}$$

$$E_{dyn,i}(x) = \sum_{j=1}^{n} \sum_{k=1}^{l} \left(\left(\frac{LCM}{T_j}\right) C_l WCEC_{ij} V_{dd,ik}^2 x_{ijk}\right) \tag{14.2b}$$

$$E_{idle,i}(x) = P_{idle,i}\, LCM \left(1 - \sum_{j=1}^{n} \sum_{k=1}^{l} \left(\frac{WCEC_{i,j}}{F_{ik} T_j} x_{ijk}\right)\right) \tag{14.2c}$$

where $E_{dyn,i}$ is the energy consumption when processor i is active, $E_{idle,i}$ is the energy consumption when processor i is idle, $\frac{WCEC_{ij}}{F_{ik} T_j}$ represents the task $\tau_j$ utilization, $u_{ijk}$, while executing in processor i at frequency $F_{ik}$ of performance state k, is the circuit capacitance constant, and $V_{dd,ik}$ is the voltage level to achieve frequency $F_{ik}$.

The term $\left(\frac{LCM}{T_j}\right) C_l WCEC_{ij} V_{dd,ik}^2 x_{ijk}$ represents the dynamic energy associated with the instances of execution of task j within the LCM. Each processor idle energy, within the LCM time window, is computed for its estimated idle time in the term $P_{idle,i} LCM\left(1 - \sum_{j=1}^{n} \sum_{k=1}^{l} \left(\frac{WCEC_{i,j}}{F_{ik} T_j} x_{ijk}\right)\right)$.

The objective function represented in Eqs. 14.2a–14.2c may still be seen as a MGAP formulation. Note that, without loss of generality, when we take the term

$P_{idle,i}$LCM out of the sum, leaving the term $mP_{idle,i}$LCM to be added to the final objective function value, we have $c_{ijk} = \left[\left(\frac{LCM}{T_j}\right) C_l WCEC_{ij} V^2_{dd,ik} - P_{idle,i} LCM\left(\frac{WCEC_{i,j}}{F_{ik} T_j}\right)\right]$.

## 14.5 Computational Techniques of Resolution

In this section, we explain the algorithmic strategy developed for the mathematical formulation of Sect. 14.4. In Sect. 14.5.1, we explain an evolutionary algorithm which produces an initial solution that can be used by the exact algorithm for finding optimal solutions, described in Sect. 14.5.2.

### *14.5.1 Approximation by Means of Evolutionary Algorithm (EA)*

We wrote an evolutionary algorithm (EA), based on genetic algorithm, for each mathematical model (Valentin 2009). We follow a similar approach as existing in the literature for other formulations on this problem (Goossens et al. 2008). The algorithm's input is the processing model H and the desired task model M (see Sect. 14.3). In our EA implementation, a solution is a chromosome that is a sequence of 0's and 1's and each gene represents one of the elements of the tri-indexed decision variable of the mathematical model. The algorithm can be simplified into two steps: (i) Initialization with random-generated individuals and (ii) Generations composed by individuals selected in tournaments and by the evolutionary operators of elitism and crossover. Algorithm 1 illustrates the overall process of our EA strategy and we describe the pieces of the EA as follows.

In the Initialization, we random-generate individuals. Random-generating individuals do not guarantee their feasibility, i.e. the generated individual may be infeasible. The process of validating or transforming individuals into feasible solution is onerous. Even then, we maintain all generations composed by feasible individuals only. We random-generate a large number of individuals, 5000, to start with a high diversity. If none of them is a feasible solution, we return the empty set $\varnothing$. If we find less than 50 feasible individuals, then we return the one with highest fitness. But when we find 50 feasible individuals, we repeat the following steps for a maximum of 100 generations, or 10 generations with same best fitness, and return the individual with best fitness. We perform the Elitism operator by always including the individual with best fitness in the next generation. We execute Selection by means of a tournament in the current population. Only 5 individuals, randomly selected, participate in the tournament. The winner of the tournament is the individual with best fitness among those participating of it. We also insert in the next generation the result of a Crossover between winners of two tournaments. The crossover operation between individuals $I_1$ and $I_2$ is done by means of selecting a pivot gene p. The genes lower than p are copied from $I_1$, the remaining genes are copied from $I_2$. When resulting individual is not feasible, we return $I_1$, if fitness($I_1$) > fitness($I_2$), or $I_2$ otherwise. We define the Fitness function to be: 1/E(individual), where the function E(individual) is the estimated energy consumption for the individual in consideration. The function E is computed using the same energy estimation as in the objective functions of the integer programming mathematical formulations.

**Algorithm 1:** Evolutionary Algorithm (EA)

```
1: Input: H, M
2: Output: best feasible solution ub found
3: /* Random-generate 5000 individuals
4: * (feasible and infeasible),
5: * to start with high diversity.
6: */
7: i ← initialization(5000);
8: /* Select 50 feasible individuals. */
9: p ← feasible(i, 50);
10: if |p|= 0 then
11:     return ∅;
12: end if
13: b ← prev ← best_individual(p);
14: if |p|< 50 then
15:     return b;
16: end if
17: g ← e ← 1;
18: while (g++ ≤ 100)and(e ≤ 10) do
19:     /* Evolve population. */
20:     tournament(p);
21:     selection(p);
22:     elitism(p);
23:     ub ← best_individual(p);
24:     if ub == prev then
25:         e++;
26:     end if
27:     prev ← ub;
28: end while
29: return (ub.structure, ub.val);
```

**Algorithm 2:** Branch-and-Cut (B&C)

```
1: Input: H, M, ub
2: Output: optimal solution (x*, v*)
3: v* ← ub.val; x* ← ub.structure;
4: L ← ILP^0;
5: while L != ∅ do
6:     n ← remove_node(L);
7:     if !schedulability(n) then
8:         continue;
9:     end if
10:    lp ← relaxation(n)
11:    (x, v) ← solve(lp);
12:    if x = infeasible then
13:        continue;
14:    end if
15:    p ← cut_planes(x, v);
16:    if p! = ∅ then
17:        add(p, lp);
18:        goto 7;
19:    end if
20:    if v ≥ v* then
21:        continue;
22:    end if
23:    if x is integer then
24:        v* ← v;x* ← x;
25:        continue;
26:    end if
27:    partition(lp);
28: end while
29: return (x*, v*);
```

### 14.5.2 Finding Optimal Solutions

We use a general branch-and-cut method combined with schedulability tests to conduct the process of finding optimal solutions. A branch-and-cut is a branch-and-bound with cut generation strategies. The algorithm's input is the processing model H, The desired task model M, and a possible upper bound ub, with objective function value and the solution structure found by the EA. The algorithm outputs the optimal distribution of hard real-time tasks among the processors that consumes less power among the possible assignments, informing as well in which frequency each tasks may be executed, and the total system estimated energy. The general solving strategy is listed in Algorithm 2.

The algorithm starts by denoting the set L of active problem nodes to contain only the initial Integer Linear Problem. When the EA returns a feasible solution, the upper bound $v^*$ and the optimal solution $x^*$ are set to match the output of the EA, otherwise they are set to $+\infty$ and to NULL, respectively. The algorithm iteratively evaluates each element of the set L. Each problem node is initially tested against the schedulability test that fits for the problem scheduling policy. In the case the schedulability test accepts the node, then a regular branch-and-cut is followed. The linear relaxation of the node is then computed and solved. When the linear relaxation is feasible, a procedure of generation of cutting planes is performed and followed by a fathoming and pruning process. The problem node is then partitioned and new restricted problem nodes are derived and incorporated into L. The iterative process repeats until the set L is empty.

## 14.6 Analysis on EA Parameters

We have tuned the EA algorithm based on an analysis of five of its parameters: number of generations, size of population, number of individuals in the tournament, the use of elitism, and percentage of mutation. We considered the CPU time needed to solve an instance with 30 tasks and 50% of estimated target CPU utilization. In Fig. 14.1 we present some graphics in which the left column shows the average CPU time and in the right column we present the average solution energy consumption, for each analysed EA parameter. We plot only observations that could be collected within an execution of less than one minute of CPU time.

As we can observe in Fig. 14.1, as expected, the execution time of the EA increases with the number of generations used, but we have noticed almost no change in the energy consumption. Similar pattern is seen for the number of individuals participating in the tournaments. We see an improvement in the energy consumption when the size of the population is higher than 20, but increasing the size of the population also increases the EA execution time. We have decided to set the parameters population and generation to 50 and the parameter tournament to 5, to avoid increasing the EA execution time, but still finding solutions with lower energy consumption. We have noticed that when we enable mutation, specially with a rate higher than 7% the execution time of the EA increases considerably, reaching more than 1 min in this analysis, and therefore, we decided to disable mutation. We have not noticed any major difference in the convergence time when enabling or disabling elitism for this particular analysis, but we decided to keep it enabled to avoid loosing promising solutions found across generations.

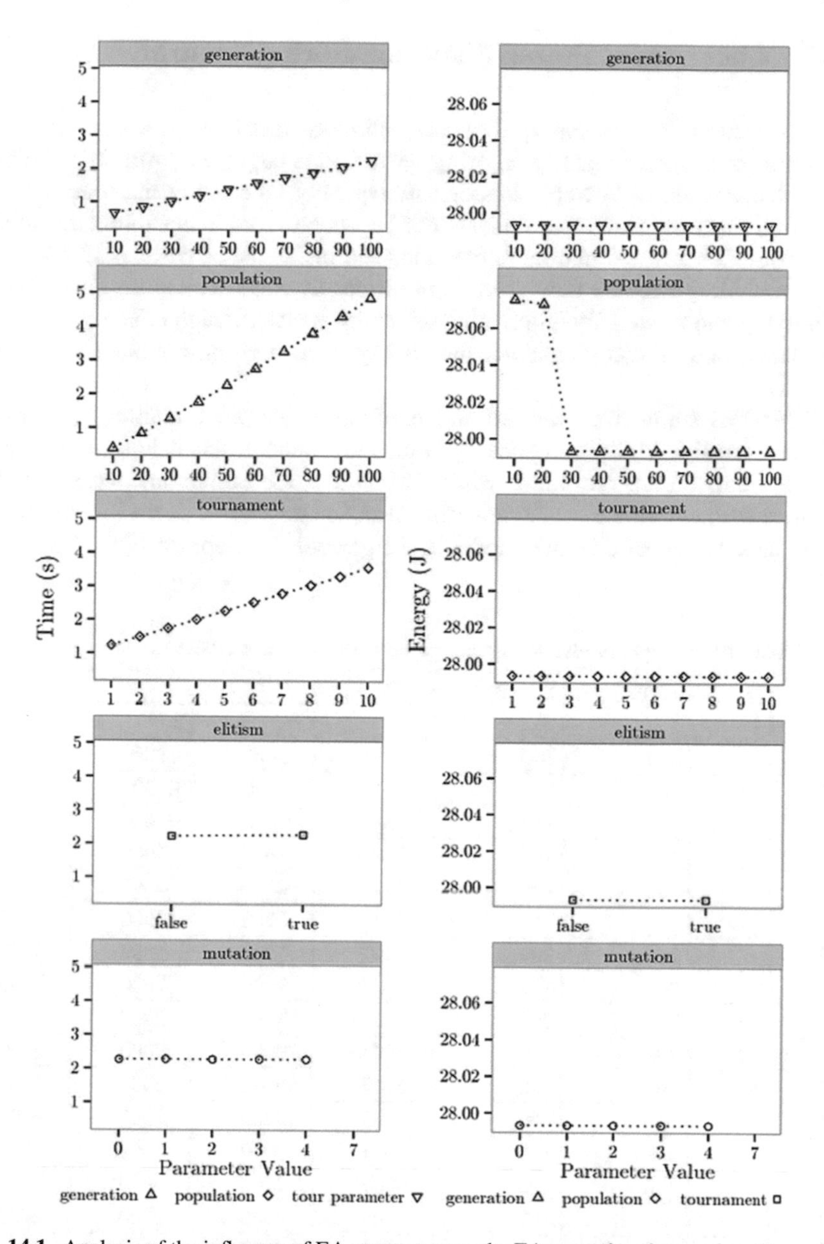


**Fig. 14.1** Analysis of the influence of EA parameters on the EA execution time and on the quality of the objective function (energy). Parameters: number of generations (generation), size of population (population), number of individuals in the tournament (tournament), the use of elitism (elitism), and percentage of mutation (mutation)

## 14.7 Case Study: Power Plant Monitoring Control

In this section, we exemplify how to optimally distribute the hard real-time workload of a power plant monitoring control in a target platform with multiple heterogeneous cores. Power plants depend typically on rotating machines such as steam turbines or generators and should be operated with maximum reliability, capacity, efficiency and minimum operating and maintenance costs. A shutdown of such machinery may be very costly and ideally avoided. Therefore, investing on identifying and potentially eliminating reliability issues through effective condition monitoring and predictive maintenance is key to modern power plant monitoring systems.

Table 14.1 summarizes the task model of the power plant monitoring example. We are considering precedence, preemption, mutual exclusion, temperature, capacity, and timing constraint while distributing the computing workload. We illustrate the precedence constraints (thin arrows) and mutual exclusion constraints (dark thick edges) of this task model in the precedence graph of Fig. 14.2.

**Table 14.1** An example of monitoring control hard real-time task model

| $\tau_i$ | $p_i$ | $T_i$ (ms) | $D_i$ (ms) | $WCEC_i$ ($\times 10^3$) | | | |
|---|---|---|---|---|---|---|---|
| | | | | 1 | 2 | 3 | 4 |
| 1 | 1 | 200 | 100 | 3000 | 3000 | 3000 | 3000 |
| 13 | 2 | 200 | 100 | 15,000 | 15,000 | 15,000 | 15,000 |
| 14 | 3 | 200 | 100 | 10,000 | 10,000 | 10,000 | 10,000 |
| 15 | 4 | 200 | 100 | 1000 | 1000 | 1000 | 1000 |
| 7 | 5 | 200 | 100 | 2000 | 2000 | 2000 | 2000 |
| 3 | 6 | 200 | 40 | 3000 | 3000 | 3000 | 3000 |
| 16 | 7 | 200 | 200 | 5000 | 5000 | 5000 | 5000 |
| 17 | 8 | 200 | 100 | 7000 | 7000 | 7000 | 7000 |
| 2 | 9 | 200 | 200 | 3000 | 3000 | 3000 | 3000 |
| 11 | 10 | 200 | 200 | 2000 | 2000 | 2000 | 2000 |
| 18 | 11 | 200 | 40 | 6000 | 6000 | 6000 | 6000 |
| 10 | 12 | 200 | 40 | 2000 | 2000 | 2000 | 2000 |
| 4 | 13 | 200 | 100 | 3000 | 3000 | 3000 | 3000 |
| 5 | 14 | 200 | 100 | 3000 | 3000 | 3000 | 3000 |
| 20 | 15 | 200 | 100 | 2000 | 2000 | 2000 | 2000 |
| 22 | 16 | 200 | 100 | 7000 | 7000 | 7000 | 7000 |
| 21 | 17 | 200 | 100 | 1000 | 1000 | 1000 | 1000 |
| 9 | 18 | 200 | 100 | 2000 | 2000 | 2000 | 2000 |
| 8 | 19 | 200 | 100 | 2000 | 2000 | 2000 | 2000 |
| 6 | 20 | 200 | 200 | 3000 | 3000 | 3000 | 3000 |
| 19 | 21 | 200 | 200 | 10,000 | 10,000 | 10,000 | 10,000 |
| 12 | 22 | 200 | 200 | 2000 | 2000 | 2000 | 2000 |

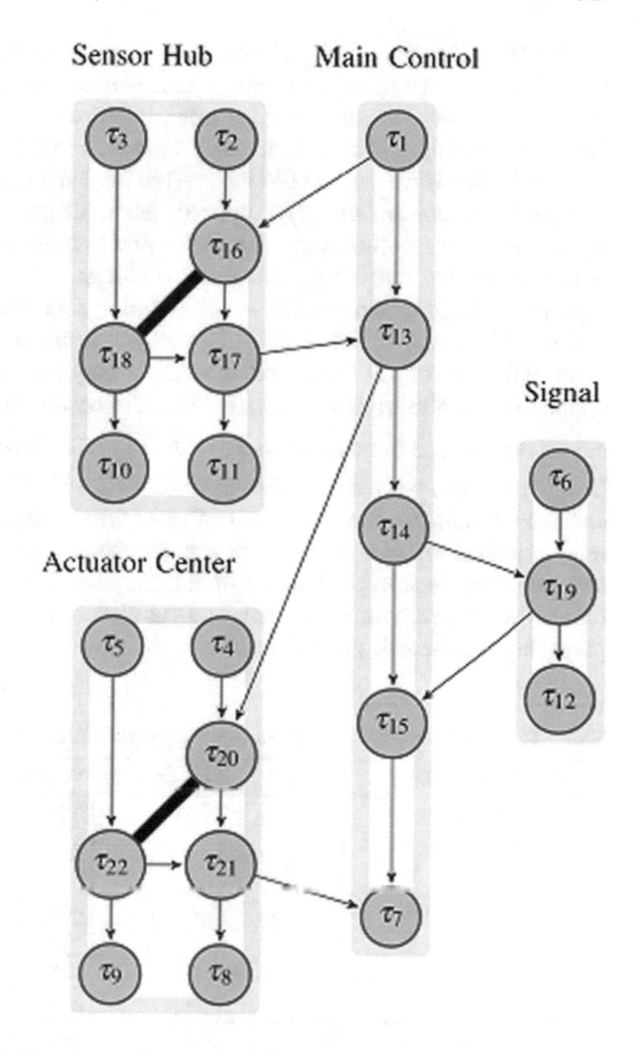


**Fig. 14.2** Precedence graph of power plant monitoring control. Arrows represent a precedence constraint, for example, $\tau_1$ precedes $\tau_{13}$. Dark thick edges represent mutual exclusion constraint, for example, $\tau_{16}$ shares a resource with $\tau_{18}$

The application example we consider, the Power Plant Monitoring Control, is composed of four logical activities that communicate among themselves: the Main Control activity, the Sensor Hub activity, the Actuator Center activity, and the Signal activity. The Main Control activity is responsible for managing the overall control system and communicating with the other activities. The Sensor Hub activity monitors environment, the Actuator Center activity is in charge of performing actions on the event of detection of failure or reliability issues, and the Signal activity reports and records any significant event detected in the system.

The Main Control activity always starts by requesting ($\tau_1$) data from the Sensor Hub. Current environment condition data is then sent back to the Main Control ($\tau_{17}$). The Main Control computes trend based on current and past environment data extrapolating and forecasting any equipment failure or reliability issues and communicates with Actuator Center ($\tau_{13}$) to implement any failure mitigation ($\tau_{20}$) or equipment adjustment ($\tau_{22}$) needed. Main Control also sends ($\tau_{14}$) regular reports of the events that happen in the control system to the Signal activity, which is responsible for activating alarms and warnings.

As an example platform, we are considering four processors: two ARM A57's and two ARM A53's. The ARM A57's may operate on seven different frequencies from 500 MHz to 1.9 GHz, and the A53's may operate on seven different frequencies from 400 MHz to 1.2 GHz. The idle power consumption is 50 mW. The circuit capacitance constant $C_1$ is $1e - 9W\frac{V^2}{Hz}$. The thermal resistance $\rho$ is $0.11\frac{C}{W}$. In this platform, we are considering the DVFS switching latency as an operation executed within the context switch of tasks with a cost of 30 ms, included in the release jitter $J_j$ of each task. More robust response time analysis considering the switching overhead in clusters and architecture influence (Valentin et al. 2015) may be also combined with the branch-and-cut algorithm when necessary. We list the platform characteristics in Table 14.2.

**Table 14.2** Architecture characteristics of a typical multi-core heterogeneous platform

| CPU | $C_1(WV^2/Hz)$ | $\kappa_i^{max}(C)$ | $\rho(C/W)$ | Voltages (V) | Frequencies (GHz) |
|---|---|---|---|---|---|
| 0 | 1e-09 | 125 | 0.11 | 0.94 | 1.9 |
| | | | | 0.86 | 1.8 |
| | | | | 0.86 | 1.7 |
| | | | | 0.78 | 1.6 |
| | | | | 0.77 | 1.5 |
| | | | | 0.77 | 1.0 |
| | | | | 0.77 | 0.5 |
| 1 | 1e-09 | 125 | 0.11 | 0.94 | 1.9 |
| | | | | 0.86 | 1.8 |
| | | | | 0.86 | 1.7 |
| | | | | 0.78 | 1.6 |
| | | | | 0.77 | 1.5 |
| | | | | 0.77 | 1.0 |
| | | | | 0.77 | 0.5 |
| 2 | 1e-09 | 125 | 0.11 | 0.82 | 1.2 |
| | | | | 0.82 | 1.1 |
| | | | | 0.7825 | 1.0 |
| | | | | 0.7575 | 0.9 |
| | | | | 0.7075 | 0.8 |
| | | | | 0.6825 | 0.7 |
| | | | | 0.6575 | 0.4 |

(continued)

**Table 14.2** (continued)

| CPU | $C_l(WV^2/Hz)$ | $\kappa_i^{max}(C)$ | $\rho(C/W)$ | Voltages (V) | Frequencies (GHz) |
|---|---|---|---|---|---|
| 3 | 1e-09 | 125 | 0.11 | 0.82 | 1.2 |
| | | | | 0.82 | 1.1 |
| | | | | 0.7825 | 1.0 |
| | | | | 0.7575 | 0.9 |
| | | | | 0.7075 | 0.8 |
| | | | | 0.6825 | 0.7 |
| | | | | 0.6575 | 0.4 |

Even though temperature is a constraint left for mechanical engineering, it can play a role while distributing the system workload. The ambient temperature in such machineries will typically be higher than the regular room temperature (25 °C) because the system is exposed to heat flowing from the mechanical engines, reaching as high as 85 °C. The common silicon junction temperature is 125 °C.

After executing the branch-and-cut optimization algorithm considering the task model of Table 14.1 and the target computing model of Table 14.2, we obtain the optimal distribution listed in Table 14.3. The precedence graph with the allocation is also illustrated in Fig. 14.3. For this case study, the optimal energy consumption is 0.1049 J for the duration of the LCM (200 ms) of tasks periods. We initialized the algorithm with the solution structure and an upper bound for the objective function extracted from the configuration found by the evolutionary algorithm. Utilizing this initial upper bound, the full optimization process took less than 1.5 h to finish and the final optimal solution differs from the logical initial distribution. Even though this case study has a set of 22 tasks, this algorithm has a reasonable performance on task models with up to 50 tasks, finishing in less than 30 min with a feasible solution for independent tasks (Valentin et al. 2016a, 2017).

As seen in Table 14.3, the schedulability analysis shows that the computed response time of each task is less than their respective deadline, meeting all timing, precedence, and mutual exclusion constraints. It is worth noting that Table 14.3 includes the inter-processor communication cost of tasks $\tau_4, \tau_5, \tau_6, \tau_{13}, \tau_{14}, \tau_{17}, \tau_{18}$, and $\tau_{19}$. The optimization process converged to an optimal solution in which tasks sharing resources are allocated in the same processor, avoiding remote blocking delays. The optimal configuration for this case study uses only three of the four available processors. The total utilization of the active processors (6.05, 16.98, and 25.00%) is well within their respective theoretical values (100%), safely respecting the capacity constraint. This configuration with low utilization is selected by the algorithm because it consumes the least energy, although it is common practice to design real-time systems with high utilization. Also, the estimated temperature of each ARM processors is less than 87 °C, in the thermal stabilization, giving enough room in the temperature constraint.

We highlight, for example, that the logical distribution setting each application activity to one processor is also feasible. This configuration uses all four processors

**Table 14.3** Optimal workload distribution result of the optimization process

| Processor: 0 | Utilization: 6.05% | Temperature: 86.85 °C | | | | |
|---|---|---|---|---|---|---|
| $\tau_i$ | WCEC ($\times 10^3$) | Frequency (GHz) | Computation (ms) | $T_i$ (ms) | $D_i$ (ms) | $R_i$ (ms) |
| 15 | 1000 | 1.9 | 0.526 | 200 | 100 | 97.331 |
| 20 | 2000 | 1.9 | 1.053 | 200 | 100 | 48.684 |
| 22 | 7000 | 1.9 | 3.684 | 200 | 100 | 13.219 |
| 21 | 1000 | 1.9 | 0.526 | 200 | 100 | 49.797 |
| 7 | 2000 | 1.9 | 1.053 | 200 | 100 | 98.443 |
| 8 | 2000 | 1.9 | 1.053 | 200 | 100 | 52.458 |
| 9 | 2000 | 1.9 | 1.053 | 200 | 100 | 18.512 |
| 10 | 2000 | 1.9 | 1.053 | 200 | 40 | 28.513 |
| 11 | 2000 | 1.9 | 1.053 | 200 | 200 | 38.019 |
| 12 | 2000 | 1.9 | 1.053 | 200 | 200 | 108.909 |
| Processor: 1 | Utilization: 16.98% | Temperature: 86.73 °C | | | | |
| $\tau_i$ | WCEC ($\times 10^3$) | Frequency (GHz) | Computation (ms) | $T_i$ (ms) | $D_i$ (ms) | $R_i$ (ms) |
| 1 | 3000 | 1.9 | 1.579 | 200 | 100 | 1.609 |
| 2 | 3000 | 1.9 | 1.579 | 200 | 200 | 3.188 |
| 3 | 3000 | 1.9 | 1.579 | 200 | 40 | 4.767 |
| 4 | 3001 | 1.9 | 1.579 | 200 | 100 | 6.346 |
| 5 | 3001 | 1.9 | 1.579 | 200 | 100 | 7.926 |
| 6 | 3001 | 1.9 | 1.579 | 200 | 200 | 9.505 |
| 16 | 5000 | 1.9 | 2.632 | 200 | 200 | 12.198 |
| 18 | 6001 | 1.9 | 3.158 | 200 | 40 | 18.483 |
| 17 | 7001 | 1.9 | 3.685 | 200 | 100 | 26.936 |
| 13 | 15,002 | 1.0 | 15.002 | 200 | 100 | 46.707 |
| Processor: 2 | Utilization: 25% | Temperature: 85.04 °C | | | | |
| $\tau_i$ | WCEC ($\times 10^3$) | Frequency (GHz) | Computation (ms) | $T_i$ (ms) | $D_i$ (ms) | $R_i$ (ms) |
| 14 | 10,001 | 0.4 | 25.003 | 200 | 100 | 71.739 |
| 19 | 10,002 | 0.4 | 25.005 | 200 | 200 | 96.774 |
| Processor: 3 | Utilization: 0.0% | Temperature: 85.005 °C | | | | |
| $\tau_i$ | WCEC ($\times 10^3$) | Frequency (GHz) | Computation (ms) | $T_i$ (ms) | $D_i$ (ms) | $R_i$ (ms) |

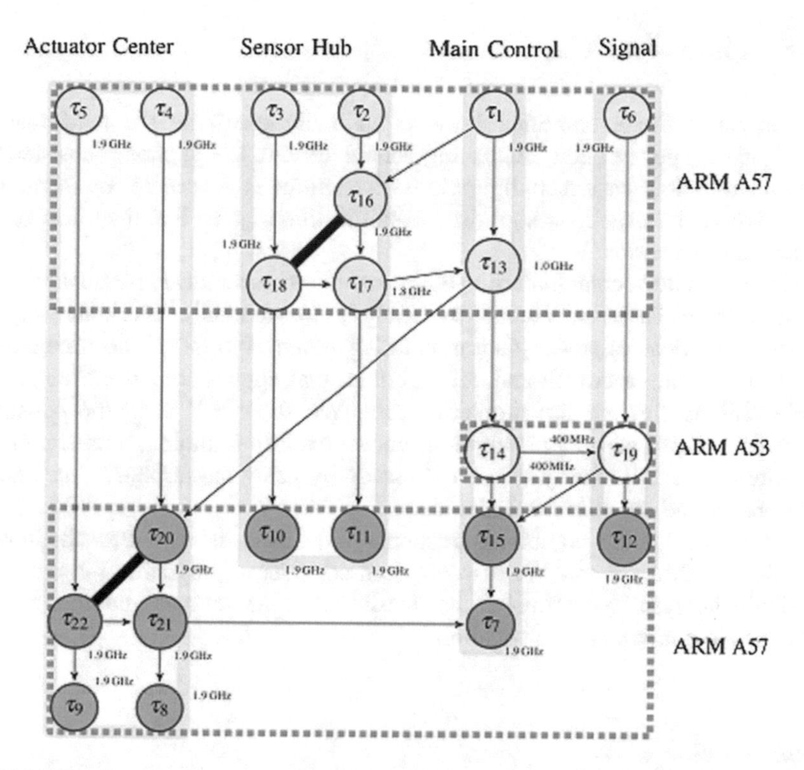


**Fig. 14.3** Precedence graph and task distribution of power plant monitoring control. White nodes are allocated in one ARM A53. Light gray nodes are allocated in one ARM A57. Dark gray nodes are allocated in the other ARM A57. The frequency that each task executes is represented close to each respective node in the graph, for example, $\tau_{19}$ executes at 400 MHz

at their respective maximum frequency. The timing, preemption, precedence, and mutual exclusion constraints are met, given that each task response time is less than their respective deadline. The capacity and temperature constraints are also met. However, this configuration's estimated total system energy is 0.1127 J for the LCM (200 ms) of tasks periods, being at least 7.4% higher than the optimal.

An intuitive approach would be to target a low power configuration, having all tasks allocated to a single ARM A53 CPU, executing at the lowest frequency of 400 MHz. That, however, is not a feasible configuration, given that the capacity constraint is not met because the total CPU utilization would be 117.5% and several tasks would not meet their deadlines in this situation.

Another intuitive approach would be to use again the logical distribution of one activity to one processor, but locking the lowest available frequency, as the utilization of each processor is not high. In this configuration, each processor utilization is less than 32%, but the system is not schedulable because the response time analysis indicates that tasks $\tau_7, \tau_8, \tau_{10}, \tau_{14}, \tau_{15}, \tau_{18}$, and $\tau_{21}$ miss their respective deadlines basically due to the accumulated precedence.

## 14.8 Final Remarks

In this paper, we exemplified how to optimally distribute the hard real-time workload of a power plant monitoring control system. We applied robust methods to avoid infeasible system configurations. Even though they can be computationally expensive, their usage in design time is still justified, given that they help prevent catastrophic scenarios.

We associated combinatorial optimization mathematical formulations and response time based schedulability analysis to optimally distribute the hard real-time workload of power plant monitoring system. We solved the combinatorial problem by using a branch-and-cut algorithm that applies response time analysis while walking through the problem nodes. We showed that all the considered constraints of precedence, preemption, mutual exclusion, timing, temperature, and capacity were met properly in our case study by using the response time analysis with branch-and-cut combined method.

We are evaluating combining response time analysis in a Branch-Cut-Price algorithm as future work. We also envision considering migration by performing sensibility analysis to determine other feasible and optimal configurations to allow for dynamic configuration switching.

## References

Audsley, N., Burns, A., Richardson, M., Tindell, K., & Wellings, A. J. (1993). Applying new scheduling theory to static priority pre-emptive scheduling. *Software Engineering Journal, 8* (5), 284–292.

Department of Energy (DoE). (2014). *Top ten exascale research challenges*. Visited in Jan 2016. URL: https://science.energy.gov/~/media/ascr/ascac/pdf/meetings/20140210/Top10report FEB14.pdf.

Glover, F., Hultz, T. J., & Klingnian, D. (1979). *Improved computer-based planning techniques*, part ii. Interfaces 9/4.

Goossens, J., Milojevic, D., & N′elis, V. (2008). Power-aware real-time scheduling upon dual cpu type multi-processor platforms. In *Proceedings of the 12th International Conference on Principles of Distributed Systems, OPODIS'08*, pp. 388–407. Berlin, Heidelberg: Springer.

He, D., & Mueller, W. (2012). Enhanced schedulability analysis of hard real-time systems on power manageable multi-core platforms. In *Proceedings of the 14th IEEE International Conference on HPCC—9th IEEE ICESS,* pp. 1748–1753, Liverpool.

Lehoczky, J., Sha, L., & Ding, Y. (1989). The rate monotonic scheduling algorithm: Exact characterization and average case behavior. In *Proceedings of the IEEE Real Time Systems Symposium*. pp. 166–171.

Markoff, J., & Lohr, S. (2002). *Intel's huge bet turns iffy*.

Valentin, E. (2009). *Github—Hydra*. Visited in Feb 2016. URL: https://github.com/toolshydra/ Hydra.

Valentin, E., de Freitas, R., & Barreto, R. (2016a). Reaching optimum solutions for the low power hard real-time task allocation on multiple heterogeneous processors problem. In *2016 VI SBESC*, pp. 128–135.

Valentin, E. B., de Freitas, R., & Barreto, R. (2016b). Applying MGAP modeling to the hard real-time task allocation on multiple heterogeneous processors problem. *Procedia Computer Science, 80,* 1135–1146. In *International Conference on Computational Science 2016*, ICCS 2016, 6–8 June 2016, San Diego, California, USA.

Valentin, E., de Freitas, R., & Barreto, R. (2017). Towards optimal solutions for the low power hard real-time task allocation on multiple heterogeneous processors. *Science of Computer Programming*.

Valentin, E., Salvatierra, M., de Freitas, R., & Barreto, R. (2015). *Response time schedulability analysis for hard real-time systems accounting dvfs latency on heterogeneous cluster-based platform* (pp. 1–8). Optimization and Simulation (PATMOS): Power and Timing Modeling.

# Chapter 15
# Operational Planning in Energy Systems: A Literature Review

**Cengiz Kahraman, Sezi Çevik Onar, Başar Öztayşi and Ali Karaşan**

**Abstract** Operational planning is the process of planning and organizing the resources to achieve organization's strategic plan. Planning the supply chain, maintenance, marketing, and production operations are the main parts of operational planning. The operational planning for energy investments is crucial since these investments are costly and the efficiency of the investments necessitates an ample planning process. Also, the type of energy source changes the operational planning need. Understanding these needs and the research gaps can enhance the efficiency of the energy systems. The objective of this chapter is to reveal the primary needs and research focuses on operation planning in energy systems. A comprehensive literature review is conducted to identify the research focuses and the gaps in operational planning in energy systems.

## 15.1 Introduction

Planning levels can be divided into three: operational level, tactical level, and strategic level. Operational planning is the processes linking strategic objectives to tactical objectives. Operational planning must answer where we are now, where we want to be, how we can get there and, how we can measure our progress. Tactical planning is the kind of planning emphasizing the present operations of various departments of an organization within one year. Managers use tactical planning to reveal what the departments of their organizations must do in order to be successful within one year. Strategic planning tries to define the strategy or direction of an organization and to make decisions on allocating its resources to pursue this

C. Kahraman · S. Çevik Onar (✉) · B. Öztayşi · A. Karaşan
Industrial Engineering Department, Management Faculty,
Istanbul Technical University, 34367 Macka, Istanbul, Turkey
e-mail: cevikse@itu.edu.tr

A. Karaşan
Institute of Natural and Applied Sciences, Yildiz Technical University,
34220 Esenler, Istanbul, Turkey

© Springer International Publishing AG, part of Springer Nature 2018

C. Kahraman and G. Kayakutlu (eds.), *Energy Management—Collective and Computational Intelligence with Theory and Applications*, Studies in Systems, Decision and Control 149, https://doi.org/10.1007/978-3-319-75690-5_15

strategy. Strategic planning deals with the whole business and tries to answer what we do, for whom we do it, and how we can success it. Energy firms generally face with operational risks arising from their systems, processes, and personnel. Operational risks result from insufficient internal processes, personnel, and systems, and/or from external events. They affect economic losses, injuries of personnel, and environmental damages.

Operational planning in an energy firm includes sourcing operations, production operations, after-sales service operations, marketing operations, and production planning operations such as routing, scheduling, dispatching, monitoring, and workforce planning. Each of these operations will be explained in Sect. 15.3 in detail.

Theory and application of intelligence in energy systems have been reviewed by many researchers in the literature. Zahraee et al. (2016) analyze artificial intelligence optimum plans in the literature, making the contribution of penetrating extensively the renewable energy aspects for improving the functioning of the systems economically. Jha et al. (2017) summarize the state-of-the-art research outcomes of renewable energy alternatives. They conclude that artificial intelligence could assist in achieving the future goals of renewable energy. Statistical and biologically inspired artificial intelligence methods have been implemented in several studies to achieve their future aims. Daut et al. (2017) provide a review of the building electrical energy consumption forecasting methods which include the conventional and artificial intelligence methods. They review, recognize, and analyse the performance of both methods for forecasting of electrical energy consumption. Cuadra et al. (2016) reviewed computational intelligence techniques used in wave energy applications, both in the resource estimation and in the design and control of wave energy converters.

Intelligent techniques have been often applied to the solutions of operational planning problems in energy sector in the literature. They have been successfully used in a large variety of energy problems, including solar systems (thermal and PV), wind energy systems, biomass energy systems, etc. Artificial Intelligence (AI) methods and tools can be used to operational planning problems related to managing the whole lifecycle of energy (Kayakutlu and Mercier-Laurent 2017). For instance, Artificial Neural Networks (ANN) was used in Photovoltaic systems and in Wind Energy Systems for maximum power point tracking of photovoltaic generators and wind energy resource assessment (Thiaw et al. 2014). Ferrari et al. (2016) challenged several computational intelligence paradigms, namely, Fuzzy C-Means, Radial Basis Function Networks, *k*-Nearest Neighbor, and Feed-forward Neural Networks, in the task of identifying the maximum power point from the working condition directly measurable from the solar panel. Mohamed et al. (2017) present a proposed particle swarm optimization (PSO) algorithm for an optimized design of grid-dependent hybrid photovoltaic-wind energy systems. This algorithm uses the actual hourly data of wind speeds, solar radiation, temperature, and electricity demand in a certain location.

The remaining of the chapter is organized as follows. In Sect. 15.2, Energy systems are briefly introduced. In Sect. 15.3, operational planning techniques are presented. In Sect. 15.4, operational planning in energy management is explained. In Sect. 15.5, conclusions and suggestions for further research are given.

## 15.2 Energy Systems

Energy is defined as "a fundamental entity of nature that is transferred between parts of a system in the production of physical change within the system and usually regarded as the capacity for doing work the capacity to do work." (Merriam-Webster). Energy can be in several forms, including heat, light, electrical or nuclear energy. Researchers define two main groups of energy, primary and secondary. The energy captured from the environment is called primary energy, the energy is later transferred to secondary energy such as electricity or fuel (Belyaev et al. 2002). The primary energy sources can be classified into three main groups: fossil (nonrenewable) energy, renewable energy, and waste (Demirel 2012).

The remains of natural sources such as dead plants and animals are transformed to energy source under the effect of heat and pressure. Coal, petroleum, natural gas, and nuclear energy are the most commonly known nonrenewable energy sources. Nonrenewable energy sources have high ratios of carbon and comprise mostly coal, petroleum, and natural gas. The characteristics such as boiling point may vary among these energy sources. While natural gas, has a shallow boiling point and gaseous components, gasoline has a much higher boiling point. Density, melting point, viscosity, and boiling point are formed based on the mixture of hydrocarbons. Nonrenewable energy sources are commonly criticised on environmental issues such as air pollution with harmful gases. Nuclear energy is different from the others because the energy generation process does not provide any harmful gases since it is based on fission of nuclear fuel (Bodansky 2004).

The energy source which is gathered from natural resources and which can naturally replenish are called renewable energy sources. Hydroelectric, solar energy, biomass, wind energy, geothermal heat, and ocean energy are among the most commonly used renewable energy types. Renewable energy sources fulfill nearly 20% of total electricity generation worldwide. The global trend is to make investments in renewable energy because of climate change concerns and high oil prices (EIA 2011). One of the most popular renewable energy is hydro energy. A water tribune or generator converts the potential energy of dammed water to kinetic energy by getting the force of moving water. Another renewable energy is a solar energy which collects the energy from solar radiation. Photovoltaics and heat engines are commonly used for solar powered electrical generation. The third group of renewable energy is biomass, which is based on microorganisms and animals. Biomass is considered as renewable energy, based on the carbon cycle since plants absorb the sun's energy, processes and produce biomass. Wood, crops, and algae are among the biomass energy sources. Another well-known renewable energy source is the wind. Wind tribunes are used to generate electric. When airflow passes through the turbines electric is generated by the tribune. Another renewable energy type is geothermal energy, which is the heat originating from the original formation of the planet. The main sources of geothermal energy are from radioactive decay of minerals, volcanic activity, and solar energy absorbed at the surface. The final renewable energy type is the ocean energy. The ocean energy focus on providing electricity using the energy passed by ocean waves, tides, salinity, and ocean temperature differences. The kinetic energy provided by the oceans is transferred to tribunes to generate electricity.

Energy systems are designed and optimized by integrating abovementioned energy sources. Liu et al. (2010) provide a brief overview of engineering techniques used in energy systems including superstructure base modeling, mixed integer programming, multiobjective optimization, and optimization under uncertainty. Superstructure based modeling (ABM) uses mathematical modeling to represent a complex system and define the optimal formation of a process (Yeomans and Grossmann 1999). To apply ABM for process engineering, all possible combination of equipment, sequences of flows, and dependencies amongst them are defined and characterized via mathematical programming. After the process design is mathematically represented, the optimal process design is obtained by solving the optimization problem. SBM has been used to solve many problems in energy planning. The most recent fields include heat energy efficiency (Cui et al. 2017), renewable energy supply system (Kwon et al. 2016), and distributed energy supply system (Voll et al. 2013).

An optimization problem is called mixed-integer programming when there are continuous and integer type of variables in the decision model. Since integer variables can be restricted to be 0 or 1, mixed-integer problems can handle binary variables. Some of the recent application areas of mixed integer programming in energy systems planning include performance comparison of energy systems (Yokoyama et al. 2017) and polygeneration energy systems design (Liu et al. 2009).

In most of the optimization problems, there are more than on the objective. The optimization models which aims to optimize a problem according to various criteria simultaneously is called multi-objective optimization. When there are trade-offs among the objective functions, multi-objective optimization can be used. A typical application of the problem in energy system design focuses on maximizing profitability while minimizing the environmental influences. The literature provides some examples of multi-objective optimization on hydrogen infrastructure strategic planning (Hugo et al. 2005), polygeneration energy systems design (Liu et al. 2009), and energy efficiency (Zhou et al. 2015).

Most of the times, uncertainty is unavoidable and unpredictable in long-term energy planning. Due to various factors such as nature of the involved tasks and high variability, the parameters of the mathematical model cannot be precisely known. The methods which focus on optimization under uncertainty can handle this kind of uncertain parameters. Most commonly used methods for optimization under uncertainty involves stochastic programming and fuzzy logic. The literature provides sample applications of uncertainty modeling in energy planning. The recent examples can be listed as; investment optimization (Cunico et al. 2017), design of solar photovoltaic supply chain (Dehghani 2018), variability reduction in energy planning (Murilo et al. 2017), microgrid energy management systems (Hu et al. 2016), and hydroelectric production planning (Zéphyr et al. 2017).

## 15.3 Operational Planning Methods

The operational plan can be described as a specific plan for the regulation of the organization's resources in pursuit of the strategic plan. It involves details about the specific activities and events to carry out of strategies. It is planned as the day-to-day management of the establishment. This planning process is created and monitored by the chief executive and board of directors of the establishment.

Today many companies that have been established for different purposes are applying operational planning in different fields in line with their strategic goals.

- **Supply chain management** enables the strategic plan to be achieved with enhancing efficiency. A supply chain strategy is an agreement of channels, and it is based on data sharing and management of data and operations within the related firms (Chopra and Meindl 2007). These operations include managerial continuums that extended to functional spaces within specific corporations and link suppliers, factories, partners and consumers across organizational boundaries (Wang and Song 2017). (Iakovou et al. 2010) presented generic system components and their unique characteristics for waste in biomass energy supply chains that differentiate them from traditional supply chains. (Balaman and Selim 2014) studied a fuzzy hybrid mathematical modelling composed of goal programming and mixed integer linear programming that is used to find the optimal design management for anaerobic digestion of bioenergy supply chains. (Agusdinata et al. 2014) presented an agent-based simulation modeling framework with artificial intelligence on uncovering system behaviors for designing the network of biofuels supply chain. (Castillo-Villar 2014) presented a study to review of applications on supply chain systems on energy with their theories, challenges and possible future studies.
- **Sourcing operations** is a procural establishment process that aims to improves continuously and re-appraises the purchasing goods from the suppliers of cooperation. In a services industry, these operations assign to a service solution which is specifically customized through the client's wishes. In a production atmosphere, it is often considered as a supply chain process of a component. These processes are day-to-day tactical transactions and aim to help to purchase orders from the suppliers. (Frackowiak and Beguin 2001) presented a study that regards the electrochemical storage of energy in various carbon materials determined as capacitor electrodes for the sourcing.
- **Production operations** aim to add value to product or service which will create a long-lasting and robust customer relationship or association. Moreover, this

can be accomplished by well and productive association between marketing and production system. Production is a component of supply chain and not only affects the suppliers' system but also customer's experience. (Rulkens 2007) presented a study that production of biogas from sewage sludge on small, medium, and large scales.

- **Build to Stock** is a way of production approach in which production schedules created upon the sales foreseen and historical demands of the customers.
- **Build to Order** or make to order is another way of production approach where products are started to manufacture until an order for those products is accepted.
- **Engineer to Order** is also manufacturing process that most complicated and customized way of production. In this way of production, demands are designed, engineered, and built in line with all customer requests.

- **After-sales service operations** refer to different products and services that are sold by cooperation, and this cooperation is responsible for the customers' satisfaction through all those products' and services' life cycle. The demands of the customers must be satisfied for them to make a positive word of mouth in the market. This effect makes an extraordinary support for the cooperation's brands and campaign. Thus, aftersales services are a crucial aspect of marketing management and must not be ignored. (Mont 2002) indicated in his study that the substitution of energy and materials with efficient services may influence overall resource consumption.

**Maintenance management** deals with the continuity and availability of resources in production plants for reaching the sustainability of the manufacturing (Ben-Daya et al. 2009). Beeftink et al. (1990) presented a model that describes energy for maintenance purposes as being obtained simultaneously from biomass degradation as well as from substrate degradation more than growth requirements. Medidi and Zhou (2006) studied maintaining an energy-efficient Bluetooth with scatternet distributed scatter net formation algorithm with simulation results. Cristaldi et al. (2011) presented a study for monitoring of a photovoltaic system to analyze its efficiency via simulation methods. Also, there are types of maintenance in the literature that are given below in detail:

- **Corrective maintenance** aims to correct the defects that can be discovered in the different components and to communicate with the maintenance department by users of the same equipment. So, this type of maintenance notices the defects after the production phases and after the delivery to the customer is completed.
- **Preventive** maintenance is maintaining the components at a certain level and scheduling the interferences of their vulnerabilities in the most well-timed. In other words, the equipment is examined even if there is no evidence of any problem. Also, periodic maintenance based on the total productive maintenance

is also called as time-based maintenance a whole range of basic tasks such as operating, monitoring, cleaning, etc.

- **Predictive** maintenance aims to continually identify and report the status and functional capacity of the installments by recollecting the values of specific variables, which substitute for the basic and operational abilities. It is a requirement to identify physical variables such as heat, vibration, consumption of energy, etc. to utilize this maintenance. On the other hand, zero hours maintenance is also a type of maintenance that inspects the components at the scheduled time intervals without considering appearing of any failure. These two type of maintenance are the most complicated and advanced type of maintenance since it needs to put together mathematical, physical and technical knowledge.

Maintenaning operations diversify with the type of factory and its range of manufacturing, but it has a significant role in production management since they may prevent loss of production, material wastage, overtime. They also regulate the production and workforce utilization. Hence, the absence of planned maintenance service causes problems for the factory. So, it should be a primary operation for the manufacturing plant by the cost-benefit analysis. Since maintenance systems also cause the stopping the production, purchasing of spare parts, etc. they may have high costs.

**Marketing operations** aim to make transparent, efficient, and accountable view of marketing for a one step further than their competitor (Jauhari and Dutta 2009). Marketing operations are often used in the energy sector (Mydock III et al. 2017). Drummond and Hanna (2001) discussed the marketing of energy in a deregulated environment. Rodriguez and Anders (2004) introduced a hybrid method for forecasting energy prices composed of artificial neural networks and fuzzy logic. Most used marketing operations in the literature are given as below:

- **Lead management** is a set of methodologies, systems, and practices that are created for generating new potential business, generally used in marketing campaigns or programs.
- **Reporting and analyzing** can be expressed separately. The process of exploring data and analyzing it is the most complex operations in the businesses. Reporting is the operation that converts raw data into knowledge. Analyzing is the operation that helps to make inferences from knowledge.
- **Data management** aims to develop architectures, and policies and to execute practices and procedures for handling data correctly for needs of an enterprise.
- **Campaign development** is in need of planning and creativity for earning substantial monetary resources. Campaign development takes time and aware planning for a sustainable competition in the market.
- **Content development** is a technique that aims to create and distribute valuable content to attract, acquire, and engage to the customer.

**Production planning** operations are the planning of production and manufacturing systems in an industry (Chapman 2006). It utilizes the raw materials, materials, and capacity of the production lines to serve the aims of factory management. Ni et al. (2006) reviewed the current developments on hydrogen production technologies to make an overview of renewable energy sources and represented potential practices for renewable hydrogen production in Hong Kong. Zhang et al. (2017) introduced a study that for a hybrid computational approach composed of simplex method and ant colony optimization to schedule products in a pipeline with multi-plant pump stations. Teeuwsen et al. (2005) presented computational intelligence methods for fast eigenvalue prediction in large interconnected power systems. Schaffner et al. (2017) introduced a study that monitors to people behaviors of moving into energy-efficient homes by using a dynamic approach based on phased model. Production planning is composed of operations which are given below:

- **Routing** is a specific way to a product or material in the manufacturing phase. This specific way ends in the employment of all the applications take place until the final product is developed.
- **Scheduling** is the process that aims to reach the most appropriate ordering, controlling and optimizing for work and workloads for the manufacturing systems. This process consists of machinery resources, workforce, raw materials and final product.
- **Dispatching** indicates to control all processes in the manufacturing system from supply part to customer shipment. Primary goal of this operation is date management and controlling capacity at specified time intervals.
- **Monitoring** means inspections of manufacturing site for any moment the production is made that is made by the inspector. An inspector checks the reproducibility of the factory, enforces terms of references, and also controls the products in case of any defects.
- **Workforce planning** is a perpetual process that aligns the demands and precedences of the cooperation for its workforce through the strategic objectives that have confliction with law, regulations and the humanitarian aspect.

## 15.4 Operational Planning in Energy Management

The operation planning need of energy systems varies with the characteristic of energy systems. For instance, for the bioenergy systems sourcing is the main problem whereas predictive and corrective maintenance is crucial for nuclear energy systems. In order to reveal the operation planning focuses on energy systems, we conduct a comprehensive literature review using "Scopus" database. We summarize the literature review results by using graphical and tabular analysis.

### *15.4.1 Supply Chain Management in Energy Systems*

Many researchers have evaluated the supply chain operations in energy systems. Supply chain management involves sourcing, production, and aftersales operations. We list down the studies in Scopus database that use both "Supply chain" and "Energy" in the title, abstract and keywords. A total of 4668 studies in Scopus database use both "supply chain" and "energy" in the title, abstract and keywords fields (Fig. 15.1).

### *15.4.2 Sourcing Operations in Energy Management*

A total of 64 studies in Scopus database use both "supply chain", "energy" and "sourcing" in the "title, abstract and keywords" fields. Figure 15.2 shows the most used keywords in these studies namely, environmental impact, biomass, sustainable development, and bioenergy. Notably, the sourcing in bioenergy is one of the primary area of interest. Richard (2010), Mirabella et al. (2013), Sikkema et al. (2014), Meadows et al. (2014), Cambero et al. (2015), Paulo et al. (2015) and Flodén and Williamsson (2016) highlighted the importance of sourcing strategy in biofuel industry. Only a few studies focus on the sourcing problems in wind energy production (Sarja 2012; Ghaffari and Venkatesh 2015).

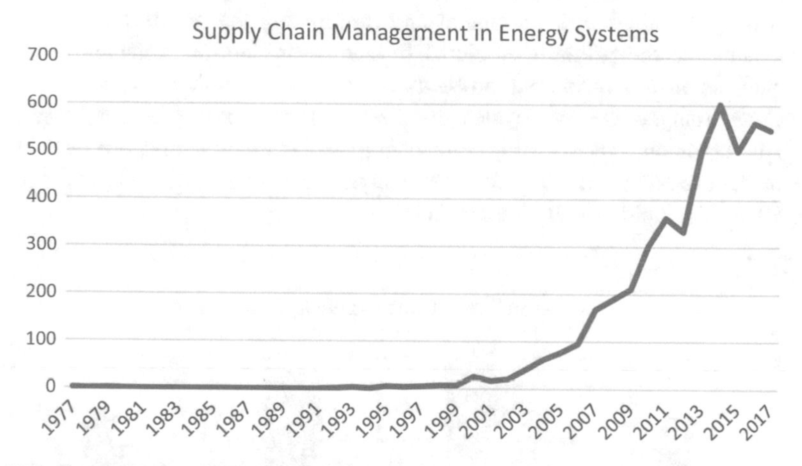


**Fig. 15.1** Supply chain management in energy systems

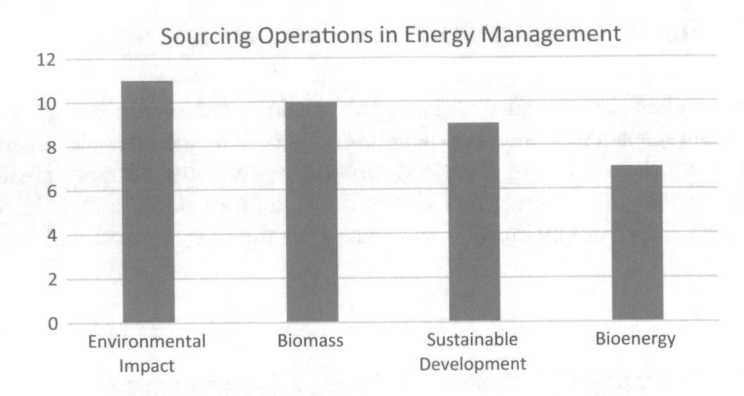


**Fig. 15.2** Sourcing operations in energy management focused areas

### *15.4.3 Production Operations in Energy Management*

A total of 1808 studies in Scopus database use both "supply chain," "energy" and "production" in the "title, abstract and keywords" fields. Energy production considered as one of the leading operations management research area in the energy field.

Similar to the sourcing operations, environmental impact, biomass, and sustainable development are the leading research areas. Figure 15.3 shows the frequency of usage of these keywords.

Providing sustainable development and minimizing the harmful impact on the environment are the primary problems in energy production. Managing biomass production is also a principal problem for the production operations. Various researchers analyze the life cycles of energy investments. Not only traditional energy investments but also the wind energy, solar energy, and biomass energy investments are analyzed with life cycle analysis (Elia et al. 2011; García-Valverde et al. 2010; Tryfonidou and Wagner 2004).

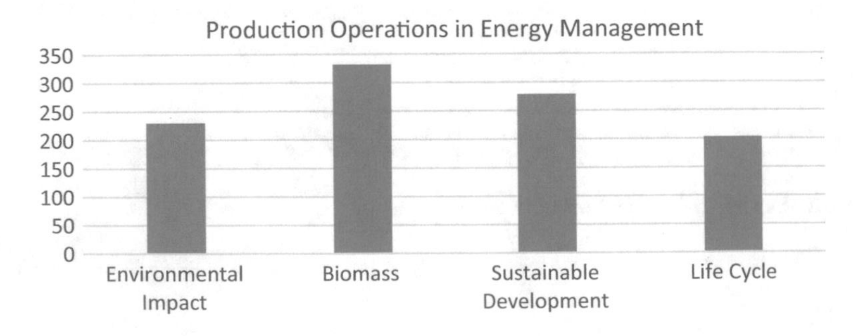


**Fig. 15.3** Production operations in energy management focused areas

### *15.4.4 Aftersales Operations in Energy Management*

A total of 44 studies in Scopus database use both "energy" and "aftersales" in the title, abstract and keywords fields. Among these studies, the primary focus is on aftersales' effect on sales and costs (Fig. 15.4). Also, the impact of sales operations in developing countries are analyzed by five studies such as Gupta (1999), Rogers (1999) and Kebede (2014). Several researchers focus on the aftersales operations in solar energy and wind energy.

## 15.5 Maintenance Management in Energy Systems

Achieving reliability, increasing cost and energy efficiency are crucial for energy systems. Maintenance planning and management enhance the performance of energy systems. In literature, some studies analyze maintenance management in energy systems.

### *15.5.1 Corrective Maintenance*

A total of 79 studies in Scopus database use both "corrective maintenance" and "energy" in the "title, abstract and keywords" fields. Figure 15.5 shows the most used keywords in these studies namely, reliability, wind power, electric utilities and nuclear energy.

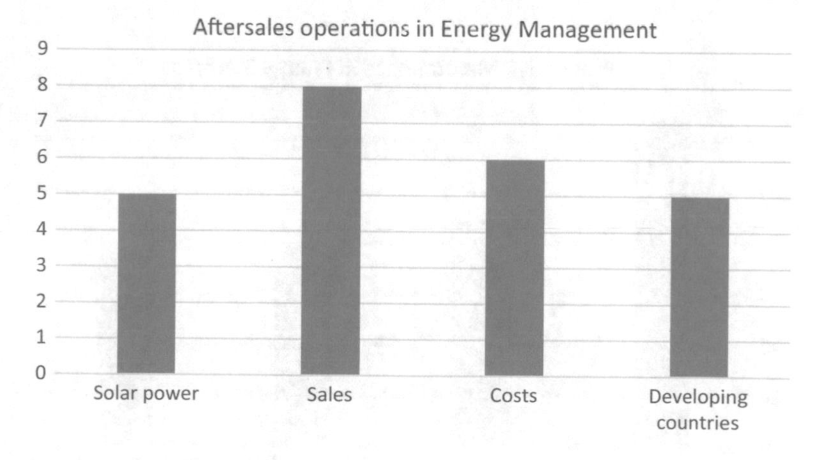


**Fig. 15.4** Aftersales operations in energy management focused areas

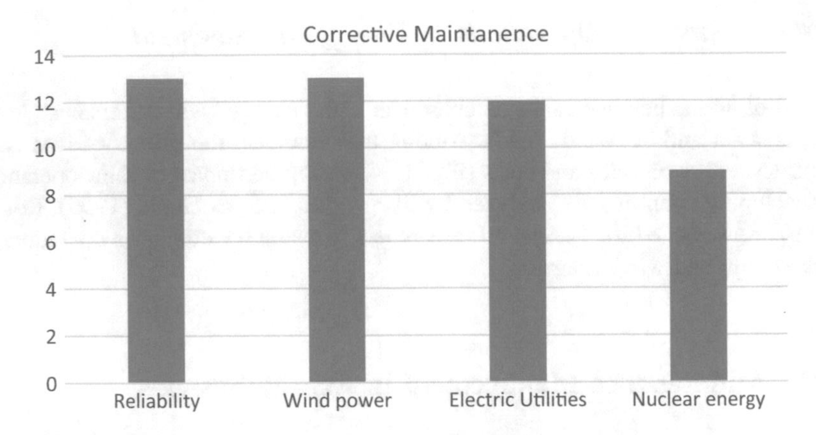


**Fig. 15.5** Corrective maintenance in energy systems focused areas

Analyzing the risks and applying corrective maintenance for wind power utilities is an important research area (Wiggelinkhuizen et al. 2008; Presencia 2017; Zhang 2017; Lei and Sandborn 2018).

### *15.5.2 Preventive Maintenance*

A total of 783 studies in Scopus database use both "preventive maintenance" and "energy" in the "title, abstract and keywords" fields. Figure 15.6 shows the most used keywords in these studies namely, costs and cost efficiency, reliability, energy efficiency and nuclear energy. Maintaining reliability by increasing both cost and energy efficiencies is the primary objective of preventive maintenance.

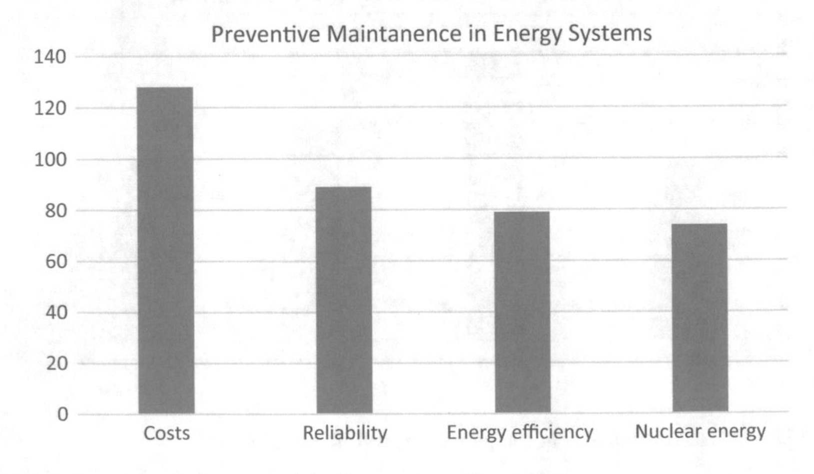


**Fig. 15.6** Preventive maintenance in energy systems focused areas

Due to the critically of the risks, the preventive maintenance has a crucial role in nuclear energy (Pereira et al. 2010; Lusby et al. 2013; Yang 2015).

### 15.5.3 *Predictive Maintenance*

A total of 288 studies in Scopus database use both "predictive maintenance" and "energy" in the "title, abstract and keywords" fields. Figure 15.7 shows the most used keywords in these studies namely, reliability, wind power, electric utilities and nuclear energy. Many researchers focus on the monitoring technologies such as sensors, signal processing and telecommunication systems (Daneshi-Far 2010; Hashemian 2011; Hashemian and Bean 2011; Zhu 2014).

Predictive maintenance is essential both for nuclear energy and wind energy systems. Therefore, many researchers conduct studies for enhancing predictive maintenance in these systems (Lin and Holbert 2009; Daneshi-Far et al. 2010; Hashemian 2011; Hashemian and Bean 2011; Hashemian et al. 2011; Yang et al. 2014).

## 15.6 Marketing Operations in Energy Systems

The changing characteristics of the available resources and needs in different locations make marketing of energy systems crucial. In Scopus database, we searched for "marketing" and "energy systems" in the "title, abstract and keywords" fields. 244 studies focus on marketing operations in energy systems (see Fig. 15.8). The marketing of renewable energy systems is the main topic of these studies.

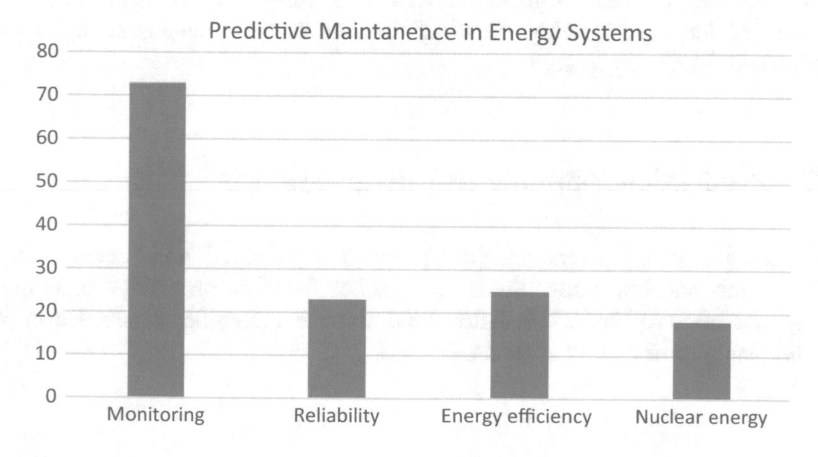


**Fig. 15.7** Predictive maintenance in Energy Systems Focused Areas

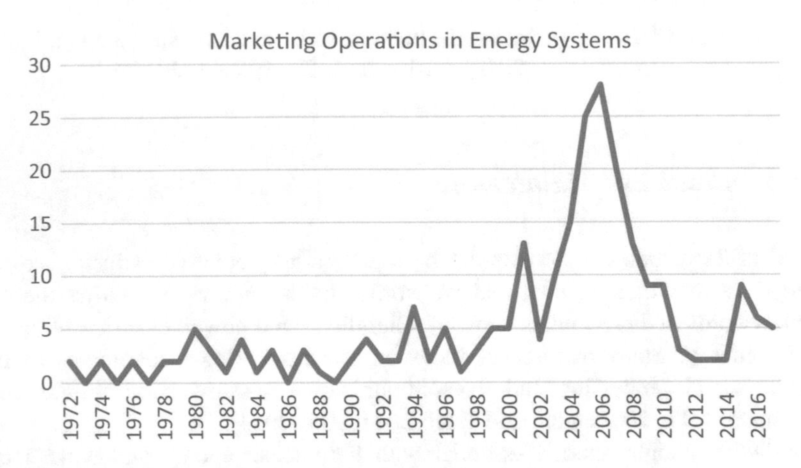


**Fig. 15.8** Marketing operating in energy systems

Between 2002 and 2010, there is a significant increase in energy marketing studies but with the full acceptance of renewables, these numbers have decreased.

We searched for "marketing," "data management" and "energy" in the "title, abstract and keywords" fields. 18 studies focus on marketing data management in energy systems (Felzien et al. 2003; Verdú et al. 2006; Gabaldón 2008; Ceci 2015).

We searched for "marketing", "analytics" and "energy" in the "title, abstract and keywords" fields. Only five studies focus on marketing analytics in energy systems (Wirl 1989; Kejariwal 2016; Akoka et al. 2017; Mueller 2017).

We searched for "marketing", "campaign development" and "energy" in the "title, abstract and keywords" fields. 51 studies focus on campaign development in energy systems. Most of these studies focus on increasing the awareness and penetration of renewable energy sources (Martinot et al. 2001; Tsoutsos 2002; Gossling et al. 2005; Zorić and Hrovatin 2012).

We searched for "content management", "marketing" and "energy" in the "title, abstract and keywords" fields. None of the studies focus on marketing content development in energy systems.

## 15.7 Production Planning in Energy Systems

In Scopus database, we searched for "production planning" and "energy" in the "title, abstract and keywords" fields. 474 studies focus on production planning in energy systems (see Fig. 15.9). After 2000, there is a significant increase in production planning in energy systems.

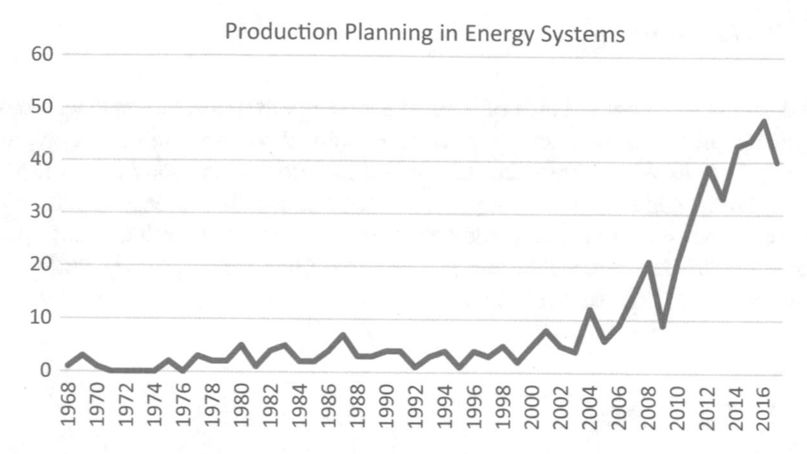


**Fig. 15.9** Production planning in energy systems

## *15.7.1 Routing*

197 studies focus on routing problems in energy systems. Solar energy, sensors, energy efficiency and energy utilization are the leading research areas (See Fig. 15.10). The routing problems are considered as a part of solar energy harvesting optimization (Alippi and Galperti 2008; Dondi et al. 2008; Ismail 2008). The energy optimization of sensor networks via the routing systems is another important research area (Voigt et al. 2003; Bergonzini et al. 2009; Chen et al. 2012). 24 studies focused on the routing problems in wind energy systems (Gupta et al. 2007; Phillips and Middleton 2012; Shafiee2015).

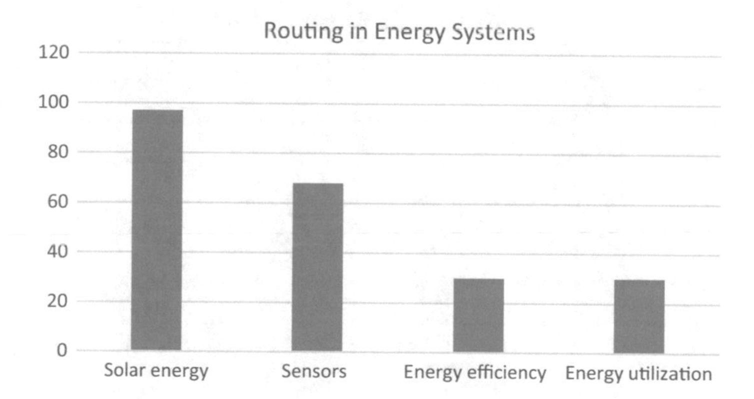


**Fig. 15.10** Routing energy systems

### 15.7.2 Scheduling

Scheduling is an important area of interest for energy systems. 424 studies focus on scheduling problems in energy systems. Optimization, renewable energy, wind power and smart power grids are the leading research areas (See Fig. 15.11).

The power generation and supply of many renewable energy resources are uncertain. Therefore, optimizing the generation and supply by using smart systems become crucial for renewable energy systems (Gupta et al. 2007; Phillips and Middleton 2012; Shafiee et al. 2015).

### 15.7.3 Dispatching

In the literature, 180 studies focus on energy dispatching problems. Electric load dispatching, optimization, renewable energy resources and wind power are the leading research areas in these studies (Xie and Ilić 2008, 2009; Tascikaraoglu et al. 2014). Figure 15.12 illustrates these research areas.

### 15.7.4 Workforce Planning

Only eight articles focus on workforce planning in energy systems. Several of these studies focus on workforce planning in nuclear energy (Irizarry and Seemer 1986; Sherrard and Horner 2007).

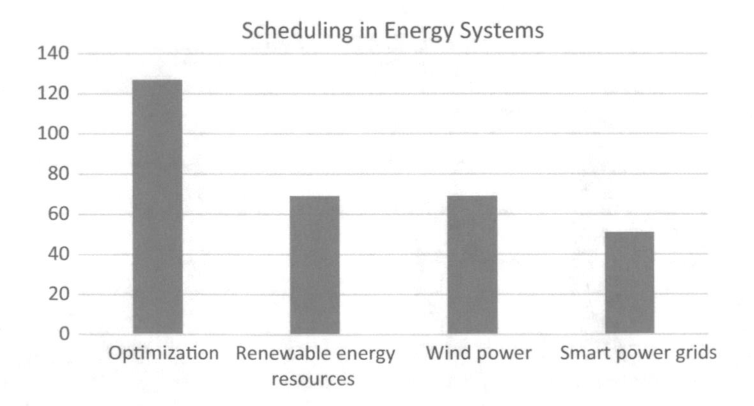


**Fig. 15.11** Scheduling energy systems

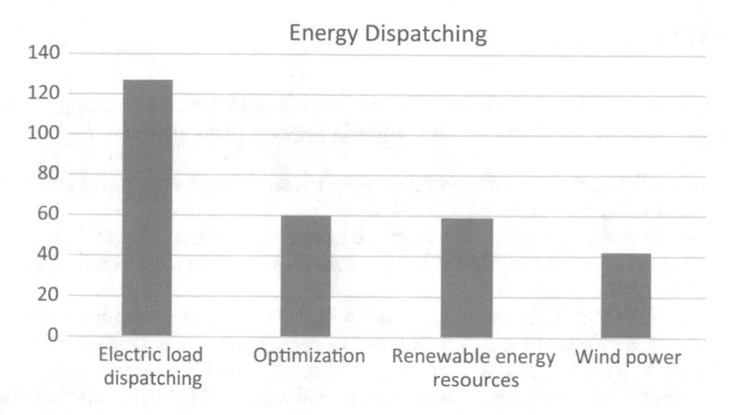


**Fig. 15.12** Energy dispatching research areas

## 15.8 Conclusions and Further Suggestions

In this chapter, we conduct a comprehensive literature review to reveal the state of the art operational planning problems in energy systems. We classify the problems and analyze the focus areas in the "Scopus" database. The result of the analysis shows the main problems in different energy types.

In literature, there is a growing interest in solving operational planning problems in renewable energy sources. Planning the supply chain operations are crucial for bioenergy fuels. Achieving a sustainable development and minimizing the environmental damage is the other focus areas for supply chain planning in energy systems. Planning the aftersales operations for solar power and energy investments in developing countries is crucial. Managing the maintenance is very important for nuclear energy and wind energy. The researchers can analyze corrective maintenance for solar energy. At the beginning of 2000 the marketing planning for renewable energy resources considered as an essential problem but with the penetration of renewables this need has been satisfied. Routing is important both for solar energy and for wind energy. Supplying energy for the wireless networks and designing the routes of these energy sources is another widely focused research problem. Scheduling is crucial for all the renewable energy sources.

In the literature, there is a need for understanding the state of the art problems of operational planning in energy systems. This chapter fulfills this need by defining the energy systems and operational planning problems and can guide future studies by showing the gaps. For the future, a qualitative study can be implemented to show the correlations among the key focus areas.

## References


Agusdinata, D. B., Lee, S., Zhao, F., & Thissen, W. (2014). Simulation modeling framework for uncovering system behaviors in the biofuels supply chain network. *Simulation, 90*(9), 1103–1116.

Akoka, J., et al. (2017). Research on big data—A systematic mapping study. *Computer Standards and Interfaces, 54,* 105–115.

Alippi, C., & Galperti, C. (2008). An adaptive system for opimal solar energy harvesting in wireless sensor network nodes. *IEEE Transactions on Circuits and Systems I: Regular Papers, 55*(6), 1742–1750.

Balaman, Ş. Y., & Selim, H. (2014). A fuzzy multiobjective linear programming model for design and management of anaerobic digestion based bioenergy supply chains. *Energy, 74,* 928–940.

Beeftink, H. H., Van der Heijden, R. T. J. M., & Heijnen, J. J. (1990). Maintenance requirements: energy supply from simultaneous endogenous respiration and substrate consumption. *FEMS Microbiology Letters, 73*(3), 203–209.

Belyaev, L. S., Marchenko, O. V., Filippov, S. P., Solomin, S. V., Stepanova, T. B., & Kokorin, A. L. (2002). *World energy and transition to sustainable development*. Boston: Kluwer.

Ben-Daya, M., Ait-Kadi, D., Duffuaa, S. O., Knezevic, J., & Raouf, A. (2009). *Handbook of maintenance management and engineering* (Vol. 7). London: Springer.

Bergonzini, C., et al. (2009). Algorithms for harvested energy prediction in batteryless wireless sensor networks. In *3rd International Workshop on Advances in Sensors and Interfaces, IWASI 2009*.

Bodansky, D. (2004). *Nuclear energy: Principles practices and prospects*. Oxford: Springer.

Cambero, C., et al. (2015). Strategic optimization of forest residues to bioenergy and biofuel supply chain. *International Journal of Energy Research, 39*(4), 439–452.

Castillo-Villar, K. K. (2014). Metaheuristic algorithms applied to bioenergy supply chain problems: theory, review, challenges, and future. *Energies, 7*(11), 7640–7672.

Ceci, M., et al. (2015). Big data techniques for supporting accurate predictions of energy production from renewable sources. In *ACM International Conference Proceeding Series*.

Chapman, S. N. (2006). *The fundamentals of production planning and control*. Prentice Hall.

Chen, S., et al. (2012). A simple asymptotically optimal energy allocation and routing scheme in rechargeable sensor networks. In *Proceedings—IEEE INFOCOM*.

Chopra, S., & Meindl, P. (2007). *Supply chain management. Strategy, planning & operation*. Das summa summarum des management (pp. 265–275).

Cristaldi, L., Faifer, M., Rossi, M., & Ponci, F. (2011). Monitoring of a PV system: The role of the panel model. In *Applied Measurements for Power Systems* (s. 90–95). IEEE.

Cuadra, L., Salcedo-Sanz, S., Nieto-Borge, J. C., Alexandre, E., & Rodríguez, G. (2016). Computational intelligence in wave energy: Comprehensive review and case study. *Renewable and Sustainable Energy Reviews, 58,* 1223–1246.

Cui, C., Li, X., Sui, H., & Sun, J. (2017). Optimization of coal-based methanol distillation scheme using process superstructure method to maximize energy efficiency. *Energy, 119,* 110–120.

Cunico, M. C., Flores, J. R., & Vecchietti, A. (2017). Investment in the energy sector: An optimization model that contemplates several uncertain parameters. *Energy, 138,* 831–845. (1 Nov 2017).

Daneshi-Far, Z., et al. (2010). Review of failures and condition monitoring in wind turbine generators. In *19th International Conference on Electrical Machines, ICEM 2010*.

Daut, M. A. M., Hassan, M. Y., Abdullah, H., Rahman, H. A., Abdullahab, M. P., & Hussin, F. (2017). Building electrical energy consumption forecasting analysis using conventional and artificial intelligence methods: A review. *Renewable and Sustainable Energy Reviews, 70,* 1108–1118.

Dehghani, E., Jabalameli, M. S., & Jabbarzadeh, A. (2018). Robust design and optimization of solar photovoltaic supply chain in an uncertain environment. *Energy, 142,* 139–156.

Demirel, Y. (2012). *Green energy and technology*. Berlin: Springer.

Dondi, D., et al. (2008). Modeling and optimization of a solar energy harvester system for self-powered wireless sensor networks. *IEEE Transactions on Industrial Electronics, 55*(7), 2759–2766.

Drummond, J., & Hanna, F. (2001). Selling power: Marketing energy under deregulation.

EIA (2011) *Renewable energy consumption and electricity preliminary statistics 2010*. Release date: 28 June 2011. http://www.eia.gov/renewable/. Accessed July 2011.

Elia, J. A., et al. (2011). Optimal energy supply network determination and life cycle analysis for hybrid coal, biomass, and natural gas to liquid (CBGTL) plants using carbon-based hydrogen production. *Computers & Chemical Engineering, 35*(8), 1399–1430.

Felzien, D., et al. (2003). IT requirements for market participant interaction with ISOs/RTOs. *IEEE Transactions on Power Systems, 18*(2), 517–519.

Ferrari, S., Lazzaroni, M., Piuri, V., Salman, A., Cristaldi, L., Faifer, M., et al. (2016). Solar panel modelling through computational intelligence techniques. *Measurement, 93,* 572–580.

Flodén, J., & Williamsson, J. (2016). Business models for sustainable biofuel transport: The potential for intermodal transport. *Journal of Cleaner Production, 113,* 426–437.

Frackowiak, E., & Beguin, F. (2001). Carbon materials for the electrochemical storage of energy in capacitors. *Carbon, 39*(6), 937–950.

Gabaldón, A., et al. (2008). Development of a methodology for improving the effectiveness of customer response policies through electricity-price patterns. In *IEEE Power and Energy Society 2008 General Meeting: Conversion and Delivery of Electrical Energy in the 21st Century*, PES.

García-Valverde, R., et al. (2010). Life cycle analysis of organic photovoltaic technologies. *Progress in Photovoltaics: Research and Applications, 18*(7), 535–538.

Ghaffari, R., & Venkatesh, B. (2015). Network constrained model for options based reserve procurement by wind generators using binomial tree. *Renewable Energy, 80,* 348–358.

Gossling, S., et al. (2005). A target group-specific approach to "green" power retailing: Students as consumers of renewable energy. *Renewable and Sustainable Energy Reviews, 9*(1), 69–83.

Gupta, P. K. (1999). Renewable energy sources-a longway to go in India. *Renewable Energy, 16* (1–4), 1216–1219.

Gupta, S. C., et al. (2007). Optimal sizing of solar-wind hybrid system. In *IET Seminar Digest*.

Hashemian, H. M. (2011). On-line monitoring applications in nuclear power plants. *Progress in Nuclear Energy, 53*(2), 167–181.

Hashemian, H. M., & Bean, W. C. (2011). State-of-the-art predictive maintenance techniques. *IEEE Transactions on Instrumentation and Measurement, 60*(10), 3480–3492.

Hashemian, H. M., et al. (2011). Wireless sensor applications in nuclear power plants. *Nuclear Technology, 173*(1), 8–16.

Hu, M.-C., Lu, S.-Y., Chen, Y.-H. (2016). Stochastic programming and market equilibrium analysis of microgrids energy management systems. *Energy, 113,* 662–670, ISSN 0360–5442.

Hugo, A., Rutter, P., Pistikopoulos, E. N., Amorelli, A., & Zoia, G. (2005). Hydrogen infrastructure strategic planning using multi-objective optimization. *International Journal of Hydrogen Energy, 30*(15), 1523–1534.

Iakovou, E., Karagiannidis, A., Vlachos, D., Toka, A., & Malamakis, A. (2010). Waste biomass-to-energy supply chain management: A critical synthesis. *Waste Management, 30*(10), 1860–1870.

Irizarry, F., & Seemer, R. H. (1986). Human resource planning: A nuclear application. In *Proceedings—Fall Industrial Engineering Conference (Institute of Industrial Engineers)*.

Ismail, M., & Sanavullah, M. Y. (2008). Security topology in wireless sensor networks with routing optimisation. In *Proceedings of the 4th International Conference on Wireless Communication and Sensor Networks, WCSN 2008*.

Jauhari, V., & Dutta, K. (2009). Services: Marketing, operations, and managment. Oxford University Press.

Jha, S. K., Bilalovic, J., Jha, A., Patel, N., & Zhang, H. (2017). Renewable energy: Present research and future scope of artificial intelligence. *Renewable and Sustainable Energy Reviews, 77,* 297–317.

Kayakutlu, G., & Mercier-Laurent, E. (2017). Intelligence for Energy. In *Intelligence in energy* (pp. 79–116).

Kebede, K. Y., et al. (2014). After-sales service and local presence: Key factors for solar energy innovations diffusion in developing countries. In *PICMET 2014—Portland International Center for Management of Engineering and Technology, Proceedings: Infrastructure and Service Integration.*

Kejariwal, A., & Orsini, F. (2016). On the definition of real-time: Applications and systems. In *Proceedings—15th IEEE International Conference on Trust, Security and Privacy in Computing and Communications, 10th IEEE International Conference on Big Data Science and Engineering and 14th IEEE International Symposium on Parallel and Distributed Processing with Applications*, IEEE TrustCom/BigDataSE/ISPA 2016.

Kwon, S., Won, W., & Kim, J. (2016). A superstructure model of an isolated power supply system using renewable energy: Development and application to Jeju Island, Korea. *Renewable Energy, 97,* 177–188.

Lei, X., & Sandborn, P. A. (2018). Maintenance scheduling based on remaining useful life predictions for wind farms managed using power purchase agreements. *Renewable Energy, 116,* 188–198.

Lin, K., & Holbert, K. E. (2009). Blockage diagnostics for nuclear power plant pressure transmitter sensing lines. *Nuclear Engineering and Design, 239*(2), 365–372.

Liu, P., Pistikopoulos, E. N., & Li, Z. (2009a). A mixed-integer optimization approach for polygeneration energy systems design. *Computers & Chemical Engineering, 33*(3), 759–768.

Liu, P., Pistikopoulos, E. N., & Li, Z. (2009b). A multi-objective optimization approach to polygeneration energy systems design. *AIChE Journal, 56*(5), 1218–1234.

Liu, P., Psitikopoulos, N. E., & Li, Z. (2010). Energy systems engineering: methodologies and applications. *Frontiers in Energy, 4*(2), 131–142.

Lusby, R., et al. (2013). A solution approach based on Benders decomposition for the preventive maintenance scheduling problem of a stochastic large-scale energy system. *Journal of Scheduling, 16*(6), 605–628.

Martinot, E., et al. (2001). World Bank/GEF solar home system projects: Experiences and lessons learned 1993–2000. *Renewable and Sustainable Energy Reviews, 5*(1), 39–57.

Meadows, J., et al. (2014). The potential supply of biomass for energy from Hardwood Plantations in the Sunshine Coast Council Region of South-East Queensland, Australia. *Small-scale Forestry, 13*(4), 461–481.

Medidi, M., & Zhou, Y. (2006). Maintaining an energy-efficient bluetooth scatternet. In *Performance, Computing, and Communications Conference* (s. 8). IEEE.

Mirabella, N., et al. (2013). Life cycle assessment of bio-based products: A disposable diaper case study. *International Journal of Life Cycle Assessment, 18*(5), 1036–1047.

Mohamed, M. A., Eltamaly, A. M., & Alolah, A. I. (2017). Swarm intelligence-based optimization of grid-dependent hybrid renewable energy systems. *Renewable and Sustainable Energy Reviews, 77,* 515–524.

Mont, O. K. (2002). Clarifying the concept of product–service system. *Journal of Cleaner Production, 10*(3), 237–245.

Mueller, T. S. (2017). Consumer perceptions of electric utilities: Insights from the center for Analytics Research & Education Project in the United States. *Energy Research and Social Science, 26,* 34–39.

Murilo, P. S., Alexandre, S., & Davi, M. V. (2017). On the solution variability reduction of stochastic dual dynamic programming applied to energy planning. *European Journal of Operational Research, 258*(2), 743–760.

Mydock III, S., Pervan, S. J., Almubarak, A. F., Lester, J., & Kortt, M. (2017). Influence of made with renewable energy appeal on consumer behaviour. *Marketing Intelligence & Planning.* https://doi.org/10.1108/mip-06-2017-0116, artical in press.

Ni, M., Leung, M. K., Sumathy, K., & Leung, D. Y. (2006). Potential of renewable hydrogen production for energy supply in Hong Kong. *International Journal of Hydrogen Energy, 31* (10), 1401–1412.

Paulo, H., et al. (2015). Supply chain optimization of residual forestry biomass for bioenergy production: The case study of Portugal. *Biomass and Bioenergy, 83,* 245–256.
Pereira, C. M. N. A., et al. (2010). A particle swarm optimization (PSO) approach for non-periodic preventive maintenance scheduling programming. *Progress in Nuclear Energy, 52*(8), 710–714.
Phillips, B. R., & Middleton, R. S. (2012). SimWIND: A geospatial infrastructure model for optimizing wind power generation and transmission. *Energy Policy, 43,* 291–302.
Presencia, C. E., & Shafiee, M. (2017). Risk analysis of maintenance ship collisions with offshore wind turbines. *International Journal of Sustainable Energy,* 1–21.
Richard, T. L. (2010). Challenges in scaling up biofuels infrastructure. *Science, 329*(5993), 793–796.
Rodriguez, C. P., & Anders, G. J. (2004). Energy price forecasting in the Ontario competitive power system market. *IEEE Transactions on Power Systems, 19*(1), 366–374.
Rogers, J. H. (1999). Learning reliability lessons from PV leasing. *Progress in Photovoltaics: Research and Applications, 7*(3), 235–241.
Rulkens, W. (2007). Sewage sludge as a biomass resource for the production of energy: Overview and assessment of the various options. *Energy & Fuels, 22*(1), 9–15.
Sarja, J., & Halonen, V. (2012). Case study of wind turbine sourcing: Manufacturer selection criteria. In *2012 IEEE Electrical Power and Energy Conference, EPEC 2012*.
Schaffner, D., Ohnmacht, T., Weibel, C., & Mahrer, M. (2017). Moving into energy-efficient homes: A dynamic approach to understanding residents' decision-making. *Building and Environment, 123,* 211–222.
Shafiee, M. (2015). Maintenance logistics organization for offshore wind energy: Current progress and future perspectives. *Renewable Energy, 77*(1), 182–193.
Sherrard, J. R., & Horner, T. A. (2007). Developing a quality entry level technician workforce for the twenty-first century commercial nuclear industry. In *CONTE 2007: Conference on Nuclear Training and Education*.
Sikkema, R., et al. (2014). Legal harvesting, sustainable sourcing and cascaded use of wood for bioenergy: Their coverage through existing certification frameworks for sustainable forest management. *Forests, 5*(9), 2163–2211.
Tascikaraoglu, A., et al. (2014). An adaptive load dispatching and forecasting strategy for a virtual power plant including renewable energy conversion units. *Applied Energy, 119,* 445–453.
Teeuwsen, S. P., Erlich, I., & El-Sharkawi, M. A. (2005). Fast eigenvalue assessment for large interconnected powers systems. In *Power Engineering Society General Meeting* (s. 1727–1733). IEEE.
Thiaw, L., Sow, G., & Fall, S. (2014). Application of neural networks technique in renewable energy systems. In *First International Conference on Systems Informatics, Modelling and Simulation, IEEE Computer Society*, Sheffield, United Kingdom, 29 April–1 May 2014.
Tryfonidou, R., & Wagner, H. J. (2004). Multi-megawatt wind turbines for offshore use: Aspects of Life Cycle Assessment. *International Journal of Global Energy Issues, 21*(3), 255–262.
Tsoutsos, T. D. (2002). Marketing solar thermal technologies: Strategies in Europe, experience in Greece. *Renewable Energy, 26*(1), 33–46.
Verdú, S. V., et al. (2006). Classification, filtering, and identification of electrical customer load patterns through the use of self-organizing maps. *IEEE Transactions on Power Systems, 21*(4), 1672–1682.
Voigt, T., et al. (2003). Utilizing solar power in wireless sensor networks. In *Proceedings—Conference on Local Computer Networks, LCN*.
Voll, P., Klaffke, C., Hennen, M., & Bardow, A. (2013). Automated superstructure-based synthesis and optimization of distributed energy supply systems. *Energy, 50,* 374–388.
Wang, S., & Song, M. (2017). Influences of reverse outsourcing on green technological progress from the perspective of a global supply chain. *Science of the Total Environment, 595,* 201–208.
Wiggelinkhuizen, E., et al. (2008). Assessment of condition monitoring techniques for offshore wind farms. *Journal of Solar Energy Engineering, Transactions of the ASME, 130*(3), 0310041–0310049.

Wirl, F. (1989). Analytics of demand-side conservation programs. *Energy systems and policy, 13* (4), 285–300.

Xie, L., & Ilić, M. D. (2008). Model predictive dispatch in electric energy systems with intermittent resources. In *Conference Proceedings—IEEE International Conference on Systems, Man and Cybernetics*.

Xie, L., & Ilić, M. D. (2009). Model predictive economic/environmental dispatch of power systems with intermittent resources. In *2009 IEEE Power and Energy Society General Meeting, PES '09*.

Yang, R., et al. (2015). An enhanced preventive maintenance optimization model based on a three-stage failure process. In *Science and Technology of Nuclear Installations 2015*.

Yang, W., et al. (2014). Wind turbine condition monitoring: Technical and commercial challenges. *Wind Energy, 17*(5), 673–693.

Yeomans, H., & Grossmann, I. E. (1999). A systematic modeling framework of superstructure optimization in process synthesis. *Computers & Chemical Engineering, 23*(6), 709–731.

Yokoyama, R., Nakamura, R., & Wakui, T. (2017). Performance comparison of energy supply systems under uncertain energy demands based on a mixed-integer linear model. *In Energy, 137,* 878–887.

Zahraee, S. M., Assadi, M. K., & Saidur, R. (2016). Application of artificial intelligence methods for hybrid energy system optimization. *Renewable and Sustainable Energy Reviews, 66,* 617–630.

Zéphyr, L., Lang, P., Lamond, B. F., & Côté, P. (2017). Approximate stochastic dynamic programming for hydroelectric production planning. *European Journal of Operational Research, 262*(2), 586–601.

Zhang, C., et al. (2017). A spare parts demand prediction method for wind farm based on periodic maintenance strategy. In *American Society of Mechanical Engineers, Power Division (Publication) POWER*.

Zhang, H., Liang, Y., Liao, Q., Wu, M., & Yan, X. (2017). A hybrid computational approach for detailed scheduling of products in a pipeline with multiple pump stations. *Energy, 119,* 612–628.

Zhou, W., Xia, X., & Wang, B. (2015). Improving building energy efficiency by multiobjective neighborhood field optimization. *Energy and Buildings, 87,* 45–56.

Zhu, J., et al. (2014). Survey of condition indicators for condition monitoring systems. In *PHM 2014—Proceedings of the Annual Conference of the Prognostics and Health Management Society 2014*.

Zorić, J., & Hrovatin, N. (2012). Household willingness to pay for green electricity in Slovenia. *Energy Policy, 47,* 180–187.

# Chapter 16
# Electrical Vehicle Charging Coordination Algorithms Framework

**Nhan-Quy Nguyen, Farouk Yalaoui, Lionel Amodeo, Hicham Chehade and Pascal Toggenburger**

**Abstract** The coordination of the electrical vehicles (EV) charging becomes an important research subject in the actual context with the growth of the EV usage. This is due to the harmful impacts of the grid and the overspending price of uncoordinated charging procedure. This work tries to provide a framework to configure and formulate the EV charging problem by the theoretical research on scheduling problem with an additional resource. Given the numerous works in the both domains, this would be advantageous to address such a general algorithm framework. This chapter also introduces our configurations, named ACPF/ACPV, to formulate and solve an actual EV charging problem for residential parking—our case study. The purpose of this case study is to illustrate how the framework would be implemented for real-life cases.

## 16.1 Introduction

Electrical vehicles (EV) have become an active topic that interested many researching domains (Cazzola and Gorner 2016; Sperling 2013). One can notice an extensive growth in term of EV sales: the milestone of two million electric cars has been met by the end of 2016 (Leech 2017). Besides the economic and ecological benefits of the EV deployment (Sperling 2013), one has to take into account its disadvantages. Since the EV charging is a high-power consumption task, it can cause much turbulence to the power grid. Also, uncoordinated charging procedure could be costly for EV users because of the personal fix cost of power subscription

N.-Q. Nguyen (✉) · F. Yalaoui · L. Amodeo · H. Chehade
ICD, LOSI, Université de Technologie de Troyes, France, UMR 6281, CNRS, 12 Rue Marie Curie, CS 42060, 10004 Troyes Cedex, France
e-mail: nhan_quy.nguyen@utt.fr

P. Toggenburger
Parkn'Plug, 7 Rue de Vanves, 92130 Issy-les-Moulineaux, France

© Springer International Publishing AG, part of Springer Nature 2018 
C. Kahraman and G. Kayakutlu (eds.), *Energy Management—Collective and Computational Intelligence with Theory and Applications*, Studies in Systems, Decision and Control 149, https://doi.org/10.1007/978-3-319-75690-5_16

and the uncontrolled load during peak hours (Nguyen et al. 2017b). One can find numerous works on the optimal charging scheduling. Most of them can be classified into two approaches: local scheduling (Deilami et al. 2011; Galus et al. 2012; Ma et al. 2013; Mohsenian-Rad et al. 2010; Musardo et al. 2005) and global scheduling (Adika and Wang 2014; Iversen et al. 2014; Li et al. 2011). Also, there are several works on scheduling problems which concern the scheduling of EV charging notably the parallel task scheduling problems (Blazewicz et al. 2011; Nguyen et al. 2016a, b; Sadykov 2012) and scheduling problems under single additional renewable resources (Blazewicz et al. 2002; Hartmann and Briskorn 2010; Józefowska et al. 2002; Nabrzyski et al. 2012; Waligóra 2009). The abundance of works on the domains has, however, some disadvantages. First, the scheduling problems developed for the EV charging, especially the control-theoretic approach, are designed for a very specific problem. Second, general configurations and formulations could be found in theoretical scheduling problems with an additional resource. Yet, as far as we know, there are no frameworks which have been developed to use those resolution methods for the EV charging coordination problem. For that reason, we tend to present in this chapter a framework that can link all those concerning works. This framework has to be both general and implementable. The generality means that it can be configurable and applicable to different real-life constraints and objectives. The implementability can be translated in a sense that it can be constructed with engineered equipment and software.

The first section is, as we have previously done, to introduce the problem addressed in this chapter and the research motivation. In the second section, we will make more details on literature reviews. Then, we can classify the similar research with real-life problems. Hence, in the third section, we can propose a charging coordination configurations family named ACP which includes ACPF and ACPV with its variation (ACPF 1, APCF 2, ACPV 1, ACPV 2.0, ACPV 2.1…). Those configurations are a part of our research for an actual industrial project on the EV charging management. In Sect. 16.4, we also present the predictive-reactive framework for the real-time EV charging management. In Sect. 16.4.1, we introduce a specific case study to illustrate how we can implement a configuration in an industrial environment with the predictive-reactive framework. Finally, we will draw conclusions and perspectives in Sect. 16.5.

## 16.2 Literature Review and Methods Classification

The literature review would be divided into three main parts. The first part is dedicated to the EV optimal scheduling problems. The second part reviews the classical scheduling problems under additional resources and the parallel tasks scheduling problem. The final one classifies existed works into groups of constraints, objectives and approaches. This classification tries to link works in different domains with the common purpose of managing the charging procedure of electric cars.

First, we review the two mains approaches of optimal EV charging scheduling problem: local approach and the global scheduling approach. In the local (or implicit) approach, the EV charging coordination usually based on a "what if" decision (causal properties), driven by some based-rule. For that reason, it can cope quickly with real-time situations. However, when the problem is not treated wholly with all its inputs, then we may find the bad result on cost/time minimization criteria and may lead to infeasible solutions. One can find many theoretical-control oriented works developed in this way. Both (Maasmann et al. 2014) and (Faddel et al. 2017) developed a feedback controlled fuzzy algorithm. The former controller (Maasmann et al. 2014) tend to level the energy consumed while the latter controller (Faddel et al. 2017) aims to maximise the parking lot profit. (Al-Awami et al. 2016) introduced a voltage-based controller to reduce the charging rate with respect to the end-time of charge reference. Álvarez et al. (2016) presented four variations of decentralised controller served the power balancing objective. The second main approach to the EVCC problem is called global (or explicit) scheduling. This scheduling approach takes into account the full input and tries to find global optimal. The advantage of this scheduling is that the solution found has better quality, and by taking into account all the constraints, feasible solutions can be found in more extreme cases. Still, this approach finds difficulties when dealing with randomness. Also, the computational cost of this method is indeed an obstacle to deal with real-time scheduling. In the literature, we can find many works related to the explicit approach, in the limit of this chapter, some featured works are cited. Using the meta-heuristic resolution method, (Alonso et al. 2014) and (Lee et al. 2012) developed genetic algorithms to deal with the charging schedule problem to minimise the total cost. To solve the coordination of EV charging and other electrical equipment, (Adika and Wang 2014) proposed a demand side management algorithm while (Karbasioun et al. 2013) introduced a power strip-packing algorithm. Both algorithms aimed to flatten the power load (Iversen et al. 2014) proposed a Markov chain formulation to deal with randomness behaviours of EV client, then propose a stochastic optimisation schedule for the EVCC problem. We resume all the mentioned works on Table 16.1.

In Table 16.1, we can rarely find works dealing with the time criteria objective (total completion time/makespan minimization). The existing works privileged the efficiency (cost) and the stability (resource levelling) to productivity (time criteria minimization) and service's degree of satisfaction. That is the reason why we would introduce more works on the scheduling field as a complementary with the intention to fill in the research gap. In the second part, we present two major approaches to the scheduling problem: the scheduling problem with controllable processing under single additional resource and the parallel task scheduling problem. With the first approach, one can find two stand-out problems, namely *Discrete-Continuous Scheduling Problem* (DCSP) and Cumulative Scheduling Problem (CuSP). In both of the methods, the more resource a job consumes at a given time, the faster it can reach its final state (completion state), and the total available resource is supposed to be constant. The main difference lies in the resource consumption function, i.e., how can the resource consumption speed-up the jobs' processes. The DCSP is first

**Table 16.1** Classification of some EVCC approaches on the literature

| Coordination method | Resolution approach | Works | Objective | | |
|---|---|---|---|---|---|
| | | | Cost minimization | Power leveling/ load flattening | Time criteria minimization |
| Local (implicit) scheduling | Feedback controlled fuzzy algorithm | Maasmann et al. (2014) | | x | |
| | | (Faddel et al. 2017) | x | | |
| | Voltage-based controller | Al-Awami et al. (2016) | | x | |
| | Decentralized controller | (Álvarez et al. 2016) | | x | |
| Global (explicit) scheduling | Genetic algorithm | Alonso et al. (2014; Lee et al. 2012) | | x | |
| | Demand side management | Adika and Wang (2014) | | x | |
| | Stochastic scheduling | Iversen et al. (2014) | x | | |
| | Power strip packing | Karbasioun et al. (2013) | | x | |

introduced by Weglarz (2012). In this problem, the processing time of a job is controlled by its resource consumption rate which is defined by a *processing rate function*. The DCSP assumes the *processing rate function* to be a convex function which is not greater than linear. In the second problem, CuSP, the resource is accumulated by a linear processing rate function (Baptiste et al. 1999). The DCSP has an advantage in considering a more generic processing rate function. However, most of the concerning works do not take into account of temporal constraints, i.e., release dates and deadlines of tasks, which is considered by CuSP. For the parallel task scheduling problem, we will discuss two specifics types of tasks, namely: moldable tasks and malleable tasks. Parallel tasks have processing time varied to the number of operators assigned to them at a time whereas their surfaces are constant. The number of operators assigned to the moldable task can be decided only once. They can be seen as rectangles with chosen height to be packed (Blazewicz et al. 2011). By default, moldable tasks are non-preemptable. The malleable task has a number of operator varying with time (Beaumont et al. 2012; Karbasioun et al. 2013). The malleable tasks can be interruptible or non-interruptible. The majority of the works studying the scheduling problem under additional resource and parallel task scheduling problem focus on the time criteria optimisation: total or total weighted completion time minimization, makespan minimization. Hence, those aspects fit well as a complementary to the research void of the EVCC. The concerned work is resumed in Table 16.2.

**Table 16.2** Classification of some featured approaches on the theoretical scheduling problem under resource constraint

| Approach | Method | Works | Constant resource | Time-varying resource | Release date | Deadline | Objective |
|---|---|---|---|---|---|---|---|
| Discrete-continuous scheduling problem | Tabu search | Waligóra (2009) | x | | | | Makespan minimization |
| | Dynamic programming and heuristic | Shabtay and Kaspi (2004) | x | | | | Total weighted flowtime |
| | Polynomial-time algorithm | Shabtay and Steiner (2007) | x | | | x | The weighted number of tardy jobs/due date assignment/makespan/ total resource consumption costs |
| Cumulative scheduling problem | Time-bound adjustment algorithms | Baptiste et al. (1999) | x | | x | x | Satisfiability tests |
| | Adjustments of heads (Jackson's pseudo-preemptive schedule) | Carlier and Pinson (2004) | x | | x | | Makespan minimization |
| Moldable task scheduling problem | Suboptimal algorithm | Blazewicz et al. (2011) | x | | | | Makespan minimization |
| Malleable task scheduling problem | Dominant class of schedules | Sadykov (2012) | x | | | | Total weighted completion time |
| | Rectangle packing algorithm | Blazewicz et al. (2006) | x | | | | Makespan minimization |

## 16.3 ACPF-ACPV: The EV Charging Coordination Configurations

After analysing the literature, we would like to classify the electrical vehicle charging coordination (*EVCC*) into specific configurations. With that kind of classification, one can decide which researching works can be used to solve each specific problem. First, we identify three groups of constraints of the EVCC issue: the human constraints, the technical constraints and the system capacity (or resource availability) constraints. Their segregation is shown in Fig. 16.1.

We would explain in detail our classification in the order of numbered constraints. The system capacity constraints group contains two types of parking with its corresponding power supply. In a dedicated parking, i.e. the power supply is subscribed for only the charging of EV, the power bandwidth is equal to the maximum power subscription, then it is considered to be constant. Let us take the residential parking case as an example. In this case, the other devices sharing the power with EV charging are the common facilities such as lightings and ventilations. The sharing power is considered to be too little to have an impact on the total power bandwidth. Hence the total available power supply is deemed to be constant. In the second constraint, the power bandwidth is time-varying. This restriction is critical to take into account because it makes the corresponding problem much harder to solve (Nguyen et al. 2016a). It happens in a case where the power bandwidth has to be shared with high-power consumption devices. The problem is very common in the professional sector where the electrical supply for the EV charging in the office parking is shared with industrial machines or at least computers and office devices (US Department of Energy (DOE), North Carolina State Energy Office 2014).

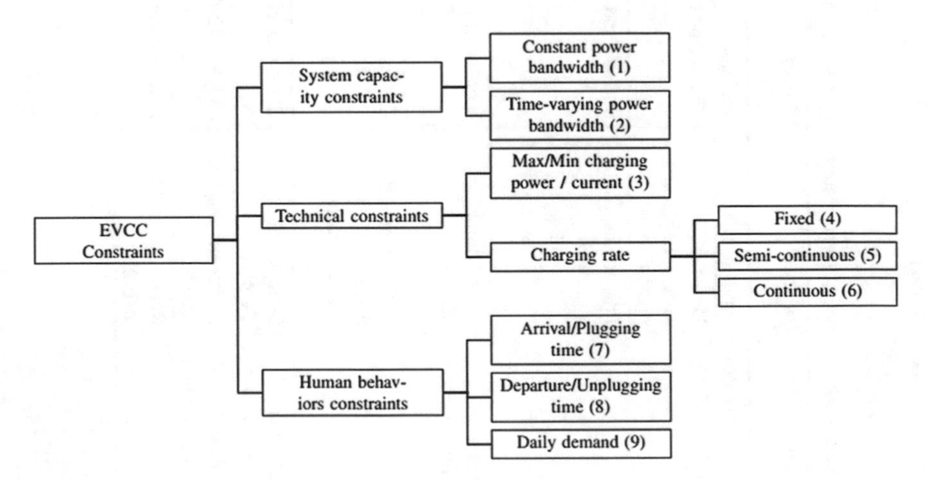


**Fig. 16.1** A classification of EVCC constraints

Despite the fact that EV chargers and electric vehicle supply equipment (EVSE) have many characteristics (Yilmaz and Krein 2013) we can classify the technical constraints into two subgroups. First, whatever the charging level is, or, it is a fast or slow charger, it always has a maximum and minimum charging power (or current) limits. The development of new protocol and charge-point technologies permits the to change the power provided on EVSE (Schmutzler et al. 2013). Depend on what kind of change can be made, we classify the charging power of EV into three categories: fixed, semi-continuous and continuous rates. To explain those constraints, we introduce some notification: let the charging job indexed by $i$, the scheduling horizon is divided into $H$ interval indexed by $k$. The charging rate of job $i$ at interval $k$ is noted by $u_{ik}$. Charging rate of job $i$ is bounded by $\bar{u}_i$ and $\underline{u}_i$. If charging rate can only be fixed at the beginning of the charging process by a value, $u_i^0 \in [\bar{u}_i, \underline{u}_i]$ then it is called fixed charging rate. With the parallel task problem notation, charging task with fixed rate can be considered to be *moldable*. In the other cases, the charging rate is whether continuous or semi-continuous. If at a given interval $k$, and the charging task has started to process: $u_{ik} \in [\bar{u}_i, \underline{u}_i]$ the charging rate is called semi-continuous or continuous depending on the lower bound $\bar{u}_i$. If $\bar{u}_i > 0$ thenthe charging rate is called semi-continuous and the charging task is non-preemptive. In the contrary, If $\bar{u}_i = 0$ then the charging rate is called continuous, the charging task is then preemptive. The segregation of charging rate is important because it defines our two configuration *ACPF* and *ACPV*. The *ACPF* configuration (French: *Algorithme de Charge de Puissance Fixe, Charging Algorithm with Fixed Power*) corresponds to the constraint (4) and the *ACPV* (French: *Algorithme de Charge de Puissance Variable, Charging Algorithm with Varying Power*) corresponds to the constraints (5) and (6). Figure 16.2 illustrates examples of the power allocation of the five configurations.

The last group of constraints is created by the behaviours of EV users. Each of them has a daily parking (i.e. plugging) time and a departure time (unplugging) time that formulates the so-called *time-windows constraint*. With the scheduling problem notation, plugging time can be seen as the release date of tasks and unplugging time can be considered as the strict deadline of task (Yalaoui and Chu 2002). Every charging task has a demanded energy to satisfy, called daily demand satisfaction constraint. This group of constraints constitutes the stochastic characteristic of the considered scheduling problem.

In Table 16.3, we specify the two configurations with their variations, accompanied by the constraints. According to each configuration, we cite the similar *EVCC* problems and the similar theoretical scheduling problem resolution approaches.

In Table 16.3, the complexity is illustrated by three objectives: feasibility test, makespan minimization and total weighted completion times minimization. In all configurations, the human behaviours constraints group are the same. Configurations having index increasing with the complexity so that one can estimate the complexity of a specific EVCC problem. According to each pair of configurations—complexity, one can choose a resolution method with reasonable computational costs and researching effort. All the settings, except ACPF 2, have their own resolution approaches.

**Table 16.3** The ACPF—ACPV configurations

| Configs. | Vers. | Power bandwidth | | Charging rate | | | Complexity | | | Scheduling algorithms |
|---|---|---|---|---|---|---|---|---|---|---|
| | | Constant | Time varying | Fixed | Continuous | Semi-continuous | Feasibility | min $C_{max}$ | min $\sum w_i C_i$ | |
| ACPF | 1 | X | | X | | | / | NP-Hard (Blazewicz et al. 2011) | / | Moldable task scheduling (Blazewicz et al. 2011; Dutot et al. 2004), Rectangle packing (Huang et al. 2007), Strip packing (Nguyen et al. 2017b) |
| | 2 | | X | X | | | / | / | / | / |
| **ACPV** | 1a | X | | | X | | Poly-nomial (Beaumont et al. 2012) | Poly-nomial (Blazewicz et al. 2006) | Poly-nomial (Beaumont et al. 2012) | Preemtable malleable task scheduling (Beaumont et al. 2012; Blazewicz et al. 2006), Discrete-continuous scheduling (Józefowska and Weglarz 1998) |
| | 1b | X | | | | X | NP-Complete (Strong) (Baptiste et al. 1999) | NP-Hard (Strong) (Carlier and Pinson 2004) | | Malleable task scheduling (Beaumont et al. 2012; Sadykov 2012), Cumulative scheduling (Baptiste et al. 1999), Jackson pseudo-preemptive scheduling (Carlier and Pinson 2004) |
| | 2 | | X | | | X | NP-Hard (Nguyen et al. 2016b) | NP-Hard (Strong) (Nguyen et al. 2016b; Nobibon et al. 2015) | | Discrete malleable task scheduling (Nguyen et al. 2016b; Nguyen et al. 2017a) |

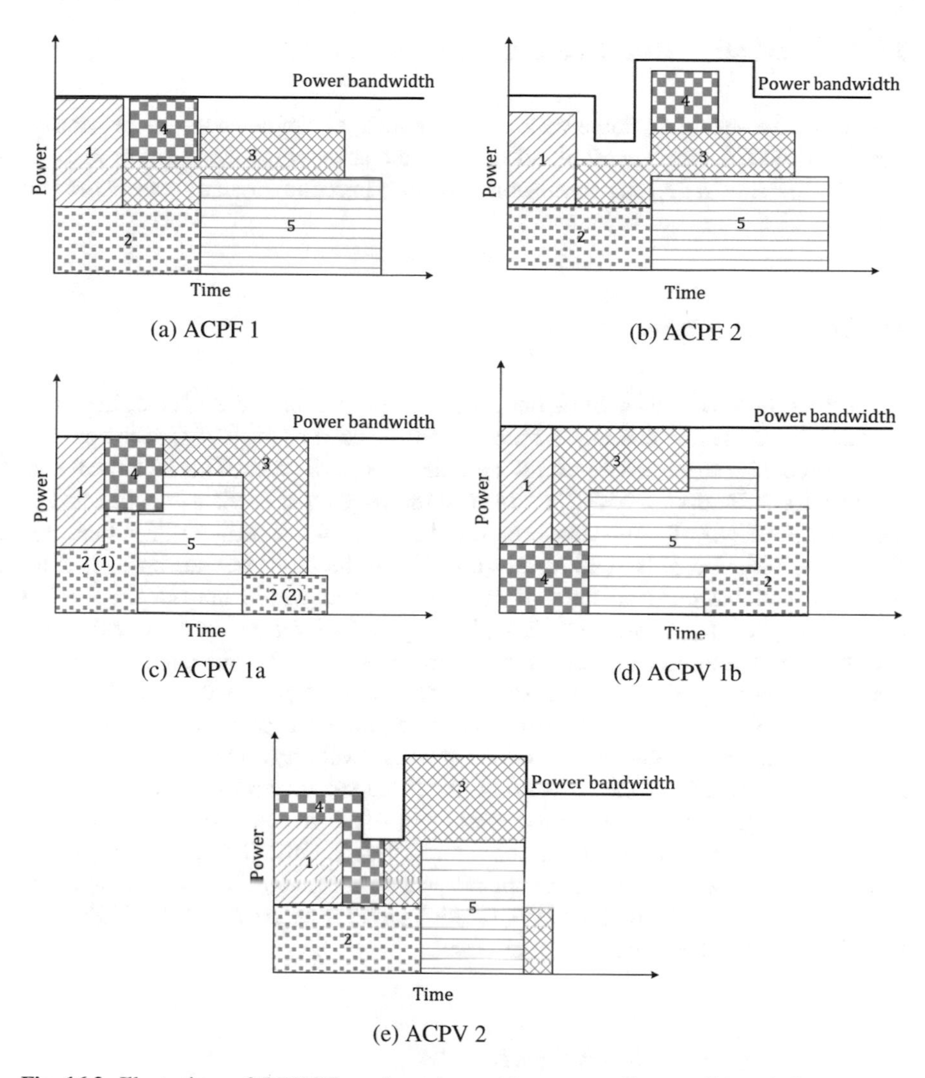


**Fig. 16.2** Illustrations of 5 EVCC configurations with corresponding possible solutions

The lack of work on *ACPF 2* is caused by the complexity of the time-varying resource constraint. This variation prevents the adaptation of other existed works on this specific problem. We can also find that the *ACPV 2* has been the most general configuration so far. For that reason, we have developed in our previous works two approaches to solve these settings: the exact and approximate ones (Nguyen et al. 2016b; 2017a). More details on the mathematical formulation of the ACPV2 configuration can be found on (Nguyen et al. 2016b). An efficient heuristic named Adapting and Layering to deal with the ACPV 2 can be found in (Nguyen et al. 2016a). An exact method to solve the problem by a Branch-and-Price algorithm is introduced in (Nguyen et al. 2017a).

## 16.4 Predictive-Reactive Framework

The predictive-reactive framework is introduced to implement the previously mentioned EVCC configuration and to apply the scheduling algorithm to real life case. Before illustrating the framework, we first introduce a study case extracted from our EVCC project in France.

### *16.4.1 Case Study*

The case study takes data from our industrial project of the EV charging management for residential parking in France. The objective of the EVCC is to optimise total weighted end-of-charge times and the electrical cost. The electrical cost includes the subscription cost (fix cost) and the usage cost (variable cost). Most of the French electrical suppliers propose the on-peak/off-peak tariff where the off-peak electrical cost is significantly lower than the on-peak one. Thus, to minimise the variable cost, all the EV must be charged within the off-peak period (overnight schedule) subject to the EV plugging times at the end of the working day and the expected unplugging times in the next morning. The histogram of arrivals time and charging demand extracted from 125 charging events is shown in Fig. 16.3a, b (Nguyen et al. 2017b). Concerning the fixed cost, there is a trade-off between the power bandwidth subscribed and the feasibility rate of the EV charging scheduling. Precisely, if the scheduling algorithm cannot find a feasible schedule baseline for a given input, the manager unit will decide to start the charging process earlier than the off-peak duration, i.e. before-off-peak shift. Thus, to re-assure the feasibility, the charging cost before the off-peak becomes costlier. According to the study case, the corresponding EVCC configuration is *ACPF 1* or *ACPV 1b*, depending on the type of the chargers used.

### *16.4.2 The Framework Conception*

The predictive-reactive framework for the EVCC problem is based on the rescheduling manufacturing framework introduced by (Vieira et al. 2003). The predictive-reactive policy (PRP) has a purpose to deal with uncertainties in scheduling, given that we have some degree of knowledge about the input. The schedule is called static if the number of input jobs is known in advance, whereas it is called dynamic when the input set is unknown. In this case study, the set of jobs is finite and known; the problem is then static. In addition, historical information of each job is logged in the database so that we can assure a certain level of information for the planning. The first part of the PRP, the predictive schedule (or the pre-schedule), intends to use that known information to generate a schedule baseline. Theoretically, any solving process to find a solution for the deterministic problem can be applied to produce a schedule baseline. Then, the

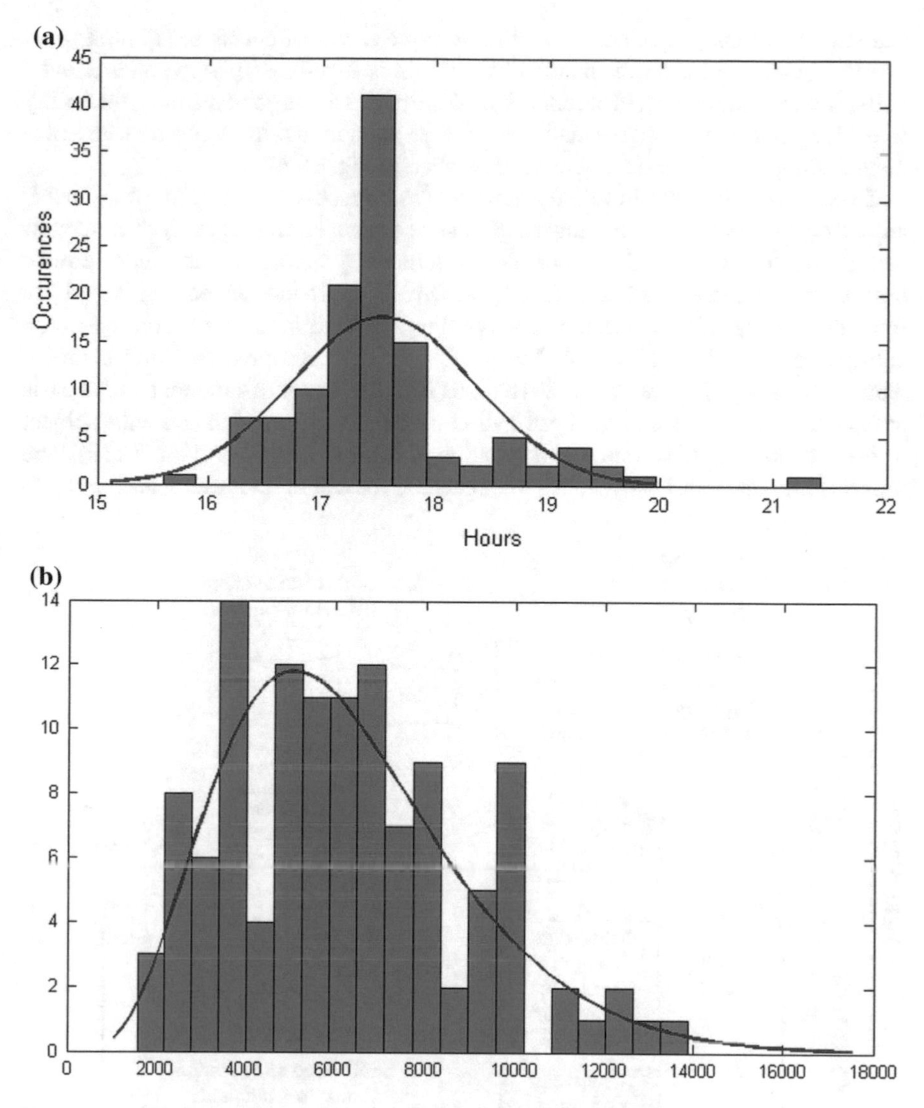


**Fig. 16.3** **a** Histogram with gamma distribution (a = 4:5; b = 1442) for the daily consumption over 120 charging events. **b** Histogram with normal distribution ($\mu = 17.5\,\text{h}; \sigma = 0.8\text{h}$) for the daily arrival times over 125 charging events

second part of the PRP, the reactive schedule (or the reschedule), is designed to treat the real-time events which are different to the baseline generated by prescheduling. More strategies and policies on this framework can be found on (Pinedo 2015; Vieira et al. 2003). The policy chosen for our predictive-reactive framework can be resumed as follows: (1) scheduling environment: static stochastic; (2) rescheduling strategy: predictive-reactive with event driven (i.e. new schedule is updated every time there are new temporal events); (3) reschedule method: partial rescheduling. The purpose of this framework is three folds, corresponding to three levels of decision. (1) Strategical level: it decides a good power-bandwidth to minimise fixed cost. (2) Operational level: daily, it decides the expected before-off-peak shift to assure a good planning feasibility rate. (3) Applicable level: it manages the charge point in real-time, coping with every new-coming event. The framework is described in Fig. 16.4.

**Predictive schedule**: In this framework, the pre-schedule work in a way that it repeatedly solves many deterministic problems generated from a random generator. The parameter of the random generator is defined by historical data, short-term or long-term, using the maximum likelihood (MLE) approach. Scheduling algorithm from the library will be selected to solve the generated instances according to the configuration of the parking: ACPF, APCV or only resource levelling heuristic namely *Layering* (Nguyen et al. 2016a, 2017b). The result of random simulators is tracked to formulate a statistical estimation on the behaviour of the baseline subject to the total power subscription and scheduling horizon (Nguyen et al. 2017b). The preschedule generates two levels of schedule baseline: the strategical and the

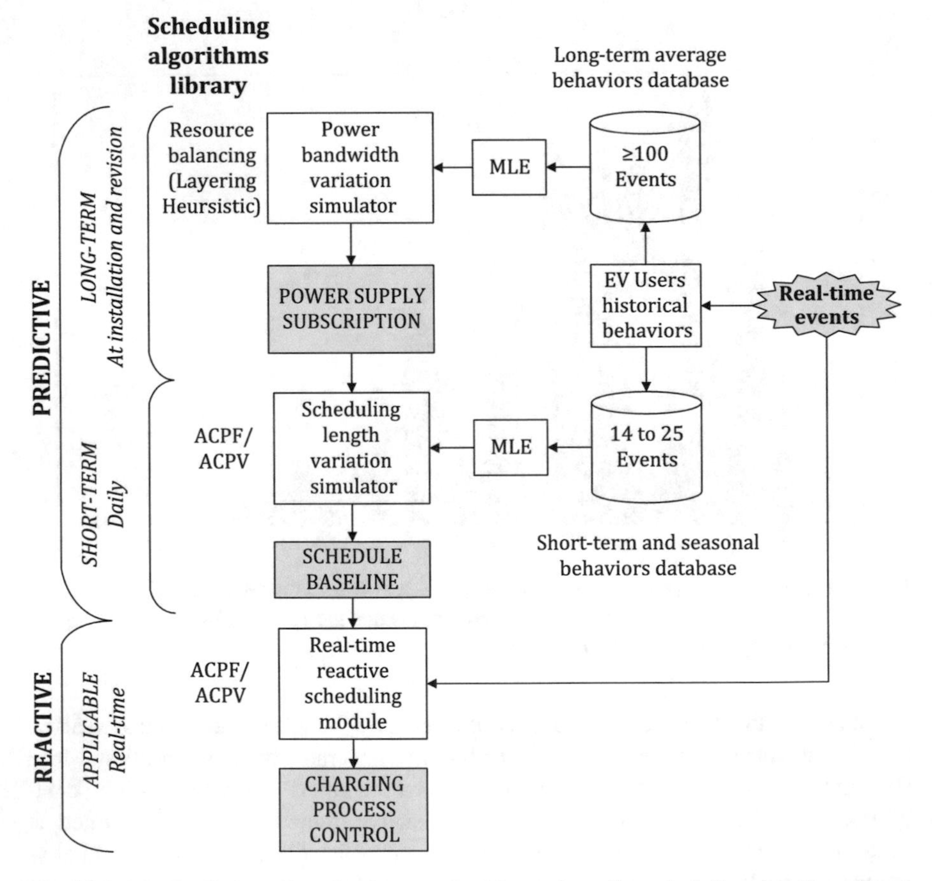


**Fig. 16.4** The Predictive—Reactive framework with corresponding scheduling algorithms library and simulators

operational level. The strategical schedule baseline, in this case, is the power subscription for the EV charging parking. The total value of power subscribed is a long-term decision. Therefore, it is calculated at the installation of the parking. This sum of power can also be revised at every trimester. The estimation of power subscription is got from the formula (16.1) presented by Nguyen et al. (2017b):

$$z(U) = f_0(U) + P_{BW}(x > U).f_p(U) \tag{16.1}$$

where:

- $P_{BW}(x > U)$ is the probability of necessary power bandwidth to assure feasibility and must be greater than the subscribed power $U$.
- $f_0(U)$ is the fixed cost of power subscription U which is usually linear: $f_0(U) = U \times c_0$, $c_0$ is the subscription cost per power unit.
- $f_p(U)$ is the penalty cost while using total power $U$ due to earlier start before off-peak hours to satisfy all energy demands.

Optimal power subscription should minimize the cost $U^* = \arg\min_u(z(U))$. Moreover, the penalty cost is calculated by $f_p(U) = U\boldsymbol{E}_U[\Delta St](c_{peak} - c_{off-peak}) = F_E(U)U\Delta c$ where $\boldsymbol{E}_U[\Delta St]$ is the expected duration of the charging operation before the off-peak hours (i.e. $\Delta St$) when the total bandwidth is $U$. $\Delta c$ is the difference of price per kW between peak and off-peak hours. $P_{BW}(x > U)$ and can be found by two simulators: the bandwidth variation and the schedule length variation (Nguyen et al. 2017b). The real-time events will be accumulated in the database. The predictive module then extracted from the database a fixed number of most recent events to simulate. For the long-term baseline estimation, the module will extract numerous events ($\geq 100$ events for each EV, corresponding to the charging behaviours in around a trimester) to parameterize the simulator. Hence one can fix the power subscription. Then, the schedule length variation will take short-term events (about 25 events each EV corresponds to 1 month) to tune the random generator. This simulator outputs the distribution of the schedule length according to the power subscribe and the recent customer behaviour. Hence, before the beginning of the charging process, one has to decide whether we have to start all the charging earlier before the off-peak hours to assure feasibility.

**Reactive schedule**: With the long-term and short-term planning starting point, the reschedule module has the information about the number of jobs it has to deal, also about the total power it can use for the charging and does it have to start earlier this day. The reactive schedule is driven by real-time events and has partial rescheduling method. Precisely, at each event: arrival and plugging of an EV, charging completion, changing of total power…, a new partial schedule will be recalculated based on the present (plugged) EV in the parking. This planning is partial due to the non-preemption properties of jobs. All the charging in the process is untouched; only the new job should be rescheduled.

For the case study, the framework gives a sharp reduction: 89% of the total power used with an expected earlier start of 4.6 min before off-peak hours. Figure 16.5 (Nguyen et al. 2017b) would justify this result.

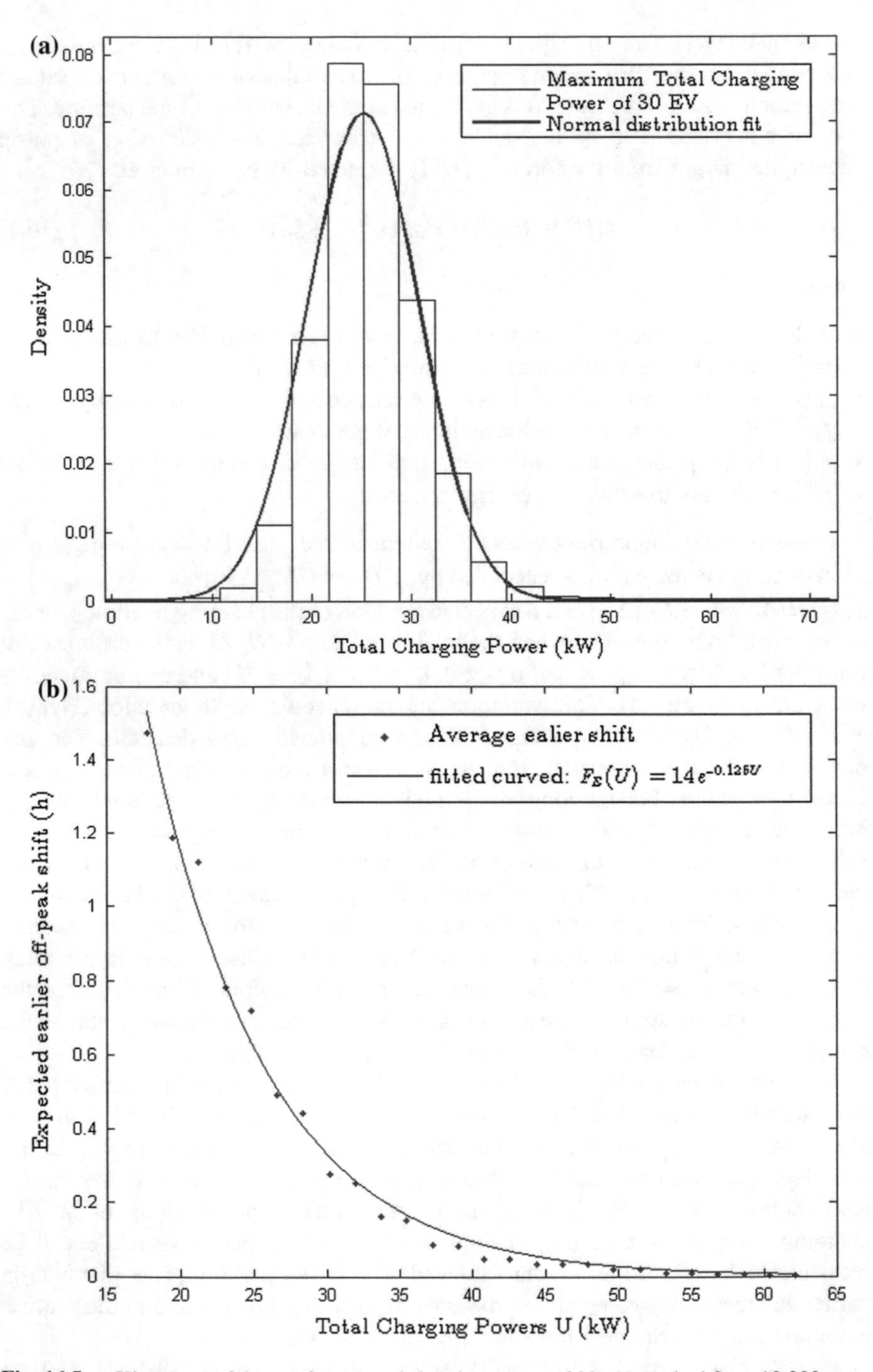


**Fig. 16.5** **a** Histogram of the maximum total charging power of 30 eV resulted from 10,000 tests. **b** Expected shift (h) to start before off-peak hours to assure feasibility according to power bandwidth used ($n_{EV}$ = 30 and $n_{tests}$ = 10,000)

## 16.5 Conclusion

In this chapter, we introduce a framework for the EV charging coordination problem. Pointing out the need of bridging the research on EVCC and theoretical scheduling problems with additional resources, we introduce five generic configurations for EVCC problems with the potential concerned scheduling algorithm. The complexity classification of each configuration suggests the corresponding amount of computational efforts should be carried out, then define the hardware requirements for the parking installation. To apply the settings to real-life problems, we introduced a predictive framework based on the manufacturing scheduling system strategy. The case study shows good time criteria and cost optimisation for the specific case study on the EVCC of residential parking. Our research points out a minor and major research gaps. Concerning the minor one, a scheduling method developed for configuration ACPF 2 is still missing. For the major one, since our EVCC problem is classified static and stochastic, a robust scheduling method could be considered.

**Acknowledgements** This research has been supported by ANRT (Association Nationale de la Recherche et de la Technologie, France).

## References

Adika, C. O., & Wang, L. (2014). Smart charging and appliance scheduling approaches to demand side management. *International Journal of Electrical Power & Energy Systems, 57,* 232–240.

Al-Awami, A. T., Sortomme, E., Akhtar, G. M. A., & Faddel, S. (2016). A voltage-based controller for an electric-vehicle charger. *IEEE Transactions on Vehicular Technology, 65*(6), 4185–4196.

Alonso, M., Amaris, H., Germain, J. G., & Galan, J. M. (2014). Optimal charging scheduling of electric vehicles in smart grids by heuristic algorithms. *Energies, 7*(4), 2449–2475.

Álvarez, JN., Knezović, K., & Marinelli, M. (2016). Analysis and comparison of voltage dependent charging strategies for single-phase electric vehicles in an unbalanced danish distribution grid. In: *Proceedings of the 51st International Universities Power Engineering Conference, IEEE.*

Baptiste, P., Le Pape, C., & Nuijten, W. (1999). Satisfiability tests and time-bound adjustments for cumulative scheduling problems. *Annals of Operations Research, 92,* 305–333.

Beaumont, O., Bonichon, N., Eyraud-Dubois, L., & Marchal, L. (2012). Minimizing weighted mean completion time for malleable tasks scheduling. In: 2012 IEEE 26th International, IEEE Parallel & Distributed Processing Symposium (IPDPS), pp. 273–284.

Blazewicz, J., Ecker, K. H., Pesch, E., Schmidt, G., & Weglarz, J. (2002). *Scheduling computer and manufacturing processes*. Newyork: Springer.

Blazewicz, J., Kovalyov, M. Y., Machowiak, M., Trystram, D., & Weglarz, J. (2006). Preemptable malleable task scheduling problem. *IEEE Transactions on Computers, 55*(4), 486–490.

Blazewicz, J., Cheng, T. E., Machowiak, M., & Oguz, C. (2011). Berth and quay crane allocation: a moldable task scheduling model. *Journal of the Operational Research Society, 62*(7), 1189–1197.

Carlier, J., & Pinson, E. (2004). Jackson's pseudo-preemptive schedule and cumulative scheduling problems. *Discrete Applied Mathematics, 145*(1), 80–94.

Cazzola, P., & Gorner, M. (2016). *Global ev outlook 2016—beyond one million electric cars*. France, Paris: International Energy Agency.

Deilami, S., Masoum, A. S., Moses, P. S., & Masoum, M. A. (2011). Real-time coordination of plug-in electric vehicle charging in smart grids to minimize power losses and improve voltage profile. *IEEE Transactions on Smart Grid, 2*(3), 456–467.

Dutot, PF., Mounié, G., & Trystram, D. (2004). Scheduling parallel tasks: Approximation algorithms.

Faddel, S., Al-Awami, A. T., & Abido, M. (2017). Fuzzy optimization for the operation of electric vehicle parking lots. *Electric Power Systems Research, 145,* 166–174.

Galus, M. D., Waraich, R. A., Noembrini, F., Steurs, K., Georges, G., Boulouchos, K., et al. (2012). Integrating power systems, transport systems and vehicle technology for electric mobility impact assessment and efficient control. *IEEE Transactions on Smart Grid, 3*(2), 934–949.

Hartmann, S., & Briskorn, D. (2010). A survey of variants and extensions of the resource-constrained project scheduling problem. *European Journal of Operational Research, 207*(1), 1–14.

Huang, W., Chen, D., & Xu, R. (2007). A new heuristic algorithm for rectangle packing. *Computers & Operations Research, 34*(11), 3270–3280.

Iversen, E. B., Morales, J. M., & Madsen, H. (2014). Optimal charging of an electric vehicle using a markov decision process. *Applied Energy, 123,* 1–12.

Józefowska, J., & Weglarz, J. (1998). On a methodology for discrete–continuous scheduling. *European Journal of Operational Research, 107*(2), 338–353.

Józefowska, J., Waligóra, G., & Węglarz, J. (2002). Tabu list management methods for a discrete–continuous scheduling problem. *European Journal of Operational Research, 137*(2), 288–302.

Karbasioun, M. M., Shaikhet, G., Kranakis, E., & Lambadaris, I. (2013). Power strip packing of malleable demands in smart grid. In: *2013 IEEE International Conference on Communications* (*ICC*), pp. 4261–4265.

Lee, J., Kim, H. J., Park, G. L., & Jeon, H. (2012). Genetic algorithm-based charging task scheduler for electric vehicles in smart transportation. In: *Asian Conference on Intelligent Information and Database Systems*, (pp. 208–217). Springer.

Leech, S. (2017). The economics of electric cars. The Market Mogul.

Li, Y., Li, L., Yong, J., Yao, Y., & Li, Z. (2011). Layout planning of electrical vehicle charging stations based on genetic algorithm. In: *Electrical Power Systems and Computers*, (pp. 661–668). New York: Springer.

Ma, Z., Callaway, D. S., & Hiskens, I. A. (2013). Decentralized charging control of large populations of plug-in electric vehicles. *IEEE Transactions on Control Systems Technology, 21* (1), 67–78.

Maasmann, J., Aldejohann, C., Horenkamp, W., Kaliwoda, M., & Rehtanz, C. (2014). Charging optimization due to a fuzzy feedback controlled charging algorithm. In: *2014 49th International Universities, IEEE Power Engineering Conference* (*UPEC*), pp. 1–6.

Mohsenian-Rad, A. H., Wong, V. W., Jatskevich, J., Schober, R., & Leon-Garcia, A. (2010). Autonomous demand-side management based on game-theoretic energy consumption scheduling for the future smart grid. *IEEE transactions on Smart Grid, 1*(3), 320–331.

Musardo, C., Rizzoni, G., Guezennec, Y., & Staccia, B. (2005). A-ecms: An adaptive algorithm for hybrid electric vehicle energy management. *European Journal of Control, 11*(4–5), 509–524.

Nabrzyski, J., Schopf, J. M., & Weglarz, J. (2012). Grid resource management: State of the art and future trends.

Nguyen, N. Q., Yalaoui, F., Amodeo, L., Chehade, H., & Toggenburger, P. (2016a). Solving a malleable jobs scheduling problem to minimize total weighted completion times by mixed integer linear programming models. In: *Asian Conference on Intelligent Information and Database Systems*, (pp. 286–295). Springer.

Nguyen, N. Q., Yalaoui, F., Amodeo, L., Chehade, H., & Toggenburger, P. (2016b). Total completion time minimization for machine scheduling problem under time windows constraints with jobs' linear processing rate function. *Computer Operational Research*, manuscript submitted.

Nguyen, N. Q., Yalaoui, F., Amodeo, L., Chehade, H., & Toggenburger, P. (2017a). A branch-and-price approach to solving a discrete malleable jobs scheduling problem with time-varying resource constraints. *European Journal of Operational Research*, manuscript submitted.

Nguyen, N. Q., Yalaoui, F., Amodeo, L., Chehade, H., & Toggenburger, P. (2017b). Predictive baseline schedule for electrical vehicles charging in dedicated residential zone parking.

Nobibon, F. T., Leus, R., Nip, K., & Wang, Z. (2015). Resource loading with time windows. *European Journal of Operational Research, 244*(2), 404–416.

Pinedo, M. (2015). Scheduling. Springer, Chap. 9.

Sadykov, R. (2012). A dominant class of schedules for malleable jobs in the problem to minimize the total weighted completion time. *Computers & Operations Research, 39*(6), 1265–1270.

Schmutzler, J., Andersen, C. A., Wietfeld, C. (2013). Evaluation of ocpp and iec 61850 for smart charging electric vehicles. In: 2013 World, IEEE Electric Vehicle Symposium and Exhibition (EVS27), pp. 1–12.

Shabtay, D., & Kaspi, M. (2004). Minimizing the total weighted flow time in a single machine with controllable processing times. *Computers & Operations Research, 31*(13), 2279–2289.

Shabtay, D., & Steiner, G. (2007). Optimal due date assignment and resource allocation to minimize the weighted number of tardy jobs on a single machine. *Manufacturing & Service Operations Management, 9*(3), 332–350.

Sperling, D. (2013). Future drive: Electric vehicles and sustainable transportation. Island Press.

US Department of Energy (DOE), North Carolina State Energy Office. (2014). Real-World Charging Behavior at the Workplace. Plug-in Electric Vehicle Consumer Usage Study, Advanced Energy leveraged a U.S. Department of Energy (DOE).

Vieira, G. E., Herrmann, J. W., & Lin, E. (2003). Rescheduling manufacturing systems: a framework of strategies, policies, and methods. *Journal of Scheduling, 6*(1), 39–62.

Waligóra, G. (2009). Tabu search for discrete–continuous scheduling problems with heuristic continuous resource allocation. *European Journal of Operational Research, 193*(3), 849–856.

Weglarz, J. (2012). *Project scheduling: recent models, algorithms and applications* (Vol. 14). Springer Science & Business Media.

Yalaoui, F., & Chu, C. (2002). Parallel machine scheduling to minimize total tardiness. *International Journal of Production Economics, 76*(3), 265–279.

Yilmaz, M., & Krein, P. T. (2013). Review of battery charger topologies, charging power levels, and infrastructure for plug-in electric and hybrid vehicles. *IEEE Transactions on Power Electronics, 28*(5), 2151–2169.

# Chapter 17
# Multi-Site Energy Use Management in the Absence of Smart Grids

**Zeynep Bektas, Gülgün Kayakutlu and M. Özgür Kayalica**

**Abstract** Demand-side management (DSM) allows an energy load to be balanced across multiple consumers. Energy consumption fluctuations cause important costs based on the alternating energy price tariffs. DSM creates opportunities for consumers to reduce their energy consumption costs by smoothing the daily load curve. An MINLP model is constructed based on power consumption, which aligns with the production schedules of the industrial units. Then, these feasible schedules are used as an input for a cooperative Bayesian game that is designed to balance the hourly loads. A case study of three factories, where the demand-side manager tries to minimize the instability of purchasing electricity from the general grid through load balancing, is considered.

## 17.1 Introduction

As industrial power consumption increases, all processes encompassing energy generation, transmission, distribution, and consumption are studied from different angles. Indirect control from the demand side, in an energy market with multiple suppliers and multiple consumers, has been studied for a long time. In countries where residential consumption comprises a major share of the energy market, smart grid applications are accompanied by incentives to implement demand-side management (DSM). DSM is a tool used by power system managers to provide balanced consumption of power across the different hours of the day. It has been shown in the literature that DSM usage balances electric energy consumption Marzband et al. (2013).

Z. Bektas (✉)
Department of Industrial Engineering, Istanbul University, Avcilar, Istanbul, Turkey
e-mail: zeynep.bektas@istanbul.edu.tr

G. Kayakutlu
Energy Institute, Istanbul Technical University, Macka, Istanbul, Turkey

M. Ö. Kayalica
Faculty of Management, Istanbul Technical University, Macka, Istanbul, Turkey

© Springer International Publishing AG, part of Springer Nature 2018

C. Kahraman and G. Kayakutlu (eds.), *Energy Management—Collective and Computational Intelligence with Theory and Applications*, Studies in Systems, Decision and Control 149, https://doi.org/10.1007/978-3-319-75690-5_17

DSM control is not as rigid as control on the generation side since it does not actually change the demand. DSM can be applied either to shift loads without decreasing total power consumption by studying the energy efficiency, or to balance hourly load flows under consumer constraints (Lopez et al. 2015).

The proliferation of smart grids represents a large share of the development of DSM programs. These smart systems enable two-way digital communication between power generators and consumers, which leads to participation that is more active on the demand side (He et al. 2015).

The main goal of this study is to construct a new load management approach from the perspective of demand-side management for multi-site manufacturing conglomerates. The proposed load management is modeled by optimizing the power usage and cooperation of consumers in order to reduce grid instability in the absence of "smart" measures. The originality of the approach rests on developing an optimized schedule of power consumption for multiple consumers and flattening the load flows using a cooperative Bayesian game approach in the absence of smart grids. Thus, when large numbers of industrial consumers put high power demands on the general grid, this approach could minimize negative effects on the stability of the grid in these districts.

When there is more than one consumer in a load management problem, the problem can be approached as a game because there is more than one independent decision maker. A recent development in DSM considers using game theory for smart grid implementations. In developing countries, two-way digital communication infrastructures are not yet in place (Atikol 2013). This study proposes a model to remove that deficiency by using Bayesian games to account for incomplete information cases through time-of-use (TOU) pricing rates. This allows modeling the distribution of the load across multiple rates.

To strengthen our model framework, an industrial case comprising three factories (consumers) owned by a single entity, which purchases energy from the general grid and distributes it to the factories, is studied with the aim of minimizing grid instability. The entity, which is assumed to own all three factories in a particular industrial zone, is taken as a demand-side manager (DSMr), whose objective is to reduce the load instability of the transmission lines and the power costs simply by optimizing the time schedules in concert with the factories in a cooperative fashion that does not change daily total electricity consumption. The DSMr is neither a power generator nor a supplier/dealer trading electricity in the market.

This paper uses a mixed integer nonlinear programming (MINLP) model proposed for DSM. Bektas and Kayalica also provide a similar framework (Bektas and Kayalica 2015). However, our model incorporates a game stage. The proposed model consists of two stages with an MINLP model giving input data for the Bayesian game. The originality of the study is enhanced by proposing a two-step model to encourage cooperative use of energy resources in developing countries, until they can catch up to the application of smart grids.

This paper is organized as follows. In Sect. 17.2, we explain the DSM approach. The methodology is given in Sect. 17.3, followed by the problem in Sect. 17.4. The case is explored in Sect. 17.5. The last section is reserved for the conclusion and

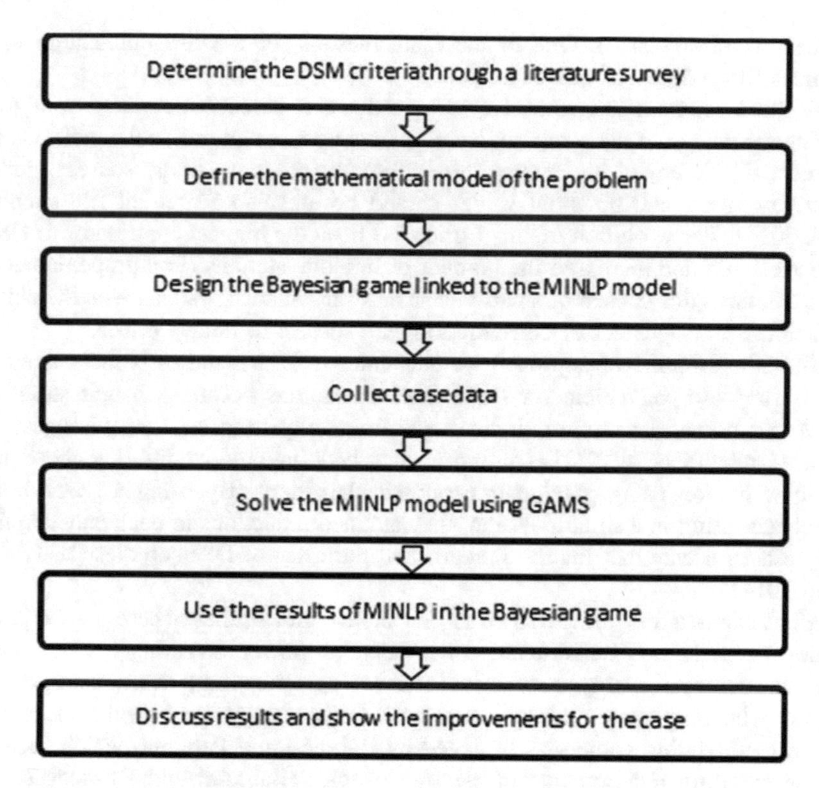


**Fig. 17.1** Flow diagram of the study

future recommendations (Sect. 17.6). A flow diagram of the study is shown in Fig. 17.1.

## 17.2 Demand-Side Management

Demand-Side Management (DSM) aims to use power efficiently by monitoring and controlling the consumer demand. DSM is comprised of policies and measures encompassing a wide range, from long-term energy efficiency programs to real-time control of distributed energy sources (Lampropoulos et al. 2013). Hence DSM programs either reduce or shift power consumption. A reduction in consumption is enabled by energy efficiency studies (Mohsenian-Rad et al. 2010). Whereas, time dependent pricing studies are focused on shaving the peaks by changing the hourly consumption levels without changing total daily electricity consumption. Power request stability is considered very important for the power suppliers; generating energy without peaks increases the efficiency (Yang et al. 2012). The supply side has raised this point for years and thus they played an important role in early DSM studies (Tembine 2016). One of the main reasons for a DSM application is to augment the stability of power grids.

A DSM program is specifically designed for a particular country or region. During the design of the program, technical, economic, and political conditions are considered. Moreover, in the program application process, energy sources, generation capacities, and potential values should be taken into account (Bergaentzle et al. 2014). The evolution of smart grids has been the biggest contributor to DSM improvements and increased the number of implementations (Lampropoulos et al. 2013). Smart grids consist of smart meters and surveillance systems with the ability to facilitate management of consumption with respect to hourly prices.

The other significant approach we encounter in DSM studies is that of microgrids. They are convenient for DSM implementations because of their structure. DSM and power scheduling studies have frequently been seen in microgrids as well. Marzband et al. (2013) proposed a scheduling model for day-ahead and real-time market using previously proposed algorithms. By using a gravitational search algorithm in a similar system, they attained reductions in peak consumption and costs in a way that fits the fundamental purposes of DSM studies (Marzband et al. 2014).

Mohsenian-Rad et al. (2010) was one of the first studies where a scheduling game approach was constructed for residential power consumers. The model assumes multiple energy consumers and a single supplier; the consumers are players who employ a strategy based on a daily energy consumption schedule. Another scheduling game was designed by Bahrami and Parniani, which focused on the charging requirements of electric vehicles (Bahrami and Parniani 2014). Chen et al. (2014) searched for the uniqueness of Nash equilibrium by modeling a cooperative game to minimize the personal energy costs of selfish players Mangiatordi et al. (2013) proposed a multi-objective particle swarm optimization algorithm to lead to the equilibrium point of a non-cooperative game for a dynamic electricity pricing structure. Yang et al. (2012) designed a game with separate utility functions for consumers and generators; this study resulted in an identification of user satisfaction levels and found the effect of demand variability on costs. Su and Huang set up a game theory model to discuss the "Energy Internet" concept. Where, three different models are proposed for different kinds of energy cells.

Unlike other studies, the Nash equilibrium equation is turned into an optimization problem using the Nikaido–Isoda function (Su and Huang 2014). Wu et al. (2011) used dynamic potential game theory by setting up a model similar to the model of (Mohsenian-Rad et al. 2010). Chen et al. (2011) used a Stackelberg game in which the power generator is the leader while the consumers are followers. Sheikhi et al. (2015) modeled a game in which the minimization of electric costs is considered in order to be able to adapt current energy management units into smart units. Nwulu and Xia determined hourly electric prices that are to be offered to customers by using multi-targeted optimization and game theory. (Nwulu and Xia 2015). In an analytically strong paper using an oligopolistic setting, Allaz and Villa utilize a Cournot-like approach and the prisoner's dilemma to show that the introduction of a single forward market reduces the supplier's market power (Allaz and Vila 1993).

In most of the above articles, two-way communication between power generators and consumers is possible. Most of them use real-time pricing.

Self-power generating case in a developing country is different from the above studies in three areas:

1. There is no smart grid infrastructure, therefore, no two way communication;
2. Holdings own the chemical product producer; the paper & pulp manufacturer and woodwork at the same site and only care the total cost of energy;
3. Prices are set by the regulatory organization for 24 h and differentiations from the committed generation are fined by penalties.

## 17.3 Methodology

Marzband et al. used several algorithms in different kinds of microgrid systems in order to attain optimal scheduling for the purpose of optimal energy management. They used the imperialist competitive algorithm and the Taguchi algorithm in Marzband et al. (2015), artificial bee colony optimization in Marzband et al. (2015), the imperialist competitive algorithm in Marzband et al. (2015), and finally, ant colony optimization in Marzband et al. (2016). It is observed that when two-way communication does not exist, it is no more deterministic and to be handled by Bayesian approach.

In this study, we propose a combination of mixed integer nonlinear programming (MINLP) and a Bayesian game, an approach that—to our knowledge—has never been used in the DSM literature.

When only some of the variables in a mathematical model are constrained as integers, we call this mixed integer programming. For example, $max\, z = x_1 + x_2\, subject\ to, x_1 + 2x_2 \leq 10\ and\ x_1 \geq 0, x_2\, integer$. On the other hand, when at least either the objective function or the constraint function is nonlinear, the mathematical model is called nonlinear. Therefore, when one or more functions of a mixed integer mathematical model are nonlinear, it is called mixed integer nonlinear programming (Winston 2004). For instance, when the objective function in the above model is $\mathrm{z} = \mathrm{x}_1 * \mathrm{x}_2$, it becomes an MINLP model.

As the second method, Bayesian game is used to model cases with incomplete information. The important novelty of Bayesian game is the notion of "type," which was introduced by Harsanyi in 1967. The impact of the game components on the utility of the players can be seen as natural effects. Nevertheless, type has to do with the willpower of the player and it reflects his decisions (Lasaulce and Tembine 2011).

According to Harsanyi's model (1967), a Bayesian game is expressed as follows.

$$G = \left(N, \{T_i\}_{i \in N}, \{S_i(\Theta_i)\}_{i \in N}, p(\Theta_{-i}|\Theta_i), \{u_i\}_{i \in N}\right) \tag{17.1}$$

where,

| | |
|---|---|
| $N = \{1, 2, \ldots, n\}$ | is a set of players; |
| $T_i = \{\theta_{i1}, \theta_{i2}, \ldots\}$ | is the type set for every player i; |
| $S_i(\Theta_i)$ | is the possible strategies set for a player i who has chosen type $\Theta_i$; |
| $p(\Theta_{-i}|\Theta_i)$ | is the belief distribution of player i with type $\Theta_i$ in that the types of others are $\Theta_{-i}$; $u_i$ is the objective function of player i. |

The Bayesian–Nash equilibrium of game G is an optimal strategy vector of $s^*\left(s_1^*, s_2^*, \ldots, s_n^*\right)$, if and only if Eq. (17.2) is valid for all i players and $\Theta_i$ types (Lasaulce and Tembine 2011).

$$s_i^*(\Theta_i) \in \arg maks_{s_i} \sum_{\Theta_{-i}} p(\Theta_{-i}|\Theta_i) \cdot u_i(s_i, s_{-i}(\Theta_{-i}), \Theta_i, \Theta_{-i}) \tag{17.2}$$

The intersection of Bayes rules and game theory appears to be in p belief functions. The Bayes rule says that $P(A|B) = [P(B|A) . P(A)]/[P(B)]$ is valid for events “A” and “B” (Doya et al. 2011). The beliefs of players are also updated by that rule in Bayesian games.

## 17.4 The Solution Proposed for Load Balancing in the Absence of Smart Grids

### *17.4.1 Problem and Proposed Model*

The problem under consideration and the proposed solution are aimed at a DSM in Turkey, where there is not yet a smart grid implementation. Nevertheless, there are organized industrial centers trying to redesign their power consumption schedules through manufacturing processes to obtain reductions in power costs and to contribute to the stability of power lines in the general grid. For instance, factories producing stock such as paper and pulp, and chemical and metal sites commonly use immense amount of power. The time scheduling for energy use among these factories can be negotiated to improve load balancing.

In the problem, grid conditions using three-time TOU rates are applied. Multi-time rate terms and current unit electricity prices in Turkey are given in Table 17.1.

For a district with high demand, developing schedules for cooperative consumption will remove grid instability, and improve the efficiency of power usage.

As long as the case is applying stock based production, without any change in demand on the manufacturing side and without changing the total daily electricity consumption, schedules can be planned to switch the power utilization times of

**Table 17.1** Time-of-use rate terms and prices in Turkey (National rates. Republic of Turkey Energy Market Regulatory Authority 2015)

| | Hour interval | Unit price ($/kWh) |
|---|---|---|
| Normal | 06:00–17:00 | 0.07 |
| Peak | 17:00–22:00 | 0.12 |
| Off peak | 22:00–06:00 | 0.03 |

certain electrical devices. To participate in such a study, production-planning teams must be flexible enough to shift their load usage (Alvarez et al. 2004). Furthermore, no load schedules are ignored during cases where energy generation is run to respond 24 h a day, 7 days a week. Synchronized machine loading is still critical.

The primary objective is to flatten the peaks in the load curves where power is consumed. Thus, the electrical devices in the factories must be inspected to differentiate the shiftable and the non-shiftable devices. This will allow the working hours of a device to be shifted to another hour during the day. For example, lighting equipment is non-shiftable whereas an oven can be operated at different hours in line with the production schedules.

The general approach with two-time rate plans in the load shifting programs is to change consumption for a whole day ratio, or a peak-to-average (PAR) consumption ratio (Lopez et al. 2015). Unlike generally accepted ratios, which are prepared for two-time rates, we need at least two ratios for a three-time rate structure.

The objective function pursued is constructed to distribute the total daily load equally. Each factory has to design the individual load curve while affecting the general load curve of the district. In our case, there is no smart grid structure; therefore, factories can only communicate once every 24 h, not on an hourly basis. This fact necessitates the use of a holistic point of view to study the district. The usage of a multi-directional communication network is replaced by a Bayesian game, undertaking a mutual optimization with a central control system.

In the first step of a two-stage method, we obtain alternative solutions for the MINLP model. Later, in the second stage, these alternative solutions will be evaluated using a Bayesian game in order to achieve a cooperative solution that is best for every player.

### 17.4.2 First Stage: Model for Power Consumers

The first step is to construct an indicator that will include the changes in machine usage rates and to reflect the transmission between the rate terms as indicated in Zhao et al. (2014). Minimizing this indicator will satisfy our goals. The indices, variables and parameters of the model are given in Table 17.2.

In this model, $l_{ijt}$ is the power consumption of the jth machine of the ith factory at hour t if the machine is used which is represented by $y_{ijt}$ being 1. For a clear

**Table 17.2** Nomenclature used in the MINLP model

| |
|---|
| $i$: Factories |
| $j$: Electrical devices for each factory |
| $t$: Hours of day (t = 1, 2, …, 24) (i.e.; t = 2 means the hour between 01:00 and 02:00) |
| $l_{i,j,t}$: The load used by device j of factory i during the hour t (kWh) |
| $y_{i,j,t}$: The binary variable shows that device j of factory i works or not during the hour t |
| $L_{i,t}$: The total load used by factory i during the hour t (kWh) |
| $L_{.t}$: The total load used in the district during the hour t (kWh) |
| $L^n$: The average hourly load used in the district during a normal period (kWh) |
| $L_i^n$: The average hourly load used in factory i during a normal period (kWh) |
| $L^p$: The average hourly load used in the district during the peak period (kWh) |
| $L_i^p$: The average hourly load used in factory i during the peak period (kWh) |
| $L^o$: The average hourly load used in the district during the off-peak period (kWh) |
| $L_i^o$: The average hourly load used in factory i during the off-peak period (kWh) |
| $\tau_{i,j}$: The daily required time to work device j of factory i (hour) |
| $w_{i,j}$: The load used by device j of factory i for one hour of work (kWh) |
| $\delta_{i,j}$: Setup time for device j of factory i to start to work (hour) |
| $dot_p$: Number of hours in the peak period |
| $dot_n$: Number of hours in the normal period |
| $dot_o$: Number of hours in the off-peak period |

**Table 17.3** Daily electrical energy consumption schedule for factory i

| | 0:00–1:00 | 1:00–2:00 | … | 23:00–0:00 |
|---|---|---|---|---|
| Device 1 | $l_{i,1,1}$ | $l_{i,1,2}$ | … | $l_{i,1,24}$ |
| Device 2 | $l_{i,2,1}$ | $l_{i,2,2}$ | … | $l_{i,2,24}$ |
| ⋮ | ⋮ | ⋮ | | ⋮ |
| Hourly total loads of factory i | $L_{i,1}$ | $L_{i,2}$ | … | $L_{i,24}$ |

understanding of the consumption schedule that is obtained by $l_{ijt}$, Table 17.3 is given in a matrix form for factory i. The aim is to reduce the total values of the columns, without changing the total values of each row.

The assumptions are summarized below.

1. Every factory in the district works 24 h in three 8-hour shifts.
2. The factories use three-time TOU rates in the presence of smart meters, which enable saving data with a time dimension.
3. The production activities of the factories are not interrupted during shift changes. The machines continue to work, so every machine has one setup process in a day. From start of work to the end, $l_{i,j,t}$ is greater than 0 and $y_{i,j,t}$ is equal to 1 for the jth machine of the ith factory.

4. The total power consumption for each device, each factory, and the district are constant each day.

For the three-time rate structure in Turkey, the argument is that besides reducing peak usage, it would be very beneficial to also use models that encourage and therefore increase off-peak use. That is why we decided to determine an indicator that captures interphase transfers, which consists of changes to all rate terms. Minimizing such an indicator in the objective function will represent the real objective better and also be different than previous works Table 17.2.

Since the objective is to flatten the daily load curve, the intention is to make the rate terms, which are sorted from largest to smallest, converge with each other. This can be achieved by both decreasing the peak-to-normal ratio and the normal-to-off-peak ratio. The best way to minimize these two ratios in the same objective function is to use a specific form of these variables by summation. Mathematically speaking, the benefit is two-fold. First, it prevents the variables from disappearing due to simplification. Second, it preserves the structures of the ratios and preserves what they represent. Therefore, a new indicator is proposed that is suitable for the three-time rate structure as shown in Eq. (17.3).

$$\frac{L^p}{L^n} + \frac{L^n}{L^o} = \frac{L^p.L^o + (L^n)^2}{L^n.L^o} \tag{17.3}$$

By minimizing this indicator, transitions are smoothened and the load curve is flattened. Thus, $L^p$ for peak, $L^n$ for normal, and $L^o$ for off peak are formulated below with respect to the three-time in the rate periods for Turkey; i.e., $dot_p$, $dot_n$, $dot_o$ are equal to 5, 11, and 8 h, respectively.

$$L^p = \frac{L_{.18} + L_{.19} + L_{.20} + L_{.21} + L_{.22}}{dot_p} \tag{17.4}$$

$$L^n = \frac{L_{.7} + L_{.8} + L_{.9} + L_{.10} + L_{.11} + L_{.12} + L_{.13} + L_{.14} + L_{.15} + L_{.16} + L_{.17}}{dot_n} \tag{17.5}$$

$$L^o = \frac{L_{.23} + L_{.24} + L_{.1} + L_{.2} + L_{.3} + L_{.4} + L_{.5} + L_{.6}}{dot_o} \tag{17.6}$$

The objective function can be rearranged to adapt to various rate structures. The summation relations between the main decision variable of the model $l_{i,j,t}$ and the related variables must be added to the model as constraints (17.7) and (17.8). The variable $L_{.t}$ will be used to examine hourly total loads of the district, after obtaining the results.

$$L_{i,t} = \sum_j l_{i,j,t} \quad \forall i, t \tag{17.7}$$

$$L_{.t} = \sum_i L_{i,t} \quad \forall t \tag{17.8}$$

Because there is no change in the daily total power consumption in the district, daily consumption of each factory and electrical device is defined as constants. They are expressed in the constraints (17.9), (17.10), and (17.11) in which $c_{2_i}$ is a vector and $c_{3_{i,j}}$ is a matrix.

$$\sum_{t=1}^{24} L_{.t} = c_1 \tag{17.9}$$

$$\sum_{t=1}^{24} L_{i,t} = \boldsymbol{c}_{2_i} \quad \forall i \tag{17.10}$$

$$\sum_{t=1}^{24} l_{i,j,t} = \boldsymbol{c}_{3_{i,j}} \quad \forall i,j \tag{17.11}$$

Equation (17.11) expresses each shiftable or non-shiftable device of a given factory as having a constant daily total load usage. However, for non-shiftable ones, $l_{i,j,t}$ is equal to a $c_{4_{i,j,t}}$ three-dimensional constant matrix that will be known in advance for these cases. The shiftable ones constitute the most important part of the problem; only Eq. (17.11) is valid for these shiftable devices. The hourly required load of a device is always the same with respect to its technical features denoted by $w_{i,j}$. Therefore, $l_{i,j,t}$ is equal to 0 or $w_{i,j}$ for these devices theoretically. It is assured by Eq. (17.12).

$$l_{i,j,t} = y_{i,j,t} \cdot w_{i,j} \quad \forall i,j,t \tag{17.12}$$

As an example, a shiftable device is the first device in the second factory and it uses 100 W in an hour. If the total consumption for that device is foreseen in the second row, the first column of the matrix is 400 W. The device has to work for 4 h a day. That is why $l_{2,1,t}$ will be equal to 0 at any 20 t value, and 100 at other 4 t values in all cases. This is shown by constraints (17.13) and (17.14).

$$\sum_{t=1}^{24} l_{i,j,t} \le \tau_{i,j} \cdot w_{i,j} \quad \forall i,\ j \tag{17.13}$$

$$\sum_{t=1}^{24} y_{i,j,t} \le \tau_{i,j} \qquad \forall i,\ j \tag{17.14}$$

On the other hand, the usable hours of a shiftable device are not completely flexible. Flexibility occurs according to the needs of the factory and the work sequences of the machines. For these devices, working intervals will be decided by solving the model. The work sequences of the machines shape these intervals. There may be more than one work sequence for electricity consuming devices in a factory. For instance, there may be a work sequence between machines 1-2-3-4, and separate from this; there may be a different sequence between machines 5-6-7. Work sequences and their setup times for each can be obtained as data.

Every work sequence can start after a certain setup time has passed at the beginning of the 24-hour period. This, however, does not mean that every sequence is going to start exactly at that time. It only means that the sequences are not going to start before that time. Therefore, the constraints for the work sequences consist of shiftable devices as written in (17.15) and (17.16).

$$\textit{If } t \leq \delta_{i,j} \textit{ then } y_{i,j,t} = 0 \tag{17.15}$$

For machine $j$ which is the first element of a sequence for factory $i$;

$$\textit{If } \sum_{t^* \geq \delta_{i,j}}^{t^* + \tau_{i,j} - 1} y_{i,j,t} = \tau_{i,j}, \textit{ then } \sum_{1}^{t^* + \tau_{i,j} - 1} y_{i,j+1,t} = 0 \textit{ for following devices} \tag{17.16}$$

Here, the index of $j + 1$ expresses the device that follows $j$ in the work sequence. Eventually, the system constraints of the model are given in (17.17) and (17.18).

$$l_{i,j,t} \geq 0 \quad \forall i, j, t \tag{17.17}$$

$$y_{i,j,t} \in \{0, 1\} \quad \forall i, j, t \tag{17.18}$$

This model is solved for each factory by using the GAMS (General Algebraic Modeling System) software.

### 17.4.3 Second Stage: Bayesian Games in a Cooperative Multi-Consumer System

A load management study is run for a single power line with multiple consumers without a smart grid infrastructure. A Bayesian game is structured to obtain mutual optimal load curves for the factories that collaborate to accept the necessary load changes. Let us explain why we use a Bayesian structure. There are two types of management in the proposed DSM system; the first one is the DSM manager (i.e., DSMr) who answers to the conglomerate; and the second one is the set of the managers of the respective factories, who again answer to the conglomerate. The factory managers are responsible for continuous production, and therefore, they may impose one of the appropriate schedules that fit in with their production plans. The DSMr, on the other hand, is responsible for the load balance in the district, which accommodates the factories. Since there is no two-way digital communication infrastructure and since each factory manager choses his schedule independently (and taking the choices of other factory managers' decisions as given),

**Table 17.4** The variables and parameters used in the game model

| |
|---|
| $i$: Players $(i = 1, 2, \ldots, n)$ |
| $j_i$: Index of player i's types |
| $T_i$: Set of player i's types |
| $\theta_{ij_i}$: $j_i$ th possible type of player i $(\theta_{ij_i} \in T_i)$ |
| $\Theta_i$: Chosen type by player i $(\Theta_i \in T_i)$ |
| $\Theta_{-i}$: Chosen types by players except i |
| $p$ : Belief function |
| $u_i$: Objective function of player i |
| $s_i^*$: Optimal strategy of player i |
| $s^*$: Optimal strategy vector |

there is incomplete information. Regarding the strategies (here, the schedules) of the factory managers, the extent of information that the DSMr has is only stochastic. This, needless to say, fits quite well within the Bayesian game approach.

The game model starts with the solutions for the MINLP model for each factory. These solutions are electricity consumption schedules. Like (Chen et al. 2014), not only the optimal solutions, but also all the feasible ones are taken into consideration for scheduling. Load flow is the concern of the DSMr, which is separate from all the factories. The Bayesian game model uses the variables and the parameters given in Table 17.4.

Only a certain number of feasible schedules appear for each factory. In the game theory model, each feasible schedule is a vector of types for the given factory. The schedule that a factory chooses is private information and not known to the other players. That is, the schedule for each factory is taken as a given for the other factories. The values of each player's belief function can be given in Eq. (17.19). Therefore, $n(.)$ is a function that gives the number of elements of a set.

$$p\left(\Theta_i = \theta_{ij_i}\right) = \frac{1}{n(T_i)} \quad \forall i, j_i \tag{17.19}$$

The objective functions of players, $u_i(\Theta_i, \Theta_{-i})$, are generated in Eq. (17.20). The indicators are included in the objective function with negative coefficients, because the game has the goal of minimization (not maximizing the benefits). Here, $Ind_i$ means the indicator value for player i, and $Ind_d$ means the indicator value for the whole district.

$$\begin{gathered} \text{If } Ind_i = \frac{L_i^p \cdot L_i^o + \left(L_i^n\right)^2}{L_i^n \cdot L_i^o} \text{ and } Ind_d = \frac{L^p \cdot L^o + (L^n)^2}{L^n \cdot L^o}; \\ u_i(\Theta_i, \Theta_{-i}) = -(Ind_i(\Theta_i) + Ind_d(\Theta_i, \Theta_{-i})) \end{gathered} \tag{17.20}$$

To solve the problem, the Bayesian–Nash equilibrium is constructed. The $s^*(s_1^*, s_2^*, \ldots, s_n^*)$ optimal strategy shows the equilibrium point defined by the equation system below. Each strategy represents one action.

$$s_i^* \in \arg max \sum_{\Theta_{-i}} p(\Theta_{-i}|\Theta_i).u_i(\Theta_i, \Theta_{-i}) \quad \forall i = 1,2,\ldots,n \tag{17.21}$$

When the belief functions are transformed with respect to the rule, the equation system for equilibrium becomes (17.22).

$$s_i^* \in \arg max \sum_{\Theta_{-i}} \left[ \frac{p(\Theta_i|\Theta_{-i}).p(\Theta_{-i})}{p(\Theta_i)} .u_i(\Theta_i, \Theta_{-i}) \right] \quad \forall i = 1,2,\ldots,n \tag{17.22}$$

Unlike other applications, the $p(\Theta_i)$ are not independent since the DSMr can make moves according to the belief of one player.

## 17.5 Case Study

In this section, we study the model framework described above to see how load balancing can benefit the players when they cooperate. It is simply assumed that, when the DSMr does not lead, each factory manager acts in his individual self-interest. This is considered a non-cooperative behavior. In the "single DSMr with multiple energy consumers" cooperative game, the first step is to collect the data for the factories. The proposed MINLP model is adapted to each factory, and feasible solutions are determined. Finally, the Bayesian game is constructed and mutual optimal consumption schedules are determined.

### 17.5.1 *Initial State of the Factories*

In the district for which the case study will be realized, there are three factories (a paper manufacturer, a detergent manufacturer, and a hygienic product manufacturer), which work 24 h, 365 days a year. The same conglomerate owns all the factories, including the district. The district buys all of its power from the general grid. The general grid pricing strategy for the district is a specific one, i.e., a three-time electricity rate. The factories produce different goods using different machines. The assumptions in Sect. 17.4.2 are valid for the factories. The shiftable devices in the factories used in the manufacturing processes with their work sequences and the data for hourly load requirements and device working hour requirements are given in Figs. 17.2, 17.3, and 17.4.

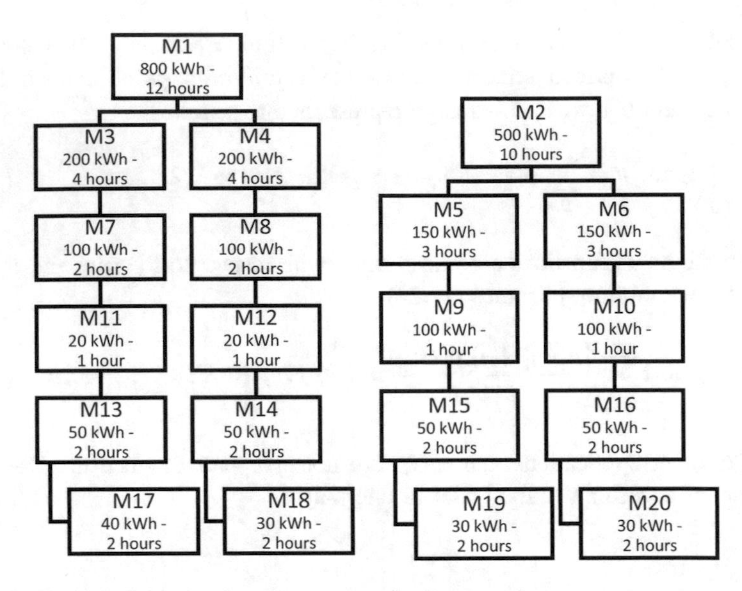


**Fig. 17.2** Shiftable devices and work sequences for the paper factory

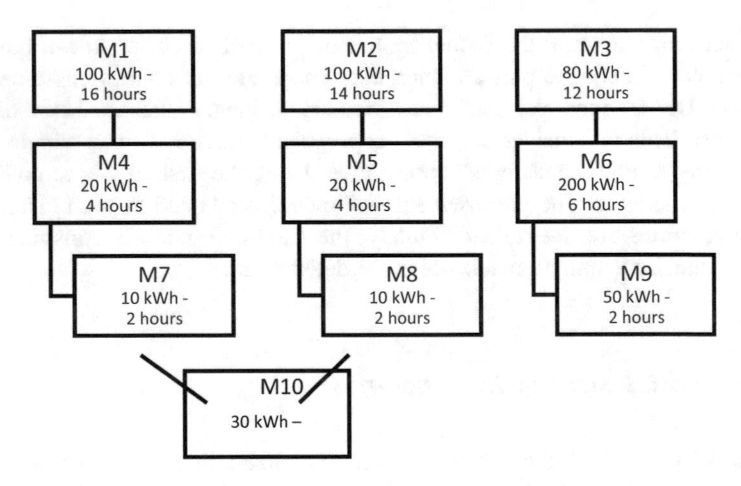


**Fig. 17.3** Shiftable devices and work sequences for the detergent factory

There are also non-shiftable devices in all three factories for which the working hours cannot be allowed to change. Generally, there are six entries: lighting, power usage for the buildings, electricity expenses for the dining hall for three shifts separately, and electricity expenses for the management offices. The lighting and general expenses over 24 h, including usage of the dining halls on the 5th, 13th and 21st hours of the shifts, and offices at the work hours of 08:00 and 18:00 are taken to be active and fixed. The power consumption amounts for non-shiftable devices in each factory are also given in Table 17.5.

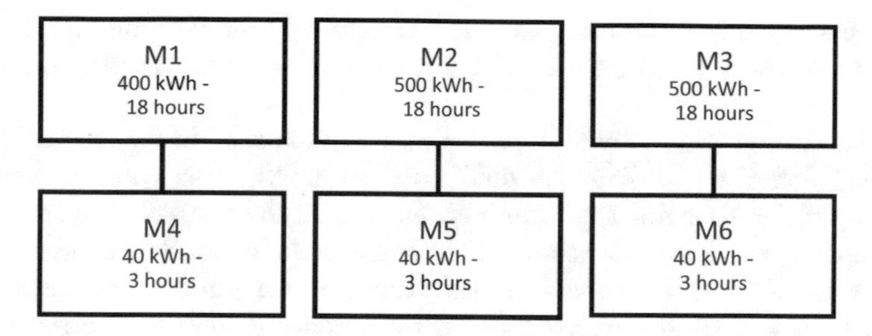


**Fig. 17.4** Shiftable devices and work sequences for the hygienic products factory

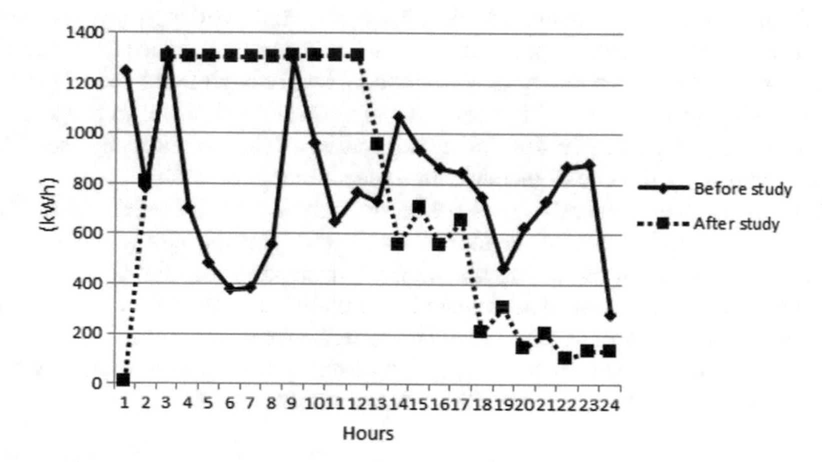


**Fig. 17.5** Daily load curves for the paper factory

**Table 17.5** Hourly load consumption for non-shiftable devices in the factories

| | Paper factory | Detergent factory | Hygienic products factory |
|---|---|---|---|
| Lighting (kWh) | 3 | 2.5 | 2 |
| General expenses of the building (kWh) | 2 | 1.5 | 1 |
| Dining hall—1st shift (kWh) | 0.5 | 0.5 | 0.5 |
| Dining hall—2nd shift (kWh) | 0.5 | 0.5 | 0.5 |
| Dining hall—3rd shift (kWh) | 0.5 | 0.5 | 0.5 |
| Management offices (kWh) | 2 | 2 | 2 |

The model also requires the setup times for the first elements in the feasible sequences. The setup times for the first elements in the work sequences in the factories are as follows: $\delta_{1,1} = 1, \delta_{1,2} = 2, \delta_{2,1} = 1, \delta_{2,2} = 1, \delta_{2,3} = 1, \delta_{3,1} = 1, \delta_{3,2} = 2$ and $\delta_{3,3} = 2$. All of the other $\delta_{i,j}$ values are equal to zero.

Considering all of these factors, the DSM study is run with the intent of converging the hourly loads for a district that buys all of its power from the general grid.

In this section, we present the power-consumption state of the district (and the factories) under the timing schedule prior to DSM. This can be seen as a non-cooperative behavior application on the part of the decision makers, i.e., the consumption scheduling decisions of the three factories are based on individual self-interest. Although all the factories belong to the same conglomerate, each factory is independently managed on its own, and this includes the power management. However, they all have to account for their management. In the absence of a DSMr, each factory individually determines its production schedule and consumes an exogenous amount of electricity within a working day. However, although the power consumption is fixed, its cost may be reduced significantly simply by optimizing the timing of consumption, i.e., through DSM and a DSMr. In the following sections, we will execute the aforementioned model to propose that under DSMr, which can be seen as a cooperative attempt initiated by the DSMr (and therefore, by the conglomerate), time scheduling is beneficial.

The initial state depends on the historical data. The potential values vary depending on the number of devices in the factory and the hourly load of each machine, are set when producing the hourly consumption data for a factory.

Two months of 24 h per day data are accumulated to be used for calculating the average indicator and the daily costs. The current values are given in Table 17.6. These cost values are the net costs originating from electricity consumption without considering the tax, line loss, and illegal usage rates.

### 17.5.2 DSMr Efficiency

In this section, the benefits of DSMr will be examined. This can be seen as an attempt to cooperatively determine the optimal power consumption schedule that will reduce power costs.

The main aim of this stage is to find all feasible consumption schedules that satisfy the constraints of our model. Thus, the sub-models for the factories are built and coded. For instance, the total number of shiftable and non-shiftable devices for the paper factory is 26. Therefore, the index of j is from 1 to 26. The sub-model for the paper factory is given in (17.23).

**Table 17.6** The current states for all three factories according to the district data

| | Paper factory | Detergent factory | Hygienic products factory | The district |
|---|---|---|---|---|
| Indicator | 2.12 | 2.35 | 2.61 | 2.33 |
| Cost | $1339.49 | $385.18 | $1816.87 | $3541.54 |

$$Min\, z = \frac{L_1^p.L_1^o + \left(L_1^n\right)^2}{L_1^n.L_1^o} \tag{17.23}$$

Subject to,

- $$For\ total\ hourly\ load,\ L_{1,t} = \sum_{j=1}^{26} l_{1,j,t} \quad \forall t$$

- $$For\ total\ daily\ load,\ \sum_{t=1}^{24} L_{1,t} = c_{21} \quad c_{21} = 18541.5$$

- $$For\ non-shiftable\ devices,$$

$$y_{1,21,t} = 1\ and\ y_{1,22,t} = 1 \quad \forall t$$

$$y_{1,23,t} = \begin{cases} 1; & if\ t = 5 \\ 0; & otherwise \end{cases} \quad y_{1,24,t} = \begin{cases} 1; & if\ t = 13 \\ 0; & otherwise \end{cases}$$

$$y_{1,25,t} = \begin{cases} 1; & if\ t = 21 \\ 0; & otherwise \end{cases} \quad y_{1,26,t} = \begin{cases} 1; & if\ 9 \le t \le 18 \\ 0; & otherwise \end{cases}$$

- *For hourly load of devices wrt whether a device works or not,* $l_{1,j,t} = y_{1,j,t}.w_{1,j}$ $\forall j, t$

- $$For\ each\ device\ to\ complete\ the\ daily\ load,\ \sum_{t=1}^{24} l_{1,j,t} = \tau_{1,j}\, w_{1,j} \quad \forall j$$

- $$For\ each\ device\ to\ complete\ the\ required\ hours,\ \sum_{t=1}^{24} y_{1,j,t} = \tau_{1,j} \quad \forall j$$

- *Defining the setup times,* $y_{1,1,1} = 0$, $y_{1,2,1} = 0$ *and* $y_{1,2,2} = 0$
- *For the work sequence of M1-M3-M7-M11-M13-M17,*

- $$If\ y_{1,1,t} = 1,\ \sum_{t}^{t+\tau_{1,1}-1} y_{1,1,t} = \tau_{1,1}\ and\ \sum_{1}^{t+\tau_{1,1}-1} y_{1,3,t} = 0\quad \forall t$$

- $$If\ y_{1,3,t} = 1,\ \sum_{t}^{t+\tau_{1,3}-1} y_{1,3,t} = \tau_{1,3}\ and\ \sum_{1}^{t+\tau_{1,3}-1} y_{1,7,t} = 0\quad \forall t$$

- $$If\ y_{1,7,t} = 1,\ \sum_{t}^{t+\tau_{1,7}-1} y_{1,7,t} = \tau_{1,7}\ and\ \sum_{1}^{t+\tau_{1,7}-1} y_{1,11,t} = 0\quad \forall t$$

- $$If\ y_{1,11,t} = 1,\ \sum_{t}^{t+\tau_{1,11}-1} y_{1,11,t} = \tau_{1,11}\ and\ \sum_{1}^{t+\tau_{1,11}-1} y_{1,13,t} = 0\quad \forall t$$

- $$If\ y_{1,13,t} = 1,\ \sum_{t}^{t+\tau_{1,13}-1} y_{1,13,t} = \tau_{1,13}\ and\ \sum_{1}^{t+\tau_{1,13}-1} y_{1,17,t} = 0\quad \forall t$$

- $$If\ y_{1,17,t} = 1,\ \sum_{t}^{t+\tau_{1,17}-1} y_{1,17,t} = \tau_{1,17}\quad \forall t$$

- *For other work sequences M1-M4-M8-M12-M14-M18, M2-M5-M9-M15-M19, and M2-M6-M10-M16-M20, the constraint sets are written in the same way.*
- *System constraints are:* $l_{1,j,t} \geq 0$ *and* $y_{1,j,t} \in \{0,1\}\ \forall j,t$.

The sub-models for the 2nd and 3rd factories are also implemented similarly. For the 2nd factory, the index of j is from 1 to 16, and for the 3rd factory, it is from 1 to 12.

The solutions are achieved using the GAMS Distribution 21.6 program, with the CPLEX solver. Four different feasible consumption schedule alternatives are found for the paper factory and hygienic products factory. Only three alternative schedules are feasible for the detergent factory. Alternative schedules achieved through the execution of the MINLP model are shown in Table 17.7.

The types and beliefs can be determined using the alternative schedules shown in Table 17.7. Each alternative is given the equal chance. If there are four alternatives,

**Table 17.7** All feasible electricity consumption schedules for the factories

| | | t = 1 | t = 2 | t = 3 | t = 4 | t = 5 | t = 6 | t = 7 | t = 8 | t = 9 | t = 10 | t = 11 | t = 12 |
|---|---|---|---|---|---|---|---|---|---|---|---|---|---|
| Paper factory (kWh) | 1st sch. | 5 | 805 | 1305 | 1305 | 1305.5 | 1305 | 1305 | 1305 | 1307 | 1307 | 1307 | 1307 |
| | 2nd sch. | 115 | 805 | 805 | 805 | 805.5 | 805 | 1305 | 1305 | 1307 | 1307 | 1307 | 1307 |
| | 3rd sch. | 5 | 805 | 805 | 805 | 1305.5 | 1305 | 1305 | 1305 | 1307 | 1307 | 1307 | 1307 |
| | 4th sch. | 5 | 805 | 805 | 1305 | 1305.5 | 1305 | 1305 | 1305 | 1307 | 1307 | 1307 | 1307 |
| Detergent factory (kWh) | 1st sch. | 4 | 184 | 284 | 284 | 284.5 | 284 | 284 | 284 | 286 | 286 | 286 | 286 |
| | 2nd sch. | 4 | 284 | 284 | 284 | 284.5 | 284 | 284 | 284 | 286 | 286 | 286 | 286 |
| | 3rd sch. | 4 | 184 | 184 | 284 | 284.5 | 284 | 284 | 284 | 286 | 286 | 286 | 286 |
| Hygienic products factory (kWh) | 1st sch. | 3 | 403 | 903 | 1403 | 1403.5 | 1403 | 1403 | 1403 | 1405 | 1405 | 1405 | 1405 |
| | 2nd sch. | 3 | 3 | 503 | 1403 | 1403.5 | 1403 | 1403 | 1403 | 1405 | 1405 | 1405 | 1405 |
| | 3rd sch. | 3 | 403 | 903 | 1403 | 1403.5 | 1403 | 1403 | 1403 | 1405 | 1405 | 1405 | 1405 |
| | 4th sch. | 3 | 3 | 1003 | 1403 | 1403.5 | 1403 | 1403 | 1403 | 1405 | 1405 | 1405 | 1405 |

| | | t = 13 | t = 14 | t = 15 | t = 16 | t = 17 | t = 18 | t = 19 | t = 20 | t = 21 | t = 22 | t = 23 | t = 24 |
|---|---|---|---|---|---|---|---|---|---|---|---|---|---|
| Paper factory (kWh) | 1st sch. | 957.5 | 557 | 707 | 557 | 657 | 207 | 305 | 145 | 205.5 | 105 | 135 | 135 |
| | 2nd sch. | 1307.5 | 907 | 907 | 907 | 707 | 507 | 505 | 245 | 105.5 | 105 | 135 | 135 |
| | 3rd sch. | 1307.5 | 907 | 557 | 557 | 707 | 357 | 355 | 245 | 205.5 | 205 | 135 | 135 |
| | 4th sch. | 1307.5 | 557 | 707 | 707 | 657 | 307 | 205 | 145 | 205.5 | 105 | 135 | 135 |
| Detergent factory (kWh) | 1st sch. | 286.5 | 406 | 406 | 406 | 326 | 246 | 244 | 94 | 84.5 | 24 | 14 | 34 |
| | 2nd sch. | 286.5 | 406 | 406 | 306 | 306 | 246 | 244 | 94 | 94.5 | 24 | 24 | 34 |
| | 3rd sch. | 286.5 | 406 | 406 | 406 | 406 | 246 | 244 | 94 | 94.5 | 24 | 24 | 34 |
| Hygienic products factory (kWh) | 1st sch. | 1405.5 | 1405 | 1405 | 1405 | 1405 | 1405 | 1403 | 1043 | 583.5 | 123 | 83 | 43 |
| | 2nd sch. | 1405.5 | 1405 | 1405 | 1405 | 1405 | 1405 | 1403 | 1403 | 943.5 | 123 | 123 | 83 |
| | 3rd sch. | 1405.5 | 1405 | 1405 | 1405 | 1405 | 1405 | 1403 | 1003 | 503.5 | 123 | 123 | 123 |
| | 4th sch. | 1405.5 | 1405 | 1405 | 1405 | 1405 | 1405 | 1403 | 1403 | 403.5 | 123 | 123 | 123 |

**Table 17.8** The objective values table for player 1

| $\theta_{21}$ | $\theta_{31}$ | $\theta_{32}$ | $\theta_{33}$ | $\theta_{34}$ |
|---|---|---|---|---|
| $\theta_{11}$ | −3.595 | −3.742 | −3.571 | −3.631 |
| $\theta_{12}$ | −4.807 | −4.992 | −4.778 | −4.852 |
| $\theta_{13}$ | −4.177 | −4.342 | −4.151 | −4.217 |
| $\theta_{14}$ | −3.847 | −4.003 | −3.823 | −3.885 |
| $\theta_{22}$ | $\theta_{31}$ | $\theta_{32}$ | $\theta_{33}$ | $\theta_{34}$ |
| $\theta_{11}$ | −3.577 | −3.722 | −3.554 | −3.613 |
| $\theta_{12}$ | −4.783 | −4.965 | −4.754 | −4.827 |
| $\theta_{13}$ | −4.157 | −4.319 | −4.131 | −4.196 |
| $\theta_{14}$ | −3.828 | −3.982 | −3.804 | −3.865 |
| $\theta_{23}$ | $\theta_{31}$ | $\theta_{32}$ | $\theta_{33}$ | $\theta_{34}$ |
| $\theta_{11}$ | −3.610 | −3.758 | −3.586 | −3.646 |
| $\theta_{12}$ | −4.828 | −5.015 | −4.798 | −4.873 |
| $\theta_{13}$ | −4.195 | −4.361 | −4.168 | −4.235 |
| $\theta_{14}$ | −3.864 | −4.021 | −3.839 | −3.902 |

the probability of choosing one of the schedules is 0.25, which is the case for both player 1 (paper factory) and player 3 (hygienic products factory). For player 2 (detergent factory) there are only three alternatives, thus, for each, the probability is 0.33.

Under these conditions, the following sets are defined for the three-player cooperative Bayesian game.

$$\begin{gathered} i = 1,2,3 \\ j_1 \in T_1 = \{\theta_{11}, \theta_{12}, \theta_{13}, \theta_{14}\} \\ j_2 \in T_2 = \{\theta_{21}, \theta_{22}, \theta_{23}\} \\ j_3 \in T_3 = \{\theta_{31}, \theta_{32}, \theta_{33}, \theta_{34}\} \end{gathered} \tag{17.24}$$

At this point, it should be noted that each type of $\theta_{1j_1}$, $\theta_{1j_2}$, $\theta_{1j_3}$ are vectors with 24 elements. The objective Eq. (17.20) and the belief vectors are used to calculate values of the objective functions in all combinations. The objective values that are composed are 4 × 3 × 4 size for each player. Table 17.8 shows the example for player 1. The values are calculated similarly for players 2 and 3.

The equilibrium equations of the game are written according to the vectors of the objective values. Since n = 3 is used in Eq. (17.21), the triple equation system takes the form of (17.25), (17.26), and (17.27).

$$s_1^* \in \arg max \sum_{\Theta_{-i}} p(\theta_{2j_2} \cap \theta_{3j_3} | \theta_{1j_1}).u_1(\theta_{1j_1}, \theta_{2j_2}, \theta_{3j_3}) \tag{17.25}$$

$$s_2^* \in \arg max \sum_{\Theta_{-i}} p(\theta_{1j_1} \cap \theta_{3j_3} | \theta_{2j_2}).u_2(\theta_{1j_1}, \theta_{2j_2}, \theta_{3j_3}) \tag{17.26}$$

$$s_3^* \in \arg max \sum_{\Theta_{-i}} p(\theta_{1j_1} \cap \theta_{2j_2} | \theta_{3j_3}).u_3(\theta_{1j_1}, \theta_{2j_2}, \theta_{3j_3}) \tag{17.27}$$

The belief functions are transformed with respect to the Bayes rule and Eq. (17.22) is modified according to the triple system, and the sets in (17.28), (17.29), and (17.30) are achieved.

$$s_1^* \in \arg max \sum_{\Theta_{-i}} \left[ \frac{p(\theta_{1j_1} | \theta_{2j_2} \cap \theta_{3j_3}).p(\theta_{2j_2} \cap \theta_{3j_3})}{p(\theta_{1j_1})} .u_1(\theta_{1j_1}, \theta_{2j_2}, \theta_{3j_3}) \right] \tag{17.28}$$

$$s_2^* \in \arg max \sum_{\Theta_{-i}} \left[ \frac{p(\theta_{2j_2} | \theta_{1j_1} \cap \theta_{3j_3}).p(\theta_{1j_1} \cap \theta_{3j_3})}{p(\theta_{2j_2})} .u_2(\theta_{1j_1}, \theta_{2j_2}, \theta_{3j_3}) \right] \tag{17.29}$$

$$s_3^* \in \arg max \sum_{\Theta_{-i}} \left[ \frac{p(\theta_{3j_3} | \theta_{1j_1} \cap \theta_{2j_2}).p(\theta_{1j_1} \cap \theta_{2j_2})}{p(\theta_{3j_3})} .u_3(\theta_{1j_1}, \theta_{2j_2}, \theta_{3j_3}) \right] \tag{17.30}$$

When the given equilibrium equations are solved, for each type vector, the initial values for the beliefs are calculated and updated using the Bayes rule. The starting point for creating the probability of the beliefs is the fact that the players will play for the strongest advantage; thus, probability for the strongest advantage will be 1 and the probability for the weakest advantage will be zero. When three players are considered, using the above concept for finding the conditional probability $p(\theta_{11}|\theta_{21} \cap \theta_{31})$ with totally different beliefs will result in the probabilities calculated in Table 17.8.

The beliefs are updated as to DSMr, where the DSMr is assigned the highest return with belief 1 and the rest are assigned belief 0.

A sample scenario is designed as player 1, player 2, and player 3 choosing type 1 schedules of their own: The conditional probability of $p(\theta_{11}|\theta_{21} \cap \theta_{31})$ reflects the decision of player 1. In this case, as seen in the objective values tables, player 1 can gain −3.595, −4.807, −4.177, or −3.847, respectively. From these values, choosing the first type by player 1 provides the highest benefit, i.e., −3.595. Therefore, the DSMr assigns 1 to the situation where player 1 chooses its first type, and assigns 0 to the other three situations for which player 1 chooses one of the other types by acting rationally. When all the actions are evaluated in a similar way, the following result is obtained.

$$\frac{p(\theta_{11}|\theta_{21} \cap \theta_{31}).p(\theta_{21} \cap \theta_{31})}{p(\theta_{11})} \cdot u_1(\theta_{11}, \theta_{21}, \theta_{31}) = \frac{1.(0.25.0.33)}{0.25} \cdot (-3.595)$$
$$= -1.18635$$

All possible scenarios for the type choices of the three players (e.g., player 1 chooses type 2 for itself, player 2 and player 3 choose type 3 for themselves; player 1 and player 3 choose type 4 for themselves and player 2 chooses type 3 for itself, etc.,) are mathematically expressed in the same way. When Eqs. (17.28), (17.29), (17.30) are solved by replacing all the unknowns with the values obtained, the equilibrium point of the Bayesian game is the set $s^*\left(s_1^* = \theta_{11}, s_2^* = \theta_{22}, s_3^* = \theta_{33}\right)$. To express the solution with respect to a definition of the Bayesian–Nash equilibrium, since Eq. (17.2) is valid for all players and all types of each, the set $\mathbf{s}^*\left(\mathrm{s}_1^* = \theta_{11}, \mathrm{s}_2^* = \theta_{22}, \mathrm{s}_3^* = \theta_{33}\right)$ is the optimal strategy for the Bayesian–Nash equilibrium. This means that the paper factory chooses its first type, the detergent factory chooses its second type, and the hygienic products factory chooses its third type.

When the obtained $\mathbf{s}^*\left(\mathrm{s}_1^* = \theta_{11}, \mathrm{s}_2^* = \theta_{22}, \mathrm{s}_3^* = \theta_{33}\right)$ equilibrium point is evaluated in terms of each player, it is seen that it gives the best strategy even in the case where some players have acted selfishly. Thus, the objective values are 1.17282 for the paper factory, 0.79449 for the detergent factory, and 1.53285 for the hygienic products factory. It can be observed in the objective values tables that none of the players can achieve a better advantage by applying a different strategy. One optimum strategy is obtained as stated in an existence theorem that says, "if the sets of players, actions and types are finite, then there is at least one Bayesian–Nash equilibrium in the game" (Lasaulce and Tembine 2011).

### 17.5.3 Results and Discussions

The Bayesian–Nash equilibrium leads us to the optimal strategy whereby the paper factory chooses its first alternative schedule, the detergent factory chooses its second alternative schedule, and the hygienic products factory chooses its third alternative schedule. These choices contribute to cost reductions for both the factories and the district management. The improvements on the current state—the non-cooperative game before the DSM study—are shown in Table 17.9.

As shown in Table 17.9, the indicators are minimized for all the factories and the district using the LP model. The number of reductions varies based on the number of machines used, the amount of power used by each machine, and the flexibility in

**Table 17.9** Values before and after the study

| | Indicator value (initial state) | Indicator value (with DSMr) | Cost value per day (initial state) | Cost value per day (with DSMr) |
|---|---|---|---|---|
| Paper factory | 2.12 | 1.49 | $1339.49 | $1049.82 |
| Detergent factory | 2.35 | 2.13 | $385.18 | $354.69 |
| Hygienic products factory | 2.61 | 2.58 | $1816.87 | $1724.52 |
| The district | 2.33 | 2.06 | $3541.54 | $3129.03 |

shifting the work hours. The value of the indicator is decreased 29.72% in the first factory, 9.36% in the second factory, and only 1.15% in the last one, but the total reduction is 11.59% for the district. The indicator value is a variable without unit, and because of that, it is obtained by proportioning the average hourly loads (kWh/kWh).

A remarkable saving is observed when the cost of daily power consumption is calculated according to the alternative schedules. The cost of consumed electrical energy for a day decreased by 21.63% in the first factory, 7.92% in the second factory, 5.09% in the third one, and almost 11.66% for the whole district.

There are a few shiftable devices in the third factory with large levels of consumption. Moreover, the power usage of those machines is not too flexible; that is why the decrease in the indicator value of the third factory is less than the others. The observations show that the indicator may be limited as the result of our model, however, further cost reductions can be achieved in the long run.

The main goal of the study was to flatten the load curves by balancing hourly electricity consumption. It can be seen from the load curves of the factories compared in Figs. 17.5, 17.6, 17.7, and 17.8, that the proposed model allows for the achievement of the main goal.

Figure 17.5 shows, in the load balance of the first factory, that the peak demand is only at about 9:30 and there are numerous rises and falls in the hours following. When the proposed model is applied and the schedules are chosen according to the Bayesian game results, the regular flow is between 02:00 and 12:00 h at a lower cost. It is also observed that the continuous rises and drops are smaller sizes after the study. During the "peak rate" hours (between 17:00 and 22:00), the size of the demand has quite diminished relative to the previous hours.

A similar trend is seen in the load balance graph for the second factory in Fig. 17.6. Although the peak is observed at 14:00, a peak rate time, it drops immediately and continues below the previous usage until the 24th hour.

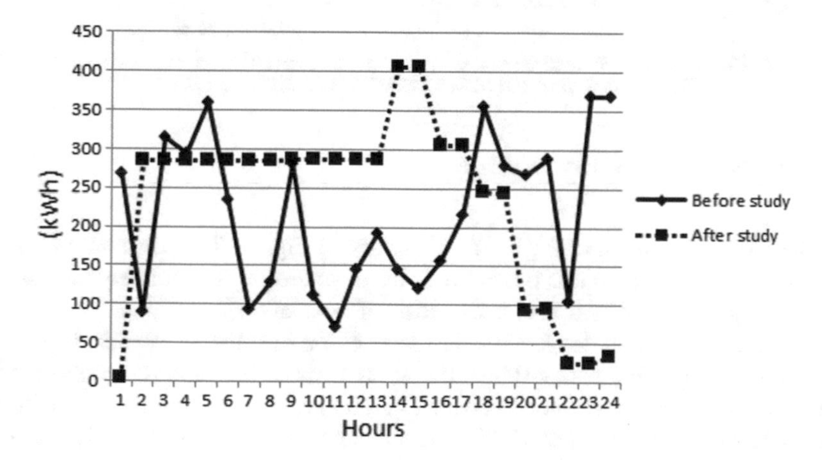


Fig. 17.6 Daily load curves for the detergent factory

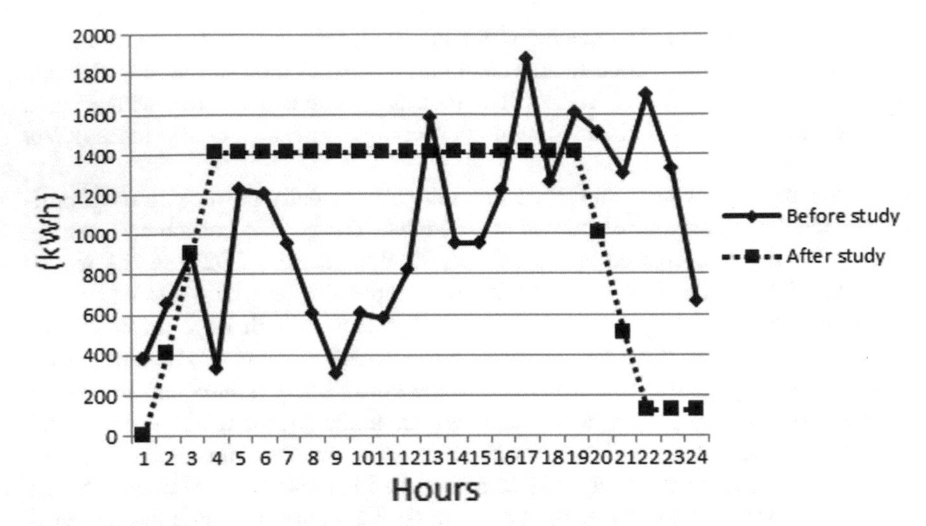


**Fig. 17.7** Daily load curves for the hygienic products factory

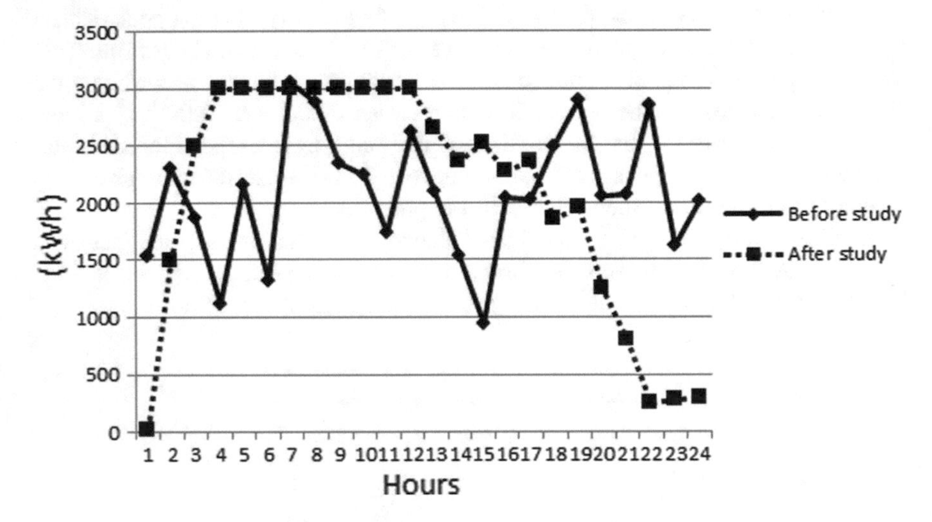


**Fig. 17.8** Daily load curves for the district

The load balance curve for the third factory in Fig. 17.7 is somewhat different than the other two due to less flexibility in the machines. It is kept in equilibrium for quite a long period, which avoids the costs of fluctuations.

The most important load curve is the one for the district, which is seen in Fig. 17.8. The graphic demonstrates the advantages of the proposed DSM model. The peak demands are replaced between 03:00 and noon, where the rate with the lower unit cost is used. After 12:00, a nearly smooth downward trend is obtained. When the after-study graph is compared with the before-study graph, the entire smoothing out action is clearly observed since there had been a continuous fluctuation beforehand, and peaks can be seen in the time ranges for the expensive rates. It is clear that the proposed model also decreases the negative effects on grid instability.

It is observed that only by using the optimized machine sequences in a cooperative way that the electricity costs for all the factories can be reduced. The price is taken only for three-time rate structures; therefore, prices are not affected by the changes in peak hours in each factory. The national grid declares prices for the three different rate times the day before.

As expected, the number of machines that can be flexibly used has an effect on the model. Therefore, the factory with the largest number of machines that could be run at different hours achieved the biggest benefit in cost reduction by smoothing the load. In the case of the DSMr, the overall benefits obtained are more important than the single factory gains. In our case study, there seems to be a strong advantage to using the DSMr. Thus, the primary goal of the proposed model is achieved. The proposed model does not make any price prediction. In cases where there is a single rate with fluctuating prices, the hourly prediction of prices should be included in the model.

After obtaining the case study results, we can articulate the following policies for manufacturing enterprises with multiple sites:

1. Power use priority should not be given to sites owning the maximum number of machines. The site with the maximum number of machines could have the maximum flexibility in machine use if the manufacturing site works to produce to stock, not produce to order.
2. Smoothing the daily power load curves reduces the cost of energy in manufacturing; thus, energy efficiency considerations are not only valuable for saving energy but also for saving money.
3. The most efficient use of power can be achieved at the Nash equilibrium point. Total energy use can be smoother and the cost can be reduced if the scheduling arrangements of the sites are made according to the Nash equilibrium, which can be created depending on demand and machine capacities.

## 17.6 Conclusion and Suggestions

As technology advances, shares in total power demand and the importance of electricity for human life have increased. Enhancements to new energy resources and new technologies that promote efficient energy use have spread globally. Research shows that the best use of electrical energy is only through improvements in the efficiency of all the processes, namely, transmission, distribution, and consumption. Smart grids and isolated grids are designed in developed countries to satisfy efficiency improvements across all three stages.

In our paper, a DSM-based load management model is proposed using the rate model of a developing country. This research is original in the sense that load shifting and flattening of the total district power curves is provided using a Bayesian game where a smart grid or isolated grid is not used.

The study is framed for an industrial district with a single DSMr with three factories (energy consumers). Recall that the district, the DSMr, and the factories belong to the same entity. A DSM model is developed with the aim of flattening the consumer load curves to minimize the grid instability caused by a district that does not use two-way digital communication. Since there is no smart grid in the system, the model is designed with the suggestion that every factory uses power consumption schedules at the equilibrium point, so that no factory can obtain priority or privilege. This can be designed since no option is given for power sales. The preference is that cost reduction in the district is offered to the factories according to their power consumption rate.

An indicator is created and calculated using an MINLP model. By solving the model, the indicator values were decreased by almost 30%. Afterwards, a Bayesian game is constructed to use the feasible schedules, creating reductions in daily power costs up to 20%.

The objectives of the proposed DSM model are satisfied in the case study experiments. The smoothing out and balancing of the energy load is achieved without violating the existing production plans and changing total daily power consumption of the district. Furthermore, optimum power-consumption schedules are recommended for the factories. By applying the proposed two-stage DSM model, load line instability is minimized while the power consumption costs of the district are reduced.

The case study results and achievements allow the design of new policies for energy efficiency considerations in developing countries.

Designing a new objective function for one-time rate users could develop the model further. Varying usage of flexible machines will be a great improvement to combine this model with the makespan studies. The case is taken for equally important factories but the model can be improved based on conglomerate given importances.

Our research will be continued to show the impact of Bayesian game structure on isolated grids with the advantage of smart features. Moreover, future studies are planned for the adoption of the proposed two-stage model in non-cooperative environments in industrial centers, where multiple suppliers and multiple consumers exist. Intraday law that permit buying from alternative suppliers and even online will be interpreted in our future studies and the Energy Internet will certainly be considered in an extension to this study.

**Acknowledgements** We would like to express our gratitude to the Hayat Kimya managers, who helped us to find a case study district and provided the necessary data and information to implement our model.

## References

Allaz, B., & Vila, J. L. (1993). Cournot competition, forward markets and efficiency. *Journal of Economic Theory, 59,* 1–16.

Alvarez, C., Gabaldon, A., & Molina, A. (2004). Assessment and simulation of the responsive demand potential in end user facilities: Application to an university customer. *IEEE Transactions on Power Systems, 19*(2), 1223–1231.

Atikol, U. (2013). A simple peak shifting DSM (demand-side management) strategy for residential water heaters. *Energy, 62,* 435–440.

Bahrami, S., & Parniani, M. (2014). Game theoretic based charging strategy for plug-in hybrid electric vehicles. *IEEE Transactions on Smart Grid, 5*(5), 2368–2375.

Bektas, Z., & Kayalica, M. O. (2015). Energy demand side management in the lack of smart grids. In F. Cucchiella & L. Koh (Eds.), *Sustainable future energy technology and supply chains* (pp. 157–170). Switzerland: Springer International Publishing.

Bergaentzle, C., Clastres, C., & Khalfallah, H. (2014). Demand-side management and European environmental and energy goals: An optimal complementary approach. *Energy Policy, 67,* 858–869.

Chen, C., Kishore, S., & Snyder, L. V. (2011). An innovative RTP-based residential power scheduling scheme for smart grids. In: IEEE book group authors (Eds.), *Proceeding Book of the International Conference on Acoustics, Speech and Signal Processing*, (pp. 5956–5959). New York: IEEE.

Chen, H., Li, Y., Louie, R. H. Y., & Vucetic, B. (2014). Autonomous demand side management based on energy consumption scheduling and instantaneous load billing: An aggregative game approach. *IEEE Transactions on Smart Grid, 5*(4), 1744–1754.

Doya, K., Ishii, S., Pouget, A., & Rao, R. P. N. (2011). *Bayesian brain: Probabilistic approaches to neural coding*. Boston: MIT Press.

He, Y., Wang, B., Li, D., Du, M., Huang, K., & Xia, T. (2015). China's electricity transmission and distribution tariff mechanism based on sustainable development. *International Journal of Electrical Power & Energy Systems, 64,* 902–910.

Lampropoulos, I., Kling, W. L., Ribeiro, P. F., Berg van den, J. (2013). History of demand side management and classification of demand response control schemes. In: IEEE book group authors (Eds.), *Proceeding Book of the Power and Energy Society General Meeting*. New York: IEEE.

Lasaulce, S., & Tembine, H. (2011). *Game theory and learning for wireless networks*. London: Elsevier.

Lopez, M. A., de la Torre, S., Martin, S., & Aguado, J. A. (2015). Demand-side management in smart grid operation considering electric vehicles load shifting and vehicle-to-grid support. *International Journal of Electrical Power & Energy Systems, 64,* 689–698.

Mangiatordi, F., Pallotti, E., & Del Vecchio, P. (2013). A non-cooperative game theoretic approach for energy management in MV grid. In: IEEE Book Group Authors (Eds.), *Proceeding book of the 13th international conference on environment and electrical engineering* (pp. 266–271). New York: IEEE.

Marzband, M., Sumper, A., Dominguez-Garcia, J. L., & Gumara-Ferret, R. (2013a). Experimental validation of a real time energy management system for microgrids in islanded mode using a local day-ahead electricity market and MINLP. *Energy Conversion and Management, 76,* 314–322.

Marzband, M., Sumper, A., Ruiz-Alvarez, A., Dominguez-Garcia, J. L., & Tomoiaga, B. (2013b). Experimental evaluation of a real time energy management system for stand-alone microgrids in day-ahead markets. *Applied Energy, 106,* 365–376.

Marzband, M., Ghadimi, M., Sumper, A., & Dominguez-Garcia, J. L. (2014). Experimental validation of a real-time energy management system using multi-period gravitational search algorithm for microgrids in islanded mode. *Applied Energy, 128,* 164–174.

Marzband, M., Azarinejadian, F., Savaghebi, M., & Guerrero, J. M. (2015a). An optimal energy management system for islanded microgrids based on multiperiod artificial bee colony combined with markov chain. *IEEE Systems Journal, 99,* 1–11.
Marzband, M., Parhizi, N., & Adabi, J. (2015b). Optimal energy management for stand-alone microgrids based on multi-period imperialist competition algorithm considering uncertainties: Experimental validation. *International Transactions on Electrical Energy Systems.*
Marzband, M., Parhizi, N., Savaghebi, M., & Guerrero, J. M. (2015c). Distributed smart decision-making for a multimicrogrid system based on a hierarchical interactive architecture. *IEEE Transactions on Energy Conversion, 99,* 1–12.
Marzband, M., Yousefnejad, E., Sumper, A., & Dominguez-Garcia, J. L. (2016). Real time experimental implementation of optimum energy management system in standalone microgrid by using multi-layer ant colony optimization. *Electrical Power and Energy Systems, 75,* 265–274.
Mohsenian-Rad, A. H., Wong, V. W. S., Jatskevich, J., Schober, R., & Leon-Garcia, A. (2010). Autonomous demand-side management based on game-theoretic energy consumption scheduling for the future smart grid. *IEEE Transactions on Smart Grid, 1*(3), 320–331.
National Rates. Republic of Turkey Energy Market Regulatory Authority. (2015). http://www.epdk.org.tr/index.php/elektrik-piyasasi/tarifeler?id=133. Accessed July 27, 2015.
Nwulu, N. I., & Xia, X. (2015). Multi-objective dynamic economic emission dispatch of electric power generation integrated with game theory based demand response programs. *Energy Conversion and Management, 89,* 963–974.
Sheikhi, A., Rayati, M., Bahrami, S., Ranjbar, A. M., & Sattari, S. (2015). A cloud computing framework on demand side management game in smart energy hubs. *International Journal of Electrical Power & Energy Systems, 64,* 1007–1016.
Su, W., & Huang, A. Q. (2014). A game theoretic framework for a next-generation retail electricity market with high penetration of distributed residential electricity suppliers. *Applied Energy, 119,* 341–350.
Tembine, H. (2016). Mean-field-type optimization for demand-supply management under operational constraints in smart grid. *Energy Systems, 7,* 333–356.
Winston, W. L. (2004). *Operations research: Applications and algorithms.* Canada: Thomson.
Wu, C., Mohsenian-Rad, H., Huang, J., & Wang, A. Y. (2011). Demand side management for wind power integration in microgrid using dynamic potential game theory. In: IEEE Book Group Authors (Eds.), *Proceeding book of the IEEE international workshop on smart grid communications and networks* (pp. 1199–1204). New York: IEEE.
Yang, P., Tang, G., & Nehorai, A. (2012). Optimal time-of-use electricity pricing using game theory. In: IEEE Book Group Authors (Eds.), *Proceeding book of the international conference on acoustics, speech and signal processing* (pp. 3081–3084). New York: IEEE.
Zhao, L., Liang, R., Zhang, J., Ma, L., & Zhao, T. (2014). A new method for building energy consumption statistics evaluation: Ratio of real energy consumption expense to energy consumption. *Energy Systems, 5,* 627–642.

# Part VI
# Performance Analysis

# Chapter 18
# A Systematic Approach to the Analysis of Barriers and Drivers of the ESCO Market in Turkey

**Özgür Yanmaz, Cigdem Kadaifci, Umut Asan and Erhan Bozdag**

**Abstract** This chapter suggests a new systematic approach to analyze the barriers and drivers of energy service contracting. In order to identify the key barriers, the proposed approach examines both direct and indirect causal relationships among barriers and allows a systematic analysis of the actors' roles in influencing the market's development. To justify the effectiveness and applicability of the proposed approach, a case is provided where the energy service contracting market in Turkey is examined.

## 18.1 Introduction

Energy efficiency has become one of the major concerns of many industries due to the overall climate change, economic developments, fluctuating prices of energy resources, technological innovations and increasing demand for renewable energy. The increasing awareness of sustainable development in recent years has promoted this growing interest. In this study, energy efficiency will be defined as "reducing the energy consumption without causing any decline in production quality and quantity in industrial establishments" (EIE 2007). By implementing energy efficiency activities, firms may avoid waste, increase their productivity, and decrease their costs and emission levels. These activities can be either outsourced or they can be performed inside the firm by investing to the necessary assets such as workforce, equipment, technology, etc. As a form of outsourcing, energy service contracting plays a critical role in realizing energy efficiency improvements (Sorrell 2007). These contracts, offered by energy service companies (ESCOs), provide an inclusive service package to the clients that enable firms to deal with various difficulties

Ö. Yanmaz · U. Asan (✉) · E. Bozdag
Department of Industrial Engineering, Istanbul Technical University, Macka, Istanbul, Turkey
e-mail: asanu@itu.edu.tr

C. Kadaifci
Department of Industrial Engineering, Doğuş University, Kadıköy, Istanbul, Turkey

© Springer International Publishing AG, part of Springer Nature 2018

C. Kahraman and G. Kayakutlu (eds.), *Energy Management—Collective and Computational Intelligence with Theory and Applications*, Studies in Systems, Decision and Control 149, https://doi.org/10.1007/978-3-319-75690-5_18

in planning, implementing and monitoring their energy efficiency projects. The type of contract mostly preferred by ESCOs suggests an energy savings offer to commercial and industrial customers in which payments are subject to the energy savings achieved or the renewable energy produced (Okay et al. 2008).

Although a trend of growth can be observed across many ESCO markets during the past few years (Bertoldi and Boza-Kiss 2017), there are still certain factors hindering the development of the energy service industry. Especially, capital inadequacy of local firms, lack of effective risk management, transparency and disclosure problems of firms, risk-averse approach of the banking system, difficult to fulfill principles and procedures in the recent legal communiqué on ESCOs, and insufficient knowledge of end-users and firms are some of the typical barriers reported in the literature (Okay et al. 2008; Onaygil and Meylani 2007; Akman et al. 2013). Recently, a number of articles and reports have been published that identify and analyze key barriers and drivers for the development of the ESCO markets. Most of these studies rely on survey data collected by means of questionnaires (Bertoldi and Boza-Kiss 2017; Kalangos 2017; Kindström et al. 2017 (among others)), in-depth interviews (Hannon et al. 2015; Sorrell et al. 2000) or Delphi studies (Pätäri and Sinkkonen 2014). Providing either an overview of locally relevant factors or of factors within a specific sector (see (Bertoldi and Boza-Kiss 2017)), the main concern of these studies is to group and prioritize these factors. However, they do not employ any further systematic analysis to closely examine the causalities and the relationships between the actors and barriers. Only few studies suggest a systematic approach to analyze the causal relationships between factors shaping the ESCO market. Fuzzy cognitive mapping (Asan et al. 2011), fuzzy time cognitive mapping (Kadaifci et al. 2014) and DEMATEL (Basak et al. 2012) are methods employed for this purpose. These studies also ignore the roles played by the main actors of the market.

In order to address the issues mentioned above, this chapter suggests a systematic approach that allows a detailed analysis of the barriers, drivers and actors of the energy service contracting market. It does not only examine the barriers and their direct causal relationships but also reveals important indirect relationships which may change the priorities of the barriers. The proposed approach provides also an overview of the interplay between the actors. Examining the actors' plans, motivations and the balance of power among them allows a better understanding of the strategic issues of the energy service contracting market. To justify the effectiveness and applicability of the proposed approach, a case is provided where the energy service contracting market in Turkey is examined. In the recent past years the energy industry in Turkey has undergone a change in terms of restructuring and deregulation. The high growth rate of Turkey's annual energy demand, the high dependence on imports to meet the current demand, the high energy intensity of the economy and the geopolitical position of Turkey indicate the need of further energy-efficiency investments and also the necessity of planning, implementing and controlling these investments (Turkyilmaz 2013). Experts who have knowledge and experience on energy service contracting were involved in the process of determining the barriers, drivers, actors and their relationships. The findings provided the basis for developing strategies for the energy service contracting market in Turkey.

The rest of this chapter is organized as follows. In order to familiarize the reader with structural analysis and actors' behavior analysis, the theoretical foundations of the proposed approach are summarized in Sect. 18.2. In the following section the proposed approach and its stages are presented. In order to demonstrate the applicability and effectiveness of the proposed approach an application to the ESCO Market in TURKEY is provided in Sect. 18.4. Finally, the contributions and limitations of the proposed approach are summarized.

## 18.2 Theoretical Basis of the Proposed Approach

Scenarios are not intended to predict the future, instead they are stories for anticipating it (Saritas and Aylen 2010) and are used when several interdependent external and internal factors need to be considered and especially the future is somewhat uncertain (Godet 1994). It seems unrealistic to predict a single future while dealing with numerous interdependent factors, which the decision makers may or may not have control on them, and the uncertainty about the future. Thus, scenario methodologies are developed to consider all possible futures, to explore alternative ways to reach them, and to show possible consequences (Godet 1994). Among different scenario methodologies, the one belongs to the "la prospective" school, is proposed by Michel Godet in order to (i) detect the priority issues, namely key variables, (ii) determine the relationships between these variables and the main actors related to the system under study, and how powerful these actors to bring their projects into reality, and finally (iii) describe the system in form of scenarios under some specific assumptions (Godet 1994).

In this study, the first two steps of the aforementioned scenario methodology, which are two separate qualitative approaches, respectively called as MICMAC (*Matrice d'Impacts Croisés Multiplication Appliquéé a un Classement/Matrix of Crossed Impact Multiplications Applied to a Classification*) and MACTOR (*Matrix of Alliances and Conflicts: Tactics, Objectives and Recommendations*), are used. MICMAC allows to detect the priority issues, in other words the key variables related to the scenario filed, and MACTOR provides a strategic point of view by exploring actor-variable relationships and the power of these actors. In the following part, the theoretical basis of these approaches are given.

Cross Impact Analysis (CIA), developed by Gordon and Helmer in 1966 (Gordon 1994), is a means of forecasting futures considered possible interrelations among future events (Chao 2008). The first application of this method was a game for an aluminum production company to give players promotional gifts (Gordon 1994). Basically, the main concern of the very first version of CIA was to explore how factors are likely to interact with each other (Porter and Hu 1990). Considering the relationships through the effects of individual factors on others provide a useful tool to analyze the structure of the scenario field. Since then, the method has been attracted attention of researchers and practitioners which result in several types of CIA (including improvements and developments) and their applications in many fields (Godet 1994; Chao 2008; Turoff 1972; Jeong and Kim 1997; Parashar et al. 1997; Bañuls et al. 2010; Turoff and Bañuls 2011; Villacorta et al. 2014). By these improvements and developments in the method, it is evolved from considering only the occurrence or not occurrence probabilities of events to taking the conditional probabilities into account (Porter and Hu 1990).

All developed versions of CIA can be classified into three groups (Asan et al. 2004): probabilistic (i.e. quantitative), deterministic (i.e. qualitative), and fuzzy. In quantitative CIA, a mathematical model considering the initial probabilities of all determined events in the selected area and the existing interrelationships is built (see (Gordon 1994)), while in qualitative one, interrelations among the events are assessed by a group of experts in form of rating values (Villacorta et al. 2014). By using qualitative techniques, one of major critics on probabilistic CIA partially disappear, which means experts do not need to identify the initial probabilities, conditional probabilities, and/or joint probabilities of events at all, they are only asked for the linguistic assessments or the strength of the interrelationships via determined scales (Weimer-Jehle 2006). Finally, fuzzy CIA enables researchers to make assessments in fuzzy or linguistic form in order to deal with the uncertainty caused by the lack of information (Asan et al. 2004).

After determining the key variables, the actors and their objectives related to the system need to be identified. Key variables provide a manageable scenario building process by dealing with a few critical factors, namely the most influential, dependent, and key factors, instead of all of them, but they constitute only one single dimension of the process. The actors in the system have different objectives either compatible or incompatible with the system's overall objectives and they may take several actions to reach their own goals. Thus, the moves of actors, their actions, and the balance of power need to be examined (Godet 1994). Examining the relationships between the actors as well as their role and power on the specific topics concerning the subject and on each other provide a significant information to build more realistic and reliable scenarios.

The MACTOR method, developed in 1985, is based on a structural analysis which identifies actors and key strategic issues (Heger et al. 2010), relationships between actors, potential alliances and conflicts, and also the balance of power (Munteanu and Apetroae 2007). As pointed out by the developer of the MACTOR method, Michel Godet, "In order to identify the most probable results, it is necessary to fully understand the actors' projects and intentions, their methods of action on one another, coupled with the constraints imposed on them" (Godet 1982). The MACTOR method produces either a graphical representation which can also be identified as a positioning map of actors based on their influence and dependence values or the convergence and divergence matrices and maps. The distances between actors show the potential alliances and conflicts by demonstrating how convergent or divergent they are. Besides observing the relative strength through convergence and divergence matrices and maps, the actors can be classified based on their position on the positioning map as dominant, dominated, independent, and key actors (Munteanu and Apetroae 2007).

This qualitative scenario methodology consisting of two distinct and also sequential methods provides scenario analysts a useful approach to deal with interrelations among numerous factors and also among considerably high number of actors. Even they are proposed as scenario tools, each can be separately or successively used in many areas and for different managerial problems. In the scope of this study, both MICMAC and MACTOR are used and a new approach to energy efficiency studies is proposed in terms of ESCOs.

## 18.3 Proposed Approach

The proposed approach consists of two methods of scenario methodology: MICMAC and MACTOR. Determining key variables in the selected area provides a more compact and manageable scenario building process, while determining particular actions corresponds to these key variables and key actors have a significant power on taking these actions and examining the relationships between them result in more coherent and reliable scenarios. Instead of identifying the drivers directly based on the scenario field analysis, literature review, and/or expert opinions, determining them by considering the key variables improves relevance of the process of thought through the right questions asked to the actors (Godet 1994). This approach aims to provide a new systematic way to recommend macro level strategies to the leading actors who have a significant impact on the actions determined. The proposed approach given in Fig. 18.1 is explained below in detail.

The MICMAC method proposed by Duperrin and Godet (1973) in order to examine the key variables in the system by questioning the interrelationships among them is a qualitative CIA technique and consists in the following steps:

1. **Problem Analysis**: The extent of the analysis and the scenario field are identified and described. The scenario analysts determine in which field the scenarios developed and in what extent the required data is collected. Thereby, the initial data is collected in this step.
2. **Variable Definition**: Based on the collected data, variables related to the system are determined and described. In this study, the barriers in ESCO market are selected as variables, and from this point on, the term variable is used to refer to the barriers.
3. **Relationship Analysis**: A committee of experts is formed for assessing the relationships. These experts can either have a profession on the related scenario field or be academicians and/or practitioners. Variables under consideration are classified into two types of relationships: direct and indirect.
   - Direct Relationships: A cross-impact matrix is formed to represent all possible relationships between the variables and determine their strengths. A pairwise comparison considers all interrelations by asking: “if

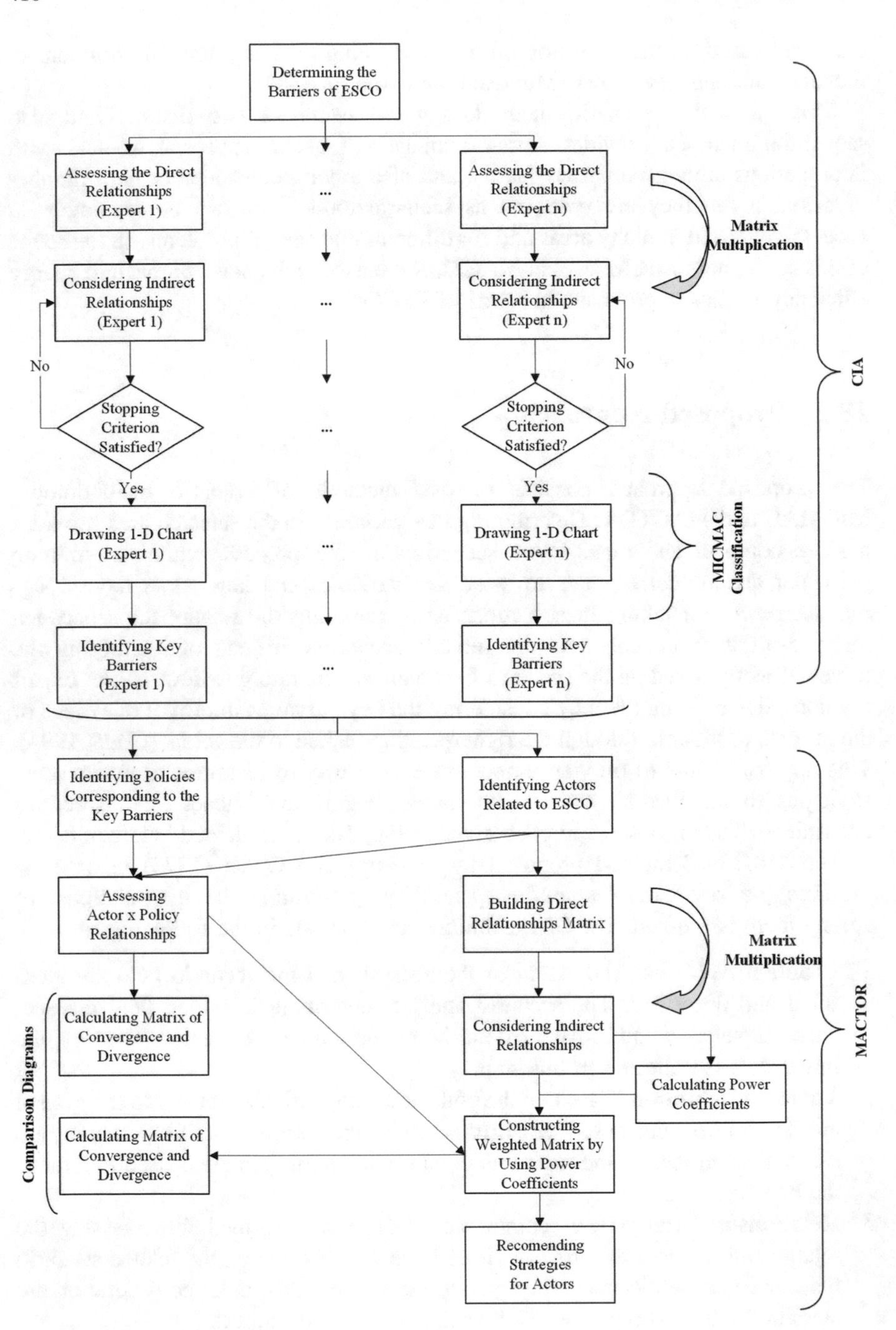


**Fig. 18.1** The proposed approach

variable A changed, what would be its direct impact on variable B?”. These assessments, performed directly by the experts, represent the direct relationships between the variables and used to classify variables based on their sum of rows and sum of columns.
  - Indirect Relationships: This type of relationships are not directly assessed by the experts. Instead, the cross-impact matrix is multiplied by its consecutive powers “to study the diffusion of impacts through reaction paths and loops” (Godet 1994). To be more specific, after the assessments are completed, the cross-impact matrix is raised to a certain power until the order of the sum of columns (and also the sum of rows) remains same in the consecutive iterations. By doing that, all possible ways from a particular variable to others are examined. This final matrix, which satisfies the stopping criteria, becomes the basis of the classification of variables based on their influence and dependence values (Godet 1994). Indirect relationship can be analyzed for the assessments of experts reached consensus, for the average or for the individual assessments.

4. **Chart Analysis**: To interpret the results, an influence-dependence chart, which is a two-dimensional graph drawn based on the influence and dependence values, is prepared (Godet 1994). On the influence-dependence chart, the vertical axis represents the degree of influence and the horizontal axis represents the degree of dependence (see Fig. 18.2). The chart, prepared for both the direct and indirect relationships analysis, includes four regions which are used to classify variables as influential, key, dependent, and excluded variables.

For a classified variable, being influential means that the relevant variable has a considerable impact on the other variables in the system and despite its high impact, other variables have almost negligible impact on it. On the contrary, dependent variables are affected by influential and key variables. Highly influential and highly

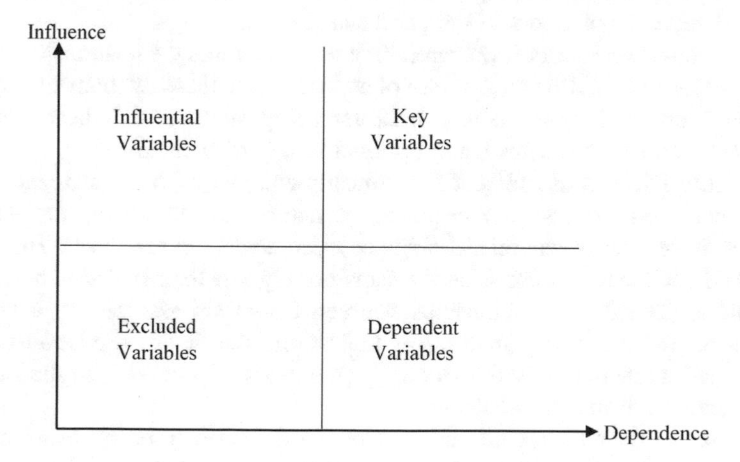


**Fig. 18.2** Influence-Dependence chart. Adapted from Godet (1994)

dependent variables are key variables which cause a reaction of others in the system in case of any change in these variables will occur. Lastly, excluded variables are those which are not considered in the further scenario process.

The cut-off points of the influence-dependence chart can be calculated by using the average ranks of the variables (Mariconda and Lurati 2015). But, in most cases, the nature of the problem necessitates the experts to determine these imaginary separation lines by considering the circumstances related to the system.

5. **Key Variable Determination**: The variables with both high dependence and high influence are identified as key variables.

Based on the key variables (i.e. barriers) determined in CIA, the drivers correspond to these variables and the actors have a considerable role in the system are identified and relationships between them are analyzed to reveal the importance of drivers and also the actual roles of actors on them to take strategic actions to deal with existing barriers. In this study, the policies to be followed in ESCO market to overcome existing barriers are selected as drivers, and from this point on, the term driver is used to refer to the policies. The MACTOR method consists in the following steps (Godet 1994):

1. **Identifying Drivers Correspond to the Key Variables**: The drivers correspond to the determined key variables are identified and described. These drivers represent strategic issues on which the actors confront each other, or in other words on which the actor have convergent or divergent objectives.
2. **Identifying Actors Related to the Problem on Hand**: The key actors having a direct or indirect control over the determined drivers are identified.
3. **Assessing Actor-Driver Relationships**: In order to run the MACTOR method, there are two main inputs collected in two matrices. The first matrix is the position matrix (MAO) that gives the position of actors over issues. The second matrix (MDA) gives the direct relationships between actors. The matrix is used to obtain agreement and disagreement coefficients, in other words convergence and divergence of actors to the particular drivers.
   The actor-driver matrix is prepared in order to examine the relationship between the actors and the drivers. A group of experts assess these relationships by using a predetermined scale where 0 indicates being neutral, +1 indicates being in favor of and −1 indicates being opposed to a particular driver.
4. **Building Direct and Indirect Relationship Matrices**: To associate the impacts of actors on drivers to the power of actors and to obtain the weighted actor-driver matrix, the relationships between actors are examined. This step of MACTOR method is based on the same principle as the MICMAC relationship analysis. The direct relationships between actors are assessed by a group of experts and by using matrix multiplication, the indirect relationships are obtained. Note that only the second degree paths are checked, so the square of the actor-actor matrix is calculated.
   The actor-driver matrix and actor-actor matrix are merged by using a power coefficient. Actors' balance of power, a scalar determining the relative strength

of each actor, is used in order to provide reasonable strategic recommendations. Power coefficient for each actor is calculated by using Eq. 18.1 based on the values in the indirect relationship matrix. If the scalar is high, the actor is in a stronger position. Then, the balance of power coefficients is calculated by using Eq. 18.2 to facilitate understanding and calculation.

$$r_i = \frac{M_i}{\sum M_i} \times \frac{M_i}{M_i + D_i} \tag{18.1}$$

$$r_i^* = \frac{r_i}{\bar{r}_i} \tag{18.2}$$

where $M_i$ is the total direct and indirect influence value of actor $i$, $D_i$ is the total direct and indirect dependence value of actor $i$, $r_i$ is the power coefficient of actor $i$ and $r_i^*$ is the balance of power coefficient of actor $i$.

5. **Calculating the Weighted Matrix**: The weighted matrix is obtained by multiplying the rows of actor-driver matrix by corresponding power coefficient to show the convergence and divergence that reflect the strength of the actors.
6. **Calculating Matrices and the Graphs of Convergence and Divergence**: The convergence and divergence matrices are calculated by using the weighted matrix. To obtain the convergence (or divergence) values of each actor pair on a particular policy, the values associated with these two actors are compared. If these actors have the same position towards any policy (i.e. the sign of the assessment is same), the absolute sum of the relevant assessments are calculated and then it is divided by two. This absolute arithmetic average demonstrates the common approach of these actors to the relevant policy. The average has a positive sign to represent the convergence and a negative sign to represent the divergence values. The convergence and divergence maps are prepared by using the values in these matrices separately where actors are represented by nodes connected by arcs, the thickness of which is proportionate to the intensity of the convergence (or divergence) between pairs of actors (Bendahan et al. 2004).

## 18.4 Application

### 18.4.1 The ESCO Market in TURKEY

During the last decade, the energy industry in Turkey has been greatly affected by the deregulation and restructuring activities. The global competitiveness has become very critical and to become a powerful player in this competitive environment, Turkey has to make difference with its technological innovations, effective and efficient energy activities, production and service quality (Kadaifci et al. 2014).

According to the report of International Energy Agency in 2005, ESCOs were not yet operating in the Turkish market, however energy efficiency activities have actually been being performed since 1995 in different sectors such as pharmaceuticals, chemicals, automotive, agriculture, paint, food and beverages, airport, hospital, and building by several medium-scale energy companies (Akman et al. 2013). In Turkey, ESCOs (which are known as Energy Efficiency Consulting Companies) operate as consulting firms which are state authorized and obliged to obey the Energy Efficiency Law (EEL) and its communiqué, which primarily aims to increase competitive advantage of Turkey in the global market (Akman et al. 2013). Although the EEL has been developed in 2007 as a result of adapting with the European Union accession period, the communiqué of the EEL related to ESCOs was released in 2012. During the development process of this communiqué, ESCO licenses have been suspended by the Ministry of Energy and Natural Resources (MENR) and candidate companies had to wait until the regulations are approved (Akman et al. 2013). Today, MENR, Energy Efficiency Coordination Board (EECB) and General Directorate of Renewable Energy (GDRE) are responsible for energy efficiency activities. The EEL promotes the efficient use of energy and covers administrative structuring, energy auditing, financial instruments and incentives, awareness raising and the establishment of an ESCO market for energy efficiency services. With the EEL it is aimed to end the state monopoly and allow private-sector participation in energy industries, aiming at cost-effective pricing through competition (Okay et al. 2008).

A strategic plan for energy efficiency covering the years 2012–2023 have been developed by General Directorate of Renewable Energy (EIE 2012). According to this plan, the decrease in the amount of energy consumed per GDP of Turkey in the year 2023 is targeted as 20%. The strategies reported in this plan, which are critical for the energy service contracting market, can be summarized as follows (i) to reduce energy intensity and energy losses in industry and services sectors, (ii) to decrease energy demand and carbon emissions of the buildings; to promote sustainable environment friendly buildings using renewable energy sources, (iii) to provide market transformation of energy efficient products, (iv) to increase efficiency in production, transmission and distribution of electricity, to decrease energy losses and harmful environment emissions, (v) to reduce unit fossil fuel consumption of motorized vehicles, to increase share of public transportation in highway, sea road and railroad and to prevent unnecessary fuel consumption in urban transportation, (vi) to use energy effectively and efficiently in public sector, and (vii) to strengthen institutional capacities and collaborations, to increase use of state of the art technology and awareness activities, to develop financial mechanisms except public financial institutions (see EIE 2012). These strategies can only be achieved in collaboration with ESCOs.

In the literature, there are a number of informative studies about the ESCO market in Turkey which helped us identifying the potential barriers and drivers influencing the market's development. Onaygil and Meylani (2007) give an overview of energy service contracting and ESCOs and provide policy suggestions for the forthcoming Turkish ESCO market. Okay et al. (2008) present views with regard to the funding and related risks that are likely to be associated with the forthcoming Turkish ESCO market. Also, the current situation of the

Turkish ESCO market is analyzed through the latest communiqué by Akman et al. (2013) from the 2013 perspective by considering the present barriers and opportunities. In another recent study, Kalangos (2017) investigates the barriers and policy drivers to energy efficiency (EE) in specific sectors in Turkey (automotive, chemicals and textile industries). Results point to the need for a policy structure that tackles the recorded poor behavioral and managerial practices on energy efficiency (EE), the lack of private EE capital funds, the inadequate energy service companies' marketplace and energy suppliers' loose EE practices. Besides, there are studies focusing on the analysis of barriers and/or success factors of energy performance contracting projects in different countries (Bertoldi and Boza-Kiss 2017; Kindström et al. 2017).

## *18.4.2 Steps of the Proposed Approach*

The proposed approach is applied to the ESCO market in Turkey. Effective use of energy resources is a non-negligible issue for Turkey when it comes to increasing energy consumption, climate change, and trade balance deficit. While energy efficiency has such a great significance, awareness created and precautions taken about this issue are quite a little. For a good future to live, it is inevitable to take potential energy efficiency actions and give this topic the credit it deserves.

In this study, barriers that hinder development of ESCO market being important factor on energy efficiency are analyzed using CIA. Then necessary actions to overcome these barriers and actors playing a significant role to take these actions are analyzed using MACTOR method.

### 18.4.2.1 Variable Definition

According to the researches, ESCO market in Turkey faces to 15 barriers including economic market barriers related to risk, heterogeneity, hidden costs, etc. (D1), high transaction costs (D2), poor market incentives provided by banks, public institutions, and third parties (D3), lack of customer awareness and understanding (D4), lack of data on energy use provided by government (D5), lack of customer information (D6), lack of trust by the clients in ESCOs (D7), lack of well-established partnerships between ESCOs and subcontractors (D8), negative influence of failed projects on the market (D9), lack of internal financial resources (D10), organizational and behavioral barriers such as inertia, lack of interest, other priorities, and lack of time, etc. (D11), lack of a supportive regulatory framework (D12), policy makers focus on energy generation and supply rather than energy efficiency improvement (D13), taxation rules that discourage investment (D14), and economic stability of the country (D15).

#### 18.4.2.2 Relationship Analysis

The direct relationships between the barriers are assessed by a group consisting of four experts who have profession on energy efficiency and ESCOs and three academicians who have researches on ESCO market. Each expert assesses the relationships and fills the cross-impact matrix separately based on a predetermined scale consists of six linguistic expressions (there is no relationship between barriers (0) and there is very little (1), little (2), medium (3), strong (4), and very strong impact (5), respectively). The cross-impact matrix, i.e. direct relationship matrix, filled by an expert is given in Table 18.1.

#### 18.4.2.3 Matrix Multiplications

The cross impact matrix shows only direct relationships between barriers. Thus, indirect impacts are needed to be determined using matrix multiplication. Sum of rows for each variable represents influence level while sum of columns for each variable represents dependence level of that barrier. The matrix multiplications are performed until order of row sum and column sum are identical in consecutive iterations. To avoid the information loss, the indirect relationships are explored for each expert and the dependence and influence rankings are calculated and sorted separately. All rankings corresponding to the barriers based on the individual expert assessments are given in Table 18.2.

**Table 18.1** Direct relationship assessments of one of the experts

| | D1 | D2 | D3 | D4 | D5 | D6 | D7 | D8 | D9 | D10 | D11 | D12 | D13 | D14 | D15 |
|---|---|---|---|---|---|---|---|---|---|---|---|---|---|---|---|
| D1 | 0 | 0 | 5 | 0 | 0 | 0 | 3 | 0 | 0 | 3 | 0 | 0 | 0 | 1 | 0 |
| D2 | 5 | 0 | 0 | 0 | 0 | 0 | 0 | 0 | 0 | 3 | 0 | 0 | 0 | 0 | 0 |
| D3 | 5 | 0 | 0 | 0 | 0 | 0 | 1 | 0 | 0 | 5 | 1 | 0 | 0 | 0 | 0 |
| D4 | 0 | 3 | 0 | 0 | 0 | 5 | 5 | 0 | 1 | 0 | 5 | 0 | 0 | 0 | 0 |
| D5 | 3 | 0 | 0 | 0 | 0 | 5 | 3 | 0 | 1 | 0 | 0 | 3 | 0 | 0 | 0 |
| D6 | 0 | 5 | 0 | 5 | 0 | 0 | 3 | 0 | 1 | 0 | 5 | 0 | 0 | 0 | 0 |
| D7 | 0 | 1 | 0 | 0 | 0 | 0 | 0 | 0 | 5 | 0 | 3 | 0 | 0 | 0 | 0 |
| D8 | 3 | 5 | 0 | 0 | 0 | 0 | 1 | 0 | 0 | 0 | 0 | 0 | 0 | 0 | 0 |
| D9 | 1 | 0 | 3 | 0 | 0 | 0 | 5 | 0 | 0 | 0 | 1 | 0 | 0 | 0 | 0 |
| D10 | 3 | 0 | 0 | 0 | 0 | 0 | 1 | 0 | 0 | 0 | 1 | 0 | 0 | 0 | 0 |
| D11 | 0 | 1 | 0 | 3 | 0 | 5 | 0 | 3 | 0 | 0 | 0 | 0 | 0 | 0 | 0 |
| D12 | 5 | 0 | 3 | 0 | 3 | 0 | 3 | 3 | 1 | 3 | 0 | 0 | 1 | 5 | 0 |
| D13 | 0 | 0 | 3 | 0 | 0 | 0 | 0 | 1 | 0 | 0 | 3 | 5 | 0 | 3 | 0 |
| D14 | 5 | 0 | 5 | 0 | 0 | 0 | 0 | 0 | 0 | 0 | 0 | 0 | 0 | 0 | 0 |
| D15 | 5 | 1 | 5 | 0 | 0 | 0 | 0 | 0 | 0 | 1 | 0 | 0 | 0 | 1 | 0 |

**Table 18.2** Rankings based on the expert assessments

| | Expert 1 | | Expert 2 | | Expert 3 | | Expert 4 | | Expert 5 | | Expert 6 | | Expert 7 | |
|---|---|---|---|---|---|---|---|---|---|---|---|---|---|---|
| | I | D | I | D | I | D | I | D | I | D | I | D | I | D |
| D1 | 8 | 11 | 4 | 15 | 13 | 9 | 5 | 10 | 5 | 13 | 8 | 13 | 5 | 13 |
| D2 | 1 | 10 | 5 | 13 | 4 | 10 | 3 | 13 | 9 | 4 | 7 | 10 | 2 | 12 |
| D3 | 10 | 7 | 8 | 12 | 8 | 6 | 10 | 7 | 8 | 8 | 10 | 8 | 6 | 9 |
| D4 | 5 | 9 | 2.5 | 9 | 14 | 3 | 8 | 9 | 2 | 9 | 2 | 9 | 15 | 7 |
| D5 | 13 | 2 | 13 | 6 | 3 | 9 | 13 | 2 | 7 | 3 | 14 | 4 | 13 | 3.5 |
| D6 | 6 | 8 | 2.5 | 8 | 9 | 3 | 6 | 8 | 2 | 5 | 2 | 7 | 14 | 10 |
| D7 | 2 | 15 | 1 | 10 | 15 | 3 | 2 | 14 | 2 | 7 | 2 | 14 | 9 | 15 |
| D8 | 4 | 12 | 9 | 14 | 12 | 13 | 7 | 12 | 11 | 14 | 6 | 15 | 3 | 6 |
| D9 | 12 | 5 | 6 | 2 | 1.5 | 11 | 14 | 5 | 14 | 1.5 | 9 | 1.5 | 8 | 8 |
| D10 | 3 | 13 | 7 | 7 | 10 | 3 | 1 | 11 | 4 | 10 | 4 | 12 | 1 | 11 |
| D11 | 7 | 14 | 10 | 11 | 11 | 3 | 4 | 15 | 6 | 15 | 5 | 11 | 11 | 14 |
| D12 | 15 | 4 | 14 | 5 | 5 | 12 | 15 | 4 | 12 | 6 | 12 | 6 | 12 | 3.5 |
| D13 | 14 | 3 | 15 | 3 | 6 | 14 | 12 | 6 | 15 | 12 | 15 | 3 | 10 | 2 |
| D14 | 9 | 6 | 11 | 4 | 7 | 7 | 9 | 3 | 10 | 11 | 11 | 5 | 4 | 5 |
| D15 | 11 | 1 | 12 | 1 | 1.5 | 15 | 11 | 1 | 13 | 1.5 | 13 | 1.5 | 7 | 1 |

*I* ranks of influence
*D* ranks of dependence

#### 18.4.2.4 Chart Analysis

After being determined direct and indirect impacts between barriers using matrix multiplications, the influence-dependence charts are prepared for seven experts to determine which barriers are key, influential, dependent, and excluded. The cut-off points of the influence-dependence chart is calculated by using the average ranks of the barriers. A sample influence-dependence chart representing one of the expert's assessments is given in Fig. 18.3.

#### 18.4.2.5 Key Barrier Determination

After the influence-dependence charts are prepared, the roles of the barriers are examined based on their positions on the chart. As mentioned before, all calculations in the CIA methodology are performed separately for each expert due to considering individual assessments in key barrier determination step. Interpreting each assessment provides a broader perspective even though there are quite similar classifications in terms of key barriers. Still, a considerable diversity of assessments exists and should be represented in the further analysis. Key barriers determined to be based on chart analysis considering differences and similarities are listed below and also shown in Table 18.3.

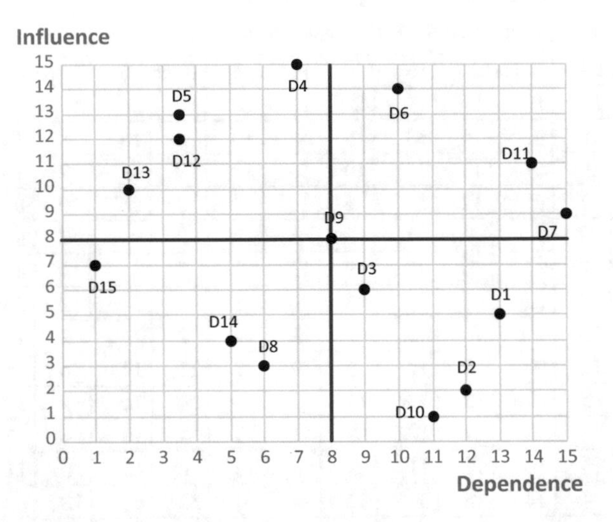


**Fig. 18.3** Influence-dependence chart of one of the experts

**Table 18.3** Key barriers determined based on individual expert assessments

| Expert 1 | Expert 2 | Expert 3 | Expert 4 | Expert 5 | Expert 6 | Expert 7 |
|---|---|---|---|---|---|---|
| D1 | D3 | D1 | D4 | D3 | D1 | D6 |
| | D8 | D8 | | D8 | D3 | D7 |
| | D11 | | | D13 | | D9 |
| | | | | D14 | | D11 |

- D1: Economic market barriers
- D3: Poor market incentives
- D4: Lack of customer awareness and understanding
- D6: Lack of customer information
- D7: Lack of trust by the clients in ESCOs
- D8: Lack of well-established partnerships between ESCOs and subcontractors
- D9: Negative influence of failed projects on the market
- D11: Organizational and behavioral barriers
- D14: Taxation rules that discourage investment

Improvement of ESCO market depends on overcoming these key barriers obtained by using Cross-Impact Analysis. In order to deal with these hindering factors, some significant actions are needed to be taken. Not only the actions but also the key actors who are responsible for or have impact on these actions are crucial in terms of the future progress of the ESCO market. The actor-related part of the proposed approach produces positioning map of actors, the conflicts and potential strategic alliances among them, and consequently the strategic actions recommended to the practitioners.

#### 18.4.2.6 Identifying Policies and Actors

Policies to overcome key barriers and actors having direct or indirect effects on these barriers are identified. Here, the main questions are who are the actors in ESCO market, how many actors should be taken into account, and which policies are considered correspond to the key barriers determined in the previous methodology. Since the purpose of this study is to propose a new systematic approach for recommending macro level strategies in general to any market, but in particular to the ESCO market in Turkey, the leading actors who have a significant impact on the policies are selected. Besides, the policies are identified based on a detailed literature survey including the academic articles and industry reports, and also on the opinions of experts in the related area.

Policies cover six issues including increasing financial incentives (O1), enabling regulations about the market such as energy price, project development processes, etc. (O2), aligning strategies and priorities of ESCOs with the market (O3), providing information and awareness for customers (O4), establishing national policies and legislations to improve the sector (O5), forming documentation of market analysis and sectoral memory (O6). Eight actors which are Governments/Ministries (A1), Local Authorities (i.e. Municipalities) (A2), Energy Efficiency Coordination Board (covers TÜBİTAK, TMMOB, etc.) (A3), European Union (A4), ESCOs (A5), Technology Suppliers (A6), Customers/End Users (A7), and Organizations Providing Financial Support (A8) are identified based on their role in the ESCO market.

#### 18.4.2.7 Assessing Actor-Policy Relationships

Two types of relationships including actor-policy and actor-actor are assessed by the same group of experts formed to assess the interrelationships among barriers. First, the actor-policy relationships are assessed by using a predetermined scale where +1 indicates being in favor of a particular policy, −1 indicates being opposed to that policy, and 0 means being neutral. In contrast to the previous methodology, the average assessments of experts are used for the further analysis. In Table 18.4, the overall assessments are shown.

**Table 18.4** The overall assessments of expert w.r.t. the actors' position towards policies

| | O1 | O2 | O3 | O4 | O5 | O6 |
|---|---|---|---|---|---|---|
| A1 | −1 | −1 | 0 | 1 | 1 | −1 |
| A2 | 0 | 0 | 0 | 1 | 0 | 0 |
| A3 | 1 | 1 | 0 | 1 | 1 | 1 |
| A4 | 1 | 1 | 0 | 1 | 1 | 1 |
| A5 | 0 | 1 | 1 | 1 | 1 | −1 |
| A6 | 0 | 0 | −1 | 0 | 0 | 0 |
| A7 | 0 | 0 | −1 | 0 | 0 | 0 |
| A8 | −1 | 0 | 0 | 0 | 0 | 1 |

#### 18.4.2.8 Building Direct and Indirect Relationship Matrices

While it is difficult to make actors talk about their strategies and priorities, and ask them to reveal their own purposes as well as their strengths and weaknesses, it is considerably easy to direct them to talk about other actors in the system (Godet 1994). Thus, after identifying their divergence and convergence to the particular policies, examining their relationships with others and seeking for their power on others will show the whole picture.

Actor-actor direct relationship matrix is filled based on the influence of one actor on another by using a scale from 0 to 3 (0: No, 1: Weak, 2: Average, and 3: Strong influence). The overall assessments of experts related to the actor-actor relationships is given in Table 18.5.

Similar to the CIA, indirect relationships are examined through matrix multiplications. At this point, only the second degree paths are checked whether there is an indirect influence of a particular actor on others. This matrix, given in Table 18.6, shows overall dependence and influence values.

**Table 18.5** Overall assessments on actor-actor relationships

| | A1 | A2 | A3 | A4 | A5 | A6 | A7 | A8 |
|---|---|---|---|---|---|---|---|---|
| A1 | 0 | 2.33 | 2.33 | 1.00 | 2.67 | 1.83 | 1.33 | 1.83 |
| A2 | 0.83 | 0 | 0.50 | 0.17 | 1.17 | 0.33 | 1.00 | 0.50 |
| A3 | 1.67 | 1.33 | 0 | 0.17 | 1.67 | 0.17 | 0.50 | 0.33 |
| A4 | 1.83 | 0.50 | 1.33 | 0 | 1.83 | 0.67 | 0.00 | 0.83 |
| A5 | 1.33 | 0.83 | 1.17 | 0.33 | 0 | 1.83 | 2.83 | 1.17 |
| A6 | 0.50 | 0.17 | 0.33 | 0.33 | 2.33 | 0 | 0.50 | 0.50 |
| A7 | 0.17 | 0.50 | 0.17 | 0.17 | 2.83 | 0.33 | 0 | 0.33 |
| A8 | 0.67 | 0.00 | 0.17 | 0.17 | 2.67 | 1.33 | 0.67 | 0 |

**Table 18.6** Indirect relationships between actors

| | A1 | A2 | A3 | A4 | A5 | A6 | A7 | A8 | I |
|---|---|---|---|---|---|---|---|---|---|
| A1 | 0.00 | 6.81 | 6.75 | 2.81 | 21.39 | 9.61 | 13.19 | 7.25 | 67.81 |
| A2 | 3.36 | 0.00 | 3.89 | 1.67 | 8.31 | 4.86 | 5.17 | 3.69 | 30.94 |
| A3 | 4.03 | 5.64 | 0.00 | 2.64 | 9.00 | 7.28 | 8.58 | 6.06 | 43.22 |
| A4 | 5.97 | 7.69 | 7.03 | 0.00 | 11.47 | 8.22 | 9.69 | 6.53 | 56.61 |
| A5 | 5.42 | 6.56 | 5.25 | 2.94 | 0.00 | 5.64 | 4.89 | 5.39 | 36.08 |
| A6 | 4.83 | 3.97 | 4.58 | 1.53 | 5.44 | 0.00 | 7.94 | 4.28 | 32.58 |
| A7 | 5.17 | 3.11 | 4.33 | 1.39 | 3.28 | 6.25 | 0.00 | 4.22 | 27.75 |
| A8 | 4.92 | 4.64 | 5.44 | 2.14 | 7.36 | 6.47 | 9.19 | 0.00 | 40.17 |
| D | 33.69 | 38.42 | 37.28 | 15.11 | 66.25 | 48.33 | 58.67 | 37.42 | |

#### 18.4.2.9 Constructing the Weighted Matrix

In order to provide reasonable strategic recommendations, actors' power coefficients are calculated and by using these, the balance of power coefficients is obtained to represent actors' relative strength. The calculated coefficients are given in Table 18.7. According to the findings, while Government/Ministries (A1) has the strongest position of power, Customer/End Users appears to be in the weakest position in ESCO market.

The balance of power coefficients and actor-policy matrix are used to construct weighted matrix. The weighted actor-policy matrix, given in Table 18.8, is obtained by multiplying each row of actor-policy matrix by power coefficient of the related actor corresponds to that row.

#### 18.4.2.10 Calculating Matrices and the Graphs of Convergence and Divergence

Two types of output are obtained as a result of MACTOR method: convergence and divergence matrices (Table 18.9) and maps. While convergence matrix identifies actors which have common positions over policies, divergence matrix identifies actors which have conflicts over them. Convergence and divergence maps in

**Table 18.7** Balance of power coefficients

| Actors | r* |
|---|---|
| A1 | 1.99 |
| A2 | 0.61 |
| A3 | 1.02 |
| A4 | 1.96 |
| A5 | 0.56 |
| A6 | 0.58 |
| A7 | 0.39 |
| A8 | 0.91 |

**Table 18.8** The weighted actor-policy matrix

| | O1 | O2 | O3 | O4 | O5 | O6 |
|---|---|---|---|---|---|---|
| A1 | −1.99 | −1.99 | 0.00 | 1.99 | 1.99 | −1.99 |
| A2 | 0.00 | 0.00 | 0.00 | 0.61 | 0.00 | 0.00 |
| A3 | 1.02 | 1.02 | 0.00 | 1.02 | 1.02 | 1.02 |
| A4 | 1.96 | 1.96 | 0.00 | 1.96 | 1.96 | 1.96 |
| A5 | 0.00 | 0.56 | 0.56 | 0.56 | 0.56 | −0.56 |
| A6 | 0.00 | 0.00 | −0.58 | 0.00 | 0.00 | 0.00 |
| A7 | 0.00 | 0.00 | −0.39 | 0.00 | 0.00 | 0.00 |
| A8 | −0.91 | 0.00 | 0.00 | 0.00 | 0.00 | 0.91 |

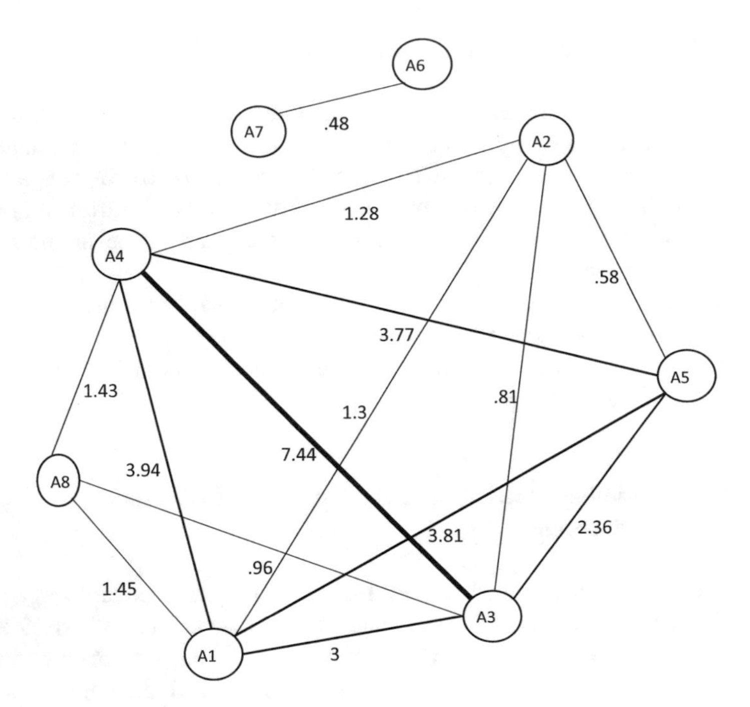


**Fig. 18.4** The convergence map

Figs. 18.4 and 18.5 are prepared by using these matrices to show the position of actors on a map. The more the actors are distant one to another, the more the intensity of their convergence (or divergence) is important. The thickness of the lines between actors indicates how powerful the alliance or the conflict between the actors are.

Convergence map shows that the Energy Efficiency Coordination Board and European Union have strong relationship and they are carrying out similar works about policies determined. Government/Ministries and ESCOs have common objectives. There are some allies and conflicts between Government/Ministries, Energy Efficiency Coordination Board, and European Union. Government/ Ministries have common policies either with the Energy Efficiency Coordination Board and European Union. On the other hand, divergence map shows Government/Ministries strongly conflicts on some policies with these actors being allies. This shows coherent works are not carried out among these actors. Local Authorities, Technology Suppliers, Customers/End Users, and Organizations providing Financial Supports have not considerable common benefits with the other actors. Local Authorities, Technology Suppliers, ESCOs, and Customers/End Users have negligible conflicts on drivers with the other actors.

The 2023 strategies associated with policies developed to improve ESCO market are demonstrated in Table 18.10. In order to accomplish these strategies, required actions need to be taken for corresponding policies.

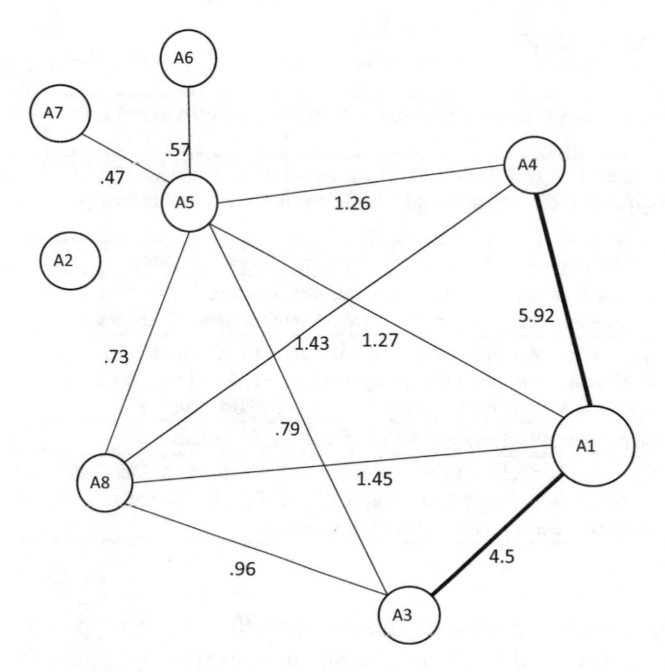


**Fig. 18.5** The divergence map

**Table 18.9** The convergence and divergence matrices

| | A1 | A2 | A3 | A4 | A5 | A6 | A7 | A8 |
|---|---|---|---|---|---|---|---|---|
| A1 | 0.00 | 1.30 | 3.00 | 3.94 | 3.81 | 0.00 | 0.00 | 1.45 |
| | 0.00 | 0.00 | −4.50 | −5.92 | −1.27 | 0.00 | 0.00 | −1.45 |
| A2 | 1.30 | 0.00 | 0.81 | 1.28 | 0.58 | 0.00 | 0.00 | 0.00 |
| | 0.00 | 0.00 | 0.00 | 0.00 | 0.00 | 0.00 | 0.00 | 0.00 |
| A3 | 3.00 | 0.81 | 0.00 | 7.44 | 2.36 | 0.00 | 0.00 | 0.96 |
| | −4.50 | 0.00 | 0.00 | 0.00 | −0.79 | 0.00 | 0.00 | −0.96 |
| A4 | 3.94 | 1.28 | 7.44 | 0.00 | 3.77 | 0.00 | 0.00 | 1.43 |
| | −5.92 | 0.00 | 0.00 | 0.00 | −1.26 | 0.00 | 0.00 | −1.43 |
| A5 | 3.81 | 0.58 | 2.36 | 3.77 | 0.00 | 0.00 | 0.00 | 0.00 |
| | −1.27 | 0.00 | −0.79 | −1.26 | 0.00 | −0.57 | −0.47 | −0.73 |
| A6 | 0.00 | 0.00 | 0.00 | 0.00 | 0.00 | 0.00 | 0.48 | 0.00 |
| | 0.00 | 0.00 | 0.00 | 0.00 | −0.57 | 0.00 | 0.00 | 0.00 |
| A7 | 0.00 | 0.00 | 0.00 | 0.00 | 0.00 | 0.48 | 0.00 | 0.00 |
| | 0.00 | 0.00 | 0.00 | 0.00 | −0.47 | 0.00 | 0.00 | 0.00 |
| A8 | 1.45 | 0.00 | 0.96 | 1.43 | 0.00 | 0.00 | 0.00 | 0.00 |
| | −1.45 | 0.00 | −0.96 | −1.43 | −0.73 | 0.00 | 0.00 | 0.00 |

**Table 18.10** Strategies-policies

| 2023 Strategies | Policies |
|---|---|
| SP-01: To reduce energy intensity and energy losses in industry and services sectors | P2 |
| SP-02: To decrease energy demand and carbon emissions of the buildings; to promote sustainable environment friendly buildings using renewable energy sources | P2, P5 |
| SP-03: To provide market transformation of energy efficient products | P2, P3, P5 |
| SP-04: To increase efficiency in production, transmission and distribution of electricity, to decrease energy losses and harmful environment emissions | P2 |
| SP-05: To reduce unit fossil fuel consumption of motorized vehicles, to increase share of public transportation in highway, sea road and railroad and to prevent unnecessary fuel consumption in urban transportation | P2, P4, P5 |
| SP-06: To use energy effectively and efficiently in public sector | P2, P5 |
| SP-07: To strengthen institutional capacities and collaborations, to increase use of state of the art technology and awareness activities, to develop financial mechanisms except public financial institutions | P1, P2, P3, P4, P6 |

All dominant actors have common understanding of and support for P4 and P5. Thus, these policies are likely to be realized. However, a considerable difference appears when it comes to the policies P1, P2, and P6. All 2023 strategies are related to P2 on which Government/Ministries have conflicts with the European Union and Energy Efficiency Coordination Board. Unless these three actors ensure coordination on this policy, realization of the following strategies seems rather challenging. Governments/Ministries and Organizations providing financial support have different opinions with European Union and Energy Efficiency Coordination Board with respect to P1 associated with the strategy SP-07. These actors should settle the conflict to accomplish SP-07. Even it creates an additional workload to maintain systematic documentation, analysis and control of the ongoing activities, Government/Ministries and ESCOs should try to reach a common ground with other actors to provide a better understanding and awareness as well as sectoral improvement.

## 18.5 Conclusion

This chapter suggests a new approach to the analysis of the potential barriers and drivers of energy service contracting markets. It allows to closely examine causalities by analyzing both direct and indirect relationships between the barriers as well as the actors and policies. Since the proposed approach considers the actors' plans, motivations and the balance of power among them, it provides the basis of a better understanding of the strategic issues of the energy service contracting market. The effectiveness and applicability of the proposed approach is demonstrated by a case study where the energy service contracting market in Turkey is examined. The findings of the case study, verified by experts, point out the presence of potential conflicts among the actors and the risk of failure in realizing some critical policies in the future. Since the barriers and policies may vary depending on country specific factors such as managerial cultures, market structures and political-legal systems, it is difficult to generalize the empirical findings of the present study. A possible future research direction might be extending the proposed approach to consider also the uncertainty involved in the judgments.

## References

Akman, U., Okay, E., & Okay, N. (2013). Current snapshot of the Turkish ESCO market. *Energy Policy, 60,* 106–115.

Asan, U., Bozdağ, C. E., & Polat, S. (2004). A fuzzy approach to qualitative cross-impact analysis. *Omega, 32,* 443–458.

Asan, U., Kutlu, A. C., & Kadaifci, C. (2011). Analysis of critical factors in energy service contracting using fuzzy cognitive mapping. In *The 41st International Conference on Computers and Industrial Engineering, Los Angeles, USA*, October 23–26, 2011.

Bañuls, V. A., Turoff, M., & Lopez, J. (2010). Clustering scenarios using cross-impact analysis. In *Proceedings of the 7th International ISCRAM Conference, Seattle, USA*, May 2010.

Basak, E., Asan, U., & Kadaifci, C. (2012). Analysis of the energy service market in turkey using the fuzzy dematel method. In *Proceedings of The 10th International FLINS Conference on Uncertainty Modeling in Knowledge Engineering and Decision Making, Istanbul, Turkey* (pp. 58–63), August 26–29, 2012.

Bendahan, S., Camponovo, G., & Pigneur, Y. (2004). Multi-issue actor analysis: Tools and models for assessing technology environments. *Journal of Decision Systems, 13*(2), 223–253.

Bertoldi, P., & Boza-Kiss, B. (2017). Analysis of barriers and drivers for the development of the ESCO markets in Europe. *Energy Policy, 107,* 345–355.

Chao, K. (2008). A new look at the cross-impact matrix and its applications in futures studies. *Journal of Futures Studies., 12*(4), 45–52.

Duperrin, J. C., & Godet, M. (1973). Methode de hierarchisation des elements d'un systeme. Rapport economique du CEA, R-45-41.

EIE. (2007). *Energy efficiency law.* http://www.eie.gov.tr, Access January 28, 2014.

EIE. (2012). *Energy efficiency strategy paper.* http://www.eie.gov.tr. Access January 28, 2014.

Godet, M. (1982). From forecasting to 'La Prospective': A new way of looking at futures. *Journal of Forecasting, 1*(3), 293–301.

Godet, M. (1994). *From anticipation to action.* Paris: UNESCO.

Gordon, T. J. (1994). The cross-impact method. A publication of United Nations Development Program's African Futures Project in collaboration with the United Nations United Nations University's Millennium Project Feasibility Study—Phase II.

Hannon, M. J., Foxon, T. J., & Gale, W. F. (2015). 'Demand pull' government policies to support product-service system activity: The case of energy service companies (ESCOs) in the UK. *Journal of Cleaner Production, 108,* 900–915.

Heger, T., Monath, T., & Kind, M. (2010). A multi-actor analysis of the QoE environment. *In the 9th Telecommunications Internet and Media Techno Economics (CTTE)* (pp. 1–9), Ghent, Belgium, June 7–9, 2010.

Jeong, G. H., & Kim, S. H. (1997). A qualitative cross-impact approach to find the key technology. *Technological Forecasting and Social Change, 55,* 203–214.

Kadaifci, C., Kucukyazici, G., Asan, U., & Bozdag E. (2014). Dynamic modeling of critical factors in energy service contracting using fuzzy time cognitive mapping. In *Proceedings of the Global Conference on Engineering and Technology Management 2014 Istanbul, Turkey*, June 23–26, 2014.

Kalangos, C. (2017). Barriers and policy drivers to energy efficiency in energy intensive Turkish industrial sectors. *International Journal of Energy Economics and Policy, 7*(3), 110–120.

Kindström, D., Ottosson, M., & Thollander, P. (2017). Driving forces for and barriers to providing energy services—A study of local and regional energy companies in Sweden. *Energy Efficiency, 10*(1), 21–39.

Mariconda, S., & Lurati, F. (2015). Stakeholder cross-impact analysis: A segmentation method. *Corporate Communications: An International Journal, 20*(3), 276–290.

Munteanu, R., & Apetroae, M. (2007). Journal relatedness: An actor-actor and actor-objectives case study. *Scientometrics, 73*(2), 215–230.

Okay, E., Okay, N., Konukman, A. E. Ş., & Akman, U. (2008). Views on Turkey's impending ESCO market: Is it promising? *Energy Policy, 36,* 1821–1825.

Onaygil, S., & Meylani, E. A. (2007). Energy efficiency consulting companies: An overview of the current situation in the world. *Energy Efficiency Congress, 2007,* 41–54. (in Turkish).

Parashar, A., Paliwal, R., & Rambabu, P. (1997). Utility of fuzzy cross-impact simulation in environmental assessment. *Environmental Impact Assessment Review, 17,* 427–447.

Pätäri, S., & Sinkkonen, K. (2014). Energy service companies and energy performance contracting: is there a need to renew the business model? Insights from a Delphi study. *Journal of Cleaner Production, 66,* 264–271.

Porter, A., & Hu, H. (1990). Cross-impact analysis. *Project Appraisal, 5*(3), 186–188.

Saritas, O., & Aylen, J. (2010). Using scenarios for roadmapping: The case of clean production. *Technological Forecasting and Social Change, 77,* 1061–1075.

Sorrell, S. (2007). The economics of energy service contracts. *Energy Policy, 35,* 507–521.

Sorrell, S., Schleich, J., Scott, S., O'Malley, E., Trace, F., Boede, U., et al. (2000). *Barriers to energy efficiency in public and private organisations*. Final report. SPRU. http://www.sussex.ac.uk/Units/spru/publications/reports/barriers/final.html. Accessed November 5, 2017.

Turkyilmaz, O. (2013). *Turkey energy outlook*. Wec Regional Meeting For Central and Eastern Europe, Bucharest, Romania.

Turoff, M. (1972). An alternative approach to cross-impact analysis. *Technological Forecasting and Social Change, 3,* 309–339.

Turoff, M., & Bañuls, V. A. (2011). Major extensions to cross-impact analysis. In *Proceedings of the 8th International ISCRAM Conference. Lisbon, Portugal*, May 2011.

Villacorta, P. J., Masegosa, A. D., Castellanos, D., & Lamata, M. T. (2014). A new fuzzy linguistic approach to qualitative cross impact analysis. *Applied Soft Computing, 24,* 19–30.

Weimer-Jehle, W. (2006). Cross-impact balances: A system-theoretical approach to cross-impact analysis. *Technological Forecasting and Social Change, 73,* 334–361.

# Chapter 19
# Fuzzy Sets Based Performance Evaluation of Alternative Wind Energy Systems

**Başar Öztayşi, Sezi Çevik Onar, Cengiz Kahraman and Ali Karaşan**

**Abstract** Evaluation of energy systems requires linguistic terms under vague and imprecise environment. The required design parameters of energy systems and the corresponding system values of those parameters should be compared to reveal that how much the alternative energy system meets the required design parameters. One of the best methods for this comparison is multi-attribute axiomatic design method. In this chapter, we present an intuitionistic fuzzy information axiom methodology in order to select the best wind energy alternative. Information axiom is used in this chapter, which is one of the two axioms of axiomatic design (AD) methodology. Triangular intuitionistic fuzzy sets are also used in the methodology. Six wind energy alternatives are evaluated based on eight attributes. A sensitivity analysis is applied to examine the robustness of the given decisions.

## 19.1 Introduction

Renewable energy sources have an immense potential to meet the energy requirements of the world. As these resources begin to be used, the energy security of the world can be powered by modern conversion technologies by reducing the long-term price of fuels from conventional sources, and decreasing use of fossil fuels. Using renewable energies does not only impact on reducing the safety risks, air pollution, and greenhouse gas emissions in the atmosphere but also they are recycled in nature.

Wind energy, one of the most-used renewable energy sources, is a carbon-free energy source depends on average wind speeds, wind turbine hubs, and turbulence intensity. There is an inclination on wind energy especially increased rapidly in the 1970s starts with the oil embargo crisis. Since that time, be able to construct a wind

B. Öztayşi · S. Çevik Onar · C. Kahraman (✉)
Istanbul Technical University, Macka, Istanbul, Turkey
e-mail: kahraman@itu.edu.tr

A. Karaşan
Yildiz Technical University, 34220 Esenler, Istanbul, Turkey

© Springer International Publishing AG, part of Springer Nature 2018

C. Kahraman and G. Kayakutlu (eds.), *Energy Management—Collective and Computational Intelligence with Theory and Applications*, Studies in Systems, Decision and Control 149, https://doi.org/10.1007/978-3-319-75690-5_19

energy system depends on many criteria such as country policies, supply chain issues of transmission and integration with the electricity system, compatibility of social and environmental conditions to the investment, economic concerns, and regional deployment. Also, when we examine alternative energy systems for wind energy, the two most important factors emerge; the first one is related with the know-how of the system which is basically composed of engineering and performance properties of the wind turbines and the other one is related to the availability of wind resources, and the second one is suitability of the area for the installation which can be called as availability of wind resources (Bansal et al. 2002).

There are many applications in the literature not only for the evaluation of alternative wind energy systems, but also other types of alternative energy sources studied. In these studies, most of the used methods are MCDM based techniques such as interval-valued intuitionistic fuzzy analytic hierarchy process (IVIF-AHP) (Onar et al. 2015), geographical information system (GIS) based AHP method (Vasileiou et al. 2017), multi-attribute Choquet integral method (Cebi and Kahraman 2013), modified ELECTRE method based on soft computing (Mousavi et al. 2017), ordinary fuzzy VIKOR method (Wang 2017). When we examine these studies, it is observed that classical methods cannot reflect the uncertainty and indeterminacy situations well. In the light of sensitivity and comparative analyses, it is also observed that some of the used methods for the solution of energy problems are not suitable to obtain the best solution.

In order to surpass these disadvantages and to well-reflect the knowledge to the problem solution, we applied multi-criteria intuitionistic fuzzy information axiom method in this study. Multi-criteria information axiom is an MCDM method based on designing the system with common areas which are determined by design ranges and system ranges. Thus, it provides an important advantage since it does not force the decision maker to define a single numerical value for design target (Kahraman et al. 2017). Intuitionistic fuzzy sets can capture the uncertainty related to the system by employing membership, non-membership, and hesitancy functions. Hence, our proposed intuitionistic fuzzy information axiom method can handle not only the uncertainties in specifications but also the hesitancies in decision makers' preferences.

In this study, an intuitionistic fuzzy axiomatic design method is applied for a selection among wind turbine alternatives for a wind energy investment in the Aegean region of Turkey. After the application, one-at-a-time sensitivity analysis is applied to check the robustness of the given decisions. The rest of the paper is organized as follows. Section 19.2 defines the meaning of performance measurement and presents a brief literature review on performance measurement. Section 19.3 presents a literature review on the evaluation of alternative wind energy systems based on MCDM methods. In Sect. 19.4, the proposed methodology is introduced. In Sect. 19.5, an application for the Aegean region of Turkey to determine the best wind turbine is given. The paper ends with the conclusions and suggestions for further research.

## 19.2 Performance Measurement

Performance measurement (PM) shows the outcome of an activity and accomplishment of the goal. It is also related to the goals of a company. PM is a socially constructed concept that the expectation of a company should be defined, and the borders of the concept should be retained (Wholey 1996). As a result of this definition, the meaning of the term may change as the goal of a company might change or as the context and competition might change in time. In other terms, the meaning of performance may change based on the time and the place (Lebas 1995).

The process of PM involves collection and analysis of data about the performance outcomes of whole systems, a sub-system or individuals (Behn 2003). Meyer (2002) states that in PM both the executed actions and their consequences should be compared with a standard or goal value so that a level of achievement can be referenced. Folan et al. (2007) state three main issues about PM as a relation, goal and characteristics. The first term "relation" emphasizes that the performance should be measured by the effects of the actions on company's outer environment. The second concept "goal" is the perception of the environment by the company. Based on this perception, the company sets its vision, core values, strategies and finally the goals. Finally, the term "characteristics" means that PM should involve well identified numerical key performance indicators. PM is used in different levels of the company with different purposes (Meyer 2002). In the top managerial level, it is used to evaluate overall activities and to prepare for the approaching performance. In the lower levels, each team or individual can use PM to assess self-performance and compensate. Also, PM can be used to motivate teams and individuals to reach better actions. In corporate life, PM can also be used to roll up performance values from the bottom to the top, and to cascade down the goals from top to bottom, to make performance comparisons across units.

Financial ratios have been used to measure the performance of companies for many years. However, PM systems based on indicators such as return on investment, and profitability can be misleading (Kaplan and Norton 1992). So more comprehensive methods have been proposed to take account of the non-financial perspective of companies. In one of the earliest studies, Keegan et al. (1989) develop Performance Measurement Matrix (PMM) which include financial, non-financial, exterior and interior indicators. Companies may use PMM matrix, as a PM tool to detect the areas that need improvement. Strategic Measurement and Reporting Technique (SMART) system is another model proposed in Wand Laboratories. In SMART's perspective, a company is evaluated in four perspectives. The first perspective involves vision and mission statements; the second perspective contains market measures and cost measures. Customer satisfaction, elasticity and efficiency take place in the third perspective and quality, distribution, lead time and cost take place in the last level (Cross and Lynch 1989). Fitzgerald et al. (1991) propose another model namely "results–determinants framework" which classify the indicators as, results and determinants. The "results" contains the measures that are related to the outcomes of actions such as competitiveness and profitability. The latter one focuses on the determinants of the outcomes such as reliability, distribution speed, and productivity. The model is based on the causality between abovementioned results and determinants. Neely et al. (2002) propose a relatively new PM framework., Performance Prism framework proposes a stakeholder focused performance evaluation system. It defines five perspectives including, stakeholder satisfaction, stakeholder contribution, strategies, processes and capabilities. Management teams use these perspectives to design the PM.

One of the recent and most accepted PM frameworks is the Balanced Scorecard proposed by Kaplan and Norton (1992). Based on the fact that in the information age may be misleading since the companies create future value through investment in customers, suppliers, employees, processes, technology and innovation, the researchers show that financial performance measures are not sufficient. What is more, the authors show that financials oriented PM may be misleading for companies. From this point, the authors propose PM framework involving four performance perspectives that are financial, customer, process and learning & growth. Different from other methods, the balanced scorecard is a flexible methodology, it allows companies to create new perspectives if needed, and also define their performance indicators under the selected perspectives.

## 19.3 Literature Review

Du et al. (2017) proposed mutual information estimation method for parameter determination method for wind turbines. The results of the real SCADA dataset shows the effectiveness of the method with vague information. Ritter and Deckert (2017) introduced a study that calculates wind energy index to support the wind farm planners and makers by using wind speed data and true production data based on long-term and low-scale analyses. Sun and Xu (2017) developed a hybrid model composed of neural networks and particle swarm optimization for the assessment of the wind turbine generators. The results of the application indicate the verification and validation of the application for wind farms of proposed method. Rehman and Khan (2017) developed and applied an MCDM model based on weighted sum approach for the most appropriate turbine selection of the wind farms. Abdulrahman and Wood (2017) introduced an optimization problem based on commercial turbine selection for wind energy farms by using genetic algorithm. Rehman and Khan (2016) identified the relevant criteria for wind turbine types and then, applied a fuzzy MCDM method to determine the best wind turbine types for wind farm in Saudi Arabia. The results revealed the effectiveness of the method for the determination of wind energy alternatives. Kahraman et al. (2016) presented a study that evaluate the wind energy technology investments based on benefit-cost analysis with extension of interval-valued intuitionistic fuzzy sets. Petković and Shamshirband (2015) studied on the selection of wind turbine model by using adaptive neuro fuzzy inference system. The results determined that blade pitch angle is the most important feature among the selection criteria of the wind turbine model. Theotokoglou and Balokas (2015) introduced a study that selects the wind turbine blade and its materials by using finite element model. Gencturk et al. (2015) proposed a study that selects the optimal wind turbine towers in seismic areas based on the total cost by using taboo search algorithm. Onar et al. (2015) applied interval-valued intuitionistic axiomatic design method to appraise wind energy alternatives based on multi-expert environment. Montoya et al. (2014) introduced a wind turbine selection method based on various multi-objective evolutionary algorithms for the wind farms. The results determined that pareto envelope-based selection algorithm has better results among the other ones such as improved pareto evolutionary algorithm, nondominated sorting genetic algorithm-II, modified sorting pareto evolutionary algorithm-II. Bagočius et al. (2014) proposed a study that assesses the wind turbine types and location of wind farms with respect to experts' opinions by using WASPAS method. Sun and Ren (2014) presented a study that selects the wind turbine types for wind farms by using a hybrid method composed of particle swarm optimization and BP neural networks. Van Buren et al. (2014) identified the criteria that are relevant with robustness and fidelity of wind turbine blades and then, applied the finite element model for the modelling of CX-100 wind turbine blade for the most appropriate wind turbine alternatives. Bassyouni and Gutub (2013) identified the most relevant materials selection strategy for the best wind turbine blades fabrication. Chowdhury et al. (2013) introduced a study that determines the wind energy turbines for wind farms based on changeable wind conditions in North Dakota by using unrestricted wind farm layout optimization method. Dong et al. (2013) applied a hybrid method composed of particle swarm optimization, differential evolution, and genetic algorithm to evaluate wind turbines in Huitengxile of Inner Mongolia. Chowdhury et al. (2013) introduced a hybrid method composed of particle swarm and mixed-discrete optimization to arrange and to select the wind farms turbines. Sarja and Halonen (2013) described the main factors of wind turbines and then applied for a region in Finland to determine best wind turbine manufacturer. Maity and Chakraborty (2012) presented a study that determines turbine blade material selection for wind energy alternatives by using fuzzy AHP. Kahraman, et al. (2010) identified the properties of wind energy sources for the selection of renewable energy sources by using fuzzy axiomatic design. Kaya and Kahraman (2010) determined the selection criteria of wind energy sources for the renewable energy planning of Istanbul by using an integrated fuzzy VIKOR & AHP methodology. Lee et al. (2009) studied on wind farms and their critical selection criteria and then, determined the best suitable alternative of wind farms by using an MCDM model. Fotuhi-Firuzabad and Dobakhshari (2009) introduced a study that determines the most appropriate wind turbine types based on a reliability-based approach for the wind farms. Tegou et al. (2009) studied an integrated methodology composed of multi-criteria analysis and geographical information systems to select most appropriate site for the wind turbines. Herbert et al. (2007) reviewed wind energy resources' criteria, methods for designing, controlling and conversion of wind energy systems, and wind energy assessment models.

## 19.4 Methodology

In this section, the principles of information axiom under classical and fuzzy environments are first given. Then intuitionistic fuzzy IA approach is presented.

### 19.4.1 *Information Axiom*

Axiomatic Design (AD) aims at creating a new design and/or to improve an existing design based on the scientific rules (Suh 2005). Axiomatic Design (AD) methodology involves two axioms (Suh 1990). These axioms are independence axiom and information axiom. The independence axiom requires the independence of functional requirements. The information axiom requires the minimization of the information content of the considered design. The best alternative having the minimum information content among the alternatives satisfying independence axiom is preferred (Suh 1990). The information content (I) is the probability of satisfying the design goals represented by functional requirements (FRs). The information content of a design is measured by (19.1) (Suh 1990):

$$I_i = \log_2 \frac{1}{p_i} \tag{19.1}$$

where the probability of success ($p_i$) is the intersection area of the probability density functions of a system range (SR) and a design range (DR). This area is called common area ($A_c$) or common range. $p_i$ is calculated as given in (19.2):

$$p_i = \int_{DR} p_s(FR_i)dFR_i \tag{19.2}$$

where $p_s(FR_i)$ is the probability density function of a functional requirement.

When there are more than one FR, the total information content $I_s$ is calculated by (19.3) (Suh 1990);

$$I_s = \sum_{i=1}^{m} \log_2(1/p_i) \tag{19.3}$$

Under vagueness and impreciseness, ordinary fuzzy IA was developed by Kulak and Kahraman (2005a, b) and Kahraman and Çebi (2009). Figure 19.1 shows how the common range is determined when system and design ranges are defined by triangular fuzzy numbers (Kahraman et al. 2018). The information content under fuzziness is calculated as in Eq. (19.4):

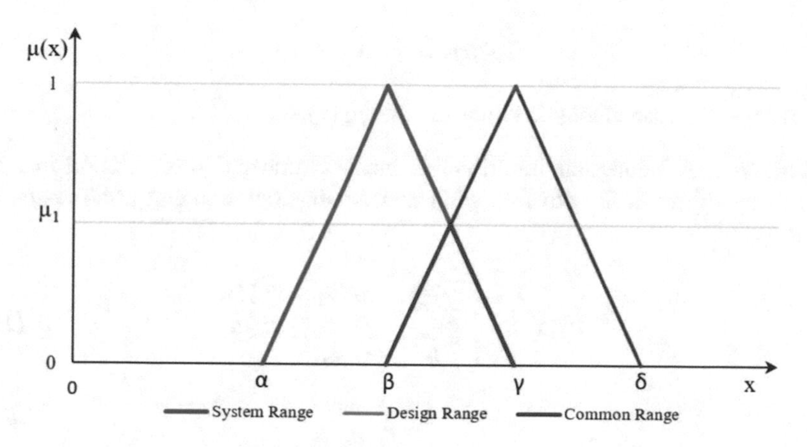


**Fig. 19.1** Triangular fuzzy SR, DR, and $A_c$

$$I_i = \log_2\left(\frac{\text{Area of System Range}}{\text{Common Area}}\right)_i \tag{19.4}$$

Based on the parameters given in Fig. 19.1, (19.4) can be given as in (19.5):

$$I_i = \log_2\left(\frac{\gamma - \alpha}{\mu_1(\gamma - \beta)}\right)_i \tag{19.5}$$

The total information content of a system involving m functional requirements under fuzziness is calculated by (19.6):

$$I_s = \sum_{i=1}^{m} \log_2\left(\frac{\gamma - \alpha}{\mu_1(\gamma - \beta)}\right)_i \tag{19.6}$$

### *19.4.2 Intuitionistic Fuzzy Sets*

An intuitionistic fuzzy sets are expressed by a membership value and a non-membership value for any x in X so that their sum is less than or equal to 1 (Atanassov 1986, 1999).

**Definition 1** Let $X \neq \emptyset$ be a given set. An intuitionistic fuzzy set in X is an object A given by

$$\tilde{A} = \{\langle x, \mu_{\tilde{A}}(x), v_{\tilde{A}}(x)\rangle; x \in X\}, \tag{19.7}$$

where $\mu_{\tilde{A}} : X \to [0, 1]$ and $v_{\tilde{A}} : X \to [0, 1]$ satisfy the condition

$$0 \le \mu_{\tilde{A}}(x) + v_{\tilde{A}}(x) \le 1, \tag{19.8}$$

for every $x \in X$. Hesitancy is equal to "$1 - (\mu_{\tilde{A}}(x) + v_{\tilde{A}}(x))$".

**Definition 2** A Triangular Intuitionistic Fuzzy Number (TIFN) $\tilde{A}$ is an intuitionistic fuzzy subset in $\mathbb{R}$ with following membership function and non-membership function:

$$\mu_{\tilde{A}}(x) = \begin{cases} \frac{x-a_1}{a_2-a_1}, & for\, a_1 \le x \le a_2 \\ \frac{a_3-x}{a_3-a_2}, & for\, a_2 \le x \le a_3 \\ 0, & otherwise \end{cases} \tag{19.9}$$

and

$$v_{\tilde{A}}(x) = \begin{cases} \frac{a_2-x}{a_2-a_1'}, & for\, a_1' \le x \le a_2 \\ \frac{x-a_2}{a_3'-a_2}, & for\, a_2 \le x \le a_3' \\ 1, & otherwise \end{cases} \tag{19.10}$$

where $a_1' \le a_1 \le a_2 \le a_3 \le a_3'$, $0 \le \mu_{\tilde{A}}(x) + v_{\tilde{A}}(x) \le 1$ and TIFN is denoted by $\tilde{A}_{TIFN} = (a_1, a_2, a_3; a_1', a_2, a_3')$.

**Definition 3** Arithmetic operations for triangular intuitionistic fuzzy numbers are as follows:

Let $\tilde{A}_{TrIFN} = (a_1, a_2, a_3; a_1', a_2, a_3')$ and $\tilde{B}_{TrIFN} = (b_1, b_2, b_3; b_1', b_2, b_3')$ be two TIFNs. Then,

**Addition:** $\tilde{C} = \tilde{A} + \tilde{B}$ is also a TIFN:

$$\tilde{C} = (a_1 + b_1,\, a_2 + b_2,\, a_3 + b_3; a_1' + b_1',\, a_2 + b_2,\, a_3' + b_3') \tag{19.11}$$

**Multiplication:** $\tilde{C} = \tilde{A} \otimes \tilde{B}$ is approximately a TIFN:

$$\tilde{C} \cong (a_1 b_1,\, a_2 b_2,\, a_3 b_3; a_1' b_1',\, a_2 b_2,\, a_3' b_3') \tag{19.12}$$

**Subtruction:** $\tilde{C} = \tilde{A} \ominus \tilde{B}$ is also a TIFN:

$$\tilde{C} = (a_1 - b_3,\, a_2 - b_2,\, a_3 - b_1; a_1' - b_3',\, a_2 - b_2,\, a_3' - b_1') \tag{19.13}$$

**Division:** $\tilde{C} = \tilde{A} ø \tilde{B}$ is approximately a TIFN:

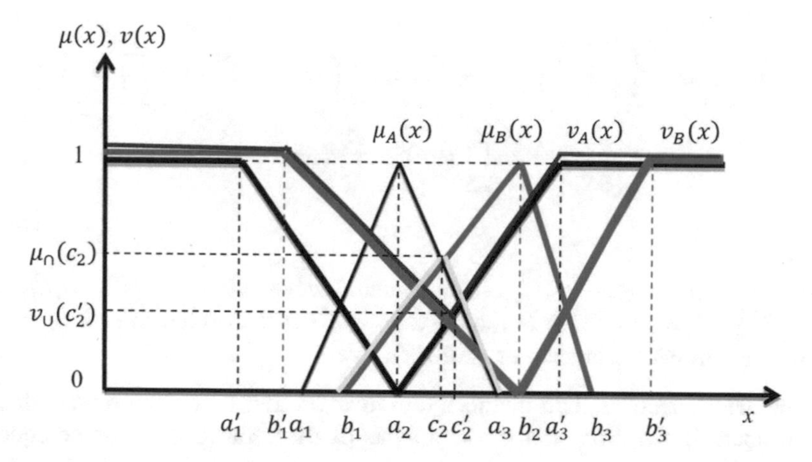


**Fig. 19.2** Intersection of two TIFNs

$$\tilde{C} \cong \left(a_1/b_3,\, a_2/b_2,\, a_3/b_1; a'_1/b'_3,\, a_2/b_2,\, a'_3/b'_1\right). \tag{19.14}$$

**Definition 4** Let $\tilde{A}$ and $\tilde{B}$ be two IFSs in the set $X$. The intersection of $\tilde{A}$ and $\tilde{B}$ is defined as in (19.15) (Bora et al. 2012).

$$\tilde{A} \cap \tilde{B} = \left\{x_i, min\left(\mu_{\tilde{A}}(x_i), \mu_{\tilde{B}}(x_i)\right), max\left(v_{\tilde{A}}(x_i), v_{\tilde{B}}(x_i)\right) \middle| x_i \in X\right\} \tag{19.15}$$

The operations given below are based on the definitions given above.

Let $\tilde{A}_{TIFN} = \left(a_1, a_2, a_3; a'_1, a_2, a'_3\right)$ and $\tilde{B}_{TIFN} = \left(b_1, b_2, b_3; b'_1, b_2, b'_3\right)$ as in Fig. 19.2 (Kahraman et al. 2017). $\mu_{\tilde{A}}(x_i)$ and $\mu_{\tilde{B}}(x_i)$ represent the membership functions of the fuzzy sets $\tilde{A}$ and $\tilde{B}$, respectively whereas $v_{\tilde{A}}(x_i)$ and $v_{\tilde{B}}(x_i)$ represent their non-membership functions, respectively. The intersection, $\tilde{A}_{TIFN} \cap \tilde{B}_{TIFN}$, is a TIFN denoted by $\tilde{C}_{TIFS} = \left(b_1, c_2, a_3; b'_1, c'_2, a'_3\right)$ where the membership values for $c_2$ and $c'_2$ are $\mu_{\cap}(c_2)$ and $v_{\cup}\left(c'_2\right)$, respectively. This TIFN can be represented by $\tilde{C}_{TIFS} = \left(b_1, (c_2, \mu_{\cap}(c_2)), a_3; b'_1, \left(c'_2, v_{\cup}\left(c'_2\right)\right), a'_3\right)$. The red colored line represents the union of the non-membership functions of $\tilde{A}$ and $\tilde{B}$ whereas the yellow colored line represents the intersection of membership functions of $\tilde{A}$ and $\tilde{B}$.

### 19.4.3 Aggregation Operators for TIFNs

Suppose $I_i = \left(\left[a_i^L, a_i^M, a_i^U\right], \left[b_i^L, b_i^M, b_i^U\right]\right)$ $(i = 1,\ 2,\ \ldots, n)$ is a set of TIFNs, then the result of the aggregation is a TIFN given by (19.16) (Zhang and Liu 2010):

$$f_m(I_1, I_2, \ldots, I_n) = \left( \left[ 1 - \prod_{i=1}^{n} (1 - a_i^L)^{w_i}, 1 - \prod_{i=1}^{n} (1 - a_i^M)^{w_i}, 1 - \prod_{i=1}^{n} (1 - a_i^U)^{w_i} \right], \right.$$
$$\left. \times \left[ \prod_{i=1}^{n} (b_i^L)^{w_i}, \prod_{i=1}^{n} (b_i^M)^{w_i}, \prod_{i=1}^{n} (b_i^U)^{w_i}, \right] \right) \tag{19.16}$$

where, $w = (w_1, w_2, \ldots, w_n)^T$ is the weight vector of $I_i (i = 1, 2, \ldots, n)$, $w_i \in [0,\ 1]$, $\sum_{i=1}^{n} w_i = 1$ (19.16) is modified as in (19.17) in order to satisfy the following conditions (Kahraman et al. 2017):

- The widespread of non-membership function must be larger than that of membership function, and the midpoints of these functions must be equal to each other.
- The sum of membership and non-membership degrees for any $x$ must be at most equal to 1.

$$f_\omega(I_1, I_2, \ldots, I_n) = \left( \left[ \max(1 - \prod_{i=1}^{n} (1 - a_i^L)^{\omega_i}, \prod_{i=1}^{n} (b_i^L)^{\omega_i}), \sqrt{(1 - \prod_{i=1}^{n} (1 - a_i^M)^{\omega_i}) \times \prod_{i=1}^{n} (b_i^M)^{\omega_i}}, \right. \right.$$
$$\left. \times \min(1 - \prod_{i=1}^{n} (1 - a_i^U)^{\omega_i}, \prod_{i=1}^{n} (b_i^U)^{\omega_i}) \right], \left[ \min(1 - \prod_{i=1}^{n} (1 - a_i^L)^{\omega_i}, \prod_{i=1}^{n} (b_i^L)^{\omega_i}), \right.$$
$$\left. \left. \times \sqrt{(1 - \prod_{i=1}^{n} (1 - a_i^M)^{\omega_i}) \times \prod_{i=1}^{n} (b_i^M)^{\omega_i}}, \max\left( 1 - \prod_{i=1}^{n} (1 - a_i^U)^{\omega_i}, \prod_{i=1}^{n} (b_i^U)^{\omega_i} \right) \right] \right) \tag{19.17}$$

where $\omega = (\omega_1, \omega_2, \ldots, \omega_n)^T$ is the weight vector of $I_i (i = 1, 2, \ldots, n)$, $w_i \in [0, 1]$, $\sum_{i=1}^{n} \omega_i = 1$.

### 19.4.4 A New Proposal for the Defuzzification of TIFNs

Chang et al. (2008) proposed the defuzzification method in (19.18) for TIFNs as in Figs. 19.3 and 19.4.

$$\bar{x}_t = \frac{a_1' + a_1 + 2a_2 + a_3' + a_3}{6} \tag{19.18}$$

In this paper, we use the defuzzification method given by (19.20) as a modification of Chang et al.'s (2008) approach.

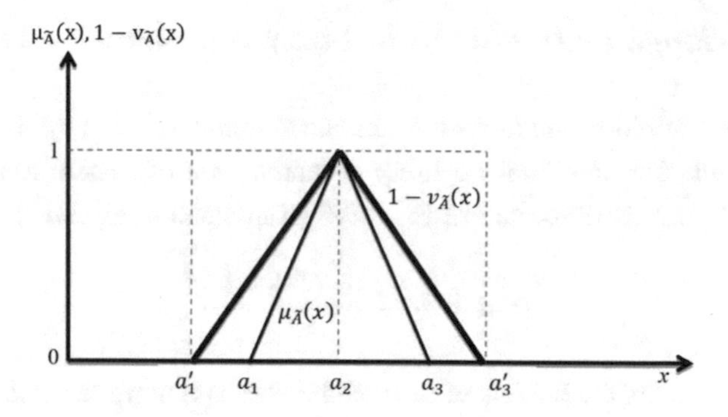


**Fig. 19.3** Membership and non-membership functions of TIFN

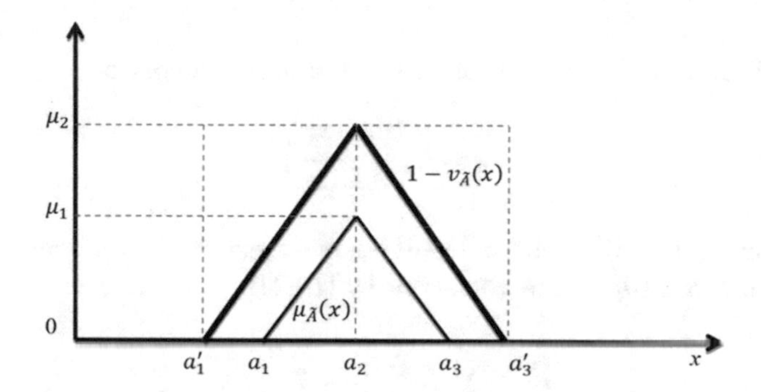


**Fig. 19.4** A TIFN with different heights for the most possible value

$$\bar{x}_t = \frac{\mu_2 \times \left(a'_1 + a_2 + a'_3\right) + \mu_1 \times (a_1 + a_2 + a_3)}{6} \tag{19.19}$$

Considering the intersection function, the defuzzification formula for $\tilde{C}_{\mathrm{TIFS}} = \left(\mathrm{b}_1, (\mathrm{c}_2, \mu_{\cap}(\mathrm{c}_2)), \mathrm{a}_3; \mathrm{b}'_1, \left(\mathrm{c}'_2, \mathrm{v}_{\cup}\left(\mathrm{c}'_2\right)\right), \mathrm{a}'_3\right)$ can be given by Eq. (19.20) (Kahraman et al. 2017):

$$\bar{x}_t = \frac{\left(1 - v_{\cup}\left(c'_2\right)\right) \times \left(b'_1 + c'_2 + a'_3\right) + \mu_{\cap}(c_2) \times (b_1 + c_2 + a_3)}{6} \tag{19.20}$$

### 19.4.5 Triangular Intuitionistic Fuzzy Information Axiom

In this chapter, intuitionistic fuzzy information axiom developed by Kahraman et al. (2017) is used. The intuitionistic fuzzy common area of system range ($\widetilde{SR}$) and design range ($\widetilde{DR}$) is illustrated in Fig. 19.5 (Kahraman et al. 2017).

$$DSR = \frac{\alpha + \alpha' + 2\xi + \kappa + \kappa'}{6} \tag{19.21}$$

(19.22) considers the heights of the membership and non-membership functions in the defuzzification process and calculates the defuzzified common area (DCA).

$$DCA = \frac{\mu_2 \times (\delta' + \rho' + \kappa') + \mu_1 \times (\delta + \rho + \kappa)}{6} \tag{19.22}$$

Thus (19.23) can be used to compute the information content

$$I_i = log_2\left(\frac{DSR}{DCA}\right)_i \tag{19.23}$$

Considering a decision matrix involving $m$ alternatives and $n$ criteria, the total information content $I_i^T$ can be calculated by (19.24).

$$I_i^T = \sum_{j=1}^{n} I_{ij} \quad i = 1, 2, \ldots, m \tag{19.24}$$

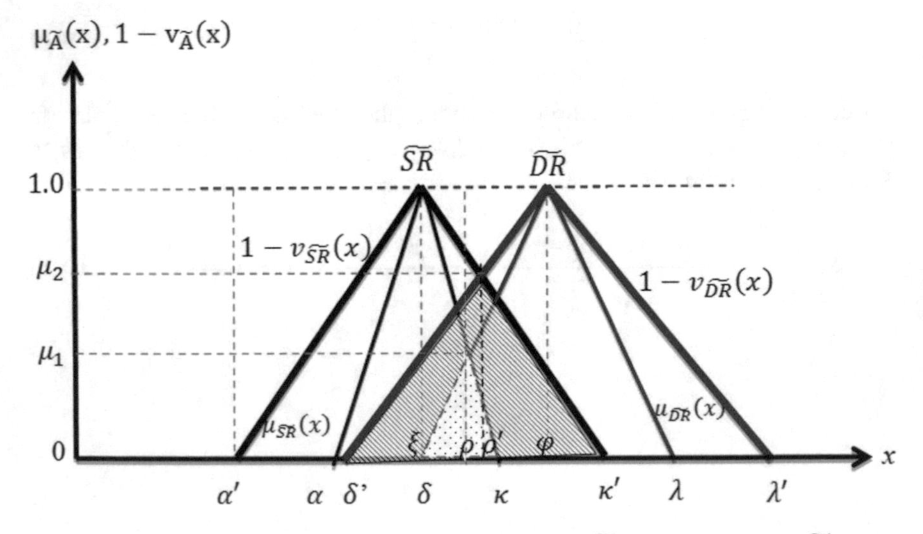


**Fig. 19.5** Intuitionistic fuzzy common area of system range ($\widetilde{SR}$) and design range ($\widetilde{DR}$)

If the criteria have different importance weights, the total weighted information content $I_i^{wT}$ can be calculated by (19.25):

$$I_i^{wT} = \sum_{j=1}^{n} I_{ij} w_j \quad i = 1, 2, \ldots, m \tag{19.25}$$

## 19.5 Numerical Application

An investor wants to select wind turbines for a wind energy investment in the Aegean region of Turkey. There are six wind turbine options, and the criteria have been determined as reliability, technical characteristics, performance, cost factors, availability, maintenance, cooperation and domesticity (Cevik et al. 2015). Each wind turbine has different characteristics. Alternative 1 has 2 MW power; its wind class is IIA and cut in/cut out speeds are 3 and 20, respectively. Alternative 2 has 2 MW power; its wind class is IIA and cut in/cut out speeds are 4 and 25, respectively. Alternative 3 has 2 MW power; its wind class is IIA and cut in/cut out speeds are 2 and 28, respectively. Alternative 4 has 2 MW power; its wind class is IIA and cut in/cut out speeds are 3 and 25, respectively. Alternative 5 has 2 MW power; its wind class is IIIA and cut in/cut out speeds are 4 and 25, respectively. Alternative 6 has 2 MW power; its wind class is IIIB and cut in/cut out speeds are 3 and 18, respectively. The criteria have equal importance. The scale in Table 19.1 is used to evaluate wind turbine alternatives (Kahraman et al. 2017). The design range for each evaluation criterion is considered as "Very Good."

Three experts evaluated the wind turbine alternatives. Two experts from the company and one academician evaluated the wind turbine options. Table 19.2 shows the evaluations of the experts.

The individual evaluations of experts are aggregated by using intuitionistic fuzzy aggregation operators, and the obtained aggregated matrix is given in Table 19.3.

The information content of wind turbines and their total information content I is calculated as in Table 19.4.

The results indicate that Alt 3 is the best alternative followed by Alt 1 and Alt 2. The worst alternative is Alt 6.

**Table 19.1** Intuitionistic fuzzy evaluation scale (Kahraman et al. 2017)

| Linguistic term | Abbreviation | Triangular intuitionistic fuzzy numbers |
|---|---|---|
| Very poor | VP | (0, 0, 0.25; 0, 0, 0.35) |
| Poor | P | (0, 0.25, 0.5; 0, 0.25, 0.6) |
| Fair | F | (0.25, 0.5, 0.75; 0.15, 0.5, 0.85) |
| Good | G | (0.5, 0.75, 1; 0.4, 0.75, 1) |
| Very good | VG | (0.75, 1, 1; 0.65, 1, 1) |

**Table 19.2** Linguistic evaluations of wind energy experts

| Reliability | | | | Technical characteristics | | | | Performance | | | | Cost factors | | | |
|---|---|---|---|---|---|---|---|---|---|---|---|---|---|---|---|
| | E1 | E2 | E3 | | E1 | E2 | E3 | | E1 | E2 | E3 | | E1 | E2 | E3 |
| Alt 1 | F | F | G | Alt 1 | F | F | F | Alt 1 | VG | G | VG | Alt 1 | F | G | F |
| Alt 2 | VG | G | VG | Alt 2 | F | G | VG | Alt 2 | G | VG | F | Alt 2 | VG | G | VG |
| Alt 3 | F | G | F | Alt 3 | G | VG | F | Alt 3 | VG | G | VG | Alt 3 | F | G | F |
| Alt 4 | G | VG | F | Alt 4 | VG | G | VG | Alt 4 | F | G | F | Alt 4 | F | F | F |
| Alt 5 | VG | G | VG | Alt 5 | F | G | F | Alt 5 | F | G | F | Alt 5 | F | G | F |
| Alt 6 | F | G | F | Alt 6 | F | F | F | Alt 6 | F | F | F | Alt 6 | F | F | F |
| Availability | | | | Maintenance | | | | Cooperation | | | | Domesticity | | | |
| | E1 | E2 | E3 | | E1 | E2 | E3 | | E1 | E2 | E3 | | E1 | E2 | E3 |
| Alt 1 | VG | G | G | Alt 1 | VG | G | VG | Alt 1 | VG | VG | G | Alt 1 | G | VG | F |
| Alt 2 | VG | G | VG | Alt 2 | VG | G | VG | Alt 2 | F | F | G | Alt 2 | VG | G | VG |
| Alt 3 | G | VG | F | Alt 3 | G | VG | F | Alt 3 | VG | G | VG | Alt 3 | F | G | F |
| Alt 4 | VG | G | VG | Alt 4 | VG | G | VG | Alt 4 | G | VG | F | Alt 4 | F | G | F |
| Alt 5 | F | G | F | Alt 5 | F | G | F | Alt 5 | VG | G | VG | Alt 5 | VG | G | VG |
| Alt 6 | F | F | F | Alt 6 | F | F | G | Alt 6 | F | G | F | Alt 6 | F | F | G |

**Table 19.3** Aggregated evaluations of wind turbines

| | Reliability | Technical characteristics | Performance | Cost factors |
|---|---|---|---|---|
| Alt 1 | (0.71, 0.97, 1; 0.59, 0.97, 1) | (0.25, 0.5, 0.75; 0.15, 0.5, 0.85) | (0.35, 0.59, 0.9; 0.21, 0.59, 1) | (0.35, 0.59, 0.9; 0.21, 0.59, 1) |
| Alt 2 | (0.35, 0.59, 0.9; 0.21, 0.59, 1) | (0.48, 0.81, 0.93; 0.28, 0.81, 1) | (0.46, 0.7, 0.97; 0.33, 0.7, 1) | (0.46, 0.7, 0.97; 0.33, 0.7, 1) |
| Alt 3 | (0.46, 0.7, 0.97; 0.33, 0.7, 1) | (0.57, 0.87, 0.97; 0.39, 0.87, 1) | (0.68, 0.95, 1; 0.55, 0.95, 1) | (0.68, 0.95, 1; 0.55, 0.95, 1) |
| Alt 4 | (0.25, 0.5, 0.75; 0.15, 0.5, 0.85) | (0.42, 0.67, 0.94; 0.28, 0.67, 1) | (0.31, 0.55, 0.88; 0.18, 0.55, 1) | (0.31, 0.55, 0.88; 0.18, 0.55, 1) |
| Alt 5 | (0.35, 0.59, 0.9; 0.21, 0.59, 1) | (0.31, 0.55, 0.88; 0.18, 0.55, 1) | (0.35, 0.59, 0.9; 0.21, 0.59, 1) | (0.35, 0.59, 0.9; 0.21, 0.59, 1) |
| Alt 6 | (0.61, 0.91, 1; 0.47, 0.91, 1) | (0.71, 0.97, 1; 0.59, 0.97, 1) | (0.25, 0.5, 0.75; 0.15, 0.5, 0.85) | (0.25, 0.5, 0.75; 0.15, 0.5, 0.85) |
| | Availability | Maintenance | Cooperation | Domesticity |
| Alt 1 | (0.63, 0.92, 1; 0.5, 0.92, 1) | (0.68, 0.95, 1; 0.55, 0.95, 1) | (0.71, 0.97, 1; 0.59, 0.97, 1) | (0.71, 0.97, 1; 0.59, 0.97, 1) |
| Alt 2 | (0.63, 0.92, 1; 0.5, 0.92, 1) | (0.42, 0.67, 0.94; 0.28, 0.67, 1) | (0.31, 0.55, 0.88; 0.18, 0.55, 1) | (0.31, 0.55, 0.88; 0.18, 0.55, 1) |
| Alt 3 | (0.38, 0.62, 0.91; 0.23, 0.62, 1) | (0.31, 0.55, 0.88; 0.18, 0.55, 1) | (0.68, 0.95, 1; 0.55, 0.95, 1) | (0.68, 0.95, 1; 0.55, 0.95, 1) |
| Alt 4 | (0.63, 0.92, 1; 0.5, 0.92, 1) | (0.56, 0.89, 1; 0.44, 0.89, 1) | (0.35, 0.59, 0.9; 0.21, 0.59, 1) | (0.35, 0.59, 0.9; 0.21, 0.59, 1) |
| Alt 5 | (0.42, 0.67, 0.94; 0.28, 0.67, 1) | (0.35, 0.59, 0.9; 0.21, 0.59, 1) | (0.68, 0.95, 1; 0.55, 0.95, 1) | (0.68, 0.95, 1; 0.55, 0.95, 1) |
| Alt 6 | (0.6, 0.89, 0.97; 0.41, 0.89, 1) | (0.31, 0.55, 0.88; 0.18, 0.55, 1) | (0.31, 0.55, 0.88; 0.18, 0.55, 1) | (0.31, 0.55, 0.88; 0.18, 0.55, 1) |

**Sensitivity Analysis**

To check the robustness of the results, we observed the changes in rankings when the weights of the criteria are changed. A one-at-a-time sensitivity analysis is applied, and the results are given in Fig. 19.6.

According to the sensitivity analysis results, the increase in the weight of performance, cooperation, cost factors, technical characteristics and domesticity does not change the wind turbine selection. Alt 3 is a robust decision. When the weights of reliability, availability and maintenance increase, Alt 1 becomes the leading wind turbine. The decision makers should carefully decide the weights of the alternatives.

**Table 19.4** Information contents of wind turbines

| | Reliability | Availability | Technical characteristics | Maintenance | Performance | Cooperation | Cost factors | Domesticity | Sum of IC |
|---|---|---|---|---|---|---|---|---|---|
| *w* | *0.125* | *0.125* | *0.125* | *0.125* | *0.125* | *0.125* | *0.125* | *0.125* | |
| Alt 1 | 0.148 | 0.366 | 3.070 | 0.247 | 1.586 | 0.148 | 1.586 | 0.148 | 0.912 |
| Alt 2 | 1.586 | 0.366 | 0.896 | 1.294 | 1.151 | 1.737 | 1.151 | 1.737 | 1.240 |
| Alt 3 | 1.151 | 1.488 | 0.605 | 1.737 | 0.247. | 0.247 | 0.247 | 0.247 | 0.746 |
| Alt 4 | 3.070 | 0.366 | 1.294 | 0.498 | 1.737 | 1.586 | 1.737 | 1.586 | 1.484 |
| Alt 5 | 1.586 | 1.294 | 1.737 | 1.586 | 1.586 | 0.247 | 1.586 | 0.247 | 1.234 |
| Alt 6 | 0.421 | 0.555 | 0.148 | 1.737 | 3.070 | 1.737 | 3.070 | 1.737 | 1.559 |

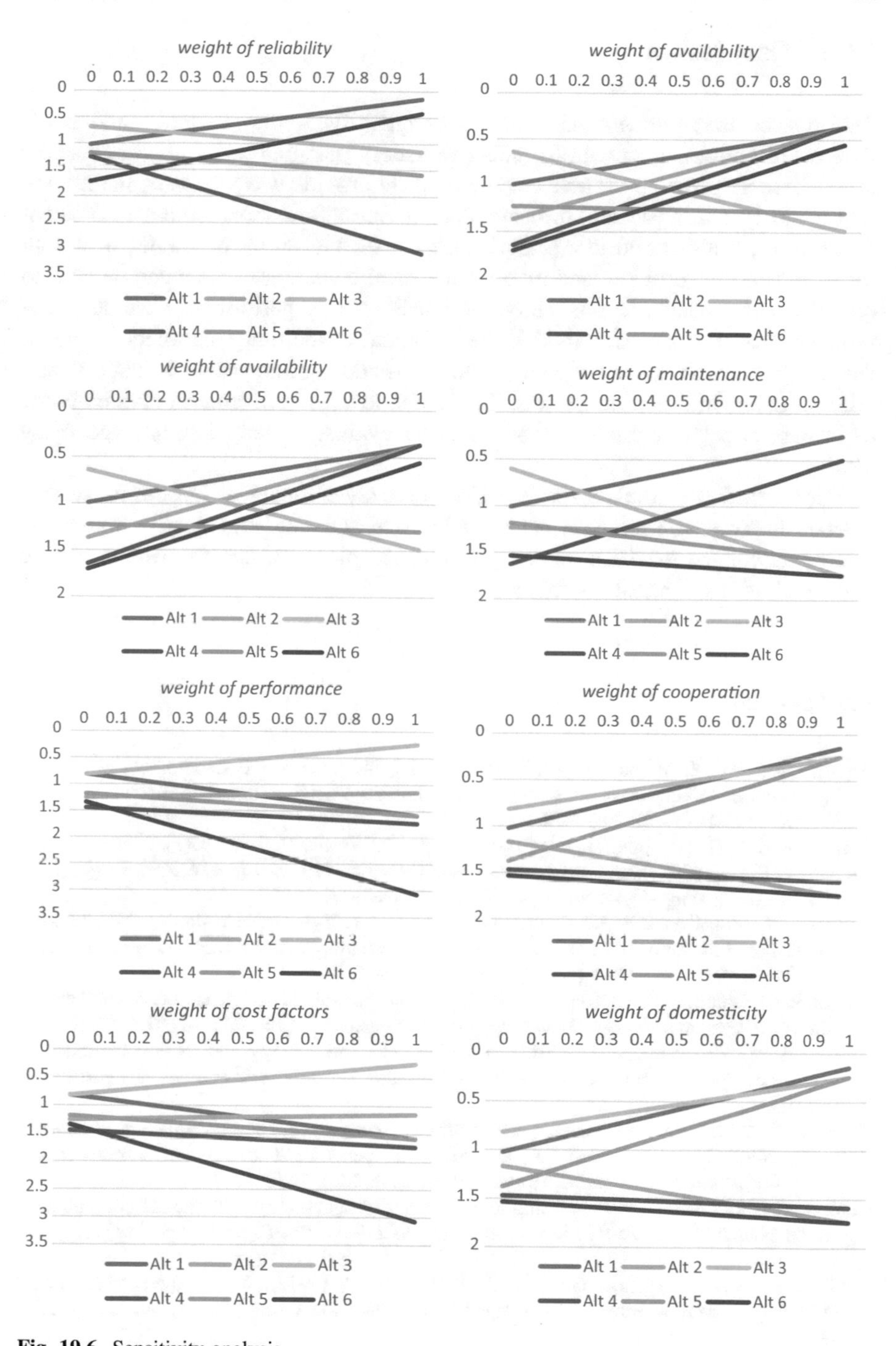


**Fig. 19.6** Sensitivity analysis

## 19.6 Conclusion

Performance measurement, the process of quantifying the efficiency and effectiveness of actions, is crucial for effective energy planning and technology selection. The literature on performance measurement provide two groups of studies; the first group deal with 'what to measure' and provide information about measurement frameworks, metrics and key performance the second group of studies deal with 'how to measure' problem and involve analytical techniques which can be used to quantify performance. In this study, we handle energy performance measurement problem as a fuzzy decision model. We use intuitionistic fuzzy axiomatic design to determine the performance of energy alternatives taking into account eight different criteria. Using information axioms, the desired level of performance can be better expressed, and the alternatives can be better evaluated using intuitionistic fuzzy sets.

For the further studies, other extensions of fuzzy sets such as Type-2 fuzzy sets, hesitant fuzzy sets, multisets, neutrosophic sets and interval-valued intuitionistic fuzzy sets can be used for energy performance problem, and the results can be compared with the results of this study.

## References

Abdulrahman, M., & Wood, D. (2017). Investigating the power-COE trade-off for wind farm layout optimization considering commercial turbine selection and hub height variation. *Renewable Energy, 102*, 267–278.

Atanassov, K.T. (1986). Intuitionistic fuzzy sets. *Fuzzy Sets and Systems, 20*(1), 87–96.

Atanassov, K.T. (1999). Intuitionistic fuzzy sets. *Theory and Applications, Studies in Fuzziness and Soft Computing* (Vol. 35). Physica-Verlag Heidelberg, Berlin.

Bagočius, V., Zavadskas, E. K., & Turskis, Z. (2014). Multi-person selection of the best wind turbine based on the multi-criteria integrated additive-multiplicative utility function. *Journal of Civil Engineering and Management, 20*(4), 590–599.

Bansal, R. C., Bhatti, T. S., & Kothari, D. P. (2002). On some of the design aspects of wind energy conversion systems. *Energy Conversion and Management, 43*(16), 2175–2187.

Bassyouni, M., & Gutub, S. A. (2013). Materials selection strategy and surface treatment of polymer composites for wind turbine blades fabricatio. *Polymers and Polymer Composites, 21* (7), 463.

Behn, R. D. (2003). *Why measure performance? Different purposes require different measures.*

Bora, M., Neog, T. J., & Kumar Sut, D. (2012). On generalized intuitionistic fuzzy soft set. *Journal of Mathematical and Computational Science*, *2*(4), 1030–1051.

Cebi, S., & Kahraman, C. (2013). Using multi attribute choquet integral in site selection of wind energy plants: The case of Turkey. *Journal of Multiple-Valued Logic & Soft Computing, 20*, 423–443.

Cevik, Onar S., Oztaysi, B., Otay, I., & Kahraman, C. (2015). Multi-expert wind energy technology selection using interval-valued intuitionistic fuzzy sets. *Energy, 90*(Part 1), 274–285.

Chang, K.-H., Cheng, C.-H., Chang, Y.-C. (2008). Reliability assessment of an aircraft propulsion system using IFS and OWA tree. *Engineering Optimization*, *40*(10), 907–921.

Chowdhury, S., Zhang, J., Messac, A., & Castillo, L. (2013). Optimizing the arrangement and the selection of turbines for wind farms subject to varying wind conditions. *Renewable Energy, 52,* 273–282.

Cross, K. F., & Lynch, R. L. (1989). The "SMART" way to define and sustain success. *National Productivity Review, 8,* 23–33. https://doi.org/10.1002/npr.4040080105.

Dong, Y., Wang, J., Jiang, H., & Shi, X. (2013). Intelligent optimized wind resource assessment and wind turbines selection in Huitengxile of Inner Mongolia, China. *Applied Energy, 109,* 239–253.

Du, M., Yi, J., Mazidi, P., Cheng, L. & Guo, J. (2017). A parameter selection method for wind turbine health management through SCADA data. *Energies, 10*(2), 253.

Fitzgerald, L., Johnston, R., Silvestro, T. J., & Voss, C. (1991). *Performance measurement in service business*. London: The Chartered Institute of Management Accountants.

Folan, P., Browne, J., & Jagdev, H. (2007). Performance: Its meaning and content for today's business research. *Computers in Industry, 58,* 605–620.

Fotuhi-Firuzabad, M., & Dobakhshari, A. S. (2009). Reliability-based selection of wind turbines for large-scale wind Farms. *World Academy of Science, Engineering and Technology, 49,* 734–740.

Gencturk, B., Attar, A., & Tort, C. (2015). Selection of an optimal lattice wind turbine tower for a seismic region based on the Cost of Energy. *KSCE Journal of Civil Engineering, 19*(7), 2179–2190.

Herbert, G. J., Iniyan, S., Sreevalsan, E., & Rajapandian, S. (2007). A review of wind energy technologies. *Renewable and Sustainable Energy Reviews, 11*(6), 1117–1145.

Kahraman, C., & Çebi, S. (2009). A new multi-attribute decision making method: Hierarchical fuzzy axiomatic design. *Expert Systems with Applications, 36*(3, Part 1), 4848–4861.

Kahraman, C., Cebi, S., Cevik Onar, S., & Oztaysi, B. (2018). A novel trapezoidal intuitionistic fuzzy information axiom approach: An application to multicriteria landfill site selection. *Engineering Applications of Artificial Intelligence, 67,* 157–172.

Kahraman, C., Cebi, S., & Kaya, I. (2010). Selection among renewable energy alternatives using fuzzy axiomatic design: The case of Turkey. *Journal of Universal Computer Science, 16*(1), 82–102.

Kahraman, C., Cevik Onar, S., & Oztaysi, B. (2016). A comparison of wind energy investment alternatives using interval-valued intuitionistic fuzzy benefit/cost analysis. *Sustainability, 8*(2), 118–136.

Kahraman, C., Onar, S. C., Cebi, S., & Oztaysi, B. (2017). Extension of information axiom from ordinary to intuitionistic fuzzy sets: An application to search algorithm selection. *Computers & Industrial Engineering, 105,* 348–361.

Kaplan, R. S., & Norton, D. P. (1992). The balanced scorecard-measures that drive performance. *Harvard Business Review, 70,* 71–79. PMID:10119714.

Kaya, T., & Kahraman, C. (2010). Multicriteria renewable energy planning using an integrated fuzzy VIKOR & AHP methodology: The case of Istanbul. *Energy, 35*(6), 2517–2527.

Keegan, D. P., Eiler, R. G., & Jones, C. (1989). Are your performance measures obsolete? *Management Accounting, 12,* 45–50.

Kulak, O., & Kahraman, C. (2005a). Fuzzy multi-attribute selection among transportation companies using axiomatic design and analytic hierarchy process. *Information Sciences, 170,* 191–210.

Kulak, O., & Kahraman, C. (2005b). Multi-attribute comparison of advanced manufacturing systems using fuzzy vs. crisp axiomatic design approach. *International Journal of Production Economics, 95,* 415–424.

Lebas, M. (1995). Performance measurement and performance management. *International Journal of Production Economics, 41,* 23–25.

Lee, A. H., Chen, H. H., & Kang, H. Y. (2009). Multi-criteria decision making on strategic selection of wind farms. *Renewable Energy, 34*(1), 120–126.

Maity, S. R., & Chakraborty, S. (2012). Turbine blade material selection using fuzzy analytic network process. *International Journal of Materials and Structural Integrity, 6*(2–4), 169189.

Meyer, M. W. (2002). *Rethinking performance measurement: Beyond the balanced scorecard*. New York: Cambridge University Press.

Montoya, F. G., Manzano-Agugliaro, F., López-Márquez, S., Hernández-Escobedo, Q., & Gil, C. (2014). Wind turbine selection for wind farm layout using multi-objective evolutionary algorithms. *Expert Systems with Applications, 41*(15), 6585–6595.

Mousavi, M., Gitinavard, H., & Mousavi, S. M. (2017). A soft computing based-modified ELECTRE model for renewable energy policy selection with unknown information. *Renewable and Sustainable Energy Reviews, 68*, 774–787.

Neely, A., Adams, C., & Kennerley, M. (2002). The performance prism: The scorecard for measuring and managing business success. London: Prentice Hall Financial Times.

Onar, S. C., Oztaysi, B., Otay, İ., & Kahraman, C. (2015). Multi-expert wind energy technology selection using interval-valued intuitionistic fuzzy sets. *Energy, 90*, 274–285. https://doi.org/10.1016/j.energy.2015.06.086.

Petković, D., & Shamshirband, S. (2015). Soft methodology selection of wind turbine parameters to large affect wind energy conversion. *International Journal of Electrical Power & Energy Systems, 69*, 98–103.

Rehman, S., & Khan, S. A. (2016). Fuzzy logic based multi-criteria wind turbine selection strategy —A case study of Qassim, Saudi Arabia. *Energies, 9*(11), 872.

Rehman, S., & Khan, S. A. (2017). Multi-criteria wind turbine selection using weighted sum approach. *International Journal of Advanced Computer Science and Applications, 8*(6), 128–132.

Ritter, M., & Deckert, L. (2017). Site assessment, turbine selection, and local feed-in tariffs through the wind energy index. *Applied Energy, 185*, 1087–1099.

Sarja, J., & Halonen, V. (2013). Wind turbine selection criteria: A customer perspective. *Journal of Energy and Power Engineering, 7*(9), 1795.

Suh, N. P. (1990). *The principles of design*. NY: Oxford University Press Inc.

Suh, N. P. (2005). *Complexity: Theory and applications*. NY: Oxford University Press.

Sun, W., & Ren, Q. (2014). BP neural network Application in wind turbine type selection based on particle swarm optimization. *Journal of Information & Computational Science, 11*(7), 2415–2423.

Sun, W., & Xu, Z. (2017). Wind turbine generator selection and comprehensive evaluation based on BPNN optimized by PSO. *International Journal of Applied Decision Sciences, 10*(4), 364–381.

Tegou, L. I., Polatidis, H., & Haralambopoulos, D. A. (2009). Wind turbines site selection on an isolated island. *WIT Transactions on Ecology and the Environment, 127*, 313–324.

Theotokoglou, E. E., & Balokas, G. A. (2015). Computational analysis and material selection in cross-section of a composite wind turbine blade. *Journal of Reinforced Plastics and Composites, 34*(2), 101–115.

Van Buren, K. L., Atamturktur, S., & Hemez, F. M. (2014). Model selection through robustness and fidelity criteria: modeling the dynamics of the CX-100 wind turbine blade. *Mechanical Systems and Signal Processing, 43*(1), 246–259.

Vasileiou, M., Loukogeorgaki, E., & Vagiona, D. G. (2017). GIS-based multi-criteria decision analysis for site selection of hybrid offshore wind and wave energy systems in Greece. *Renewable and Sustainable Energy Reviews, 73*, 745–757.

Wang, Y. (2017). A fuzzy VIKOR approach for renewable energy resources selection in China. *Revista de la Facultad de Ingeniería, 31*(10), 62–77.

Wholey, J. (1996). Formative and summative evaluation: Related issues in performance measurement. *Evaluation Practice, 17*, 145–149.

Zhang, X., & Liu, P. D. (2010). Method for aggregating triangular intuitionistic fuzzy information and its application to decision making. *Technologic and Economic Development of Economy, 16*, 280–290.

# Chapter 20
# Sustainability Performance Evaluation of Energy Generation Projects

**Yağmur Karabulut and Gülçin Büyüközkan**

**Abstract** Affordable and reliable energy is not only central to prosperity, but also to poverty reduction, local development, environmental integrity, quality of life, growth, and progress. Given the importance and wide scale of energy generation all around the world, its ever growing economic, social and environmental aspects need to be taken into better consideration. The sustainability performance of energy operations shall be assessed on a project basis, as energy generation projects may significantly vary, depending on the needs and circumstances. This chapter introduces a novel approach for evaluating energy projects from a sustainability point of view and estimates their sustainability performance as a decision-making support tool. Decision environments can sometimes be complicated for an individual decision maker (DM) to consider every aspect of the problem. Group decision making (GDM) can be advantageous to reduce the impact of biased and personal opinions on the decision process. Moreover, DMs' judgments are mostly far from being completely certain, making it more difficult to put their ratings into numerical forms. In such circumstances, the fuzzy set theory can be applied to better represent DMs' preferences. This chapter applies GDM together with the fuzzy set theory to find the importance of the selected evaluation criteria. Then, GDM and the fuzzy set theory are combined with VIKOR (Vlse Kriterijumska Optimizacija Kompromisno Resenje) technique to rank the energy project alternatives. This approach is particularly useful for its strength in dealing with actual energy projects so that it can support both researchers and business managers to compare the sustainability performance of planned or realized power plants in a balanced manner. The usability of the proposed approach is shown in a case study from Turkey, where different energy projects are evaluated for their overall sustainability performance.

Y. Karabulut
Mavi Consultants, Altunizade Mah. Kisikli Caddesi No: 28
Avrupa Is Merkezi K.1/2, 34662 Uskudar, Istanbul, Turkey

G. Büyüközkan (✉)
Department of Industrial Engineering, Galatasaray University,
Çırağan Caddesi No: 36, 34357 Ortaköy, Istanbul, Turkey
e-mail: gulcin.buyukozkan@gmail.com

© Springer International Publishing AG, part of Springer Nature 2018

C. Kahraman and G. Kayakutlu (eds.), *Energy Management—Collective and Computational Intelligence with Theory and Applications*, Studies in Systems, Decision and Control 149, https://doi.org/10.1007/978-3-319-75690-5_20

## 20.1 Introduction

Energy is essential to reduce underdevelopment, spur economic growth and ensure environmental protection. According to the World Bank (2017), 1.06 billion people today do not have access to electricity, despite its vital role in contributing to the quality of life, economic progress and social advancement. Therefore, ensuring energy supply in an affordable, reliable, low-carbon and sustainable way is crucial. In 2015, countries worldwide agreed on a set of 17 goals to protect the planet and ensure prosperity for all as part of a new sustainable development agenda. Each so-called Sustainable Development Goal has specific targets to be achieved over the next 15 years (United Nations 2015). Energy generation is related to all of these goals. Nevertheless, at this rate of investments, the rate of electrification is forecasted to only reach 92% by 2030 (World Bank 2017), still leaving 8% of global population in the dark. Given the importance of energy generation, its ever growing economic, social and environmental aspects need to be taken into spotlight.

The world is struggling to find the optimal solution to expand its clean energy base amid climate change concerns, air and water pollution of conventional power generation and consistently decreasing capital costs for renewable energy technologies. Urgent measures are required to promote people's access to energy, supported by policies, regulations, and incentives to accommodate for the growing role for the private sector to finance energy projects to assure investors to earn returns on their investments (IBRD and World Bank 2017). Deciding on energy projects solely on economic considerations has become a thing of the past, as governments and communities scrutinize social and environmental dimensions nowadays more than ever. Developing sustainably also requires preventing ecological degradation and creating decent jobs and opportunities. The divergence from fossil fuels towards renewables will, therefore, be inevitable. This, however, does not explain which concrete energy projects investors should be investing in. Ultimately, the aim is to build one or more of these most sustainable energy projects.

There are many types of energy generation options available for investors. Selecting the optimal project is important to ensure long term profitability, social acceptance, and environmental protection. These implications of energy generation do not only vary for different technologies, but also for the particular conditions of each and every energy generation project. Therefore, identifying the most sustainable energy generation project requires a careful review of alternative projects in the light of different sustainability criteria (Büyüközkan and Karabulut 2017). New methods for evaluating the level of sustainability of energy projects can help reduce environmental burden and resource use, and also contribute to the local economy, employment, and technology transfer. Widely recognized methods for selecting the most sustainable energy generation project are largely missing, which would be able to compare different energy projects with multiple criteria in differing circumstances.

This chapter introduces a novel approach for evaluating energy projects from a sustainability point of view and estimate their sustainability performance for decision making purposes. The uncertainty, subjectivity and vagueness of human cognitive processes in the area of sustainability are handled with an evaluation model based on the fuzzy multi criteria decision making (MCDM) method for measuring the performance of alternative projects with regard to criteria. MCDM is a capable and popular tool used in decision problems for assessing and ranking alternatives on the basis of multiple, usually conflicting criteria. The proposed approach is based on two analytical methods; the fuzzy set theory to determine the importance weights of evaluation criteria and VIKOR (Vlse Kriterijumska Optimizacija Kompromisno Resenje) to consolidate the ratings of feasible alternatives. These techniques are supported with a Group Decision Making (GDM) approach to reduce the level of personal bias and subjectivity of individual decision makers (DMs). The methodological workings are tested and validated on a case from Turkey, where actual energy generation projects are compared. The results intend to present how individual energy projects can be assessed in terms of their overall sustainability impacts.

The chapter continues with a summary of the literature on the selection of energy generation projects from a sustainability point of view, followed by the presentation of publications that utilize the Fuzzy VIKOR technique. Then, the proposed approach will be explained. The section after that will demonstrate the application of the approach on a real case study. The final section will present authors' conclusions.

## 20.2 Literature Review

There is extensive research on energy generation technologies and their sustainability impacts. For concrete energy generation projects, on the other hand, literature is quite limited. In terms of the proposed methodology, the Fuzzy VIKOR technique is a known method, as also discussed below.

### *20.2.1 Energy Generation Project Selection*

There is a wide range of analytical methods in the literature that compare and rank energy technologies (Kaya and Kahraman 2010). A very limited number of these papers deal with specific energy projects, while many others focus on renewables, national energy policies, optimization of energy mix or comparison of energy technologies. Analytical methods, such as MCDM and fuzzy set theory, are popular because of the variety of the factors that affect the eventual decision (Wang et al. 2009; Hsueh and Yan 2013).

Looking at the last ten years of publications, Garg et al. (2007) studied different thermal power plant alternatives with TOPSIS, another MCDM technique. Burton and Hubacek (2007) explored the question whether small-scale energy projects effectively fare better than large-scale ones with respect to social, economic and environmental targets. In another paper, Chatzimouratidis and Pilavachi (2007) evaluated 10 different energy generation technologies with regards to five environmental criteria. The same authors, Chatzimouratidis and Pilavachi (2008), studied the environmental impacts of ten different power plants on the living standard by using the Analytic Hierarchy Process (AHP) method, a popular MCDM approach. Chatzimouratidis and Pilavachi (2009) also discussed power plants in terms of technological, economic and sustainability criteria again with the help of AHP to find out that renewables fare better than thermal power plants in general. Another article by Pilavachi et al. (2009) analyzed 9 power plant types fuelled with natural gas or hydrogen. Using AHP for seven criteria, they concluded that natural gas combined cycle plants perform better overall. Kowalski et al. (2009) determined the most suitable sustainable energy technology by making use of an MCDM technique and found out that natural gas is the best fuel type. Lee et al. (2009) developed a new MCDM model, based on AHP, to select a suitable wind farm project, while Kahraman et al. (2009) employed axiomatic design and AHP under fuzzy environment to select the best possible renewable energy alternative. Wang et al. (2009) reviewed MCDM methods for sustainable energy systems, including criteria selection, criteria weighting, evaluation, and aggregation. Streimikiene et al. (2012) developed an MCDM decision support framework for choosing the most sustainable electricity production technology with the help of MULTIMOORA and TOPSIS methods. More recently, Onar et al. (2015) focused on the evaluation of wind energy investments and selected the most appropriate wind energy technology with fuzzy MCDM techniques to help investors.

Some publications approach the subject with a more regional point of view. In the past, Polatidis and Haralambopoulos (2004) developed a new multi-participatory and MCDM framework to evaluate renewable energy alternatives in Greece. Cavallaro and Ciraolo (2005) made use of selected evaluation criteria for a case study in Italy, where they selected the most appropriate site for wind energy. Another article by Krukanont and Tezuka (2007) presented an optimization model based on two-stage stochastic programming to evaluate the energy system of Japan, while Begic and Afgan (2007) evaluated different power system options in Bosnia Herzegovina with a multi-criteria sustainability assessment framework. Tsoutsos et al. (2009) identified a set of renewable energy technologies for the island of Crete and then used MCDM approach for assessing them in terms of economic, technical, social and environmental criteria, which are identified by sectoral experts. Another researcher, Cristóbal (2011), combined AHP and VIKOR to select a renewable energy project that suits national energy policies in Spain and identified the biomass plant option to be the best choice. Nixon et al. (2013) introduced an MCDM-based method for evaluating alternative technologies for generating electricity from municipal solid waste in India. In another publication, Nixon et al. (2014) presented a goal programming model to optimize the deployment of bio-energy plants in Punjab, India and demonstrated its use on two alternative case scenarios. Zhang et al. (2015) developed an improved MCDM method that is based on the fuzzy measure and integral to evaluate clean energy options for Jiangsu province, China. They ranked the solar photovoltaic technology as the first, followed by the wind, biomass, and nuclear technologies. Abdullah and Najib (2016) proposed a method based on intuitionistic fuzzy AHP to establish a preference in the sustainable energy planning decision-making problem in Malaysia. In another recent paper, Ishizaka et al. (2016) explored the best energy technologies for the United Kingdom by using GAIA (graphical analysis for interactive aid) and AHP techniques. Recently, Grilli et al. (2017) deployed MCDM methods to assess the best solutions for enhancing the production of renewable energy in the Alps.

### *20.2.2 Energy Generation Project Selection in Turkey*

There is extensive research on evaluating energy technologies in Turkey. These are summarized next. In the last decade, Köne and Büke (2007) determined the optimal fuel mixture for electricity generation in Turkey with Analytic Network Process (ANP). Önüt et al. (2008) also used ANP to assess energy resources for the Turkish manufacturing industry. Kahraman et al. (2009) used a fuzzy AHP method to select the most appropriate renewable energy alternative for Turkey and concluded that the wind power technology promises the best outcomes. Atmaca and Başar (2012) performed multi-criteria evaluations of six different energy plants in Turkey with respect to 13 criteria under 4 main clusters; technology and sustainability, economic suitability, life quality and socio-economic. They used the ANP to find out that nuclear power technology is the best solution, followed by natural gas. In a study by Kaya and Kahraman (2011), a modified fuzzy TOPSIS method is proposed for energy planning.

In more recent years, Kabak and Dağdeviren (2014) proposed a hybrid model that considers benefits, opportunities, costs and risks related to energy technologies and prioritized available renewable energy alternatives in Turkey with ANP. Büyüközkan and Güleryüz (2014) developed an evaluation model to guide planners with their critical renewable energy technology alternative selection processes. In another paper, Erdogan and Kaya (2015) applied fuzzy AHP based on interval type-2 fuzzy sets to obtain the weights of the criteria affecting energy alternatives. Then, they fuzzified the TOPSIS method by interval type-2 fuzzy sets to rank them. Şengül et al. (2015) used fuzzy TOPSIS method for ranking renewable energy supply systems in Turkey, while Büyüközkan and Güleryüz (2016) combined Decision Making Trial and Evaluation Laboratory Model (DEMATEL) technique with ANP for selecting the most appropriate energy technology in Turkey from an investor-focused perspective. Balin and Baraçli (2017) used fuzzy AHP procedure based upon type-2 fuzzy sets, and interval type-2 TOPSIS method to find the best renewable alternative energy for Turkey. A similar problem was studied by Çolak and Kaya (2017), who introduced another MCDM model for ranking renewable energy alternatives in Turkey. They integrated AHP based on interval type-2 fuzzy sets with hesitant fuzzy TOPSIS methods for this purpose. Another study in this field was published by Büyüközkan and Karabulut (2017). They proposed a novel method combining AHP with VIKOR for better selecting concretely defined energy projects from a sustainability point of view. Büyüközkan and Güleryüz (2017) utilized linguistic interval fuzzy preferences with DEMATEL, ANP and TOPSIS to find the most appropriate energy alternative for Turkey. Özkale et al. (2017) analysed the strengths, weaknesses, opportunities and threats (SWOT) of energy alternatives and applied PROMETHEE method to ranked them for the Turkish energy market. Kuleli Pak et al. (2017) employed ANP to find the sustainable energy scenario of Turkey.

This review suggests that energy technologies are researched with various MCDM methods in terms of technologies, including geographic considerations.

### 20.2.3 Fuzzy VIKOR Literature

This chapter utilizes Fuzzy VIKOR, the combination of the fuzzy set theory and the VIKOR method, to deal with the energy project selection problem from a sustainability point of view. This combination is used in the literature for different purposes. For example, Vinodh et al. (2013) made use of Fuzzy VIKOR to select the best concept among five alternative designs for adopting agile manufacturing for the speedy production of customized, high-quality products in different lot sizes. Liu et al. (2014) again used linguistic variables, expressed as TFNs, to evaluate the weights of the selection criteria with the help of Ordered Weighted Averaging (OWA) operator to convert the fuzzy decision matrix into crisp values. VIKOR is utilized to find the ranking of disposal site alternatives for municipal solid waste. Chang (2014) developed a framework based on the concept of fuzzy sets theory and the VIKOR method for evaluating hospital service quality and tested their method on an empirical case with 33 evaluation criteria and 5 medical centers in Taiwan, assessed by 18 evaluators. Mandal et al. (2015) developed a methodology for identifying failure modes of overhead crane operations with the help of fuzzy set theory and VIKOR to quantify risks of different human errors using the experts' opinions. Rostamzadeh et al. (2015) introduced a quantitative Fuzzy VIKOR evaluation model to solve the green MCDM problem with the help of TFNs to deal with the imprecision of expert evaluations. More recently, Wu et al. (2016) applied Fuzzy VIKOR to solve a CNC machine tool selection problem with the use of linguistic input. In another paper, Sofiyabadi et al. (2016) used Fuzzy VIKOR to measure the importance of key performance indicators in a successful service business of LG, a South Korean multinational conglomerate corporation. The same method was employed by Bahadır and Büyüközkan (2016) to select the most appropriate robots for warehouses. In another paper, Wang (2017) made use of this methodological combination for selecting the most suitable energy technology in

China. Also lately, Foroozesh et al. (2017) integrated hesitant fuzzy sets with VIKOR and applied it to new product selection and energy policy selection problems under uncertainty.

### 20.2.4 *Findings of Literature Review*

This extensive literature analysis indicates the following research gaps:

i. Energy literature mostly discusses energy generation technologies. Concrete projects are largely ignored. This can cause oversimplifications, as project-specific conditions can be decisive for sustainability in different decision-making problems.
ii. Energy literature frequently does not consider the different scales of sustainability impacts. Sustainability of large and small-scale technologies are frequently compared, which can lead to questionable findings. Alternative projects that are evaluated shall be of comparable size so that their economic, environmental and social implications can also be comparable.
iii. Fuzzy VIKOR is used very limited for energy related decision-making problems. So far, there is no publication which applies Fuzzy VIKOR to evaluate concretely defined energy generation projects with a sustainability perspective.

To address these literature findings, a robust and practical Fuzzy VIKOR-based framework is proposed for assessing and ranking alternative energy generation projects in terms of their sustainability aspects.

## 20.3 Proposed Fuzzy VIKOR-Based Framework for Energy Generation Project Selection

The proposed integrated Fuzzy VIKOR-based framework consists of a hierarchical sustainability criteria model, integrated with MCDM methods to provide researchers and business managers with a functional framework that can be applied to other energy generation project selection problems. For a DM, it is sometimes difficult to identify a single alternative that satisfies all the evaluation criteria at once. In such MCDM problems, a compromise solution can be sought (Büyüközkan and Ruan 2008). VIKOR is integrated to the proposed framework to identify such compromise solutions by seeking a maximum group utility for the majority and simultaneously a minimum of an individual regret for the opponent. The original VIKOR requires crisp numerical input from a DM. Linguistic information can be modelled by extending the VIKOR method with Zadeh's (1965) fuzzy logic to process such data to achieve a more comprehensive evaluation. Considering that several DMs can be involved in real world problems, who prefer to voice their opinions with words instead of exact numbers, the VIKOR method will be applied in a fuzzy environment with GDM.

Criteria weights are found with the fuzzy set theory and the ranking of energy generation projects is achieved with Fuzzy VIKOR, both with the help of GDM. The proposed evaluation procedure consists of the following consecutive phases: (i) Identify evaluation criteria and their hierarchy; and, (ii) Determine criteria weights and rank energy project alternatives with Fuzzy VIKOR.

## *20.3.1 Identify Evaluation Criteria and Their Hierarchy*

Many factors affect the selection of the most sustainable energy generation project. Each of these factors shall be addressed with usually vague and imprecise data collected from a team of experts with different backgrounds. A GDM approach is utilized with the fuzzy set theory to minimize the bias stemming from experts' fuzzy evaluations and to better manage the associated uncertainties and partiality. For this evaluation, first the evaluation criteria shall be identified.

The first phase in the proposed Fuzzy VIKOR-based framework is the determination of criteria for evaluating energy projects. These criteria, along with their structured levels and definitions, are taken from Büyüközkan and Karabulut (2017), as given in Fig. 20.1.

The objective of the decision-making problem is located at the highest level of the proposed hierarchy. The main dimensions ($C_1$, $C_2$, and $C_3$) are located on the 2nd level of the hierarchy, with their sub-assigned criteria in the 3rd and 4th levels. The lowest level belongs to the energy generation project alternatives.

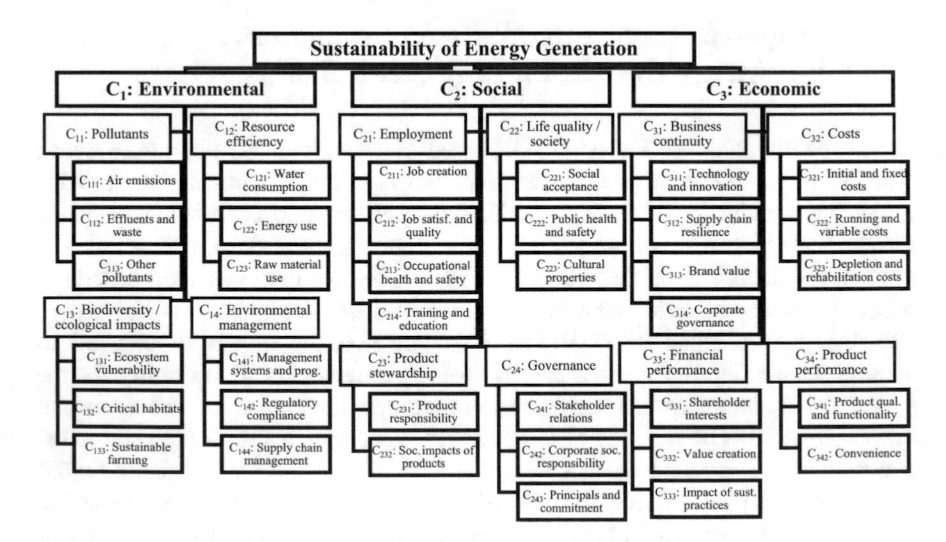


**Fig. 20.1** Evaluation criteria and their hierarchy

### *20.3.2 Determine Criteria Weights and Rank Energy Generation Project Alternatives with Fuzzy VIKOR*

In this second phase, the priority of the identified criteria will be found, with which the alternative energy projects will be compared with each other. The techniques used in this phase are introduced next.

#### 20.3.2.1 Group Decision Making

An expert might not possess sufficient knowledge of part of the problem and, as a consequence, he/she might not always provide all the needed information. Relying on a single DM can pose risks in decision-making because of individual limitations on experiences, preferences or biases. To address these difficulties, more than one DM is involved in the process. A GDM process is characterized by its involvement of two or more industry specialists who differ in their preferences but have the same access to information, where each DM brings along her/his own perceptions, attitudes, motivations, and personalities.

#### 20.3.2.2 Fuzzy Assessment of Criteria Weights

When DMs are asked about their thoughts on the sustainability aspects of an energy generation project, the responses will be mostly linguistic, which are inherently imprecise, and unquantifiable. In such settings, DMs' inputs can be collected as fuzzy numbers to adequately represent the uncertainty in human perceptions. The fuzzy set theory is a method proposed by Zadeh (1965) to handle ambiguity, uncertainty, and vagueness of decision-making problems. Fuzzy sets are a class of objects with a continuum of grades of membership from 0 to 1, implying partial membership to that set. A linguistic value that is not described explicitly can be represented mathematically in the interval [0,1] that indicates the degree of its membership.

#### 20.3.2.3 Rank Energy Generation Project Alternatives with Fuzzy VIKOR

Following the fuzzy assessment of the criteria weights, the project evaluations are collected from the DMs in fuzzy environment and used to compare and rank the available energy generation project alternatives with VIKOR, another analytical method. VIKOR as an MCDM technique is a powerful compromise ranking method that is preferred for its ability for dealing with complex decision problems. It can successfully process conflicting and non-commensurable criteria for choosing the best solution from a set of available alternatives (Opricovic 1998; Opricovic and

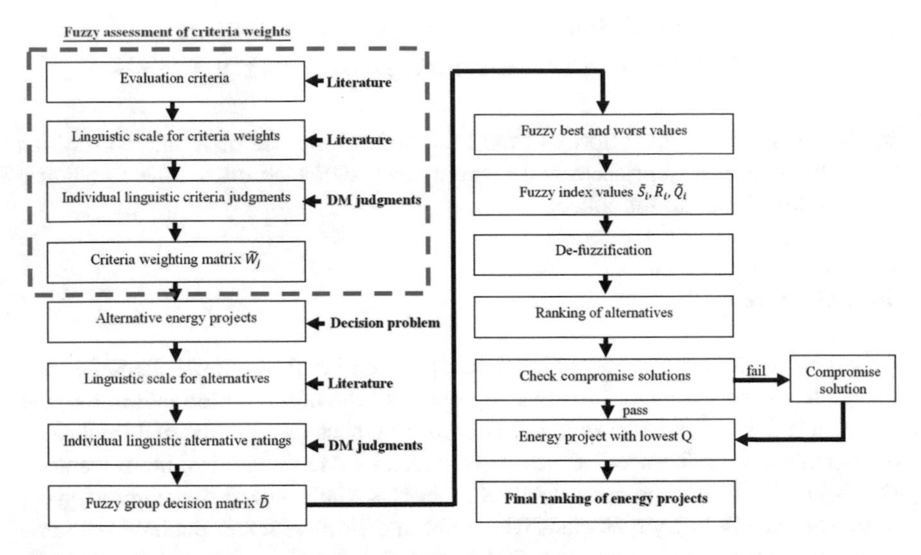


**Fig. 20.2** General structure of the proposed framework

Tzeng 2004). It ranks alternatives on the basis of closeness of each alternative to the ideal result under given conditions by providing the highest group utility for the 'majority' while maintaining the lowest individual regret of the 'opponent'. The general structure of the proposed Fuzzy VIKOR-based framework is illustrated in Fig. 20.2.

Following the presentation of the sustainability model and the individual methods that compose the integrated proposed Fuzzy VIKOR-based framework, its steps are explained next.

**Step 1**. Construct a committee of DMs with K members, determine n alternatives, and m criteria.

**Step 2**. Identify the evaluation base, in other words, the linguistic variables for weighing the DMs, criteria and alternatives, all of which are given in Tables 20.1, 20.2 and 20.3 in this order.

**Step 3**. Collect the judgments of the DMs on the DMs, criteria, and alternatives. DMs give their linguistic evaluations with the linguistic variables given in Step 2.

**Step 4**. Calculate DMs' weights, $\lambda_k$, where k = 1, 2, …, K. The $k$th DM gives his/her judgment about other DMs using the linguistic terms given in Table 20.1. Their judgments are aggregated with "$\tilde{\lambda}_k$" and de-fuzzified with "$\lambda_k'$" according to the generalized mean of fuzzy number by Opricovic and Tzeng (2003). The aggregation is done by the formula in Eq. (20.1).

$$\tilde{\lambda}_k = \frac{\sum_{t=1}^{K} a_t}{3}, \frac{\sum_{t=1}^{K} b_t}{3}, \frac{\sum_{t=1}^{K} c_t}{3} \quad k \neq t \tag{20.1}$$

**Table 20.1** Evaluation scale for DMs (Büyüközkan and Ruan 2008)

| Linguistic terms | | Fuzzy numbers |
|---|---|---|
| Extremely important | EI | {0.70 0.90 1.00} |
| Very important | VI | {0.50 0.70 0.90} |
| Important | IM | {0.30 0.50 0.70} |
| Medium importance | MI | {0.20 0.30 0.50} |
| Unimportant | U | {0.10 0.20 0.30} |
| Very unimportant | VU | {0.00 0.10 0.20} |

**Table 20.2** Evaluation scale for criteria weights (Chen 2000)

| Linguistic terms | | Fuzzy numbers |
|---|---|---|
| Very Very Low | VVL | {0.00 0.10 0.10} |
| Very Low | VL | {0.10 0.20 0.30} |
| Low | L | {0.20 0.30 0.40} |
| Medium Low | ML | {0.30 0.40 0.50} |
| Medium | M | {0.40 0.50 0.60} |
| Medium High | MH | {0.50 0.60 0.70} |
| High | H | {0.60 0.70 0.80} |
| Very High | VH | {0.70 0.80 0.90} |
| Very Very High | VVH | {0.80 0.90 1.00} |

**Table 20.3** Evaluation scale for alternative evaluations (Chen 2000)

| Linguistic terms | | Fuzzy numbers |
|---|---|---|
| Very Very Bad | VVB | {0.00 1.00 1.00} |
| Very Bad | VB | {1.00 2.00 3.00} |
| Bad | B | {2.00 3.00 4.00} |
| Medium Bad | MB | {3.00 4.00 5.00} |
| Intermediate | I | {4.00 5.00 6.00} |
| Medium Good | MG | {5.00 6.00 7.00} |
| Good | G | {6.00 7.00 8.00} |
| Very Good | VG | {7.00 8.00 9.00} |
| Very Very Good | VVG | {8.00 9.00 10.00} |

Here, the parameters a, b and c are the indices for triangular fuzzy numbers, where $1 \leq k \leq K, 1 < t \leq K$. To map a fuzzy number into a corresponding crisp number, Opricovic and Tzeng (2003) proposed the triangular fuzzy number (a, b, c) de-fuzzification formula in Eq. (20.2).

$$\lambda'_k = \left(\frac{a+b+c}{3}\right) \tag{20.2}$$

De-fuzzified values are normalized to adjust the values measured on a different scale to a common scale between [0–1]. The normalization is accomplished with Eq. (20.3).

$$\lambda_k = \frac{\lambda'_k}{\sum_{k=1}^{K} \lambda'_k} \tag{20.3}$$

**Step 5**. Determine the aggregated fuzzy weight $\tilde{w}_j$ of criterion $C_j$, with j = 1, 2, …, m with Eq. (20.4) and establish criteria weighting matrix $\tilde{W}_j$ with j criteria, as shown in Eq. (20.5). Main criteria weights and sub-criteria weights are multiplied to find the final evaluation weights of the evaluation criteria.

In the past, Yager (1988) introduced an aggregation technique based on the ordered weighted averaging operators. The so-called Ordered Weighted Averaging (OWA) operators have been discussed in a large number of papers. Here, this OWA operator in Eqs. (20.4) and (20.6) is applied.

$\lambda_k = (\lambda_1, \lambda_2, \ldots, \lambda_K)$, where $\lambda_k \in [0,\ 1],\ 1 \leq k \leq K$, and $\sum_{k=1}^{K} \lambda_k = \lambda_1 + \cdots + \lambda_k = 1$.

Furthermore,

$$\tilde{w}_j = \sum_{k=1}^{K} \lambda_k a_j^k, \sum_{k=1}^{K} \lambda_k b_j^k, \sum_{k=1}^{K} \lambda_k c_j^k \tag{20.4}$$

where $a_j$, $b_j$, and $c_j$ are the *k*th largest elements of the DMs' judgments.

$$\tilde{W}_j = \left[\tilde{w}_1, \tilde{w}_2, \ldots, \tilde{w}_j\right] \tag{20.5}$$

**Step 6**. Determine the aggregation of fuzzy rating $\tilde{f}_{ij}$ of alternative $A_i$ i = 1,2, …, n under criterion $C_j$ j = 1, 2, …, m with the help of Eq. (20.6). Establish the fuzzy matrix $\tilde{D}$ with i alternative and j criteria, as shown in Eq. (20.7).

$$\tilde{f}_{ij} = \sum_{k=1}^{K} \lambda_k a_{ij}^k, \sum_{k=1}^{K} \lambda_k b_{ij}^k, \sum_{k=1}^{K} \lambda_k c_{ij}^k \tag{20.6}$$

where $a_j$, $b_j$, $c_j$ is the *k*th largest element of the DMs judgments.

$$\tilde{D} = \begin{bmatrix} \tilde{f}_{11} & \tilde{f}_{12} & \cdots & \tilde{f}_{1j} \\ \tilde{f}_{21} & \tilde{f}_{22} & & \tilde{f}_{2j} \\ \vdots & \vdots & \ddots & \vdots \\ \tilde{f}_{i1} & \tilde{f}_{i2} & \cdots & \tilde{f}_{mn} \end{bmatrix} \tag{20.7}$$

**Step 7**. Compute the values of $\tilde{f}_j^*$, the best values of benefit criteria and worst values of cost criteria and $\tilde{f}_j^-$, the worst values of benefit criteria and best values of cost criteria, as in Eqs. (20.8) and (20.9).

$$\text{Benefit criteria} \quad \tilde{f}_j^* = \max_j \tilde{f}_{ij}, \quad \tilde{f}_j^- = \min_j \tilde{f}_{ij} \tag{20.8}$$

$$\text{Cost Criteria} \quad \tilde{f}_j^* = \min_j \tilde{f}_{ij}, \quad \tilde{f}_j^- = \max_j \tilde{f}_{ij} \tag{20.9}$$

**Step 8**. Calculate the values of $\tilde{S}_j$ and $\tilde{R}_j$ using Eqs. (20.10) and (20.11).

$$\tilde{S}_i = \sum_{j=1}^{n} \left[ \tilde{w}_j \left( \tilde{f}_j^* - \tilde{f}_{ij}^* \right) / \left( \tilde{f}_j^* - \tilde{f}_j^- \right) \right] \tag{20.10}$$

$$\tilde{R}_i = \max_j \left[ \tilde{w}_j \left( \tilde{f}_j^* - \tilde{f}_{ij}^* \right) / \left( \tilde{f}_j^* - \tilde{f}_j^- \right) \right] \tag{20.11}$$

Here, $\tilde{S}_j$ and $\tilde{R}_j$ are used for ranking "group utility" and the "individual regret", respectively.

**Step 9**. Compute the values $\tilde{Q}_i$ using the $\tilde{S}_i^*$, $\tilde{S}_i^-$ and $\tilde{R}_i^*$, $\tilde{R}_i^-$ values.

$$\tilde{S}_i^* = \min_i \tilde{S}_i, \quad \tilde{S}_i^- = \max_i \tilde{S}_i \tag{20.12}$$

$$\tilde{R}_i^* = \min_i \tilde{R}_i, \quad \tilde{R}_i^- = \max_i \tilde{R}_i \tag{20.13}$$

$$\tilde{Q}_i = \upsilon\left(\tilde{S}_i - \tilde{S}_i^*\right) / \left(\tilde{S}_i^- - \tilde{S}_i^*\right) + (1 - \upsilon)\left(\tilde{R}_i - \tilde{R}_i^*\right) / \left(\tilde{R}_i^- - \tilde{R}_i^*\right) \tag{20.14}$$

Here, "$\upsilon$" is introduced as a weight of the strategy of "the majority of criteria", as proposed in the VIKOR method. Usually $\upsilon$ value is taken as 0.5.

**Step 10**. Determine the ranking of the alternatives. De-fuzzify the triangular fuzzy numbers into crisp numbers by using Eq. (20.15), as proposed by Opricovic and Tzeng (2003).

$$\frac{a + b + c}{3} \tag{20.15}$$

Alternatives are ranked by sorting each of the de-fuzzified $S_i$, $R_i$, and $Q_i$ index values. These $S_i$, $R_i$, and $Q_i$ index values are sorted in an increasing order, as proposed in the original VIKOR method (Opricovic 1998). The result is a set of three ranking lists denoted as $S_{[\cdot]}$, $R_{[\cdot]}$ and $Q_{[\cdot]}$.

**Step 11**. The alternative $i_1$ corresponding to $Q_{[1]}$ (the smallest among $Q_i$ values) is proposed as a single compromise solution if;

1. Alternative $i_1$ has an acceptable advantage:
   $Q_{[2]} - Q_{[1]} \geq DQ$ where $DQ = 1/(n - 1)$ and n is the number of the alternatives.
2. Alternative $i_1$ is stable, i.e. it is also the best ranked in $S_{[\cdot]}$, $R_{[\cdot]}$.

If one of the above conditions is unmet, then a set of compromise solutions is proposed:

(a) Alternatives $i_1$ and $i_2$, where $Q_{i2} = Q_{[2]}$, if only the 2nd condition is not met; or,
(b) Alternatives $i_1$, …, $i_z$, if the 1st condition is not met. Here, $i_z$ is determined by the relation $Q_{[z]} - Q_{[1]} < DQ$ for the maximum z where $Q_{iz} = Q_{[z]}$.

## 20.4 Case Study

The proposed framework is applied to a case study with 3 energy projects, where experts are provided with detailed project data to prevent evaluations to be affected by prejudices or common opinions. In this case study, an energy project portfolio from Turkey is compared to understand the extent of how sustainable they are. Alternatives from Turkey with similar installed capacities, in megawatts (MW), are chosen to facilitate balanced comparison, without any other selection factor. The energy projects are presented in Table 20.4.

Four experts evaluated the weights of the evaluation criteria weights. Then, the same experts scored the three alternative projects according to the criteria. All four experts have experience in the Turkish energy sector between 5 and 20 years and are adequately qualified for this evaluation.

### *20.4.1 Implementation of the Proposed Methodology*

**Step 1**. A decision committee with 4 DMs ($\lambda_k$, $k = 1, 2, 3, 4$) are constructed and 3 alternatives ($A_i$, i = 1, 2, 3) and 3 main and 3 mid evaluation criteria with 33 sub-criteria are determined, as illustrated in Fig. 20.1.
**Step 2**. A six-point triangular fuzzy scale for DMs importance weights is used. For the alternatives and criteria, nine-point scales are preferred.
**Step 3**. The judgments of DMs are collected as linguistic evaluations for the DMs, criteria, and alternatives, as given in Tables 20.5, 20.6, and 20.7, respectively.
**Step 4**. DMs' weights, $\lambda_k$, are calculated after gathering the evaluations of the DMs. For instance, the second DM weighed the first DM as EI, the third DM as VI and the fourth DM as I. These evaluations are aggregated with Eq. (20.1), as shown below. All aggregated fuzzy values are illustrated in Table 20.8. Aggregated fuzzy values are de-fuzzified with Eq. (20.2) and normalized with Eq. (20.3).

**Table 20.4** Selected energy generation project alternatives

| # | Technology | Installed capacity (MW) | Location |
|---|---|---|---|
| $A_1$ | Wind power plant | 43.75 | Aegean region of Turkey |
| $A_2$ | Hydropower plant | 33.0 | Central Anatolia region of Turkey |
| $A_3$ | Landfill power plant | 42.3 | Marmara region of Turkey |

Table 20.5 DM's linguistic evaluations of DMs

| DM | 1 | 2 | 3 | 4 |
|---|---|---|---|---|
| 1 | | EI | EI | EI |
| 2 | VI | | VI | VI |
| 3 | VI | VI | | EI |
| 4 | MI | IM | IM | |

$$\tilde{\lambda}_2 = \begin{pmatrix} a = 0.50 = \frac{0.70+0.50+0.30}{3}, \\ b = 0.70 = \frac{0.90+0.70+0.50}{3}, \\ c = 0.87 = \frac{1.00+0.90+0.70}{3} \end{pmatrix}$$

$$\lambda_2' = \frac{0.50+0.70+0.87}{3} = 0.689$$

$$\lambda_2 = \frac{0.689}{2.767} = 0.249$$

**Step 5**. The aggregation of the fuzzy weight $\tilde{w}_j$ of criterion $C_j$, j = 1, 2, …, m, is accomplished with Eq. (20.4). An illustration of the "a" indices is shown below;

- $C_1$ = 0.30 * 0.60 + 0.25 * 0.50 + 0.25 * 0.40 + 0.20 * 0.30 = 0.465
- $C_{11}$ = 0.30 * 0.50 + 0.25 * 0.40 + 0.25 * 0.30 + 0.20 * 0.30 = 0.365
- $C_{111}$ = 0.30 * 0.80 + 0.25 * 0.60 + 0.25 * 0.50 + 0.20 * 0.50 = 0.615

The aggregated indices are multiplied to find the final weight. The first index of $C_{111}$ is calculated by 0.465 * 0.365 * 0.615 = 0.103. The final fuzzy output set of fuzzy weights are converted by de-fuzzification into crisp weights. The criteria weighting matrix, $W_j$, with its 33 criteria is established in Table 20.9.

**Step 6**. Aggregation of the fuzzy rating $\tilde{f}_{ij}$ of alternative $A_i$ under criterion $C_j$ is accomplished with Eq. (20.6). An illustration is displayed below for the first alternative's first indices;

$f_{11}$ = 0.29 * 8 + 0.25 * 8 + 0.25 * 6 + 0.21 * 6 = 7.08

**Step 7**. $\tilde{f}_j^*$ values and $\tilde{f}_j^-$ values are calculated with the Eqs. (20.8) and (20.9).

**Step 8**. The values of $\tilde{S}_i$ and $\tilde{R}_i$ are calculated with Eqs. (20.10) and (20.11), as shown in Table 20.10.

**Step 9**. Computed $\tilde{Q}_i$ values are calculated with Eq. (20.14), using the values of $\tilde{S}_i^*$, $\tilde{S}_i^-$ and $\tilde{R}_i^*$, $\tilde{R}_i^-$ found by Eqs. (20.12) and (20.13). The Tables 20.11 and 20.12 show these values.

**Step 10**. The ranking order of alternatives is determined with the de-fuzzified triangular fuzzy numbers. At this step, the generalized mean of fuzzy number technique is used for the de-fuzzification. The triangular fuzzy number (a, b, c) are de-fuzzified with Eq. (20.15). An illustration for the first alternative is given below;

**Table 20.6** DMs' evaluations of alternatives

| | DM | $C_{111}$ | $C_{112}$ | $C_{113}$ | $C_{121}$ | $C_{122}$ | $C_{123}$ | $C_{131}$ | $C_{132}$ | $C_{133}$ | $C_{141}$ | $C_{142}$ |
|---|---|---|---|---|---|---|---|---|---|---|---|---|
| $A_1$ | 1 | VVG | VVG | VVG | G | VVG | VVG | VB | VVB | VB | G | VVG |
| | 2 | G | B | I | VVG | I | G | VVG | VVG | VB | VVG | VVG |
| | 3 | VVG | VVG | VVG | VVG | VVG | VVG | G | G | G | VVG | VVG |
| | 4 | G | VVG | VVG | VVG | VVG | VVG | G | G | VVG | VVG | VVG |
| $A_2$ | 1 | I | B | G | VVB | B | VB | VB | I | G | I | VVG |
| | 2 | G | B | I | VB | VVG | G | B | B | I | VVG | B |
| | 3 | G | B | I | VB | I | VB | VB | B | VB | B | I |
| | 4 | G | G | B | B | VVG | I | B | I | VB | VVG | VVG |
| $A_3$ | 1 | VVG | VVG | VVG | I | B | VVG | I | VVG | VVG | VVG | VVG |
| | 2 | VVG | VVG | VVG | VVG | B | VVG | VVG | VVG | VB | VVG | VVG |
| | 3 | G | VVG | G | I | I | I | I | I | G | G | VVG |
| | 4 | VVG | VVG | VVG | G | G | VVG | VVG | VVG | VVG | VVG | VVG |

| | DM | $C_{143}$ | $C_{144}$ | $C_{211}$ | $C_{212}$ | $C_{213}$ | $C_{214}$ | $C_{221}$ | $C_{222}$ | $C_{223}$ | $C_{231}$ | $C_{232}$ |
|---|---|---|---|---|---|---|---|---|---|---|---|---|
| $A_1$ | 1 | G | VVG | I | I | G | I | G | G | B | B | I |
| | 2 | I | I | I | G | I | G | G | VVG | VVG | VVG | VVG |
| | 3 | G | VVG | I | G | VVG | VVG | VVG | VVG | VVG | G | G |
| | 4 | VVG | VVG | I | G | G | VVG | G | VVG | VVG | G | G |
| $A_2$ | 1 | B | G | G | I | I | B | B | VB | B | B | B |
| | 2 | I | I | I | G | I | G | B | G | G | I | VVG |
| | 3 | B | I | G | I | VVG | I | VB | B | VB | VB | VB |
| | 4 | I | VVG | G | I | VVG | VVG | VB | VVG | G | B | G |
| $A_3$ | 1 | VVG | G | VVG | G | I | G | I | I | B | I | G |
| | 2 | I | I | I | G | I | G | G | VVG | VVG | VVG | VVG |
| | 3 | VVG | G | VVG | I | I | VVG | G | VVG | I | G | G |
| | 4 | G | VVG | VVG | I | I | VVG | G | VVG | VVG | G | G |

| | DM | $C_{233}$ | $C_{311}$ | $C_{312}$ | $C_{313}$ | $C_{314}$ | $C_{321}$ | $C_{322}$ | $C_{323}$ | $C_{331}$ | $C_{332}$ | $C_{333}$ |
|---|---|---|---|---|---|---|---|---|---|---|---|---|
| $A_1$ | 1 | I | B | VVG | I | G | G | G | G | I | G | G |
| | 2 | VVG | G | VVG | VVG | VVG | I | VVG | I | G | VVG | VVG |
| | 3 | G | VVG | VVG | G | G | I | G | I | G | VVG | VVG |
| | 4 | VVG | VVG | VVG | G | G | I | VVG | VVG | VVG | VVG | VVG |
| $A_2$ | 1 | VB | B | B | I | I | G | B | VB | G | B | VVB |
| | 2 | I | G | VVG | I | G | G | VVG | G | G | I | G |
| | 3 | I | G | VVB | VB | B | B | I | I | G | B | VVB |
| | 4 | G | I | VVG | I | I | VVG | VVG | VVG | VVG | G | I |
| $A_3$ | 1 | G | VVG | VVG | I | VVG | G | I | I | I | VVG | VVG |
| | 2 | VVG | I | G | VVG | VVG | B | G | I | G | VVG | VVG |
| | 3 | G | VVG | G | G | G | G | G | I | VVG | VVG | G |
| | 4 | G | VVG | G | VVG | G | B | G | VVG | VVG | VVG | VVG |

**Table 20.7** DMs' importance weights for criteria

| DM | $C_1$ | $C_2$ | $C_3$ | $C_{11}$ | $C_{12}$ | $C_{13}$ | $C_{14}$ | $C_{21}$ | $C_{22}$ | $C_{23}$ | $C_{31}$ | $C_{32}$ | $C_{33}$ |
|---|---|---|---|---|---|---|---|---|---|---|---|---|---|
| 1 | MH | MH | M | ML | MH | VL | ML | H | H | L | VH | M | ML |
| 2 | ML | ML | VVH | L | M | M | L | ML | H | MH | VH | VL | H |
| 3 | M | L | VVH | MH | VVL | MH | L | M | VH | ML | M | ML | VH |
| 4 | H | VL | VH | M | L | M | L | VH | H | VL | MH | L | VH |

| DM | $C_{111}$ | $C_{112}$ | $C_{113}$ | $C_{121}$ | $C_{122}$ | $C_{123}$ | $C_{131}$ | $C_{132}$ | $C_{133}$ | $C_{141}$ | $C_{142}$ | $C_{143}$ | $C_{144}$ |
|---|---|---|---|---|---|---|---|---|---|---|---|---|---|
| 1 | H | MH | ML | H | ML | MH | M | VVH | L | M | L | L | M |
| 2 | VVH | H | VVL | M | H | M | VVH | ML | ML | L | L | L | H |
| 3 | MH | MH | M | H | M | M | H | MH | ML | M | VVL | VH | VL |
| 4 | MH | MH | M | VVH | VVL | H | H | H | L | M | MH | ML | VVL |

| DM | $C_{231}$ | $C_{232}$ | $C_{233}$ | $C_{221}$ | $C_{222}$ | $C_{223}$ | $C_{211}$ | $C_{212}$ | $C_{213}$ | $C_{214}$ |
|---|---|---|---|---|---|---|---|---|---|---|
| 1 | MH | ML | H | H | H | L | VL | ML | H | L |
| 2 | MH | M | MH | M | VVH | L | VL | L | VVH | VL |
| 3 | VVH | VVL | H | H | M | M | ML | VVL | H | ML |
| 4 | M | H | M | M | VII | ML | L | M | M | L |

| DM | $C_{331}$ | $C_{332}$ | $C_{333}$ | $C_{321}$ | $C_{322}$ | $C_{323}$ | $C_{311}$ | $C_{312}$ | $C_{313}$ | $C_{314}$ |
|---|---|---|---|---|---|---|---|---|---|---|
| 1 | ML | MH | H | MH | ML | H | ML | M | VL | M |
| 2 | M | M | H | VVH | H | VVL | H | M | VVL | L |
| 3 | H | M | M | H | H | L | L | VVH | VVL | L |
| 4 | VH | M | ML | VVH | M | L | M | MH | L | VL |

**Table 20.8** DMs' weights

| DM | | Weights |
|---|---|---|
| 1 | {0.40 0.57 0.77} | 0.209 |
| 2 | {0.50 0.70 0.87} | 0.249 |
| 3 | {0.50 0.70 0.87} | 0.249 |
| 4 | {0.63 0.83 0.97} | 0.293 |

$$Qi = \frac{0.454 + 0.452 + 0.451}{3} = 0.452$$

The energy project alternatives are ranked by sorting in an increasing order, as instructed in the VIKOR method. The result is a set of three ranking lists denoted as $S_{[\cdot]}$, $R_{[\cdot]}$ and $Q_{[\cdot]}$. The Table 20.13 displays the calculated values.

**Step 11**. The alternative $A_3$ corresponds to the smallest among $Q_i$ values. It is the proposed compromise solution if alternative $A_3$ had an acceptable advantage, in other words, $0.455 - 0.015 < 0.5$ where $DQ = 1/(3 - 1)$. However, $A_3$ is not

**Table 20.9** Criteria weights

| 1st level criteria | | 2nd level criteria | | | 3rd level criteria | | Overall |
|---|---|---|---|---|---|---|---|
| $C_1$ | 0.32 | $C_{11}$ | 0.27 | $C_{111}$ | Air emissions | 0.41 | 0.035 |
| | | | | $C_{112}$ | Effluents and waste | 0.36 | 0.031 |
| | | | | $C_{113}$ | Other pollutants | 0.23 | 0.020 |
| | | $C_{12}$ | 0.24 | $C_{121}$ | Water consumption | 0.41 | 0.031 |
| | | | | $C_{122}$ | Energy use | 0.26 | 0.020 |
| | | | | $C_{123}$ | Raw material use | 0.34 | 0.026 |
| | | $C_{13}$ | 0.29 | $C_{131}$ | Ecosystem vulnerability | 0.4 | 0.037 |
| | | | | $C_{132}$ | Critical habitats | 0.4 | 0.037 |
| | | | | $C_{133}$ | Sustainable farming | 0.2 | 0.019 |
| | | $C_{14}$ | 0.20 | $C_{141}$ | Man. systems and programs | 0.27 | 0.017 |
| | | | | $C_{142}$ | Regulatory compliance | 0.21 | 0.013 |
| | | | | $C_{143}$ | Environmental governance | 0.28 | 0.018 |
| | | | | $C_{144}$ | Supply chain management | 0.24 | 0.015 |
| $C_2$ | 0.23 | $C_{21}$ | 0.35 | $C_{211}$ | Job creation | 0.17 | 0.014 |
| | | | | $C_{212}$ | Job satisfaction and quality | 0.21 | 0.017 |
| | | | | $C_{213}$ | Occupational health and safety | 0.43 | 0.035 |
| | | | | $C_{214}$ | Training and education | 0.19 | 0.015 |
| | | $C_{22}$ | 0.42 | $C_{221}$ | Social acceptance | 0.35 | 0.034 |
| | | | | $C_{222}$ | Public health and safety | 0.42 | 0.041 |
| | | | | $C_{223}$ | Cultural properties | 0.23 | 0.022 |
| | | $C_{23}$ | 0.23 | $C_{231}$ | Stakeholder relations | 0.38 | 0.020 |
| | | | | $C_{232}$ | Corporate social responsibility | 0.26 | 0.014 |
| | | | | C233 | Principals and commitment | 0.36 | 0.019 |
| $C_3$ | 0.45 | $C_{31}$ | 0.39 | $C_{311}$ | Technology and innovation | 0.3 | 0.053 |
| | | | | $C_{312}$ | Supply chain resilience | 0.39 | 0.068 |
| | | | | $C_{313}$ | Brand value | 0.11 | 0.019 |
| | | | | $C_{314}$ | Corporate governance | 0.21 | 0.037 |
| | | $C_{32}$ | 0.21 | $C_{321}$ | Initial and fixed costs | 0.45 | 0.043 |
| | | | | $C_{322}$ | Running and variable costs | 0.34 | 0.032 |
| | | | | $C_{323}$ | Depletion and rehab. costs | 0.22 | 0.021 |
| | | $C_{33}$ | 0.4 | $C_{331}$ | Shareholder interests | 0.36 | 0.065 |
| | | | | $C_{332}$ | Value creation | 0.31 | 0.056 |
| | | | | $C_{333}$ | Impact of sustainable practices | 0.34 | 0.061 |

**Table 20.10** $\tilde{S}_i$ and $\tilde{R}_i$ values

| | $\tilde{S}_i$ | | | $\tilde{R}_i$ | | |
|---|---|---|---|---|---|---|
| $A_1$ | {0.591 | 1.063 | 1.745} | {0.211 | 0.338 | 0.506} |
| $A_2$ | {2.308 | 4.330 | 7.178} | {0.220 | 0.349 | 0.521} |
| $A_3$ | {0.618 | 1.166 | 1.911} | {0.125 | 0.226 | 0.366} |

**Table 20.11** $\tilde{S}_i^*$, $\tilde{S}_i^-$ and $\tilde{R}_i^*$, $\tilde{R}_i^-$ values

| | | | |
|---|---|---|---|
| $\tilde{S}_i^-$ | {0.591 1.063 1.745} | $\tilde{R}_i^-$ | {0.125 0.226 0.366} |
| $\tilde{S}_i^*$ | {2.308 4.330 7.178} | $\tilde{R}_i^*$ | {0.220 0.349 0.521} |

**Table 20.12** $\tilde{Q}_i$ values

| |
|---|
| {0.454 0.452 0.451} |
| {1.000 1.000 1.000} |
| {0.008 0.016 0.015} |

**Table 20.13** Ranking lists for Q, S, and R

| Alternative | | $Q_i$ | | $S_i$ | | $R_i$ | |
|---|---|---|---|---|---|---|---|
| | | Index | Rank | Index | Rank | Index | Rank |
| $A_1$ | Wind farm | 0.452 | 2 | 1.133 | 1 | 0.352 | 2 |
| $A_2$ | Hydropower plant | 1.000 | 3 | 4.682 | 3 | 0.364 | 3 |
| $A_3$ | Landfill power plant | 0.013 | 1 | 1.232 | 2 | 0.239 | 1 |

ranked best in $S_{[\cdot]}$. Since one of the two conditions in Step 9 is not satisfied, a set of compromise solutions consisting of the alternatives $A_3$ and $A_1$ is proposed.

## 20.5 Discussion of Results

An analytical decision support method is proposed for evaluating and ranking concretely defined energy generation projects from a sustainability perspective. This integrated method is then applied to a case in Turkey, where a group of experts assessed three actual project alternatives. The results of the case study offer powerful insight about the sustainability performance of energy generation projects, where A1, the landfill power plant, is proposed together with A3, the wind farm, as the most sustainable compromise set of project alternatives. This ranking is discussed with the DMs and it is observed that the results are in agreement with DMs' expectations. As a comparative perspective, this result is also in general agreement with energy technology selection literature. The wind, biogas and landfill technologies are frequently preferred in similar rankings over hydropower plants (Kahraman et al. 2009; Kaya and Kahraman 2010, 2011; Büyüközkan and Karabulut 2017).

According to this case study, the most important sustainability aspect is economic, followed by environmental and social aspects. Under economic aspects, $C_{33}$, Financial Performance, and $C_{31}$, Business Continuity, are deemed to be the most influential factors in deciding on the energy generation project. Located under these criteria, the overall most important sub-criteria with a global priority more than 6% are identified as $C_{312}$, Supply Chain Resilience, $C_{331}$, Shareholder interests, and $C_{333}$, Impact on sustainable practices. For the environmental criteria, $C_{111}$, Air Emissions, $C_{131}$, Ecosystem Vulnerability, and $C_{132}$, Critical Habitats, stand out with global weights exceeding 3.5% each. For the social criteria, the experts indicate that $C_{213}$, Occupational Health and Safety, and $C_{221}$, Social Acceptance, are the most important sub-criteria with global weights more than 3% each, as Table 20.9 displays. It should be noted that these weights are not limited to the alternative energy generation projects in the case study and can be used in other energy generation project selection problems as well.

The reason of using the fuzzy set theory is its ability to gather DMs' linguistic opinions, instead of exact numerical values, so that the ambiguity and uncertainty of human judgments can be better captured and reduced. Criteria weights are evaluated with this method in a GDM environment, which is advantageous over a single DM for reducing individual bias and insufficient knowledge on certain aspects of decision problems. VIKOR is preferred thanks to its strength of coming up with compromise set solutions, a preferable feature in decision problems with subjective criteria, such as sustainability. Input for VIKOR is collected from DMs as linguistic input, which are then aggregated with GDM and de-fuzzified. The case study indicates that this combined Fuzzy VIKOR-based framework generates plausible results. Another benefit of this method is that the DMs are not asked to make pair-wise comparisons, such as in AHP, which is cognitively demanding and time consuming as the number of criteria increases.

This framework can be used by energy investors for effective and data-based assessment of well-defined energy generation projects, instead of generic technology comparisons that rely on individuals' generalizations and common opinions. Energy project selection is a significant decision problem for investors, which requires the consideration of many different sustainability aspects. The proposed method successfully integrates economic, environmental and social factors into this decision process. It can guide business managers and researchers in reaching realistic and practical judgments about sustainability performance of energy projects that are in operation, as well as estimating the level of the sustainability of those projects that are under planning or construction. The criteria cover a wide range of impacts, which can also help developers of energy generation projects to take the necessary precautions for improving their projects' sustainability performance. It can also be useful to researchers for better understanding the underlying sustainability components of energy project selection problem. The energy generation project selection model presented in this chapter can be used in different geographies for smaller or larger concrete projects, as long as they are of comparable size.

## 20.6 Conclusion

Energy is essential for economic development, environmental protection and social agenda. Energy generation is expected to be affordable, reliable, economically viable, environmentally friendly and socially acceptable. Therefore, those energy projects that satisfy these expectations shall be prioritized. This raises the question of how to evaluate energy generation projects in terms of sustainability in a holistic way.

This chapter presented a combined Fuzzy VIKOR-based framework for an effective energy project selection from a sustainability point of view. The 4-level criteria structure originally proposed by Büyüközkan and Karabulut (2017) consisting of 3 main dimensions, 12 criteria, and 37 sub-criteria is integrated with two MCDM techniques, the fuzzy set theory, and VIKOR, to evaluate this set of multilevel criteria. An empirical case from Turkey is studied to demonstrate the usability of the proposed framework for identifying the most sustainable energy generation project from a set of alternative projects. Three projects are compared with the help of four industry experts. The analysis resulted in a compromise solution set, consisting of the wind farm and landfill power plant projects. This suggests that the selected hydropower plant project is not sustainable enough, while this implication shall not be simply extended and generalized to its technology, considering that this evaluation is carried out for specific projects.

The article is original for different reasons. The work by Büyüközkan and Karabulut (2017) is extended by making use of their criteria structure but integrating with another technique, fuzzy set theory, to allow for uncertain and imprecise evaluation data. This allows linguistic evaluations to be gathered from experts, which is more natural and easier for human way of thinking. Thus, the major scientific contribution of the proposed framework is its ability to select specific energy generation projects, instead of generic energy technologies, on the ground of their sustainability impacts with linguistic input.

The proposed Fuzzy VIKOR-based framework can be helpful to researchers and businesses to better understand project-based evaluation and design more sustainable energy generation projects in a more structured manner. It allows DMs to provide their opinions linguistically to mimic natural thinking, without having to voice their judgments with numerical values. The method can also be applied to smaller, larger or more projects in other geographies, as long as sufficient sustainability information is available to DMs. This approach can guide energy generation project investors to identify acceptable compromises in their project prioritization problems.

In terms of limitations, the interactions among evaluation criteria are not considered. One of the perspectives for future research is then to consider the inner dependence of criteria and their interactions. Integration of the ANP technique (Saaty 2008) can be a possible solution to extend this work.

**Acknowledgements** The authors would like to express their sincere gratitude to the experts for their invaluable support in the evaluation. This research was supported by Galatasaray University Research Fund (Projects number: 17.402.004 and 17.402.009).

## References

Abdullah, L., & Najib, L. (2016). Sustainable energy planning decision using the intuitionistic fuzzy analytic hierarchy process: Choosing energy technology in Malaysia. *International Journal of Sustainable Energy, 35,* 360–377. https://doi.org/10.1080/14786451.2014.907292.

Atmaca, E., & Basar, H. B. (2012). Evaluation of power plants in Turkey using Analytic Network Process (ANP). *Energy, 44,* 555–563. https://doi.org/10.1016/j.energy.2012.05.046.

Bahadır, B., & Büyüközkan, G. (2016). Robot selection for warehouses. In *Proceeding of LM SCM 2016 Conference*, p. 341.

Balin, A., & Baraçli, H. (2017). A fuzzy multi-criteria decision making methodology based upon the interval Type-2 fuzzy sets for evaluating renewable energy alternatives in Turkey. *Technological and Economic Development of Economy, 23,* 742–763. https://doi.org/10.3846/20294913.2015.1056276.

Begić, F., & Afgan, N. H. (2007). Sustainability assessment tool for the decision making in selection of energy system—Bosnian case. *Energy, 32,* 1979–1985. https://doi.org/10.1016/j.energy.2007.02.006.

Burton, J., & Hubacek, K. (2007). Is small beautiful? A multicriteria assessment of small-scale energy technology applications in local governments. *Energy Policy, 35,* 6402–6412. https://doi.org/10.1016/j.enpol.2007.08.002.

Büyüközkan, G., & Güleryüz, S. (2014). A new GDM based AHP framework with linguistic interval fuzzy preference relations for renewable energy planning. *Journal of Intelligent & Fuzzy Systems, 27,* 3181–3195.

Büyüközkan, G., & Güleryüz, S. (2016). An integrated DEMATEL-ANP approach for renewable energy resources selection in Turkey. *International Journal of Production Economics, 182,* 435–448. https://doi.org/10.1016/j.ijpe.2016.09.015.

Büyüközkan, G., & Güleryüz, S. (2017). Evaluation of renewable energy resources in Turkey using an integrated MCDM approach with linguistic interval fuzzy preference relations. *Energy, 123,* 149–163. https://doi.org/10.1016/j.energy.2017.01.137.

Büyüközkan, G., & Karabulut, Y. (2017). Energy project performance evaluation with sustainability perspective. *Energy, 119,* 549–560. https://doi.org/10.1016/j.energy.2016.12.087.

Büyüközkan, G., & Ruan, D. (2008). Evaluation of software development projects using a fuzzy multi-criteria decision approach. *Math Comput Simul, 77,* 464–475. https://doi.org/10.1016/j.matcom.2007.11.015.

Cavallaro, F., & Ciraolo, L. (2005). A multicriteria approach to evaluate wind energy plants on an Italian island. *Energy Policy, 33,* 235–244. https://doi.org/10.1016/S0301-4215(03)00228-3.

Chang, T.-H. (2014). Fuzzy VIKOR method: A case study of the hospital service evaluation in Taiwan. *Information Sciences, 271,* 196–212. https://doi.org/10.1016/j.ins.2014.02.118.

Chatzimouratidis, A. I., & Pilavachi, P. A. (2007). Objective and subjective evaluation of power plants and their non-radioactive emissions using the analytic hierarchy process. *Energy Policy, 35,* 4027–4038. https://doi.org/10.1016/j.enpol.2007.02.003.

Chatzimouratidis, A. I., & Pilavachi, P. A. (2008). Multicriteria evaluation of power plants impact on the living standard using the analytic hierarchy process. *Energy Policy, 36,* 1074–1089. https://doi.org/10.1016/j.enpol.2007.11.028.

Chatzimouratidis, A. I., & Pilavachi, P. A. (2009). Technological, economic and sustainability evaluation of power plants using the Analytic Hierarchy Process. *Energy Policy, 37,* 778–787. https://doi.org/10.1016/j.enpol.2008.10.009.

Chen, C.-T. (2000). Extensions of the TOPSIS for group decision-making under fuzzy environment. *Fuzzy Sets and Systems, 114,* 1–9. https://doi.org/10.1016/S0165-0114(97)00377-1.

Çolak, M., & Kaya, İ. (2017). Prioritization of renewable energy alternatives by using an integrated fuzzy MCDM model: A real case application for Turkey. *Renewable and Sustainable Energy Reviews, 80,* 840–853. https://doi.org/10.1016/j.rser.2017.05.194.

Erdogan, M., & Kaya, I. (2015). An integrated multi-criteria decision-making methodology based on type-2 fuzzy sets for selection among energy alternatives in Turkey. *Iran J Fuzzy Syst, 12,* 1–25.

Foroozesh, N., Gitinavard, H., Mousavi, S. M., & Vahdani, B. (2017). A hesitant fuzzy extension of VIKOR method for evaluation and selection problems under uncertainty. *International Journal of Applied Management Science, 9,* 95–113. https://doi.org/10.1504/IJAMS.2017.084946.

Garg, R. K., Agrawal, V. P., & Gupta, V. K. (2007). Coding, evaluation and selection of thermal power plants—A MADM approach. *International Journal of Electrical Power & Energy Systems, 29,* 657–668. https://doi.org/10.1016/j.ijepes.2006.08.002.

Grilli, G., Meo, I. D., Garegnani, G., & Paletto, A. (2017). A multi-criteria framework to assess the sustainability of renewable energy development in the Alps. *Journal of Environmental Planning and Management, 60,* 1276–1295. https://doi.org/10.1080/09640568.2016.1216398.

Hsueh, S.-L., & Yan, M.-R. (2013). A multimethodology contractor assessment model for facilitating green innovation: The view of energy and environmental protection. *Scientific World Journal, 2013,* e624340. https://doi.org/10.1155/2013/624340.

IBRD, & World Bank. (2017). *State of electricity* (Access Report 2017).

Ishizaka, A., Siraj, S., & Nemery, P. (2016). Which energy mix for the UK (United Kingdom)? An evolutive descriptive mapping with the integrated GAIA (graphical analysis for interactive aid)–AHP (analytic hierarchy process) visualization tool. *Energy, 95,* 602–611. https://doi.org/10.1016/j.energy.2015.12.009.

Kabak, M., & Dağdeviren, M. (2014). Prioritization of renewable energy sources for Turkey by using a hybrid MCDM methodology. *Energy Conversion and Management, 79,* 25–33. https://doi.org/10.1016/j.enconman.2013.11.036.

Kahraman, C., Kaya, İ., & Cebi, S. (2009). A comparative analysis for multiattribute selection among renewable energy alternatives using fuzzy axiomatic design and fuzzy analytic hierarchy process. *Energy, 34,* 1603–1616. https://doi.org/10.1016/j.energy.2009.07.008.

Kaya, T., & Kahraman, C. (2010). Multicriteria renewable energy planning using an integrated fuzzy VIKOR & AHP methodology: The case of Istanbul. *Energy, 35,* 2517–2527. https://doi.org/10.1016/j.energy.2010.02.051.

Kaya, T., & Kahraman, C. (2011). Multicriteria decision making in energy planning using a modified fuzzy TOPSIS methodology. *Expert Systems with Applications, 38,* 6577–6585. https://doi.org/10.1016/j.cswa.2010.11.081.

Köne, A. Ç., & Büke, T. (2007). An analytical network process (ANP) evaluation of alternative fuels for electricity generation in Turkey. *Energy Policy, 35,* 5220–5228. https://doi.org/10.1016/j.enpol.2007.05.014.

Kowalski, K., Stagl, S., Madlener, R., & Omann, I. (2009). Sustainable energy futures: Methodological challenges in combining scenarios and participatory multi-criteria analysis. *European Journal of Operational Research, 197,* 1063–1074. https://doi.org/10.1016/j.ejor.2007.12.049.

Krukanont, P., & Tezuka, T. (2007). Implications of capacity expansion under uncertainty and value of information: The near-term energy planning of Japan. *Energy, 32,* 1809–1824. https://doi.org/10.1016/j.energy.2007.02.003.

Kuleli Pak, B., Albayrak, Y. E., & Erensal, Y. C. (2017). Evaluation of sources for the sustainability of energy supply in Turkey. *Environment Progress & Sustainable Energy, 36,* 627–637. https://doi.org/10.1002/ep.12507.

Lee, A. H. I., Chen, H. H., & Kang, H.-Y. (2009). Multi-criteria decision making on strategic selection of wind farms. *Renewable Energy, 34,* 120–126. https://doi.org/10.1016/j.renene.2008.04.013.

Liu, H.-C., You, J.-X., Chen, Y.-Z., & Fan, X.-J. (2014). Site selection in municipal solid waste management with extended VIKOR method under fuzzy environment. *Environmental Earth Sciences, 72,* 4179–4189. https://doi.org/10.1007/s12665-014-3314-6.

Mandal, S., Singh, K., Behera, R. K., et al. (2015). Human error identification and risk prioritization in overhead crane operations using HTA, SHERPA and fuzzy VIKOR method. *Expert Systems with Applications, 42,* 7195–7206. https://doi.org/10.1016/j.eswa.2015.05.033.

Nixon, J. D., Dey, P. K., Davies, P. A., et al. (2014). Supply chain optimisation of pyrolysis plant deployment using goal programming. *Energy, 68,* 262–271. https://doi.org/10.1016/j.energy.2014.02.058.

Nixon, J. D., Dey, P. K., Ghosh, S. K., & Davies, P. A. (2013). Evaluation of options for energy recovery from municipal solid waste in India using the hierarchical analytical network process. *Energy, 59,* 215–223. https://doi.org/10.1016/j.energy.2013.06.052.

Onar, S. C., Oztaysi, B., Otay, İ., & Kahraman, C. (2015). Multi-expert wind energy technology selection using interval-valued intuitionistic fuzzy sets. *Energy, 90,* 274–285. https://doi.org/10.1016/j.energy.2015.06.086.

Önüt, S., Tuzkaya, U. R., & Saadet, N. (2008). Multiple criteria evaluation of current energy resources for Turkish manufacturing industry. *Energy Conversion and Management, 49,* 1480–1492. https://doi.org/10.1016/j.enconman.2007.12.026.

Opricovic, S. (1998). Multi-criteria optimization of civil engineering systems. Faculty of Civil Engineering, Belgrade. (Table II Perform Matrix).

Opricovic, S., & Tzeng, G.-H. (2004). Compromise solution by MCDM methods: A comparative analysis of VIKOR and TOPSIS. *European Journal of Operational Research, 156,* 445–455. https://doi.org/10.1016/S0377-2217(03)00020-1.

Opricovic, S., & Tzeng, G.-H. (2003). Defuzzification within a multicriteria decision model. *International Journal of Uncertainty, Fuzziness and Knowledge-Based Systems, 11,* 635–652. https://doi.org/10.1142/S0218488503002387.

Özkale, C., Celik, C., Turkmen, A. C., & Cakmaz, E. S. (2017). Decision analysis application intended for selection of a power plant running on renewable energy sources. *Renewable and Sustainable Energy Reviews, 70,* 1011–1021. https://doi.org/10.1016/j.rser.2016.12.006.

Pilavachi, P. A., Stephanidis, S. D., Pappas, V. A., & Afgan, N. H. (2009). Multi-criteria evaluation of hydrogen and natural gas fuelled power plant technologies. *Applied Thermal Engineering, 29,* 2228–2234. https://doi.org/10.1016/j.applthermaleng.2008.11.014.

Polatidis, H., & Haralambopoulos, D. (2004). Local renewable energy planning: A participatory multi-criteria approach. *Energy Sources, 26,* 1253–1264. https://doi.org/10.1080/00908310490441584.

Rostamzadeh, R., Govindan, K., Esmaeili, A., & Sabaghi, M. (2015). Application of fuzzy VIKOR for evaluation of green supply chain management practices. *Ecol Indic, 49,* 188–203. https://doi.org/10.1016/j.ecolind.2014.09.045.

Saaty, T. L. (2008). Decision making with the analytic hierarchy process. *International Journal of Services Sciences, 1*(1), 83–98. https://doi.org/10.1504/IJSSci.2008.01759.

San Cristóbal, J. R. (2011). Multi-criteria decision-making in the selection of a renewable energy project in Spain: The Vikor method. *Renew Energy, 36,* 498–502. https://doi.org/10.1016/j.renene.2010.07.031.

Şengül, Ü., Eren, M., Eslamian Shiraz, S., et al. (2015). Fuzzy TOPSIS method for ranking renewable energy supply systems in Turkey. *Renewable Energy, 75,* 617–625. https://doi.org/10.1016/j.renene.2014.10.045.

Sofiyabadi, J., Kolahi, B., & Valmohammadi, C. (2016). Key performance indicators measurement in service business: A fuzzy VIKOR approach. *Total Quality Management & Business Excellence, 27,* 1028–1042. https://doi.org/10.1080/14783363.2015.1059272.

Streimikiene, D., Balezentis, T., Krisciukaitienė, I., & Balezentis, A. (2012). Prioritizing sustainable electricity production technologies: MCDM approach. *Renewable and Sustainable Energy Reviews, 16,* 3302–3311. https://doi.org/10.1016/j.rser.2012.02.067.

Tsoutsos, T., Drandaki, M., Frantzeskaki, N., et al. (2009). Sustainable energy planning by using multi-criteria analysis application in the island of Crete. *Energy Policy, 37,* 1587–1600. https://doi.org/10.1016/j.enpol.2008.12.011.

United Nations. (2015). Sustainable development goals. In *U. N. sustainable development* http://www.un.org/sustainabledevelopment/sustainable-development-goals/. Accessed July 21, 2017.

Vinodh, S., Varadharajan, A. R., & Subramanian, A. (2013). Application of fuzzy VIKOR for concept selection in an agile environment. *International Journal of Advanced Manufacturing Technology, 65,* 825–832. https://doi.org/10.1007/s00170-012-4220-2.

Wang, J.-J., Jing, Y.-Y., Zhang, C.-F., & Zhao, J.-H. (2009). Review on multi-criteria decision analysis aid in sustainable energy decision-making. *Renewable and Sustainable Energy Reviews, 13,* 2263–2278. https://doi.org/10.1016/j.rser.2009.06.021.

Wang, Y. (2017). A fuzzy VIKOR approach for renewable energy resources selection in China. *Revista de la Facultad de Ingeniería, 31*(10).

World Bank. (2017). *Energy overview*. http://www.worldbank.org/en/topic/energy/overview#1. Accessed April 24, 2017.

Wu, Z., Ahmad, J., & Xu, J. (2016). A group decision making framework based on fuzzy VIKOR approach for machine tool selection with linguistic information. *Applied Soft Computing, 42,* 314–324. https://doi.org/10.1016/j.asoc.2016.02.007.

Yager, R. R. (1988). On ordered weighted averaging aggregation operators in multicriteria decisionmaking. *IEEE Trans Syst Man Cybern, 18,* 183–190. https://doi.org/10.1109/21.87068.

Zadeh, L. A. (1965). Fuzzy sets. *Information and Control, 8,* 338–353.

Zhang, L., Zhou, P., Newton, S., et al. (2015). Evaluating clean energy alternatives for Jiangsu, China: An improved multi-criteria decision making method. *Energy 90: Part, 1,* 953–964. https://doi.org/10.1016/j.energy.2015.07.124.

# Part VII
# Collective Intelligence

# Chapter 21
# Review of Collective Intelligence Used in Energy Applications

**Gülgün Kayakutlu and Secil Ercan**

**Abstract** Low-Carbon Economy policies drive Europe for an integrated approach for utility consumption and management; number of integrated distribution companies are increasing. This new trend will soon cause the need for group decisions, collective intelligence approach to the energy industry. This study aims to review the collective intelligence concepts and methods to give a summary of collective intelligence use in energy applications. It can be considered as a foundation for the future of collective intelligence in the energy industry.

## 21.1 Introduction

New benchmarks on integrated utility industry of Europe shows positive results in price reduction and energy saving (Georges et al. 2017). Solar and wind installations in Germany and extended nuclear installations in France are leading the positive impact on aggressive policies for a low-carbon economy in Europe. The more integrated the utility industries are, the more need will arise for group decisions through a variety of dimensions. We believe this will cause the wider use of collective intelligence in the energy field.

A recent development in the knowledge management field is represented by collective intelligence to find better solutions for current social and cultural challenges (Elia et al. 2014). In only a few decades the nature inspired collective intelligence methods have proven their value (Król and Lopes 2012). The growth of network economy has steered for the success of collective approaches. Wikipedia and Social media sites have been the sparkling results in information and communication technologies. These methods achieved results representing collaborative work provided by integrated but competitive brains in several industries.

G. Kayakutlu (✉)
ITU Ayazaga Campus, Energy Institute, 34469 Sariyer, İstanbul, Turkey
e-mail: kayakutlu@itu.edu.tr

S. Ercan
University of Reims Champagne Ardennes, Grand Est, France

© Springer International Publishing AG, part of Springer Nature 2018

C. Kahraman and G. Kayakutlu (eds.), *Energy Management—Collective and Computational Intelligence with Theory and Applications*, Studies in Systems, Decision and Control 149, https://doi.org/10.1007/978-3-319-75690-5_21

Using collective intelligence approaches have shown that "there could well be a solution out there somewhere, far outside of the traditional places" (Bonabeau 2009). In order to choose the right tool that fits for locating the solution needs the knowledge of concepts and the choices.

This article aims to review the collective intelligence concepts and methods besides giving the summary of using these methods for the energy applications. A brief on the future trends of using collective intelligence will also bring a new dimension for the readers. It is strongly believed that the reviews given in this chapter will give some hints for both the academicians and the practitioners in the energy field.

This paper is so organized that the definitions and the main concepts of the collective intelligence will be given in the following section. The third section is a short review of the methods and the fourth section is reserved for the energy applications. The last section will include the conclusions and results.

## 21.2 Collective Intelligence Concepts and Definitions

Collective Intelligence is generally defined as a group intelligence that emerges from collaborative efforts of individual team members for a consensus in decision making. The knowledge management specialists like Quinn and Nonaka would take collective intelligence as an organizational learning rather than representing the collection of knowledge (Merali 2000). According to Tapscott and Williams (2008) if an organisation uses the collective intelligence for learning and group decisions they should give an importance that the group has the four principals that enable the success.

- *Openness*: Members of the decision team would join meetings without thinking "I own this idea", ready to share the ideas with the group. This will create an opportunity for the review of other expert minds to clean the flaws and the bugs as well as expanding the creativity.
- *Peering*: This principal is creating the basis for an equality for the decision makers avoiding the need for a hierarchical approval. Since this basis would allow peer contributions, the desired contributions will be equally contributing to accomplish the common goals.
- *Sharing*: Though it seems similar to the openness, it is not limited to the new ideas but sharing the intellectual property with the rest of the group. It is a necessity for the flow of ideas and critics feeding in the group creativity.
- *Global Acting*: This concept is mainly introduced by the enhancement of communications. It simply means that the organization should be open to the industry and to the entire world of supply and marketing chains. This principle will allow to overcome the borders of company, industry or country based culture to enter the new markets with the new talents.

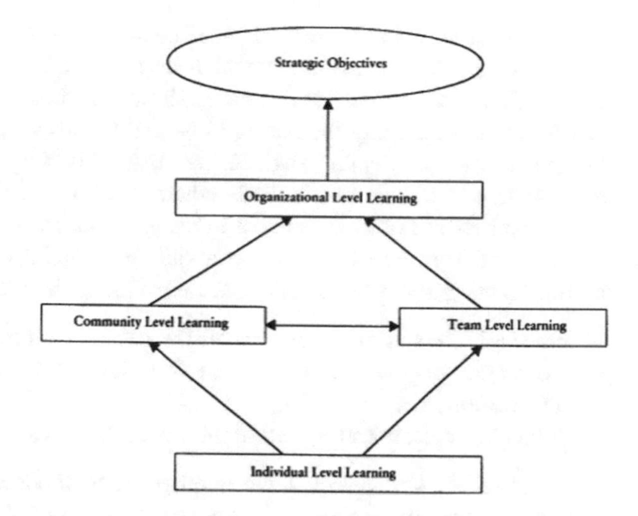


**Fig. 21.1** Organisational learning (Cross and Israelit 2000)

As a learning Cross defines collective intelligence as part of the Fig. 21.1.

Wikipedia and on-line games are shown as the best examples of using the collective intelligence for learning.

Moffet (2010) tells the story of collective intelligence growing from Wheeler's experience as follows:

> In 1911 William Morton Wheeler observed ants working in a group and saw them not as an individual but a single unit in a colony which created superorganism. A year later Emile Durkheim identified society as the sole source of human logical thought.

It is known that humans interact in several ways which are similar to animal interactions (Fisch et al. 2012):

- Individuals obey certain social rules—swarm of birds or fish obey rules to survive.
- Individuals exchange information for a common goal-ants following the pheromone for food.
- Babies learn by observing mimics of mother-monkeys copy movement of other monkeys or humans.
- Individuals learn from each other by rules or change of experience-small fish learn to escape from the big fish with rules and experiences.

Hence knowledge created in collective synergy can act timely in critical situations and moreover, may solve problems that cannot be solved individually.

It was only in the fifth generation of computer systems launched in Japan in 1980, that this concept has been accepted as the new era for social web (Kapetanios 2008). Main aim of this new approach is the absorption and spreading of information through evolving knowledge processes. This new computing method has created solutions with inspirations from the biological groups in the nature.

The dynamic pattern mechanism of neurons has helped converting the continuous time into discrete solutions, which saves execution time in computing (Conrad 1987). That is why Elsevier databases show that there have been more than 6000 researches just applying the Artificial Neural Networks (ANN) on time series data for predictions in health, engineering, materials and other fields. Planning or scheduling can be one of the fields where ANN is beneficial (García et al. 2014). Stock Markets are of great interest in using the collective intelligence because, the synthesis of a group of brains can avoid shortcomings. Kaplan (2001) has summarized it decently in three major reasons of using the group synergy in investment:

- No single person has access to all the necessary information;
- No single person can use of all the alternative approaches to process the information;
- A single investor can not all alone avoid mistakes caused by subjectivity.

Other widely known collective intelligence methods are ant colony and particle swarm optimisation besides evolutionary algorithms.

Collective intelligence also helps combining a variety of information as in this example of steel industry case given by Lazaric. The study aims to make a technical analysis of the scrap quality and internal temperature of a blast furnace. However, the approach to analysis is focused on different levels of knowledge which could be lost by communication problems of experts from different "communities of practice" (Lazaric et al. 2003).

Use of collective knowledge creates an important organisational concept of "**collective memory**". This concept is almost in conflict with the history, since, the ordinary people get concerned about present based on the considerations of the past. However, in collective memory for an organization is constructed by giving access to organizational information like mission, goals, objectives and basic policies. This information is recorded in group meetings and help to educate the new employees as well as supporting the multi-stage group tasks (Paul et al. 2004).

Literature review shows that Collective Intelligence has been the focus of a variety of scientific fields like Medical Sciences, Management Science, Sociology, Social Psychology, Information Technology and Knowledge Management. It has been observed that earliest research is in medical science (Thomas 1973), but the most crowded one is on creating the collective intelligence in the society, which is still centre of attention (Tounsi and Rais 2018). In Fig. 21.2 you can see the growing interest on collective intelligence for society, health, organizations and for computation.

Collective intelligence is mostly used in societal and organisational search because the approach is mainly used in complex environments to respond with abstract chaotic models. Szuba (1999) defines the society with thinking and non-thinking beings with "non-continuous" ways of interactions.

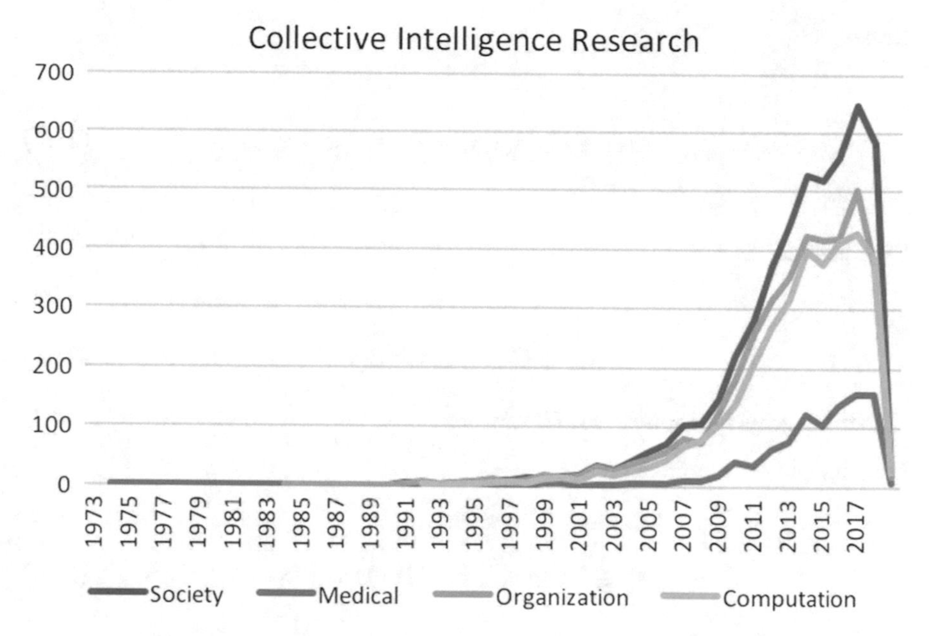


**Fig. 21.2** Distribution of research on collective intelligence. *Source* SCOPUS

## 21.3 Collective Intelligence Methods

Collective intelligence methods use a stochastic approach, since they include at least one random variable. Szuba (1999) says individual in a society move randomly because the inferences in real life are made randomly. He suggests using Brownian movements to represent social behaviour summarizing his reasons as:

- Single incidences are random;
- Incidences use resources distributed in time and space among the individuals;
- Performance of any social structure depends on the organization;
- There are inconsistencies caused by facts and rules.

By defining the above, he can express the levels and performance of collective intelligence mathematically. Maleszka and Nguyen (2015) have recently studied the computational expression of the integration of individual intelligence. The first assumption is individual intelligence is nonnegative, reflexive and symmetric so that the following expressions are used to make mathematical definitions follow.

Nonnegative:

$$\forall x, y \in U : d(x, y) \geq 0 \tag{21.1}$$

Reflexive:

$$\forall x, y \in U : d(x, y) = 0 \quad \text{if} \quad x = y \tag{21.2}$$

Symmetrical:

$$\forall x, y \in U : d(x, y) = d(y, x) \tag{21.3}$$

An Integrity function $I \in$ *Int(U)* satisfies:

Unanimity:

$$\forall n \in N \quad and \quad \forall x \in U\, I(n.x) = x \tag{21.4}$$

Simplification: Collective($X$) is a multiple of Collective($Y$)

$$I(X) = I(Y) \tag{21.5}$$

Consistency:

$$\forall x \in U (x \in I(X)) \Rightarrow (x \in I(X \cup x)) \tag{21.6}$$

Proportion:

$$\begin{gathered} \forall X_1 \in U \quad and \quad \forall X_2 \in U \\ ((X_1 \subset X_2) \wedge (x \in I(X_1)) \wedge (y \in I(X_2))) \Rightarrow (d(x, X_1) \leq d(y, X_2)) \end{gathered} \tag{21.7}$$

Optimality of consistent $X$ for any $X \in \prod (U)$

$$
\begin{aligned}
(x \in I(X)) &\Leftrightarrow (d(x,X) = \min_{y \in U} d(y,X)) \\
(x \in I(X)) &\Leftrightarrow (d^2(x,X) = \min_{y \in U} d^2(y,X))
\end{aligned}
\tag{21.8}
$$

Complexity:

$$
\text{If } X \text{ is consistent then } X \subseteq I(X) \tag{21.9}
$$

Even though the collective intelligence of human beings can be shown mathematically, computations show an exploration, exploitation dual dilemma between the individuals. Although social learning through the current social media reduce collective exploitation costs, total group exploration costs increase. Toyokawa et al. (2014) have used the similarity of human interaction dilemma with social insects approach and figured out that people can learn about the previous choices about other and solve the exploitation-exploration dilemma. This is a recent example of using the cases from the nature to solve collective intelligence problems.

A few decades ago nature inspired collective intelligence solutions were taken by the scientists with strong suspicions. The improvements in computational methods and artificial intelligence have success of those algorithms increased. Though evolutionary algorithms and neural networks are also considered in collective intelligence by some researchers, the majority accepts the early algorithms to be ant colony and particle swarm optimisation (Penalva 2006).

In Ant Colony algorithm (ACO) a population of ants are initiated for random solutions where each ant is assigned a transition probability. Then the pheromone trail is updated. At each iteration, the pheromone amount is evaporated and best ants can survive using the Eqs. (21.10)–(21.12) defined by Goss (Dorigo and Di Caro 1999) (Fig. 21.3).

Probability with heuristic information $\eta$ and S, and constants $\alpha$ and $\beta$ depends on the ratio of individual pheromone $\tau_{ij}$ at iteration t and total of $\tau_{ij}$ for all the ants:

$$
\rho_{ij}(t) = \frac{\tau_{ij}(t)^{\alpha} \eta_{ij}^{\beta}}{\sum_{j \in S} \tau_{ij}(t)^{\alpha} \eta_{ij}^{\beta}} \tag{21.10}
$$

where, pheromone is renewed by evaporation rate $\rho$:

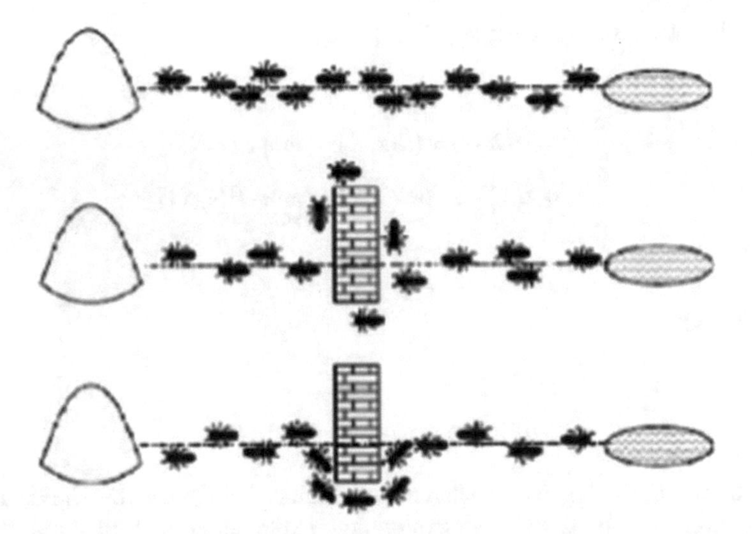

**Fig. 21.3** Typical food search of ants (Ghanbari et al. 2013)

$$\tau_{ij}(t+1) = (1-\rho)\tau_{ij}(t) \tag{21.11}$$

The trail is renewed for the global best path $\Delta\tau_{ij}$

$$\tau_{ij}(t+1) = \tau_{ij}(t) + \Delta\tau_{ij} \tag{21.12}$$

Kennedy and Eberhardt (1995) defined Particle Swarm Optimization (PSO) to find the individual best (particle best-Pbest) as well as the social best (Global Best-Gbest). It is the best representative of showing that the ability of a group to solve problems is richer than its individual members (Fig. 21.4).

This algorithm is defined by imitating simulating the flocking bird groups (Sharafi and ElMekkawy 2015). Given the position and the velocity as below the objective function is calculated for each particle to find the position and velocity for personal and global best using the random number r.

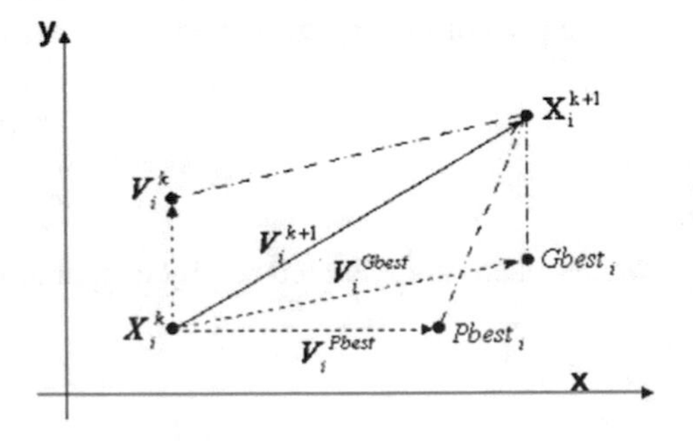


**Fig. 21.4** PSO geometry in finding particle best and the global best (Kennedy and Eberhart 1995)

$$x_{ij} = x_{\min} \times r(x_{\max} - x_{\min}) \quad \forall i = 1,\ldots,N; \forall j = 1,\ldots,n$$
$$v_{ij} = \alpha \frac{x_{\min} + r(x_{\max} - x_{\min})}{\Delta t} \tag{21.13}$$

After having found the particle and global best the velocity and positions are recalculated using the constants $c_1$ and $c_2$:

$$v_{ij} \leftarrow v_{ij} + c_1 q \left( \frac{x_{ij}^{pb} - x_{ij}}{\Delta t} \right) + c_2 r \left( \frac{x_{j.}^{sb} - x_{ij}}{\Delta t} \right) \quad \forall i = 1,\ldots,N; \forall j = 1,\ldots,n$$
$$x_{ij} \leftarrow x_{ij} + v_{ij} \times \Delta t \tag{21.14}$$

Kitamura shows a good example of a two-objective operation planning problem for a period of 24 h in industry using PSO, with the objectives being the simultaneous minimization of energy cost and $CO_2$ emissions (Kitamura et al. 2006). An operation schedule is worked out by dividing the problem into 1-h intervals.

PSO is widely used in forecasting and clustering in the energy field. Kumar and Chaturvedi (2013) gives a good example of optimizing the power flow in the grid. In this analysis, the total load of electricity is optimized considering the busses, the branches, the generators and the transformers. These widely known algorithms found applications in the areas of network security, pervasive computing, mobile and embedded systems, pattern recognition, data classification as well as the energy fields. Krol gives some examples with their impact in industry (Król and Lopes 2012):

> The artificial immune system can efficiently detect changes in the environment or deviations from the normal system behaviour via self-optimization and learning process. The concepts of intercellular information exchange can be used to learn: efficient dispatching, shortening of signalling pathways and modelling the control loop for a regulatory process in an organism. Some bio-ideas can be successfully exploited to elaborate sound strategies against cascading failures in the systems, or to provide insights into complex social phenomena such as terrorist cells.

Artificial Bee Colony (ABC) is a new algorithm that is preferred in forecasting as well. Karaboga and Akay (2007) defines the food sources, foragers as main self-organisers. The foragers can be employed or non employed. The employed foragers are associated with a particular food source, which they are exploiting or are employed at. Unemployed Foragers are continually looking for a food source to exploit. There are two types of unemployed foragers: scouts, searching the environment surrounding the nest for new food sources and onlookers waiting in the nest and establishing a food source through the information shared by employed foragers. The algorithm defines two leading modes of behaviour: either recruits at the food source or abandons the source. The important part of the algorithm is the exchange of information among the bees. Employed foragers share their information with a probability proportional to the profitability of the food source found in the dancing area. An onlooker on the dance floor, decides to employ herself at the most profitable source. But, there is a greater probability of onlookers choosing more profitable sources since more information is circulated about the more profitable sources. Creating a solution $v_{ij}$ for ith source and jth bee and find the probability for each source through the fit function associated with it $fit_{i,}$ which is a ratio of sum of all fit values for all the sources as calculated using Eq. (21.15) first.

$$\begin{aligned} & v_{ij} = x_{ij} + \phi_{ij}(x_{ij} - x_{kj}) \\ & where \\ & k : \{1, 2, \ldots number\, of\, employed\, bees\} \\ & j : \{1, 2, \ldots D\} \\ & D : number\, of\, parameters\, to\, optimize \\ & \phi_{ij} \in [-1 + 1]\, random\, number \\ & p_i = \frac{fit_i}{\sum_{l=1}^{S} fit_l} \end{aligned} \tag{21.15}$$

All these nature inspired methods can be used with the advantages in solving the Non-Polynomial; applying the deep learning; designing the IT security; discover semantic networks; financial portfolios; characteristic or demographic segmentation (Kayakutlu and Mercier-Laurent 2017). The cutting-edge research used biological or nature inspired intelligence algorithms in dynamic agent based systems. As in the case of Zhang et al. (2015) where the swarm algorithm is used to define the topology of the multi-agent dynamic intelligence.

More hybrid algorithms are constructed gradually, when an operational function is studied together with nature concerns. BeeCup to study energy efficient mobile communication clustering (Xia et al. 2014), bee swarm to design safe routing (Bitam et al. 2013) and a harmony search for implementing energy efficient sensor networks are good examples.

Zak & Zak (1994) have looked for responses to define collective brain through the probabilistic structure can be installed and how can it be improved with collective actions depending on the Newtonian dynamics and gradient descent. The research could not find the structure but defined the improvements for collective intelligence. Equation (21.16) defines the *i*th intelligent unit (neuron) to access m existing and m + 1…n predicted information with its own probability function $\varphi_I$ which will eventually converge to 0 when no coincidence occurs.

$$\begin{aligned} & \dot{x}_i = \gamma_i \sin^k\left(\frac{\sqrt{\omega}}{\alpha_i} \sum_j T_{ij} x_j\right) \sin \omega t + \varepsilon_0^2 \phi_i(x_1^i, \ldots, x_m^i, x_{m+1}^i, \ldots, x_n^i), \varepsilon_0 \to 0 \\ & L = \sum \int_0^T (x_i - x_i)^2 dt\, for\, relatively\, l \arg e\, T \end{aligned} \tag{21.16}$$

$L$ is the improvement structure depending on the synaptic (generally sigmoid) oscillations $T_{ij}$. There is no more a need for gradient descent because prediction is continued on the same direction as far as $L^1 \geq L^0$ and $L^2 \geq L^1$. If the improvement is not observed the direction is changed.

Gavrilets (2015) studied the collective action of the collaborative brain using the evolutionary algorithms trying to show the individual costs and benefits when a certain collaborative action. This research shows that collective action starts first with the neighbouring groups after a certain state of benefit they become involved in other collective actions. Evolution starts with "us versus them" and with the cultural learning, social intelligence improves to "us versus nature".

## 21.4 Collective Intelligence in Energy Applications

A review in Web of Science shows that individual algorithms like Ant colony and Particle Swarm Optimization have been used in energy forecasting for long (Kitamura et al. 2006; Dorigo et al. 2006; Toksarı 2009; Robinson 2005). However, the collective learning or collective intelligence concepts have been introduced to the industry only after the use of distributed energy, smart buildings, smart city concepts and economic concerns are integrated with the climate concerns. Non-profit services, federated operations and intellectual property rights forced the collective management of the grids (Ishida et al. 2011) and smart building networks (Rahimian et al. 2015; Nemoz 2013).

One of the early studies of collective learning applied in the energy industry is focused on responding to "Why? What? How?" questions collaboratively (Korsvold et al. 2010). The research results in collaboration on only one single question, never combined. These findings would also explain the method oriented use of collective intelligence in the energy industry, since, most of the research was only targeting the economic analysis. Use of multiple resources has made the daily life of energy generators more complex and introduced multi agent systems as a tool for operations management (Saba et al. 2015).

Following the improvements in the energy research, review of collective intelligence applications will be organised as forecasting studies, economic analysis, strategic studies, operational analysis, performance review and future studies.

### *21.4.1 Energy Forecasting*

Demand, renewable energy availabilities and the price are the main uncertainties in the energy research. Energy forecasting studies in short and long terms are motivated by these uncertainties.

Long term demand forecasts are the basis for creating policies. Early samples of creating are fuel policies in Iran (Assareh et al. 2010) and long term power demand forecasting in China (Pi et al. 2010) and in Turkey (Ünler 2008). Toksari gives the domestic power consumption forecasts for Turkey between 2014 and 2030 using a hybridized ACO algorithm (Toksari 2016). The forecast is based on population, import, export and GDP for two scenarios. A very recent study of Karadede et al. (2017) on Natural gas consumption in Turkey has also used hybridized collective intelligence to reduce the forecasting errors. However, with the analysis of He et al. (2017) it is observed that the methodology might be of second importance to improve the long term forecasts if the impact factors are well chosen. Inopportunely, majority of long term demand forecasts still only rely on GDP, Population, industrialization and resource availability where as the new factors are to be defined as shown in Fig. 21.5.

Energy demand by the transportation is a specific industry demand which is forecasted by using the artificial bee colony in Turkey (Sonmez et al. 2017). They are expecting the energy use to be doubled in 2034, that of 2013.

The short-term forecasting is highly requested for the renewable energy resource estimation, energy loading amount prediction and power price prediction. Autoregressive Integrated Moving Average (ARIMA) is the generally preferable method for short term forecasts. But as in the case of Su et al. (2014) using (PSO) reduces the prediction errors of wind speed forecasting.

If distributed wind turbines are used in the wind farms, then hybridization of PSO and ACO can be combined to suggest better schedules (Rahmani et al. 2013). In majority of the short-term forecasts Artificial Neural Network (ANN) systems are used as a black box as in the electricity generation using solar resources (Bugała et al. 2018). In Bugala et al's study, number of sunny hours, length of day, air pressure, maximum temperature, cloudiness are taken as input and daily electric energy is given as the output.

There is a remarkable change on the energy price forecasts which even makes ministries of energies change their policies. There were studies to predict which prices should be more beneficial for the power price forecasting (Melichar 2016;

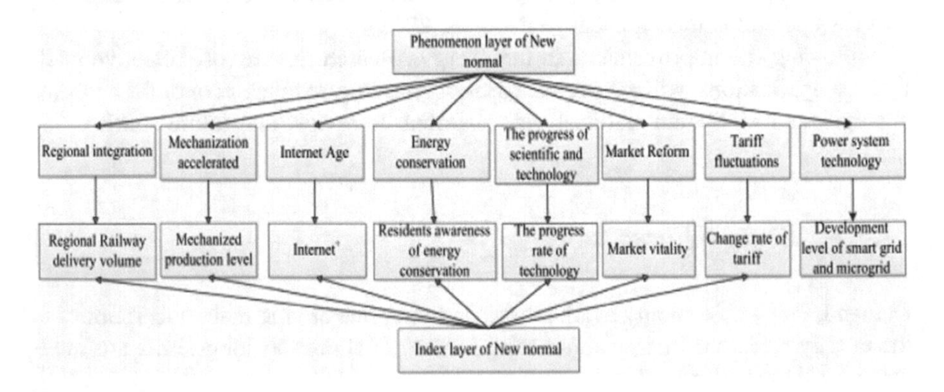


**Fig. 21.5** Sample of combining new and factors effecting the energy demand forecast (He et al. 2017)

Lahiani et al. 2017). Conventionally prices are based on petroleum but now natural gas seems the start of the market. So gas prices or both petroleum and gas prices are to be taken as historical data. Doubts due to rapid changes in prices caused discussions even on DOE forecasts (Mamatzakis and Koutsomanoli-Filippaki 2014). Sisodia ct al. (2015) Sisodia discussed the existing pricing models in accordance with the differences of the developing countries.

Raza and Khosravi (2015) give an insight to the intelligent studies on smart grids and smart buildings. But unlike the load forecasts analysed in this review, Dudek introduces a new technique for load prediction. He uses Artificial Immune System for estimating the energy load by finding the data patterns and clustering the patterns on similarities (Dudek 2015). Ma et al. (2017) reviews all the market models including the IEA designed ones (e.g. MARKAL and LEAP) and discusses the applicability of those models on district load forecasting. They construct a new hybrid modelling framework where cuckoo search, a multi-objective optimization algorithm is hybridized with some machine learning methods.

Multi Agent applications have taken off in short term prediction for the smart grid management with distributed resources. The load demand schedule is to be designed depending on the battery scheduling as in the case of Chaouachi et al. (2013).

### *21.4.2 Energy Economy*

Energy economy includes the investment analysis, cost optimization and lifecycle evaluations of energy resources as well as cost comparisons. Portfolio construction are also part of economic analysis, since the resources are operated together according to the total cost. New technologies are centre of attraction for the economic analysis. The first example is the multi objective analysis of the cost of energy generation while minimizing the carbon emission and maximizing the wind energy share. There are several economic studies alike, but Khodja et al. (2014) used ACO to perform the analysis and showed that the results are achieved in shorter computational time than the other methods.

An investment in a bioethanol is studied by using the particle swarm optimization in Redlarski et al. (2017) where the multi objective optimization is designed for the profitability of the supply chain and the investor separately. Researchers emphasize the importance of location as well as the cost of a rich variety of equipment requested. The solution is modelled by using several PSO run in parallel.

In a review of intelligent techniques used for multiple energy resource usage in energy generation mentions particle swarm optimization more for the use of multi objective optimization but ant colonies and artificial immune system in hybrid resource sizing (Zahraee et al. 2016).

A survey on Honey Bee Collective Intelligence algorithm use, shows that any economic dispatch problem in energy field can use this method (Rajasekhar et al.

2017). Actually, one of the dispatch evaluations for chiller dispatch used bee swarm optimization to save energy (Lo et al. 2016).

### 21.4.3 Energy Strategies

Critical and long-term decisions are the subject of strategic analysis. Collective intelligence techniques can well be used for the long term decisions in the energy field. As in the case of reinforcement planning of the distribution systems studied by Favuzza et al. (2007). It is a strategic plan since the costs are to be considered together with the selling and buying conditions. The solutions are to be minimized and the cables have to be installed optimally. An annual multi objective function to create consensus of all the concerns at once is solved using Ant Colony Search algorithm. The economic impact of the best strategy is minimized in minimum computational time.

Facility location has always been a strategic decision. With the distributed energy management the locations for the distribution centres are chosen through the use of bi-level ant colony algorithm by Yegane et al. (2016). In this interesting study, the case of two computing supply chain is analysed where both have the chance to locate the distribution centre in the same place. The well-known NP-hard problem by scenarios of importance for the followers. The results show that even large-scale problems can be solved with the same algorithm.

Maintenance and spare part renewals in transmissions systems are strategic and, it is well known that the mistakes may cause blackouts. Optimization problem for maintenance plans are designed as a knapsack problem and solved using ant colony optimization designed as a chain of capital and operational expenses (Rhein et al. 2017). Hazard rates are calculated with the designed new ant colony application and the reliability of the transmission system is improved by the right timing.

The last sample of strategic analysis is on the fuel cell capacitors of the electric vehicles. Strategic energy management system for the electrical vehicle is designed by Koubaa and Krichen (2017) using a layer for rule based intelligence and the second layer for the swarm optimisation. The rules are defined rules for standing/breaking, demand, decelerating, accelerating and state of charge and then the PSO is run to define the probabilities for state of charge to be more than minimum and less than maximum. The first layer served for speeding up the computational time and the second layer optimized hydrogen use, which is in fact better than the approaches using genetic algorithm.

### 21.4.4 Energy Operations

There are numerous samples of collective intelligence applications in energy supply chain operations, mainly on planning, scheduling and expansion of the transmission

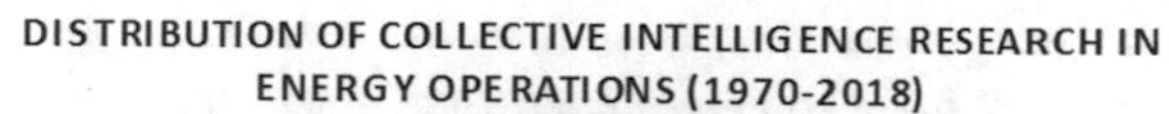


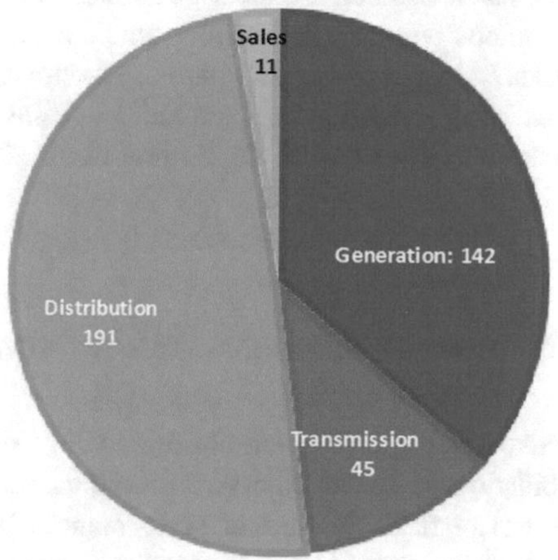


**Fig. 21.6** Research on energy operations (SCOPUS 1970–2018)

and distribution. The supply chain will be considered as energy generation, transmission, distribution, and sales. After the distributed energy providing with micro-grids, the samples are more concentrated on the energy generation operations because of the hybrid use of the resources (Fig. 21.6).

Sorensen explains the difficulties of generation, transmission and distribution of new energy sources like bio-fuels and liquid fuels which are foreseen to replace the fossil sources. Watanabe (2005) and Li et al. (2005) give good distribution or distribution expansion planning by using the ant colony optimization.

For allocation of distributed resources two features of the cat swarm are simulated: seeking slowly to exploit and tracing fast (explore) using velocity (Chu et al. 2006). This algorithm has been the preferred collective intelligence for allocating the turbines and keeping the reliability standards (Mezian et al. 2015; Selvakumar et al. 2017).

In hybrid energy resources are used for power generation, multiple objectives create an NP-Hard problem to solve. If minimizing the cost, minimizing the losses, maximizing voltage stability and minimizing the emissions are the objectives there is a need for hybridizing the solution algorithms. In the research of Kefayat et al. (2015) pareto optimal is to be created in the case of discrete non-linear problems. In fuel management of the above research the new algorithm is created to combine the ACO and the ABC, so that the local optima problem of ACO can be avoided.

When distributed energy use started the concept of smart distribution has been part of demand site management. An in depth analysis provided by Cardenas et al. (2014) gives all the details intelligent algorithms used in detailed operations of smart grids and smart distribution. A study of Zu, Mitra and Cai takes only the planning details in smart grids considering the vulnerability, reliability and economy as multiple objectives (Xu et al. 2014). Multi agent systems with PSO are used in optimizing the integration of distributed services.

Energy problems have both the supply and demand sites. Demand site management has also been a concern of smart grids due to homogeneous use of energy in the campus with supply equilibrium. The energy saving is possible by smoothing the peaks. Sustainability, fair share of resources, control of energy utilization are the complex problems that are solved by using collective intelligence techniques (Palensky and Dietrich 2011; Yanine et al. 2014; Aghajani et al. 2017; Jin et al. 2017).

### 21.4.5 Energy Management Performance

Collective intelligence methods are mainly designed to approach optimality for probabilistic non-linear objective functions, many examples of which are observed in the energy field. Technical performances like energy transfer, conversion efficiency is as important as the managerial performances as in building, grid or hub management.

Udayraj et al. (2015) compares the performances of ACO, PSO and Cuckoo Search in heat transfer. ACO is found to be better in time and step size when the heat flux problem is studied with noise and temperature robustness. In Beltramo et al. study biogas flow rate is optimised by using ANN and ACO with a performance increase by identification of the significant variables. Optimal design of a standalone wind-solar hybrid power generator needs a model for load sizing and performance optimization overcoming the hourly uncertainties. Suhane et al. (2016) has found a solution using the ant colony optimization measuring the performance by total cost of demand load and unmet demand, contribution of different resources and measuring the state of battery.

Performance of building management is improved by saving energy without a compromise of the resident comfort. Faia et al. studies the previous cases after having clustered the similar cases, uses the PSO to optimize the energy saving parameters and gives those to the building SCADA system as the reduction suggested (Faia et al. 2017). Results show that adequate savings are achieved.

Control of charge and discharge is optimized by Marzband et al. (2016) using a multi-level Ant Colony algorithm in a stand alone Micro-grid. The novel algorithm allows the ants find the best way to the food by showing them the path with maximum pheromone. Real-time scheduling of the wind and solar resources and the batteries was based on the parameters optimized by this algorithm.

Huo et al. (2018) studied the hub optimization for multi carrier energy infrastructure. Energy storage consideration while controlling the real-time changes in demand and energy generation is analysed in the mentioned research. The performance optimisation model is constructed with PSO hybridized with interior point method. The suggested model is also applied on an eleven hub system and found fast enough to be used online.

Gonzalez et al. (2018) suggests that the high cost of deep water and the uncertainties force the offshore windfarm to be located in certain areas which may be crowded. Hence, the location project needs to solve the conflict of each investor has his own self-interest but they have to collaborate for the best performance. They developed a co-evolutionary algorithm using Nash equilibrium with a sequential single project approach. Hence the algorithm generates two population to start, calculates the Net Present Values as fit function for each and runs the Nash equilibrium among the best choices before the mutation and crossover. Overall maximum benefits are achieved at the end.

## 21.5 Conclusions

This review is motivated by increasing necessity of using collective intelligence methods to find a solution for the complex energy problems. An effort is given to give collective dynamics in energy networks in knowledge clusters as defined (Liu et al. 2013). The concepts and definitions are introduced before explaining the mathematical expression of most widely known techniques. Overview of almost ninety articles from respected scientific journals allowed the classification of the energy fields where sample applications are presented.

It is observed that swarm and colony algorithms are not any more limited to PSO and ACO but being extended gradually. They are applied according to the problem specific features. Recently better results in forecasting, economic, strategic, operational and performance approaches for different energy problems are achieved by combining the methods. Method hybridization is not specific to intelligent methods like combining the PSO and ACO but also numeric optimizations like mixed integer programming, interior-point and game theory approaches.

Future trends in the energy field show complexities in risk reduction (Balo et al. 2015), modelling and optimization of integrated energy supply systems (Burke and Stephens 2017) and strategic decision making for multiple micro grids (Jalali et al. 2017). It is recommended that the new complexities proposed by these trends are to be followed through the collective intelligence dimension.

This review will be helpful both the intelligence researchers as well as energy planning, optimisation and management experts.

## References

Aghajani, G. R., Shayanfar, H. A., & Shayeghi, H. (2017). Demand side management in a smart micro-grid in the presence of renewable generation and demand response. *Energy, 126,* 622–637.

Assareh, E., Behrang, M. A., Assari, M. R., & Ghanbarzadeh, A. (2010). Application of PSO (particle swarm optimization) and GA (genetic algorithm) techniques on demand estimation of oil in Iran. *Energy, 35*(12), 5223–5229. Available from: http://dx.doi.org/10.1016/j.energy.2010.07.043.

Bale, C. S. E., Varga, L., & Foxon, T. J. (2015, January 21). Energy and complexity: New ways forward. *Appl Energy, 138*, 150–159. Available from: http://www.sciencedirect.com/science/article/pii/S0306261914011076.

Bitam, S., Mellouk, A., & Zeadally, S. (2013). HyBR: A hybrid bio-inspired bee swarm routing protocol for safety applications in Vehicular Ad hoc NETworks (VANETs). *Journal of Systems Architecture, 59*(10 PART B), 953–957.

Bonabeau, E. (2009). Decisions 2.0: The power of collective intelligence. *MIT Sloan Management Review, 50*(2), 45–52.

Bugała, A., Zaborowicz, M., Boniecki, P., Janczak, D., Koszela, K., Czekała, W., et al. (2018). Short-term forecast of generation of electric energy in photovoltaic systems. *Renewable and Sustainable Energy Reviews, 81,* 306–312.

Burke, M. J., & Stephens, J. C. (2017). Political power and renewable energy futures: A critical review. *Energy Research & Social Science.*

Cardenas, J. A., Gemoets, L., Ablanedo Rosas, J. H., & Sarfi, R. (2014). A literature survey on smart grid distribution: An analytical approach. *Journal of Cleaner Production, 65,* 202–216.

Chaouachi, A., Kamel, R. M., Andoulsi, R., & Nagasaka, K. (2013). Multiobjective intelligent energy management for a microgrid. *IEEE Transactions on Industrial Electronics, 60*(4), 1688–1699.

Chu, S-C., Tsai, P., & Pan, J-S. (2006). Cat swarm optimization. In *Pacific Rim International Conference on Artificial Intelligence* (pp. 854–858). Available from: http://link.springer.com/10.1007/978-3-540-36668-3_94.

Conrad, M. (1987). Rapprochement of artificial intelligence and dynamics. *European Journal of Operational Research, 30*(3), 280–290.

Cross, R. L., Jr., & Israelit, S. B. (2000). Strategic learning in a knowledge economy : Individual, collective and organizational learning process. In *Resources for the knowledge based economy series* (Vol. xviii, p. 348).

Dorigo, M., Birattari, M., & Stützle, T. (2006) Ant colony optimization. *IEEE Computational Intelligence Magazine, 1*(4), 28–39. Available from: http://ieeexplore.ieee.org/lpdocs/epic03/wrapper.htm?arnumber=4129846.

Dorigo, M., & Di Caro, G. (1999). The ant colony optimization meta-heuristic. *New Ideas Optimization, 2*, 11–32. Available from: http://portal.acm.org/citation.cfm?id=329055.329062.

Dudek, G. (2015). Pattern similarity-based methods for short-term load forecasting—Part 1: Principles. *Applied Soft Computing, 37,* 277–287.

Elia, G., Margherita, A., & Vella, G., et al. (2014). A conceptual model to design a collective intelligence system supporting technology entrepreneurship. In: C. Vivas, & P. Sequeira (Eds.), *Proceedings of the 15th European Conference on Knowledge Management (ECKM 2014)* (pp. 297–305).

Faia, R., Pinto, T., Abrishambaf, O., Fernandes, F., Vale, Z., & Corchado, J. M. (2017). Case based reasoning with expert system and swarm intelligence to determine energy reduction in buildings energy management. *Energy and Buildings, 155.*

Favuzza, S., Graditi, G., Ippolito, M. G., & Sanseverino, E. R. (2007). Optimal electrical distribution systems reinforcement planning using gas micro turbines by dynamic ant colony search algorithm. *IEEE Transactions on Power Systems, 22*(2), 580–587.

Fisch, D., Jänicke, M., Kalkowski, E., & Sick, B. (2012). Learning from others: Exchange of classification rules in intelligent distributed systems. *Artificial Intelligence, 187–188,* 90–114.

García, T. R., Cancelas, N. G., & Soler-Flores, F. (2014). The artificial neural networks to obtain port planning parameters. *Procedia-Social and Behavioral Sciences, 162,* 168–177. Available from: http://linkinghub.elsevier.com/retrieve/pii/S1877042814062983.

Gavrilets, S. (2015). Collective action and the collaborative brain. *Journal of the Royal Society Interface, 12*(102), 20141067–20141067. Available from: http://rsif.royalsocietypublishing.org/cgi/doi/10.1098/rsif.2014.1067.

Georges, P., Nietvelt, K., de Taisne, B., Stenqvist, A., & Schiavo, M. (2017). *European integrated utilities in 2017: Rebenchmarking the sector.* Standard & Poor's Financial Services LLC.

Available from: https://www.spratings.com/documents/20184/1481001/European+Integrated+Utilities+In+2017/a5a59071-259b-4566-a0e2-abfd1652d9fb.

Ghanbari, A., Kazemi, S. M. R., Mehmanpazir, F., & Nakhostin, M. M. (2013). A cooperative ant colony optimization-genetic algorithm approach for construction of energy demand forecasting knowledge-based expert systems. *Knowledge-Based Systems, 39*, 194–206. Available from: http://dx.doi.org/10.1016/j.knosys.2012.10.017.

González, J. S., Burgos Payán, M., & Riquelme Santos, J. M. (2018). Optimal design of neighbouring offshore wind farms: A co-evolutionary approach. *Applied Energy, 209,* 140–152.

He, Y., Jiao, J., Chen, Q., Ge, S., Chang, Y., & Xu, Y. (2017). Urban long term electricity demand forecast method based on system dynamics of the new economic normal: The case of Tianjin. *Energy, 133,* 9–22.

Huo, D., Le Blond, S., Gu, C., Wei, W., & Yu, D. (2018). Optimal operation of interconnected energy hubs by using decomposed hybrid particle swarm and interior-point approach. *International Journal of Electrical Power & Energy Systems, 95,* 36–46.

Ishida, T., Murakami, Y., Tsunokawa, E., Kubota, Y., & Sornlertlamvanich, V. (2011). Federated operation model for service grids. *Cognitive Technologies,* 279–298.

Jalali, M., Zare, K., & Seyedi, H. (2017). Strategic decision-making of distribution network operator with multi-microgrids considering demand response program. *Energy, 141,* 1059–1071.

Jin, M., Feng, W., Marnay, C., & Spanos, C. (2017). Microgrid to enable optimal distributed energy retail and end-user demand response. *Applied Energy.*

Kapetanios, E. (2008). Quo Vadis computer science: From turing to personal computer, personal content and collective intelligence. *Data & Knowledge Engineering, 67*(2), 286–292.

Kaplan, C. A. (2001). Collective intelligence: A new approach to stock price forecasting. In *2001 IEEE International Conference on Systems, Man and Cybernetics e-Systems and e-Man for Cybernetics in Cyberspace (CatNo01CH37236)* (Vol. 5, pp. 2893–2898).

Karaboga, D., & Akay, B. (2007). Artificial Bee Colony (ABC) Algorithm on training artificial neural networks. In *2007 IEEE 15th Signal Processing and Communications Applications* (pp. 1–4). Available from: http://ieeexplore.ieee.org/document/4298679/.

Karadede, Y., Ozdemir, G., & Aydemir, E. (2017). Breeder hybrid algorithm approach for natural gas demand forecasting model. *Energy, 141,* 1269–1284.

Kayakutlu, G., & Mercier-Laurent, E. (2017). *Intelligence in energy*. London: ISTE-Elsevier Ltd.

Kefayat, M., Lashkar Ara, A., & Nabavi Niaki, S. A. (2015, March 28). A hybrid of ant colony optimization and artificial bee colony algorithm for probabilistic optimal placement and sizing of distributed energy resources. *Energy Conversion and Management, 2*, 149–161. Available from: http://www.sciencedirect.com/science/article/pii/S0196890414010760.

Kennedy, J., & Eberhart, R. (1995). Particle swarm optimization. In *1995 Proceedings, IEEE International Conference on Neural Networks* (Vol. 4, pp. 1942–1948).

Khodja, F., Younes, M., Laouer, M., Kherfane, R. L., Kherfane, N. (2014). A new approach ACO for solving the compromise economic and emission with the wind energy. *Energy Procedia,* 893–906.

Kitamura, S., Mori, K., Shindo, S., & Izui, Y. (2006). Modified multiobjective particle swarm optimization method and its application to energy management system for factories. *Electrical Engineering in Japan, 156*(4), 33–42. Available from: http://search.ebscohost.com/login.aspx?direct=true&db=a9h&AN=21611108&lang=tr&site=ehost-live.

Korsvold, T, Madsen, B-E., Bremdal, B., Herbert, M. C., Nystad, E., & Danielsen, J. E., et al. (2010). Creating an intelligent energy organization through collective learning. In *SPE Intelligent Energy Conference & Exhibition* (pp. 892–923). Available from: http://www.scopus.com/inward/record.url?eid=2-s2.0-77953993993&partnerID=40&md5=abf9e61bb1848d9d215ce144eb10f513.

Koubaa, R., & Krichen, L. (2017). Double layer metaheuristic based energy management strategy for a fuel cell/ultra-capacitor hybrid electric vehicle. *Energy, 133,* 1079–1093.

Król, D., & Lopes, H. S. (2012). Nature-inspired collective intelligence in theory and practice. *Information Sciences, 182*(1), 1–2.

Kumar, S., & Chaturvedi, D. K. (2013). Optimal power flow solution using fuzzy evolutionary and swarm optimization. *International Journal of Electrical Power & Energy Systems, 47*, 416–423. Available from: http://www.sciencedirect.com/science/article/pii/S014206151200662X.

Lahiani, A., Miloudi, A., Benkraiem, R., & Shahbaz, M. (2017). Another look on the relationships between oil prices and energy prices. *Energy Policy, 102,* 318–331.

Lazaric, N., Mangolte, P. A., & Massué, M. L. (2003). Articulation and codification of collective know-how in the steel industry: Evidence from blast furnace control in France. *Research Policy, 32*(10), 1829–1847.

Li, Z., Xie, Z., & Qin, J. (2005). Application of ant colony algorithms in optimized design of gas transmission pipelines. *Tianranqi Gongye/Natural Gas Ind., 25*(9), 104–119.

Liu, X., Jiang, T., & Ma, F. (2013). Collective dynamics in knowledge networks: Emerging trends analysis. *Journal of Informetrics, 7*(2), 425–438.

Lo, C. C., Tsai, S. H., & Lin, B. S. (2016). Economic dispatch of chiller plant by improved ripple bee swarm optimization algorithm for saving energy. *Applied Thermal Engineering, 100,* 1140–1148.

Ma, W., Fang, S., Liu, G., & Zhou, R. (2017). Modeling of district load forecasting for distributed energy system. *Applied Energy,* 181–205.

Maleszka, M., & Nguyen, N. T. (2015). Integration computing and collective intelligence. *Expert Systems with Applications, 42*(1), 332–340.

Mamatzakis, E., & Koutsomanoli-Filippaki, A. (2014). Testing the rationality of DOE's energy price forecasts under asymmetric loss preferences. *Energy Policy, 68,* 567–575.

Marzband, M., Yousefnejad, E., Sumper, A., & Domínguez-García, J. L. (2016). Real time experimental implementation of optimum energy management system in standalone Microgrid by using multi-layer ant colony optimization. *International Journal of Electrical Power & Energy Systems, 75,* 265–274.

Melichar, M. (2016). Energy price shocks and economic activity: Which energy price series should we be using? *Energy Economics, 54,* 431–443.

Merali, Y. (2000). Individual and collective congruence in the knowledge management process. *The Journal of Strategic Information Systems, 9*(2–3), 213–234. Available from: http://linkinghub.elsevier.com/retrieve/pii/S0963868700000445.

Mezian, R., Boufala, S., Amara, M., & Amara, H. (2015). Cat swarm algorithm constructive method for hybrid solar gas power system reconfiguration. In: *3rd International Renewable and Sustainable Energy Conference (IRSEC)* (pp. 1–7).

Moffett, M. W. (2010). *Adventures among ants: A global Safari with a cast of trillions.* (1–280 pp). Available from: https://www.scopus.com/inward/record.uri?eid=2-s2.0-84987834827&partnerID=40&md5=5d907b41e3ab02966b2b1695c30e04d7.

Nemoz, S. (2013). Smart campus: Recent advances and future challenges for action research on territorial sustainability. *Implementing Campus Greening Initiatives,* 313–323.

Palensky, P., & Dietrich, D. (2011). Demand side management: Demand response, intelligent energy systems, and smart loads. *IEEE Transactions on Industrial Informatics, 7*(3), 381–388.

Paul, S., Haseman, W. D., & Ramamurthy, K. (2004). Collective memory support and cognitive-conflict group decision-making: An experimental investigation. *Decision Support Systems, 36*(3), 261–281.

Penalva, J. M. (2006). *Intelligence collective.* Paris: Mines.

Pi, D., Liu, J., & Qin, X. (2010). A grey prediction approach to forecasting energy demand in China. *Energy Sources, Part A: Recovery, Utilization, and Environmental Effects, 32*(16), 1517–1528.

Rahimian, M., Cardoso-Llach, D., & Iulo, L. D. (2015). Participatory energy management in building networks. *Sustainable Human-Building Ecosystems,* 27–35.

Rahmani, R., Yusof, R., Seyedmahmoudian, M., & Mekhilef, S. (2013). Hybrid technique of ant colony and particle swarm optimization for short term wind energy forecasting. *Journal of Wind Engineering and Industrial Aerodynamics, 123,* 163–170.

Rajasekhar, A., Lynn, N., Das, S., & Suganthan, P. N. (2017) Computing with the collective intelligence of honey bees—A survey. *Swarm and Evolutionary Computation,* 25–48.

Raza, M. Q., & Khosravi, A. (2015). A review on artificial intelligence based load demand forecasting techniques for smart grid and buildings. *Renewable and Sustainable Energy Reviews., 50,* 1352–1372.

Redlarski, G., Krawczuk, M., Kupczyk, A., Piechocki, J., Ambroziak, D., & Palkowski, A. (2017). Swarm-assisted investment planning of a bioethanol plant. *Polish Journal of Environmental Studies, 26*(3), 1203–1214.

Rhein, A., Balzer, G., & Renz, P. (2017). Reliability-based improvement of life-cycle maintenance and replacement strategies in transmission systems using ant colony optimization. In *2017 6th International Youth Conference on Energy (IYCE).*

Robinson, D. G. (2005). Reliability analysis of bulk power systems using swarm intelligence Year: 2005. In *Annual Reliability and Maintainability Symposium* (pp. 96–102).

Saba, D., Laallam, F. Z., Hadidi, A. E., & Berbaoui, B. (2015). Contribution to the management of energy in the systems multi renewable sources with energy by the application of the multi agents systems "MAS." *Energy Procedia,* 616–623.

Selvakumar, K., Vijayakumar, K., & Boopathi, C. S. (2017). Demand response unit commitment problem solution for maximizing generating companies' profit. *Energies, 10*(10).

Sharafi, M., & ElMekkawy, T. Y. (2015). Stochastic optimization of hybrid renewable energy systems using sampling average method. *Renewable and Sustainable Energy Reviews, 52,* 1668–1679.

Sisodia, G. S., Soares, I., Banerji, S., & Van Den Poel, D. (2015). The status of energy price modelling and its relevance to marketing in emerging economies. *Energy Procedia,* 500–505.

Sonmez, M., Akgüngör, A. P., & Bektaş, S. (2017). Estimating transportation energy demand in Turkey using the artificial bee colony algorithm. *Energy, 122,* 301–310.

Suhane, P., Rangnekar, S., Mittal, A., & Khare, A. (2016). Sizing and performance analysis of standalone wind-photovoltaic based hybrid energy system using ant colony optimisation. *IET Renewable Power Generation, 10*(7), 964–972. Available from: http://ieeexplore.ieee.org/ielx7/4159946/7514393/07514408.pdf?tp=&arnumber=7514408&isnumber=7514393%5Cnhttp://ieeexplore.ieee.org/xpl/articleDetails.jsp?tp=&arnumber=7514408&source=tocalert&dld=aG90bWFpbC5jb20=.

Su, Z., Wang, J., Lu, H., & Zhao, G. (2014). A new hybrid model optimized by an intelligent optimization algorithm for wind speed forecasting. *Energy Conversion and Management, 85,* 443–452.

Szuba, T. (1999). A formal definition of the phenomenon of collective intelligence and its IQ measure. In *Lecture notes in computer science (including subseries lecture notes in artificial intelligence and lecture notes in bioinformatics)* (pp. 165–173).

Tapscott, D., & Williams, A. D. (2008). Wikinomics: How mass collaboration changes everything. *Journal of Information Technology and Politics, 5*(2), 259–262. Available from: http://search.ebscohost.com/login.aspx?direct=true&db=lxh&AN=35157924.

Thomas, L. (1973). On various words. *The New England Journal of Medicine, 289*(19), 1024–1026.

Toksarı, M. D. (2009). Estimating the net electricity energy generation and demand using the ant colony optimization approach: Case of Turkey. *Energy Policy, 2009*(37), 1181–1187.

Toksari, M. D. (2016). A hybrid algorithm of Ant Colony Optimization (ACO) and Iterated Local Search (ILS) for estimating electricity domestic consumption: Case of Turkey. *International Journal of Electrical Power & Energy Systems, 78,* 776–782.

Tounsi, W., & Rais, H. (2018). A survey on technical threat intelligence in the age of sophisticated cyber attacks. *Computers and Security., 72,* 212–233.

Toyokawa, W., Kim, H. R., & Kameda, T. (2014). Human collective intelligence under dual exploration-exploitation dilemmas. *PLoS One, 9*(4), e95789.

Udayraj, Mulani, K., Talukdar, P., Das, A., & Alagirusamy, R. (2015). Performance analysis and feasibility study of ant colony optimization, particle swarm optimization and cuckoo search algorithms for inverse heat transfer problems. *International Journal of Heat and Mass Transfer, 89*, 359–378.

Ünler, A. (2008). Improvement of energy demand forecasts using swarm intelligence: The case of Turkey with projections to 2025. *Energy Policy, 36*(6), 1937–1944.

Watanabe, I. (2005). An ACO algorithm for service restoration in power distribution systems. In *IEEE CEC 2005 Proceedings IEEE Congress on Evolutionary Computation* (pp. 2864–2871).

Xia, F., Zhao, X., Zhang, J., Ma, J., & Kong, X. (2014). BeeCup: A bio-inspired energy-efficient clustering protocol for mobile learning. *Future Generation Computer Systems, 37,* 449–460.

Xu, X., Mitra, J., Cai, N., & Mou, L. (2014). Planning of reliable microgrids in the presence of random and catastrophic events. *International Transactions on Electrical Energy Systems, 24*(8), 1151–1167.

Yanine, F. F., Caballero, F. I., Sauma, E. E., & Córdova, F. M. (2014). Building sustainable energy systems: Homeostatic control of grid-connected microgrids, as a means to reconcile power supply and energy demand response management. *Renewable and Sustainable Energy Reviews., 40,* 1168–1191.

Yegane, B. Y, Nakhai Kamalabadi, I., & Farughi, H. (2016). A non-linear integer bi-level programming model for competitive facility location of distribution centers. *International Journal of Engineering-Transactions B: Applications, 29*(8), 1131–1140. Available from: https://www.scopus.com/inward/record.uri?eid=2-s2.0-84984637702&partnerID=40&md5=2cfe1e1faca06a0dedd3440ab0cfb226.

Zahraee, S. M., Khalaji Assadi, M., & Saidur, R. (2016). Application of artificial intelligence methods for hybrid energy system optimization. *Renewable and Sustainable Energy Reviews, 66,* 617–630. Available from: http://dx.doi.org/10.1016/j.rser.2016.08.028.

Zak, M., & Zak, A. P. (1994). Unpredictable dynamics and collective brain. *Computers & Mathematics with Applications, 27*(9110), 185–197.

Zhang, H., Li, Z., Qu, Z., & Lewis, F. L. (2015). On constructing Lyapunov functions for multi-agent systems. *Automatica, 58,* 39–42.

# Chapter 22
# Fuzzy Collective Intelligence for Performance Measurement in Energy Systems

**Cengiz Kahraman, Sezi Çevik Onar and Basar Oztaysi**

**Abstract** Collective intelligence (CI) means that a group of people or animals can solve problems efficiently and offer greater insight and a better answer than any individual could provide. Fuzzy sets have been integrated with collective intelligence techniques in order to allow uncertain, vague imprecise and incomplete information to be incorporated to the CI models. The fuzzy CI techniques have been rarely used in the solution of energy problems even they still present new research opportunities to researchers. This chapter gives the results of the literature review on fuzzy CI research for energy systems.

**Keywords** Collective intelligence · Fuzzy sets · Performance · Energy systems
Particle swarm optimization

## 22.1 Introduction

Collective Intelligence is the wisdom of crowds introduced by Francis Galton. Collective intelligence is based on the fact that groups of people can be more intelligent than an intelligent individual and that groups do not always require intelligent people to reach a smart decision or outcome. Several collective intelligence techniques have been proposed in the literature. The most efficient heuristic based Swarm Intelligence algorithms are Particle Swarm Optimization (PSO), Artificial Bee Colony (ABC), Bat algorithm (BA), Cat Swarm Optimization, Bacterial Foraging, Stochastic diffusion search, Glowworm Swarm Optimization, Gravitational search algorithm, Cuckoo Search Algorithm, Differential Evolution, and Ant Colony Optimization (ACO) algorithm.

Fuzzy set theory (FST) has often been used in collective intelligence modeling since FST allows uncertain, vague imprecise and incomplete information to be

C. Kahraman (✉) · S. Çevik Onar · B. Oztaysi
Department of Industrial Engineering, Istanbul Technical University,
34367 Besiktas, Istanbul, Turkey
e-mail: kahraman@itu.edu.tr

© Springer International Publishing AG, part of Springer Nature 2018

C. Kahraman and G. Kayakutlu (eds.), *Energy Management—Collective and Computational Intelligence with Theory and Applications*, Studies in Systems, Decision and Control 149, https://doi.org/10.1007/978-3-319-75690-5_22

incorporated to the models. Chohra et al. (2010) develop a fuzzy cognitive and social negotiation strategy for autonomous agents with incomplete information, where the characters conciliatory, neutral, or aggressive, are suggested to be integrated in negotiation behaviors. First, one-to-one bargaining process, in which a buyer agent and a seller agent negotiate over price is developed for a time-dependent strategy and for a fuzzy cognitive and social strategy. Second, experimental measures carried out for different negotiation deadlines of buyer and seller agents are detailed. Third, experimental results for both time-dependent and fuzzy cognitive and social strategies are presented, analyzed, and compared for different deadlines of agents. The suggested fuzzy cognitive and social strategy allows agents to improve the negotiation process, with regard to the time-dependent one, in terms of agent utilities, round number to reach an agreement, and percentage of agreements. We present some examples of fuzzy collective intelligence algorithms in the following.

Venayagamoorthy et al. (2009) present two new strategies for navigation of a swarm of robots for target/mission focused applications including landmine detection and firefighting. The first method presents an embedded fuzzy logic approach in the particle swarm optimization (PSO) algorithm robots and the second method presents a swarm of fuzzy logic controllers, one on each robot. The framework of both strategies has been inspired by natural swarms such as the school of fish or the flock of birds. In addition to the target search using the above methods, a hierarchy for the coordination of a swarm of robots has been proposed. The robustness of both strategies is evaluated for failures or loss in swarm members. Yazdani et al. (2011) present a multiobjective optimization method that uses a Particle Swarm Optimization algorithm enhanced with a Fuzzy Logic-based controller. Their method uses a number of fuzzy rules and dynamic membership functions to evaluate search spaces at each iteration. The method works based on Pareto dominance and is tested using standard benchmark data sets.

Based on Ant Colony Optimization (ACO), Fidanova et al. (2012) use intuitionistic fuzzy estimation of start nodes with respect to the quality of the solution. They suggest various start strategies. Sensitivity analysis of the algorithm behavior according to estimation parameters is also made.

Wang et al. (2013) propose a hybrid artificial bee colony optimization (BCO) for solving the fuzzy flexible job-shop scheduling problem. The proposed algorithm utilizes multiple strategies in a combined way to generate the initial solutions with certain quality and diversity as the food sources, and applies the left-shift decoding scheme to convert solutions to active schedules. The exploitation search procedures based on the crossover operators for machine assignment and operation sequence in the employed bee phase are designed to generate the new neighbouring food sources. To prevent premature convergence in the scout bee phase, the population is updated by the new source with an adjustable search radius. Based on the Taguchi method of design of experiment, the influence of parameter setting is investigated and suitable parameter values are suggested. Fevrier and Valdez Castillo (2016) present a modification of a bio-inspired algorithm based on the bee colony optimization for optimizing fuzzy controllers. First, the traditional BCO is tested with the optimization of fuzzy controllers. Second, a modification of the original method is presented by including fuzzy logic to dynamically change the main parameter values of the algorithm during execution. Third, the proposed modification of the BCO algorithm with the fuzzy approach is used to optimize benchmark control problems. Dell'Orco et al. (2017) develop a new metaheuristic algorithm, based on the Fuzzy Bee Colony Optimization (FBCO), which integrates the concepts of BCO with a Fuzzy Inference System. The proposed method assigns, through the multi-criteria analysis, airport gates to scheduled flights based on both passengers' total walking distance and use of remote gates, to find an optimal flight-to-gate assignment for a given schedule.

Premkumar and Manikandan (2016) present the design of fuzzy proportional derivative controller and fuzzy proportional derivative integral controller for speed control of brushless direct current drive. Optimization of the above controllers' design is carried out using nature inspired optimization algorithms such as particle swarm, cuckoo search, and bat algorithms. The precise investigation through simulation is performed using simulink toolbox. From the simulation test results, it is found that bat optimized fuzzy proportional derivative controller has superior performance than the other considered controllers. Khooban and Niknam (2015) propose a new online intelligent strategy to realize the control of multi-area load frequency systems. The proposed intelligent strategy is based on a combination of a novel heuristic algorithm named Self-Adaptive Modified Bat Algorithm and the fuzzy logic which is used to optimally tune parameters of proportional–integral controllers which are the most popular methods.

Xu and Xuesong (2014) propose a new fuzzy identification method for T-S model identification algorithm, based on cat swarm and least squares method. T-S model identification is divided into structural and parameter identification. In the structure identification using cats warm can effectively overcome the traditional clustering algorithms. Using recursive least squares method to identify parts of the model parameters, forming the fuzzy identification method is based on cat swarm optimization.

Tabatabaei and Vahidi (2011) propose a novel methodology for the optimal location and sizing of shunt capacitors in radial distribution systems. Their method is based on a fuzzy decision making which using a new evolutionary method. The capacitor placement optimization problem includes: minimizing the cost of peak power, reducing energy loss and improving voltage profile. The installation node is selected by the fuzzy reasoning supported by the fuzzy set theory in a step by step procedure. Also an evolutionary algorithm known as bacteria foraging algorithm (BFA) is utilized in solving the objective multivariable optimization problem and the optimal node for capacitor placement is determined.

Neshat and Sepidname (2015) introduce fuzzy adaptive Swallow Swarm Optimization (FASSO) method to provide advantages such as high speed of convergence, avoidance from falling into local extremum, and high level of error tolerance. González et al. (2015) use fuzzy gravitational search algorithm with dynamic parameter adaptation for optimizing the modular neural network in a particular pattern recognition application. The proposed method is applied to medical images in echocardiogram recognition. One of the most common methods for detection and analysis of diseases in the human body, by physicians and specialists, is the use of medical images. Guerrero et al. (2015) describe the enhancement of the Cuckoo Search (CS) Algorithm via Lévy flights using a fuzzy system to dynamically adapt its parameters. The original CS method is compared with the proposed method called Fuzzy Cuckoo Search (FCS) on a set of benchmark mathematical functions. Liu and Lampinen (2005) introduce a new version of the Differential Evolution algorithm with adaptive control parameters—the fuzzy adaptive differential evolution algorithm, which uses fuzzy logic controllers to adapt the search parameters for the mutation operation and crossover operation.

There are not too many fuzzy collective intelligence works in energy systems in the literature. The aim of this chapter is to summarize those published works and give ideas to the readers how collective intelligence can be used under fuzziness for the problems of energy systems.

The rest of this chapter is organized as follows. Section 22.2 presents some graphical illustrations of the data on collective intelligence in energy systems. Section 22.3 classifies fuzzy collective intelligence approaches to energy systems under subtitles forecasting, economic analysis, strategic analysis, operational analyses, and performance analyses. Finally, conclusion and suggestions for further research are presented.

## 22.2 Literature Review: Graphical Analyses on CI in Energy (SCO)

Various fuzzy collective intelligence methods and algorithms are utilized for improving the performances of the energy systems. In this study, we categorize the collective intelligence algorithms as intelligent agents, particle swarm, ant colony, genetic algorithm, honeybee, particle bee, cuckoo search, artificial bee, bee swarm, swarm intelligence, quantum particle swarm, bee colony, bat algorithm and butterfly algorithm. In Scopus database, we search for these algorithms, fuzzy and energy performance at the "abstract, title and keywords" fields. We use the following search algorithm in the Scopus database: "*(((TITLE-ABS-KEY (intelligent AND agents) OR TITLE-ABS-KEY (particle AND swarm) OR TITLE-ABS-KEY (ant AND colony) OR TITLE-ABS-KEY (genetic AND algorithm) OR TITLE-ABS-KEY (honey AND bee) OR TITLE-ABS-KEY (particle AND bee) OR TITLE-ABS-KEY (cuckoo AND search) OR TITLE-ABS-KEY (artificial AND bee) OR TITLE-ABS-KEY (bee AND swarm) OR TITLE-ABS-KEY (swarm AND intelligence) OR TITLE-ABS-KEY (quantum AND particle AND swarm) OR TITLE-ABS-KEY (bee AND colony) OR TITLE-ABS-KEY (bat AND algorithm) OR TITLE-ABS-KEY (butterfly AND algorithm) OR TITLE-ABS-KEY (collective AND intelligence))) AND (TITLE-ABS-KEY (fuzzy)) AND TITLE-ABS-KEY (energy)) AND TITLE-ABS-KEY (performance)*".

There are a total of 599 studies utilized fuzzy collective intelligence methods for performance measurement in energy systems. The studies by year are given in Fig. 22.1.

Figure 22.1 shows the increasing trend on the collective intelligence methods in energy system performance. Especially between 2006 and 2013; there is a significant rise in the field. These papers are published in various sources. These sources can be grouped into three groups. The first group is the material science journals. "Applied Mechanics and Materials" with 33 papers and "Advanced Materials Research" with 23 papers are the leading journals that publish papers in collective intelligence on energy system performance. Further analysis shows that these studies held between 2012 and 2013. The second group is the energy journals (see Fig. 22.2). "Energy" is the leading journal that publishes fuzzy collective intelligence papers. "Energy Conversion and Management," "International Journal of Hydrogen Energy," "Energy and Buildings" and "Renewable Energy" are the leading energy journals that publish fuzzy collective intelligence in energy performance measurement.

The third group is the methodological journals that focus on methods and algorithms (see Fig. 22.3). "Advances in Intelligent Systems and Computing" is the leading journal that publishes fuzzy collective intelligence papers. "Communications in Computer and Information Science," "Journal of Intelligent and Fuzzy Systems," "Mathematical Problems in Engineering," and "Soft Computing" are the leading methodological journals that publish fuzzy collective intelligence in energy performance measurement.

In literature, various algorithms such as genetic algorithm, particle swarm, artificial intelligence, ant colony, intelligent agents, bee colony, honeybee, artificial

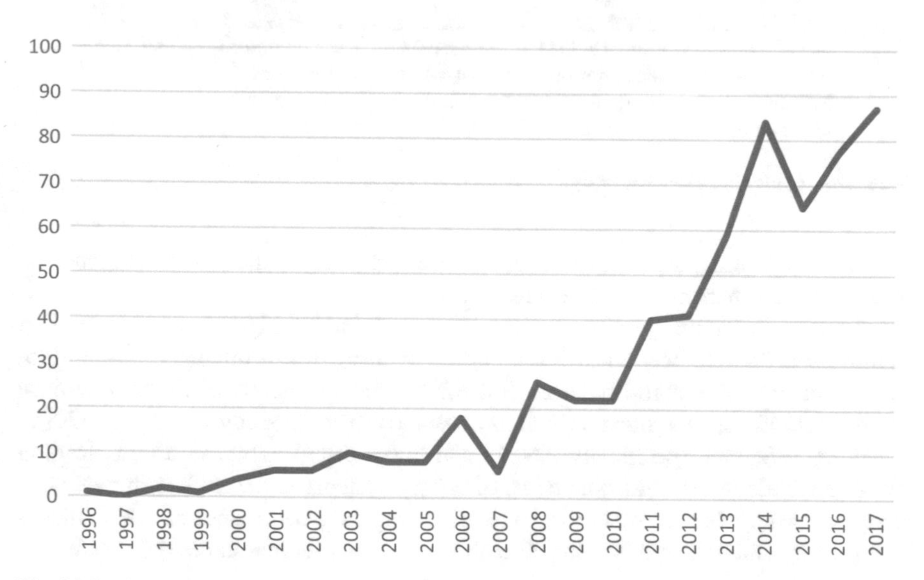


**Fig. 22.1** Collective intelligence for performance measurement in energy system studies by year

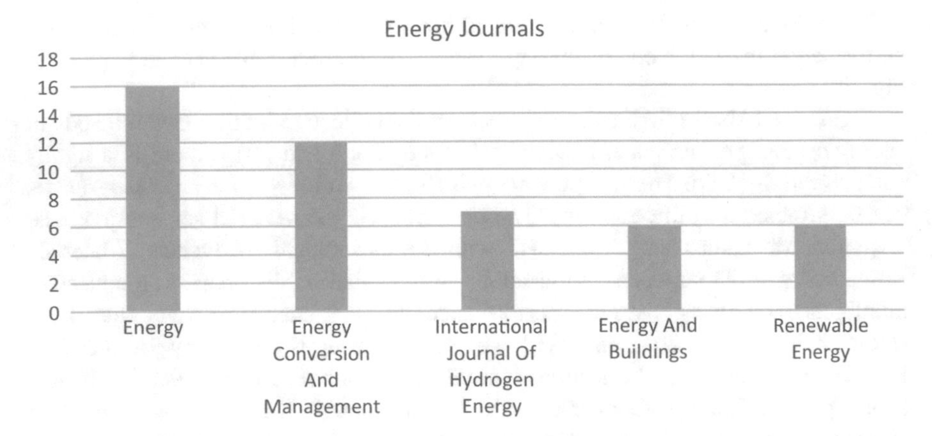


**Fig. 22.2** Energy journals

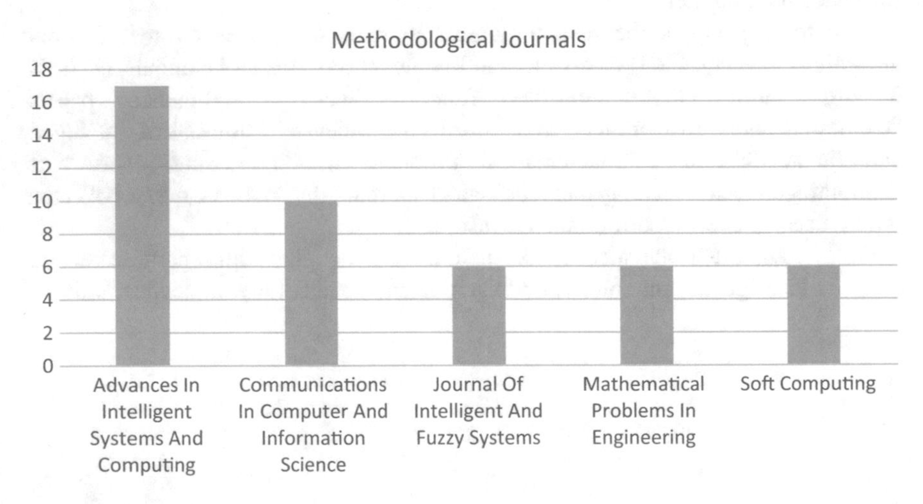


**Fig. 22.3** Methodological journals

bee and quantum particle swarm are utilized in fuzzy collective intelligence in energy system performance (see Fig. 22.4).

Genetic algorithm, particle swarm, and artificial intelligence are the leading algorithms. Particle bee, cuckoo search, bee swarm, swarm intelligence, bat algorithm and butterfly algorithms are not utilized in the literature. Fuzzy inference based methods are the most used fuzzy methods in these studies (see Fig. 22.5).

Fuzzy inference specifically ANFIS and fuzzy neural networks are the leading fuzzy methods in the field. The types of energy systems in the studies show a wide range of variety, but the measuring and improving the performance of renewables is the primary interest area. Figure 22.6 shows the most common energy problems.

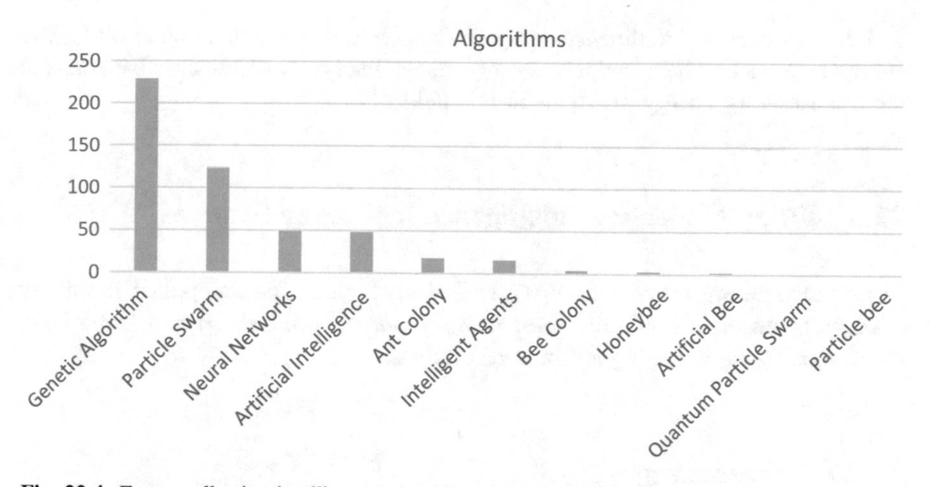


**Fig. 22.4** Fuzzy collective intelligence algorithms in energy system performance

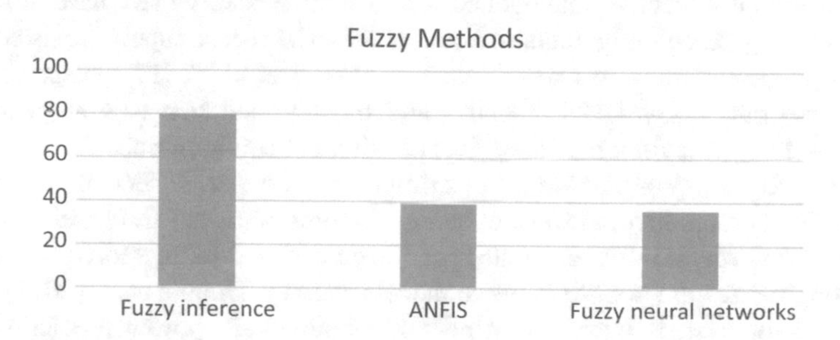


**Fig. 22.5** Fuzzy methods

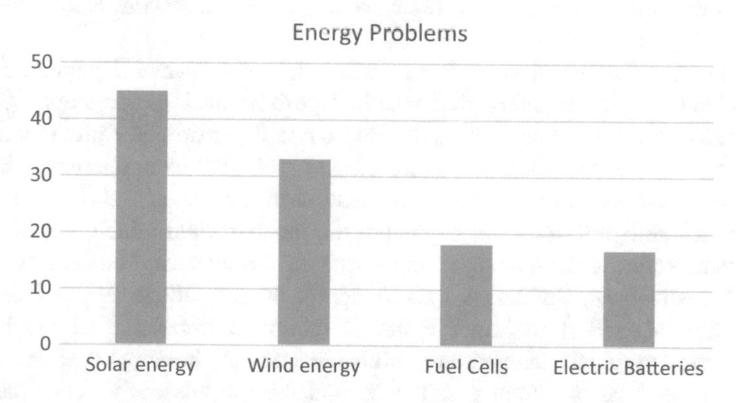


**Fig. 22.6** Energy problems

Fuzzy collective intelligence for performance measurement in solar and wind energy systems are the principal areas of interest. Fuel cells and electric batteries are the other leading energy problems in this field.

## 22.3 Fuzzy Collective Intelligence in Energy Systems

In this section, we classify fuzzy collective intelligence works applied in energy systems based on the subtitles *forecasting, economic analysis, strategic analysis, operational analyses*, and *performance analyses*.

### *22.3.1 Forecasting*

In the literature collective intelligence methods have been widely used with fuzzy logic in energy forecasting domain. Some of the most recent papers are listed in this section. In one of the most recent studies, Hugett et al. (2017) present a model which integrate fuzzy KNN (FKNN) and the artificial bee colony methods to predict weld quality for a particular friction stir weld configuration. The authors use artificial bee colony algorithm to determine the best parameters of the (F)KNN model. The authors involve wavelet energy features extracted from weld signals of X-Force, Y-Force, spindle rotational speed, feed rate, and plunge force are involved to the original feature pool to improve model accuracy. Dounia et al. (2017) present a multi-agent system based on wind and photovoltaic power prediction using artificial neural networks. The paper aims to implement a hybrid renewable energy system with generation units and storage units. To address uncertainties in the system, the authors use a fuzzy logic technique to decide the auxiliary energy source.

Wang et al. (2017) aim to optimize the thermal performance of parabolic trough solar collector systems using the genetic algorithm-back propagation (GA-BP) neural network model. The authors develop thermal performance prediction of the parabolic trough solar collector systems based on GA-BP neural network model to overcome fuzziness and incomplete information in the systems. Aziz et al. (2017) propose an intelligent approach to predict the biochar yield which is a renewable energy that produced from biomass thermochemical processes. The authors propose adaptive neuro-fuzzy inference system approach and utilize a particle swarm optimization algorithm to improve the prediction performance of the biochar. Heating rate, pyrolysis temperature, Moisture content, holding time and sample mass were used as the input parameters, and the outputs are biochar mass and biochar yield. Chen et al. (2017) focus on predicting the carbon efficiency and the comprehensive carbon ratio (CCR) is taken to be a measure of carbon efficiency. In the paper a model involving three steps is proposed. In the first step, an integrated fuzzy predictive model is proposed to predict the key state parameters based on the evaluation of current operating conditions. In the second step, a model is proposed to predict the yield using predicts the yield by using predicted values of the state parameters along with key material parameters are used as inputs for a particle swarm optimisation-based backpropagation neural network predictive model In the third step, the predicted yield is fed into the mechanism model in order to calculate the CCR.

Sadaei et al. (2017) propose a short-term load forecasting approach using a combination of Fuzzy Time Series (FTS) with Seasonal Auto Regressive Fractionally Integrated Moving Average method. The authors utilize Particle Swarm Optimization (PSO) to optimize the parameters of the method. Kassa et al. (2017) propose an artificial neuro-fuzzy inference system based approach for one-day-Ahead hourly wind power generation prediction. The proposed model is compared with a hybrid genetic algorithm based back propagation neural network model. The results show that the proposed approach outperformed the hybrid model demonstrating its favourable accuracy and reliability. Ruan et al. (2017) propose an on-line hybrid intelligent control system based on a genetic algorithm (GA) evolving fuzzy wavelet neural network software sensor to control dissolved oxygen (DO) in an anaerobic/anoxic/oxic process for treating papermaking wastewater. The outcomes specify that the reasonable forecasting and control performances are achieved when the proposed model is used.

Kaboli et al. (2017) propose a model for electrical energy consumption using optimized gene expression programming. The results are compared with those obtained from artificial neural network, support vector regression, adaptive neuro-fuzzy inference system, rule-based systems, linear and quadratic models optimized by particle swarm optimization, cuckoo search algorithm, and back-tracking search algorithm. The results reveal that the proposed method outperforms the other artificial intelligence based models. Basterrech (2017) analyse the performance of an automatic procedure for selecting the most important input features which effect the solar irradiance. The proposed method is based on a generalisation of swarm optimisation named Geometrical Particle Swarm Optimization. As a good combination of weather information is defined, the authors utilize a reservoir computing model as forecasting technique. Ahmet et al. (2017) present a new hybrid algorithm that uses Krill Herd optimization algorithm and Adaptive Neuro-Fuzzy Inference System for wind speed forecasting. In the proposed approach Krill Herd optimization is used to optimize the parameters of Adaptive Neuro-Fuzzy Inference System. The proposed model other models optimized using particle swarm optimization and genetic algorithms.

Chen et al. (2017) propose a dynamic-neighbourhood particle swarm optimization algorithm in the local optimal energy management strategy of plug-in hybrid electric vehicles. The authors also propose an online correction algorithm based on the backup control strategy and fuzzy logic. The authors also propose a predictive energy management strategy with an online correction algorithm. Son and Kim (2017) provide a precise model for the one-month-ahead forecast of electricity demand in the residential sector. In this study, a total of 20 influential variables are considered. A forecasting model based on support vector regression and fuzzy-rough feature selection with particle swarm optimization algorithms is proposed. The proposed forecasting model was validated using historical data from South Korea between January 1991 and December 2012. Alzoubi et al. (2017) focus on determining the best linear model using an artificial neural network (ANN) with the imperialist competitive algorithm (ICA-ANN) and ANN to predict the energy consumption for land levelling. In this research, effects of various soil properties such as embankment volume, soil compressibility factor, specific gravity, moisture content, slope, sand percent and soil swelling index on energy consumption are investigated. Panapakidis and Dagoumas (2017) propose a forecasting model using proposed model combines the Wavelet Transform, Genetic Algorithm, Adaptive Neuro-Fuzzy Inference System and Feed-Forward Neural Network. The authors apply the model to day-ahead natural gas demand predictions problem and show the forecasting performance of the model.

### *22.3.2 Economic Analyses*

Economics may be unique among the sciences in offering a formal mathematical proof that collective intelligence works, at least in the case of the market. a counter-hypothesis to the Efficient Markets Hypothesis can be formulated to explain the collective intelligence of the market—a counter-hypothesis that takes the natural variation of market behavior into account (Lo 2015).

Families, armies, countries, and companies all act collectively in ways that seem intelligent. Researchers have studied many fields from economics to political science based on collective intelligence. Analyzing how individual people's attitudes are determined or how they make economic choices would not be central to collective intelligence, but analyzing how different regulatory mechanisms in markets lead to more or less intelligent behavior by the markets as a whole would be central to collective intelligence (Malone et al. 2015).

Collectives or groups of individuals or animals do things that seem to be intelligent. They jointly achieve better results than a single individual could achieve. For instance, internet presents low cost communication and interaction which makes it feasible for groups to do many more things than before. Web-based collective intelligence is defined as the ability of a collective to learn or understand or to deal with new or trying situations, the skilled use of reason, the ability to apply knowledge to manipulate one's environment or to think abstractly as measured by objective criteria, based on the internet and associated technologies. These networks allow for synchronizing and parallelizing the different phases of the innovation process thus reducing transaction costs and costs of redesigning and testing products (Ickler 2010).

It has not yet much been studied on fuzzy collective intelligence based on economic analyses for energy investments. For instance, Yin et al. (2014) present a fuzzy logic controller for a parallel hybrid electric bus (PHEB). They propose a genetic-ant colony algorithm in order to reduce fuel consumption and emissions, including hydrocarbon, carbon monoxide, nitrogen oxide and particulate matter for optimum the fuzzy logic controller of PHEB, which is based on the ability of the quick global searching of genetic algorithm and the mechanism of positive feedback of ant colonies algorithm. Hence, a great potential exists in this area for the researchers who seeks for new research areas.

### 22.3.3 Strategic Analyses

Energy management strategy involves measuring energy usage, identifying the ways to decrease this usage, developing and implementing necessary systems for optimizing the energy usage and minimizing the costs and harms to the environment. Development of fuzzy control strategies for hybrid vehicles where the optimal fuel consumption is one of the primary problems in this field. Various computational intelligence techniques such as particle swarm optimization, dynamic programming, and ant colony are combined with fuzzy logic to achieve optimal control strategy for hybrid electric vehicles. Poursamad and Montazeri (2008) develop a genetic fuzzy control strategy for fuel cell based hybrid electric vehicles where the control is obtained with a fuzzy logic controller. Similarly, Zhou et al. (2013) develop a fuzzy logic control strategy for hybrid electric vehicles by using genetic algorithms. Ravey et al. (2012) used dynamic programming and fuzzy logic controller in developing control strategies for hybrid electric vehicles. Wu et al. (2008) develop fuzzy control strategies for hybrid electric vehicles where the control parameters are optimized with particle swarm optimization. Li et al. (2014) develop ant colony based control strategies for hybrid electric vehicles. Fuzzy control strategies are adapted to not only to hybrid vehicles but also to various fields such as to heating, ventilation and air-conditioning systems. Navale and Nelson (2012) showed that the performances of adaptive fuzzy logic controllers enhanced with genetic algorithms are superior to the conventional proportional, integral and derivative controllers.

One of the primary objectives of energy management strategy is to predict energy consumption and take the necessary actions for managing this need. Herrera et al. (2016) try to achieve optimal battery and supercapacitor combinations in tramways. The sizes of the energy storage systems are developed with multi-objective genetic algorithms where costs such as energy absorbed and the operating expenses are the primary determinants. Similarly, Yu et al. (2016) develop a real-time energy management strategy for electric urban busses where the objective is to achieve the optimal combination of conventional batteries and supercapacitors. The problem is considered as a multi-objective optimization problem where the cycle life of the battery, total consumption and the use of battery are taken into account. Both dynamic programming and fuzzy control strategy are used for optimizing the problem and the results show that dynamic programming gives better results. When the genetic algorithm is used to optimize the membership functions of the conventional fuzzy control strategy, the system performs almost as good as dynamic programming.

Developing the operation strategy for the energy systems is another critical research area. Moradi and Eskandari (2014) develop an operation strategy for microgrids. Initial investment costs, operational strategy costs, purchase of electricity from the utility, maintenance, and operational costs, as well as revenues including those associated with a reduction in non-delivered energy, the credit for the reduction in levels of environmental pollution, and sales of electricity back to the utility are taken into account for developing operational strategies. The quadratic programming (QP) and the particle swarm optimization (PSO) algorithms are used to define the operational strategy for the microgrid where the uncertainty in the power price is modeled with fuzzy sets. Similarly, Cha et al. (2015) develop a multi-agent system-based microgrid operation strategy.

### 22.3.4 Operational Analyses

The literature provides various energy operational management studies involving fuzzy collective intelligence. In one of the most recent studies, Amrani et al. (2017) propose a multi-agent approach for managing electric vehicle energy. The management strategy is a hybrid strategy, utilizing artificial intelligence, fuzzy logic and genetic algorithms and the multi-agent systems. The fuzzy inference system is first optimized off-line by the genetic algorithm. Then it is used during on-line checking to take into account the uncertain case. Kanellos et al. (2017) propose a cost-effective and emission-aware power management system for ships. The proposed optimization method is exploiting an interactive approach based on particle swarm optimization method and a fuzzy mechanism to improve the computational efficiency of the algorithm. The proposed fuzzy-based particle swarm optimization algorithm aims at minimizing the operation cost, limiting the greenhouse gas (GHG) emissions and satisfying the technical and operational constraints of the ship. Ravadanegh et al. (2017) propose a comprehensive method to solve a combinatorial problem consisting of distribution system reconfiguration, capacitor allocation, and renewable energy resources sizing and sitting simultaneously and to improve power system's accountability and system performance parameters. Due to finding solution which is closer to realistic characteristics, load forecasting, market price errors and the uncertainties related to the variable output power of wind-based DG units are put in consideration. The authors use NSGA-II accompanied by the fuzzy set theory to solve the aforementioned multi-objective problem. The proposed scheme finally leads to a solution with a minimum voltage deviation, a maximum voltage stability, a lower amount of pollutant and lower cost.

Xu et al. (2017) propose an improved bacterial foraging optimization (IBFO) algorithm by swarm intelligence algorithm to improve the establishment of rule library. Initially, the imperfectness of fuzzy rule library based on artificial experience induction is analyzed. Then, the improved fuzzy control system is described.

Finally, the Gaussian membership function parameters of the improved TSK fuzzy system (C-ATSKFS, constant-ameliorative TSK fuzzy system) rule library are optimized. Compared with the existing method, the improved algorithm can effectively increase the recognition accuracy of the fuzzy control system which is very critical in the fields of energy production and robot control. Abd et al. (2017) focus on managing power consumption in cloud computing infrastructure. The authors propose a DNA-based Fuzzy Genetic Algorithm (DFGA) that employs DNA-based scheduling strategies to reduce power consumption in cloud data centres. The method has a power-aware architecture for managing power consumption in the cloud computing infrastructure. The authors also identify the performances metrics that are needed to evaluate the proposed work performance. The experimental results show that DFGA reduced power consumption when compared with other algorithms. Hosseinpour et al. (2017) propose a flexible and reliable framework based on a combination of artificial neural network, genetic algorithm, and fuzzy systems for multi-objective exergetic optimization of continuous photobiohydrogen production process from syngas by *Rhodospirillum rubrum* bacterium. To this end, the artificial neural network is extended with fuzzy clustering method to model exergetic outputs by input variables. The outputs of modelling system are then fed into a novel optimization approach developed by hybridising additive linear interdependent fuzzy multi-objective optimization and the elitist non-dominated sorting genetic algorithm. The optimization is carried out to minimize the normalized exergy destruction and maximize the rational and process exergetic efficiencies, simultaneously. The solutions of the proposed approach are also compared with conventional fuzzy multi-objective optimization procedure with independent objectives.

Mohammadi et al. (2017) present a fuzzy based methodology for distribution system feeder reconfiguration considering DSTATCOM with an objective of minimizing real power loss and operating cost. Installation costs of DSTATCOM devices and the cost of system operation. In the proposed approach, the fuzzy membership function of loss sensitivity is used for the selection of weak nodes in the power system for the placement of DSTATCOM and the optimal parameter settings of the DFACTS device along with an optimal selection of tie switches in reconfiguration process are governed by a genetic algorithm. Rezvani et al. (2016) focus on grid-connected intelligent hybrid battery/photovoltaic system using the new hybrid fuzzy-neural method. To capture the maximum power point (MPP), a hybrid fuzzy-neural maximum power point tracking method is applied. Obtained results represent the effectiveness and superiority of the proposed method, and approximately two percentage points increment the average tracking efficiency of the hybrid fuzzy-neural in comparison to the conventional methods. The proposed method has the advantages of robustness, fast response and good performance. Chen et al. (2016) propose a novel optimal power management approach for plug-in hybrid electric vehicles against uncertain driving conditions. To optimize the threshold parameters of the rule-based power management strategy under a certain driving cycle, the particle swarm optimization algorithm is employed, and the optimization results are used to determine the optimal control actions. To better implement the power management strategy in real time, a driving condition recognition algorithm is proposed to identify real-time driving conditions through a fuzzy logic algorithm. To adjust the thresholds of the rule-based strategy adaptively under uncertain driving cycles, a dynamic optimal parameters algorithm has been further established accordingly, and it is helpful for avoiding the problem that the thresholds of the rule-based strategy are very sensitive to the driving cycles. Finally, in combination with the above efforts, a detailed operational flowchart of the particle swarm optimization algorithm-based optimal power management through driving cycle recognition is proposed.

Aghbashlo et al. (2016) focus on optimization of continuous photobiological hydrogen production using a hybrid approach. To this end, the authors propose a multi-objective hybrid optimization technique was developed by coupling the elitist NSGA-II with the adaptive neuro-fuzzy inference system to optimize the operational conditions of the photobioreactor. The syngas flow rate and culture agitation speed are selected as the independent variables, while rational, and process exergy efficiencies, as well as normalized exergy destruction, are dependent variables. The ANFIS is used to establish an objective function for each dependent variable individually based on the independent variables. The developed ANFIS model is then utilized by the NSGA-II approach to finding the optimal operating conditions leading to the highest rational and process exergy efficiencies and the lowest normalized exergy destruction. Hussain et al. (2015) focus on optimization based fuzzy resource allocation framework for smart grids. The authors propose a power flow control scheme using a framework of fuzzy logic and genetic algorithm to manage desired power flow levels within the smart grid efficiently. A fuzzy decision criterion is designed to choose a most suitable power source to deliver power to a certain demand. A genetic algorithm is used to choose a most suitable route from source to demand and optimize a cost function based on distance. Simulations show that the smart grid power flow can achieve the desired thresholds by incorporating the proposed approach even in the presence of unpredictable power fluctuations from renewable energy resources. Karavas et al. (2015) propose a multi-agent decentralized energy management system based on distributed intelligence for the design and control of autonomous polygeneration microgrids. The authors present the design and investigation of a decentralized energy management system for the autonomous polygeneration microgrid topology. The decentralized energy management system gives the possibility to control each unit of the microgrid independently. The most important advantage of using a decentralized architecture is that the managed microgrid has much higher chances of partial operation in cases when malfunctions occur at different parts of it, instead of a complete system breakdown. The designed system was based on a multi-agent system and employed Fuzzy Cognitive Maps for its implementation. It is then compared to a case study with an existing centralized energy management system. The technical performance of the decentralized solution performance is on par with the existing centralized one, presenting improvements in financial and operational terms for the implementation and operation of an autonomous polygeneration microgrid. Moradi et al. (2015) energy management in microgrids are addressed from different perspectives such as economic efficiency, environmental restrictions, and reliability improvement. The hybrid optimization method is used to optimize the type and capacity of distributed generation, sources, and the capacity of storage devices. Quadratic programming and particle swarm optimization algorithms are used for optimization integrated with fuzzy logic.

### *22.3.5 Performance Analyses*

Fuzzy sets have been used with collective intelligence techniques in the field of performance. In one of the most recent studies, Jia et al. (2017) focus on the performance of train operations regarding energy efficiency and service quality. The authors formulate a bi-objective train-speed trajectory optimization model to minimize the energy consumption and travel time in an inter-station section simultaneously. For obtaining an optimal train-speed trajectory which has an equal satisfactory degree on both objectives, a fuzzy linear programming approach is applied to reformulate the objectives. Also, a genetic algorithm is developed to solve the proposed train-speed trajectory optimization problem. Wang et al. (2017) aim to optimize the thermal performance of parabolic trough solar collector systems in order to improve its thermal performance, based on the genetic algorithm-back propagation neural network model. In order to deal with issues such as fuzzy or incomplete information and a complex architecture, the authors propose a system based on genetic algorithm and backpropagation neural networks. Nilashi et al. focus on energy performance prediction in residual buildings. The proposed method utilizes clustering, noise removal and prediction techniques. In this manner Expectation Maximization, Principal Component Analysis and Adaptive Neuro-Fuzzy Inference System methods are used. The authors provide experimental results on real-world dataset show the efficiency of the proposed method. Sukumar et al. (2017) present a power management system for a grid-connected PV and solid oxide fuel cell, considering variation in the load and solar radiation. The objective of the proposed system is to minimize the power drawn from the grid and operate the SOFC within a specific power range. Since the photovoltaic is operated at the maximum power point, the power management involves the control of SOFC active power where a proportional and integral (PI) controller is used. In the study, the control parameters are determined by the genetic algorithm and simplex method. Also, a fuzzy logic controller is also developed to generate appropriate control parameters for the PI controller. The performance of the controllers is evaluated by minimizing the integral of time multiplied by absolute error criterion. Wang et al. (2017) focus on optimizing the thermal performance (system output energy, thermal efficiency, and heat loss of cavity absorber) of parabolic trough solar collector (PTC) systems to improve its thermal performance, based on the genetic algorithm-back propagation (GA-BP) neural network model. There are some undefined problems, fuzzy or incomplete information and a complex thermal performance of the PTC systems. Therefore, the thermal performance prediction of the PTC systems based on GA-BP neural network model was developed. Subsequently, the metrics performances have been adopted to understand the algorithm and evaluate the prediction accuracy comprehensively.

Wang (2017) focus on performance problem of electric vehicles and present a novel double—energy fuzzy control algorithm for battery-supercapacitor based on particle swarm optimization. The proposed algorithm can avoid falling into local optimum and being over-reliance on prior knowledge by using the swarm intelligence global optimization and evolutionary operation. The simulation results show that this method can improve the vehicle performances in the large extent and verify the effectiveness of the control strategy. Huang and Chen (2017) develop a hybrid powertrain in which a magnetic flywheel system (MFS) is integrated with the fuel cells to improve the performance of fuel cells used in vehicles. The authors also propose an auto-tuning proportional–integral–derivative (PID) controller based on the controls of multiple adaptive neuro-fuzzy interference systems and particle swarm optimization. Furthermore, MATLAB/Simulink simulations considering an FTP-75 urban driving cycle are conducted, and a variability improvement of approximately 27.3% in fuel cell output is achieved. Anicic and Jovic (2016) investigate the hydrodynamic performance of a novel type of ducted tidal turbine. The authors propose an adaptive neuro-fuzzy inference system to estimate power coefficient value of the ducted tidal turbines. The backpropagation learning algorithm is used for training this network. This intelligent controller is implemented using Matlab/Simulink, and the performances are investigated. The simulation results presented in this paper show the effectiveness of the developed method.

Aghbashlo et al. (2016) focus on optimizing the performance of a continuous photobioreactor for hydrogen production from syngas via water gas shift reaction by Rhodospirillum rubrum. To this end, a new multi-objective hybrid optimization technique was developed by coupling the elitist NSGA-II with the ANFIS to optimize the operational conditions of the photobioreactor. The ANFIS was used to establish an objective function for each dependent variable individually based on the independent variables. The developed ANFIS model was then utilized by the NSGA-II approach to find the optimal operating conditions simultaneously leading to the highest rational and process exergy efficiencies and the lowest normalized exergy destruction. Aghajani et al. (2015) propose a multi-objective energy management system is proposed to optimize micro-grid performance in a short-term in the presence of Renewable Energy Sources. In the typical micro-grid, different technologies including Wind Turbine, PhotoVoltaic cell, Micro-Turbine, Full Cell, battery hybrid power source and responsive loads are used which makes the energy management problem very complex. The authors use multi-objective particle swarm optimization algorithm to handle the nonlinearity in the problem. The authors also utilize, fuzzy-based mechanism and non-linear sorting system to determine the best compromise considering the set of solutions from Pareto-front space. The numerical results represented the effect of the proposed Demand Side Management scheduling model on reducing the effect of uncertainty. Camargo et al. (2010) propose an intelligent supervision system for industrial production performance in oil wells. The proposed method combines Genetic Algorithms, Fuzzy

Classification, Neo-Fuzzy systems and Energy Mass Balance. In the study, the completion geometry and the reservoir potential is considered in order to establish the oil or gas flow that a well can produce. Calderaro et al. (2007) focus on wind generators performance optimization and propose a design for wind tribunes that generate an adaptive fuzzy model for maximum energy extraction. The proposed design involves fuzzy clustering combined with genetic algorithms and recursive least-squares optimization methods. Zhao et al. (2004) propose a multivalued fuzzy behavior control system for robot navigation using singletons instead of fuzzy set consequences. In the study, a genetic algorithm approach which improves the performance of the proposed fuzzy system with singleton consequences is proposed. The authors Show the effectiveness of the proposed system by simulation results.

## 22.4 Conclusions

In this chapter, we examined the usage of fuzzy collective intelligence techniques in energy systems. Even they have not frequently been used for energy problems in the literature; there are still great opportunities for them to be used for the solution of energy problems. Fuzzy collective intelligence techniques such as particle swarm optimization, artificial bee colony, bat algorithm, cat swarm optimization, bacterial foraging, stochastic diffusion search, glowworm swarm optimization, gravitational search algorithm, cuckoo search algorithm, differential evolution, and ant colony optimization algorithm have been already developed and used in the solutions of other problems. This may provide a reference for them to be used in the energy systems in the future.

## References


Abd, S. K., Al-Haddad, S. A. R., Hashim, F., Abdullah, A. B. H. J., & Yussof, S. (2017). An effective approach for managing power consumption in cloud computing infrastructure. *Journal of Computational Science, 21,* 349–360.

Aghbashlo, M., Hosseinpour, S., Tabatabaei, M., Younesi, H., & Najafpour, G. (2016). On the exergetic optimization of continuous photobiological hydrogen production using hybrid ANFIS-NSGA-II (adaptive neuro-fuzzy inference system-non-dominated sorting genetic algorithm-II). *Energy, 96,* 507–520.

Ahmed, K., Ewees, A. A., Abd El Aziz, M., Hassanien, A. E., Gaber, T., Tsai, P.-W., et al. (2017). A hybrid krill-ANFIS model for wind speed forecasting. *Advances in Intelligent Systems and Computing, 533,* 365–372.

Alzoubi, I., Delavar, M., Mirzaei, F., & Nadjar Arrabi, B. (2017). Integrating artificial neural network and imperialist competitive algorithm (ICA), to predict the energy consumption for land leveling. *International Journal of Energy Sector Management, 11*(4), 522–540.

Amrani, R. E., Yahyaouy, A., & Tairi, H. (2017). Hybrid strategy based on MAS for an intelligent energy management: Application to an electric vehicle. *Intelligent Systems and Computer Vision, ISCV* 2017, art. no. 8054917,

Aziz, M. A. E., Hemdan, A. M., Ewees, A. A., Elhoseny, M., Shehab, A., Hassanien, A. E., et al. (2017). Prediction of biochar yield using adaptive neuro-fuzzy inference system with particle swarm optimization. *Proceedings—2017 IEEE PES-IAS PowerAfrica Conference: Harnessing Energy, Information and Communications Technology (ICT) for Affordable Electrification of Africa*, PowerAfrica 2017, art. no. 7991209, pp. 115–120.

Basterrech, S. (2017). Geometric particle swarm optimization and reservoir computing for solar power forecasting. *Advances in Intelligent Systems and Computing, 576,* 88–97.

Cha, H. J., Won, D. J., Kim, S. H., Chung, I. Y., & Han, B. M. (2015). Multi-agent system-based microgrid operation strategy for demand response. *Energies, 8*(12), 14272–14286.

Chen, X., Chen, X., She, J., & Wu, M. (2017a). Hybrid multistep modeling for calculation of carbon efficiency of iron ore sintering process based on yield prediction. *Neural Computing and Applications, 28*(6), 1193–1207.

Chen, Z., Xiong, R., & Cao, J. (2016). Particle swarm optimization-based optimal power management of plug-in hybrid electric vehicles considering uncertain driving conditions. *Energy, 96,* 197–208.

Chen, Z., Xiong, R., Wang, C., & Cao, J. (2017b). An on-line predictive energy management strategy for plug-in hybrid electric vehicles to counter the uncertain prediction of the driving cycle. *Applied Energy, 185,* 1663–1672.

Chohra, A., Madani, K., & Kanzari, D. (2010). Fuzzy cognitive and social negotiation agent strategy for computational collective intelligence. Lecture Notes in Computer Science, Vol. 6220 LNCS, pp. 143–159.

Dell'Orco, M., Marinelli, M., & Altieri, M. G. (2017). Solving the gate assignment problem through the Fuzzy Bee Colony Optimization. *Transportation Research Part C: Emerging Technologies, 80,* 424–438.

Dounia, E. B., Ali, Y., & Jaouad, B. (2017). Multi-agent system based on the fuzzy control and extreme learning machine for intelligent management in hybrid energy system. *2017 Intelligent Systems and Computer Vision, ISCV 2017*, art. no. 8054922.

Fevrier, C. C., & Valdez Castillo, O. (2016). Optimization of fuzzy controller design using a new bee colony algorithm with fuzzy dynamic parameter adaptation. *Applied Soft Computing, 43,* 131–142.

Fidanova, S., Atanassov, K., & Marinov, P. (2012). Intuitionistic fuzzy estimation of the ant colony optimization starting points (Conference Paper), Lecture Notes in Computer Science (including subseries Lecture Notes in Artificial Intelligence and Lecture Notes in Bioinformatics), Vol. 7116 LNCS, pp. 222–229. *8th International Conference on Large-Scale Scientific Computations, LSSC 2011*; Sozopol; Bulgaria; 6 June 2011 through 10 June 2011; Code 89998.

González, B., Valdez, F., Melin, P., & Prado-Arechigab, G. (2015). Fuzzy logic in the gravitational search algorithm enhanced using fuzzy logic with dynamic alpha parameter value adaptation for the optimization of modular neural networks in echocardiogram recognition. *Applied Soft Computing, 37,* 245–254.

Guerrero, M., Castillo, O., & Mario García, M. (2015). Fuzzy dynamic parameters adaptation in the Cuckoo Search Algorithm using fuzzy logic. *2015 IEEE Congress on Evolutionary Computation (CEC),* May 25–28, 2015, Sendai, Japan.

Herrera, V., Milo, A., Gaztañaga, H., Etxeberria-Otadui, I., Villarreal, I., & Camblong, H. (2016). Adaptive energy management strategy and optimal sizing applied on a battery-supercapacitor based tramway. *Applied Energy, 169,* 831–845.

Hosseinpour, S., Aghbashlo, M., Tabatabaei, M., Younesi, H., Mehrpooya, M., & Ramakrishna, S. (2017). Multi-objective exergy-based optimization of a continuous photobioreactor applied to produce hydrogen using a novel combination of soft computing techniques. *International Journal of Hydrogen Energy, 42*(12), 8518–8529.

Huang, C. N., & Chen, Y. S. (2017). Design of magnetic flywheel control for performance improvement of fuel cells used in vehicles. *Energy, 118,* 840–852.

Huggett, D. J., Liao, T. W., Wahab, M. A., & Okeil, A. (2017). Prediction of friction stir weld quality without and with signal features. *International Journal of Advanced Manufacturing Technology*, 1–15.

Hussain, S., Al Alili, A., & Al Qubaisi, A. M. (2015). Optimization based fuzzy resource allocation framework for smart grid. *International Conference on Smart Energy Grid Engineering, SEGE 2015*, art. no. 7324627.

Ickler, H. (2010). An approach for the visual representation of business models that integrate web-based collective intelligence into value creation. In T. J. Bastiaens, U. Baumöl, & B. J. Krämer (Eds.), *On collective intelligence. Advances in intelligent and soft computing* (Vol. 76). Berlin, Heidelberg: Springer.

Jędrzejowicz, P., Nguyen, N. T., & Hoang, K. (Eds.). (2011). Computational collective intelligence: technologies and applications. *ICCCI 2011, Third International Conference on Computational Collective Intelligence*, Gdynia, Poland, September 21–23, 2011, Proceedings, Part II.

Jia, J., Yang, K., Yang, L., Gao, Y., & Li, S. (2017). Designing train-speed trajectory with energy efficiency and service quality. *Engineering Optimization*, 1–22 (Article in Press).

Kaboli, S. H. A., Fallahpour, A., Selvaraj, J., & Rahim, N. A. (2017). Long-term electrical energy consumption formulating and forecasting via optimized gene expression programming. *Energy, 126,* 144–164.

Kanellos, F. D., Anvari-Moghaddam, A., & Guerrero, J. M. (2017). A cost-effective and emission-aware power management system for ships with integrated full electric propulsion. *Electric Power Systems Research, 150,* 63–75.

Karavas, C.-S., Kyriakarakos, G., Arvanitis, K. G., & Papadakis, G. (2015). A multi-agent decentralized energy management system based on distributed intelligence for the design and control of autonomous polygeneration microgrids. *Energy Conversion and Management, 103* (art. no. 7262), 166–179.

Kassa, Y., Zhang, J. H., Zheng, D. H., & Wei, D. (2017). Short term wind power prediction using ANFIS. *2016 IEEE International Conference on Power and Renewable Energy, ICPRE 2016*, art. no. 7871238, pp. 388–393.

Khooban, M. H., & Niknam, T. (2015). A new intelligent online fuzzy tuning approach for multi-area load frequency control: Self adaptive modified Bat Algorithm. *Electrical Power and Energy Systems, 71,* 254–261.

Leonori, S., Paschero, M., Rizzi, A., & Mascioli, F. M. F. (2017). An optimized microgrid energy management system based on FIS-MO-GA paradigm. *IEEE International Conference on Fuzzy Systems*, art. no. 8015438,.

Li, L., Huang, H., Lian, J., Yao, B., Zhou, Y., Chang, J., et al. (2014). Research of Ant Colony optimized adaptive control strategy for hybrid electric vehicle. *Mathematical Problems in Engineering*.

Liu, J., & Lampinen, J. (2005). A fuzzy adaptive differential evolution algorithm. *Soft Computing, 9*(6), 448–462.

Lo, A. W. (2015). The wisdom of crowds vs. the madness of mobs. In T. W. Malone & M. S. Bernstein (Eds.), *Handbook of collective intelligence*. Massachusetts London, England: The MIT Press Cambridge.

Malone, T. W., & Bernstein, M. S. (2015). Introduction. In T. W. Malone & M. S. Bernstein (Eds.), *Handbook of collective intelligence*. Massachusetts London, England: The MIT Press Cambridge.

Mohammadi, M., Abasi, M., & Rozbahani, A. M. (2017). Fuzzy-GA based algorithm for optimal placement and sizing of distribution static compensator (DSTATCOM) for loss reduction of distribution network considering reconfiguration. *Journal of Central South University, 24*(2), 245–258.

Moradi, M. H., & Eskandari, M. (2014). A hybrid method for simultaneous optimization of DG capacity and operational strategy in microgrids considering uncertainty in electricity price forecasting. *Renewable Energy, 68*, 697–714.

Moradi, M. H., Eskandari, M., & Mahdi Hosseinian, S. (2015). Operational strategy optimization in an optimal sized smart microgrid. *IEEE Transactions on Smart Grid, 6*(3), 1087–1095 (art. no. 6891360).

Navale, R. L., & Nelson, R. M. (2012). Use of genetic algorithms and evolutionary strategies to develop an adaptive fuzzy logic controller for a cooling coil—Comparison of the AFLC with a standard PID controller. *Energy and Buildings, 45,* 169–180.

Neshat, M., & Sepidname, G. (2015). A new hybrid optimization method inspired from swarm intelligence: Fuzzy adaptive swallow swarm optimization algorithm. *Egyptian Informatics Journal, 16*(3), 339–350.

Panapakidis, I. P., & Dagoumas, A. S. (2017). Day-ahead natural gas demand forecasting based on the combination of wavelet transform and ANFIS/genetic algorithm/neural network model. *Energy, 118,* 231–245.

Poursamad, A., & Montazeri, M. (2008). Design of genetic-fuzzy control strategy for parallel hybrid electric vehicles. *Control Engineering Practice, 16*(7), 861–873.

Premkumar, K., & Manikandan, B. V. (2016). Bat algorithm optimized fuzzy PD based speed controller for brushless direct current motor. *Engineering Science and Technology, an International Journal, 19*(2), 818–840.

Ravadanegh, S. N., Oskuee, M. R. J., & Karimi, M. (2017). Multi-objective planning model for simultaneous reconfiguration of power distribution network and allocation of renewable energy resources and capacitors with considering uncertainties. *Journal of Central South University, 24*(8), 1837–1849.

Ravey, A., Blunier, B., & Miraoui, A. (2012). Control strategies for fuel-cell-based hybrid electric vehicles: From offline to online and experimental results. *IEEE Transactions on Vehicular Technology, 61*(6), 2452–2457.

Rezvani, A., Khalili, A., Mazareie, A., & Gandomkar, M. (2016). Modeling, control, and simulation of grid connected intelligent hybrid battery/photovoltaic system using new hybrid fuzzy-neural method. *ISA Transactions, 63,* 448–460.

Ruan, J., Zhang, C., Li, Y., Li, P., Yang, Z., Chen, X., et al. (2017). Improving the efficiency of dissolved oxygen control using an on-line control system based on a genetic algorithm evolving FWNN software sensor. *Journal of Environmental Management, 187,* 550–559.

Sadaei, H. J., Guimarães, F. G., José da Silva, C., Lee, M. H., & Eslami, T. (2017). Short-term load forecasting method based on fuzzy time series, seasonality and long memory process. *International Journal of Approximate Reasoning, 83,* 196–217.

Son, H., & Kim, C. (2017). Short-term forecasting of electricity demand for the residential sector using weather and social variables. *Resources, Conservation and Recycling, 123,* 200–207.

Sukumar, S., Marsadek, M., Ramasamy, A., Mokhlis, H., & Mekhilef, S. (2017). A fuzzy-based PI controller for power management of a grid-connected PV-SOFC hybrid system. *Energies, 10* (11) (art. no. 1720).

Tabatabaei, S. M., & Vahidi, B. (2011). Bacterial foraging solution based fuzzy logic decision for optimal capacitor allocation in radial distribution system. *Electric Power Systems Research, 81* (4), 1045–1050.

Venayagamoorthy, G. K., Grant, L. L., & Doctor, S. (2009). Collective robotic search using hybrid techniques: Fuzzy logic and swarm intelligence inspired by nature. *Engineering Applications of Artificial Intelligence, 22*(3), 431–441.

Wang, L., Zhou, G., Xu, Y., & Liu, M. (2013). A hybrid artificial bee colony algorithm for the fuzzy flexible job-shop scheduling problem. *International Journal of Production Research, 51* (12), 3593–3608.

Wang, W., Li, M., Hassanien, R. H. E., Ji, M. E., & Feng, Z. (2017a). Optimization of thermal performance of the parabolic trough solar collector systems based on GA-BP neural network model. *International Journal of Green Energy, 14*(10), 819–830.

Wang, W., Li, M., Hassanien, R. H. E., Ji, M. E., & Feng, Z. (2017b). Optimization of thermal performance of the parabolic trough solar collector systems based on GA-BP neural network model. *International Journal of Green Energy, 14*(10), 819–830.

Wang, X. (2017). Research on double energy fuzzy controller of electric vehicle based on particle swarm optimization of multimedia big data. *International Journal of Mobile Computing and Multimedia Communications, 8*(3), 32–43.

Wu, X., Cao, B., Wen, J., & Bian, Y. (2008). Particle swarm optimization for plug-in hybrid electric vehicle control strategy parameter. *2008 IEEE Vehicle Power and Propulsion Conference, VPPC* 2008.

Xu, C., Feng, X., Zhang, J., Li, X., & Cao, Y. (2017). The design of a fuzzy control rule library based on improved bacterial foraging optimization. *Chongqing Daxue Xuebao/Journal of Chongqing University, 40*(7), 63–71.

Xu, S., & Xuesong, X. (2014). Fuzzy identification base on cat swarm optimization algorithm. *The 26th Chinese Control and Decision Conference (CCDC 2014)*, 31 May–2 June 2014, Changsha, China.

Yazdani, H., Kwasnicka, H., & Ortiz-Arroyo, D. (2011). Multiobjective particle swarm optimization using fuzzy logic. In P. Jędrzejowicz, N. T. Nguyen, & K. Hoang (Eds.), *Computational Collective Intelligence: Technologies and Applications, ICCCI 2011, Third International Conference on Computational Collective Intelligence*, Gdynia, Poland, September 21–23, Proceedings, Part II, pp. 224–233.

Yin, A., Zhao, H., & Zhang, B. (2014). Optimisation of fuzzy control strategy for hybrid electric bus based on genetic-ant colony algorithm. *Australian Journal of Electrical and Electronics Engineering, 11*(3), 339–346.

Yu, H., Tarsitano, D., Hu, X., & Cheli, F. (2016). Real time energy management strategy for a fast charging electric urban bus powered by hybrid energy storage system. *Energy, 112,* 322–331.

Zhou, M., Lu, D., Li, W., & Xu, H. (2013). Optimized fuzzy logic control strategy for parallel hybrid electric vehicle based on genetic algorithm. *Applied Mechanics and Materials, 274,* 345–349.

# Part VIII
# Future Trends

# Chapter 23
# A Framework for the Performance Evaluation of an Energy Blockchain

**Seda Yanik and Anil Savaş Kiliç**

**Abstract** In this study, we first discuss the disruption and its impacts on the power and utilities sector. We identify the drivers of the disruption and the impact it will make on the sector. Then, as one of the drivers of disruption, we examine the blockchain technology, its features, benefits and limitations. We identify a performance evaluation framework for a power and utilities blockchain, in a distributed generation setting. Finally, we evaluate the performance factors' interrelationships on the performance of the blockchain system and investigate the importance and cause-effect groups of the factors.

## 23.1 Introduction

In a world where utilities see revenue opportunities arising from a new paradigm, which is being formed by mega trends including emerging technologies; they think that they have to change but are unsure about how to do it (Utility Dive 2016). The power generation is becoming decentralized, fragmented, and democratized. Homes and even electric vehicles will be nodes of providing electricity to the grid, on top of consuming from it. These nodes are expected to be able to sell electricity to each other autonomously in a peer-to-peer setting without an intermediary, and blockchain is going in a direction that will position itself in the heart of this new paradigm.

There is a vision that blockchain will play a significant role in every aspect of responding to disruption. It can help to modernize the physical infrastructure or to develop new business models or to manage a new type of autonomous market. However, blockchain has its limitations together with the benefits it serves. Evaluation of the performance of the blockchain in response to the requirements of the power and utilities sector is an important research topic. Power and utilities term used in this study represents the entire value chain: generation, transmission, distribution and supply.

---

S. Yanik (✉) · A. S. Kiliç
Istanbul Technical University, Macka, Istanbul, Turkey
e-mail: sedayanik@itu.edu.tr

© Springer International Publishing AG, part of Springer Nature 2018

C. Kahraman and G. Kayakutlu (eds.), *Energy Management—Collective and Computational Intelligence with Theory and Applications*, Studies in Systems, Decision and Control 149, https://doi.org/10.1007/978-3-319-75690-5_23

In this study, we envision a distributed generation context where domestic consumers can produce electricity at their homes, buy electricity from the grid or other consumers, store the electricity at their homes and/or electric vehicles, sell the surplus electricity to other consumers and/or to the grid. And the blockchain enables the peer to peer setting to buying and selling as well as managing the smart contracts, where different prices can be set autonomously based on supply and demand or other parameters.

Our final aim in this study is to evaluate the performance of such a setting of a power and utilities blockchain. It is important to understand what it is capable of and what performance criteria should be prioritized to assess current use cases and potential future applications in the power and utilities sector. To this aim, we employ fuzzy DEMATEL (decision making trial and evaluation laboratory) approach for exploring the importance of factors affecting the blockchain performance and the interrelationships among the factors. DEMATEL is a multi-criteria decision making method used to evaluate the interrelationships among factors of a system. It uses expert opinion and draws results using the group knowledge of experts. To represent the linguistic judgements of experts, we use fuzzy sets and employ a fuzzy DEMATEL method. Finally we group the factors into cause and effect groups.

The chapter is organized as follows: Sect. 23.2 discusses the changes and disruption that will transform the power and utilities sector. Section 23.3 describes blockchain as an emerging technology and a driver of disruption. Then Sect. 23.4 presents the fuzzy DEMATEL method and its application to the power and utilities blockchain and finally Sect. 23.5 concludes with the discussions and future research suggestions.

## 23.2 Upcoming Disruption and What It Means for Power and Utilities Sector

It is impossible to understand what blockchain means for power and utilities sector, without examining all of the drivers of disruption and their impact on the sector. Disruption is like the perfect storm where a number of factors has emerged to mature at the same time period, which lead to a revolutionary path that lays ahead of us for the upcoming decades. Blockchain is a significant disruption factor and its future applications needs to be discussed in that context.

### *23.2.1 Disruption and Its Impacts*

Disruption, a term coined by Clayton Christensen (1997), is fundamentally changing the way the world works. According to Christensen, disruption can equally come from innovations other than technological breakthroughs (such as new business models or new production processes). Actually, the idea of "disruptive innovation" has become so common that the number of media articles mentioning "disruptive innovation" between 2010 and 2015 have increased more than 440% (EY 2016) (Fig. 23.1).

The reason why the organizations and individuals are talking about disruption ever-increasingly is the critical alteration of the three primary forces, as stated by EY (2016): Technology, globalization, and demographics. Especially "technology", presenting the next wave of technologies such as Artificial Intelligence (AI), robotics, Internet of Things (IoT), blockchain, Mixed Reality (MR), Additive Manufacturing etc., is taking the innovation to a level it has never had the potential to create this big impact before.

EY (2016) reveals that eight mega trends are the consequence of the mentioned forces: Industry redefined, the future of smart, the future of work, behavioral revolution, empowered customer, urban world, health reimagined, and resourceful planet. All of these eight megatrends are interconnected: they are "the causes of", and "the effects that lead to" each other in a number of ways. Business articles published by a number of other professional services firms or institutions also reveal more or less the similar mega-trends (pwc 2016; Frost and Sullivan 2017; Singh 2014; Weller 2016; Vielmetter and Sell 2014; Korn Ferry|Hay Group 2017).

Industry redefined is a critical trend for utilities as it stresses on industry convergence, the blurring of two or more previously distinct industries and sets of participants, which may lead to the disappearance of the utility sector as we know it in a decade or two. For example, "smart home" or "connected home", whatever the name is given, is the target of many companies from a number of sectors: Utilities, tech giants, telecom companies, and even automotive and pharmaceutical companies.

The future of smart is the trend that connects every "thing", collects and analyses their data, and make them more autonomous and effective. Industry redefined and the future of work trends incorporate IoT, AI, robotics, blockchain etc. to displace humans and create "lights out" factories and offices.

Behavioral revolution is bringing the behavioral economics (BE) into the mainstream, using smartphones and sensors to monitor behavior and respond accordingly in real-time. Empowered customer trend is derived from the individual

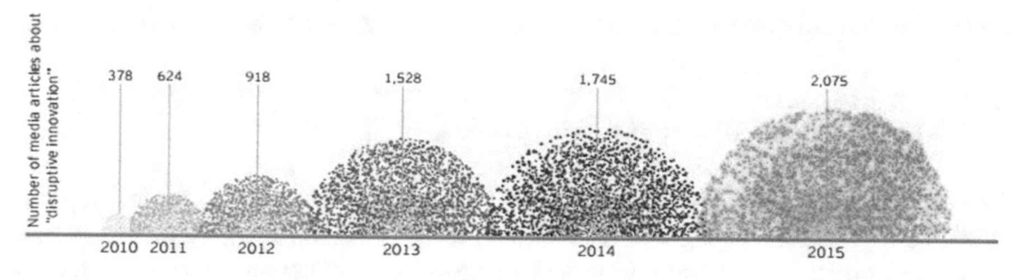


**Fig. 23.1** Number of media articles mentioning "Disruptive Innovation" (EY 2016)

customers who understand their commercial value and want to participate more in the commercial processes. Health reimagined trend focuses on the outcome of the treatments and preventive healthcare, rather than the treatment itself and the effort required for it.

As demographic trends push the world's population to 9.7 billion by 2050, natural resource constraints—whether in availability or infrastructure—challenge established modes of consumption, from the individual to global corporate supply chains (United Nations 2015), which is exactly why resourceful planet trend has emerged.

And lastly; urban world trend is identified by the milestone in 2008; when the majority of the world's population started living in cities for the first time in history. By 2050, at current rates of urbanization, the world will be two-thirds urban and one-third rural, a reversal of the global distribution pattern of 1950 (United Nations 2015). The ongoing migration to urban centers and the natural growth increases the urban population further more (EY 2017a). On top of that, it is evident that effects of disruption are beginning to extend far beyond the business world. For example, "sharing economy" start-ups such as Uber and Airbnb are already disrupting regulatory frameworks and some of the most disruptive technologies on the horizon will not only disrupt corporate business models, but also society as a whole—realigning income distribution, altering relationships between governments and citizens, and perhaps even calling into question fundamental aspects of the human experience (EY 2016). Even disrupters are facing the threat of being disrupted; as in Arcade City's effort to bypass the intermediary in mobility sharing, attacking Uber's business model with blockchain (Carmichael 2016).

### 23.2.2 *What Disruption Means for Power and Utilities Sector*

Traditional utility model has been facing disruption for a long time; resulting in a dramatic global value decline. The majors in Europe have not recovered from the significant decline in their combined market value that began in early 2008 (Robinson 2015). Barclays Bank's bond rating service has downgraded the entire U.S. electric utility sector bond market rating against the U.S. Corporate Bond Index due to the challenge from ratepayers' increasing opportunities to cut grid electricity consumption with solar and battery storage (Trabish 2014) (Fig. 23.2).

#### 23.2.2.1 Drivers of Disruption

We have identified six drivers namely prosumers, emerging technologies, regulations, de-regulations, aging workforce, and old infrastructure, which challenge the traditional utility model struggling with impairments. These drivers have led to

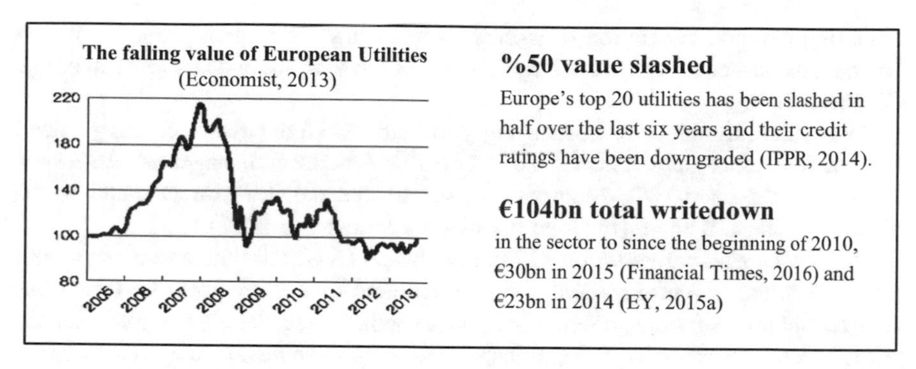


**Fig. 23.2** Utility impairments

market value decline, and new market entries have been enabled by the emerging technologies, which will be described in the following sections. Three types of responses are identified as the best options to overcome the mentioned challenges: Innovation on new customer experiences, new business models, and operational efficiency solutions (Fig. 23.3).

*Aging workforce and old infrastructure.* It is widely articulated that the aging of the workforce and old infrastructure are the most pressing challenges for the electric industry (Utility Dive 2016). A total of 71% of the utility executives surveyed cited

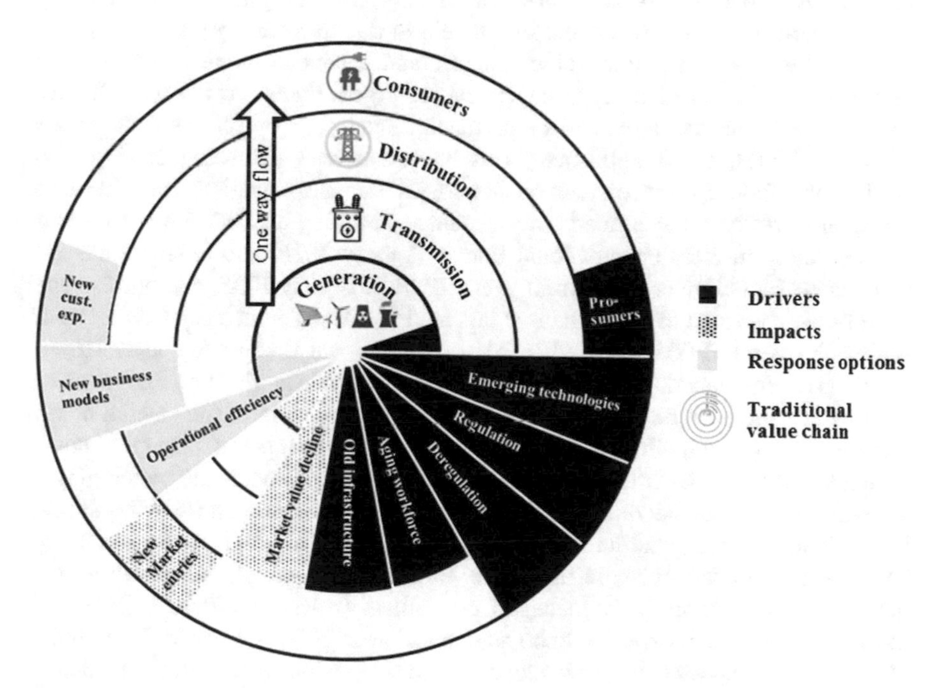


**Fig. 23.3** Disruption of the traditional electricity value chain

attracting the next generation of workers a "growing" or "urgent" concern, which has become larger recently as a large portion of the workforce nears retirement age (Engerati 2016).

In addition; the industry hosts aging grids and workforce assets, creating a need for efficient asset performance and process performance management (Ericsson 2014). $40bn/year worth resources are wasted due old infrastructure and inefficiencies in the US alone (Institute for Electric Innovation 2015).

*Regulation and deregulation.* Clean energy and $CO_2$ emission targets put a high pressure on the decentralization of supply required by decarbonization, which has created millions of small renewables. On the other hand, lagging regulations are perceived as a barrier to it: Significant regulatory changes are required to adapt disruption. 53% of utilities set forth that regulation delays investment (Engerati 2016). Lastly, liberalization processes such as unbundling and privatization are moving forward. Value chain is being unbundled and eligibility limits are converging to zero. Other players, such as telecoms, can easily climb over the barriers of vertical integration and exclusive advantages of utilities.

*Prosumers.* 69% of utility executives stated that they were excelling in customer satisfaction, confirmed by only 25% of customers (Engerati 2016), who no longer benchmark utilities against each other, but sectors such as telco and banking. The more frequent interaction customers have with utility customer services, the less satisfied they are (EY 2015c).

Moreover, customers are evolving to the prosumer, who can also produce power and challenge one-way physical flow that is only from the grid to themselves.

*Emerging technologies.* Global investment in disruptive energy technologies is growing fast, creating new value streams and business models economically compelling. Distributed energy concept has emerged, thanks to the technological improvements that reshaped energy production landscape by photovoltaic panels, electric vehicles, smart appliances, and battery storage (Munsing et al. 2017). Adding the falling costs of these technologies into the equation, they have the potential to become the primary energy source, reaching a 500 TWh solar generation capacity in 2020 (International Energy Agency 2016) and reducing the utilities market by half over the next 20 years (IPPR 2014). By 2035, microgrids could become the base load in some mature markets (EY 2015b), with a capacity at record high with over 150 GW installed in 2015 (International Energy Agency 2016).

IoT is another technological leap that is central to manage these components in a microgrid, which can be described as "a collection of prosumers (residential nodes) that are arranged within the same distribution feeder and support exchange of power between them (Laszka et al. 2017)". As utility-specific protocols are converging to TCP-IP, utilities has become one of the largest IoT spenders and will be the 3rd largest industry by expenditure in IoT, with over $69bn worldwide (Ericsson 2016). There is a significant increase in number of IoT projects in the sector with more than 1.5bn devices globally managed by utilities (Ericsson 2014). Utilities IoT market size in 2014 is expected to be quadrupled as of 2020 (European Commission 2014). IoT is expected to grow mostly in consumer-interfacing devices such as smart meters, thermostats, and appliances. Residential IoT market is expected to

| | |
|---|---|
| **93% increase in # of IoT projects** by energy & utilities adopters (Vodafone, 2016) | **Smart lighting growth: 5 times to 2.54bn** is expected **between 2014-2020** |
| **Utilities will spend $87.5bn for IoT** products and services by 2018 (Ericsson, 2016) | **€40bn utility IoT market** is expected by 2020, 4 times bigger than €10bn in 2014 (European Commission, 2014) |
| **67% of utilities** will account for overall M2M connections worldwide by 2023 (Utility Week, 2014) | **$117bn residential IoT market in 2026** is expected to grow from **$26.5bn** in 2016 (Metering & Smart Energy International, 2016) |
| **900M smart meters** is expected in 2020, creating **$60bn efficiency** (Greenough, 2015) | |

**Fig. 23.4** Selected IoT market projections

grow more than four times the size of 2016 (Metering and Smart Energy International 2016a). Utilities' use of IoT is still not sophisticated, yet the perceived business value delivered by existing IoT use cases is high (Fig. 23.4).

US Utilities have mostly invested in utility-scale renewables, demand side management, distributed generation and natural gas power plants. In the future, respondents indicated that their companies should invest more in energy storage, distributed generation and utility-scale renewables (Utility Dive 2016).

#### 23.2.2.2 Impacts of Disruption

Very immediate, tangible and ultimate impact of disruption is the market value decline, as explained previously.

New market entries is another consequence of disruption. Today's customers demand transparent and competitive pricing, as well as energy-efficient, environmentally friendly solutions. Within the utilities industry, innovation sit on the periphery rather than infiltrating the core business. Therefore, disruption is being perpetrated by new entrants. Retailers from industries such as telecommunications, consumer products, security services, and technology have moved into the smart home market. Technology giants such as Google, Apple, and Amazon have made major investments in the smart home. Hardware providers, such as Google's NEST business, software platform providers, such as EnergyHub, and aggregators, such as Comverge, are all seeking opportunities in a space traditionally served solely by the utilities. Consortiums have emerged as well: In March 2016, South Korea's SK Telecom signed a MoU with equipment provider Kocom, followed by a partnership with Samsung Electronics to transform the South Korean city of Daegu into a IoT test hub. SKT previously signed agreements with home automation company Commax as well as Hyundai Telecom in 2015 (Metering and Smart Energy International 2016c). Npower has launched an interactive trial for 300 homes using the latest smart home technology, working with D-Link and Yale with Nest integration (Moore 2015). German utility RWE has announced plans to acquire grid-scale solar and storage provider Belectric through its renewable energy subsidiary, Innogy SE (Metering and Smart Energy International 2016b).

The growth of new entrants, though halting at first, has captured close to 20% market share (Utility Week 2016). Traditional utilities face the risk of becoming mere infrastructure managers and seeing value formerly captured by their production units transferred to other players in the value chain. This could lead to further impairment of their traditional generation assets.

#### 23.2.2.3 How to Respond Disruption

Three main ways are identified to help utilities fit into the quickly evolving energy ecosystem: Creating a new customer experience on current business models, developing new business models, and building efficient operations are the key actions to be taken as a response to the disrupted market.

*New customer experience.* Customers expect free and accurate billing, of which they are getting a sophisticated version from companies like Amazon. It's not enough to aim for neither modest incremental change nor best in the industry. From the customers' point of view, all they do is buy energy and get billed; and they wonder why their bills are still wrong or why moving house is such a hassle. Utilities need to (1) optimize Customer and Billing costs (2) understand their customers (3) create loyalty (4) master the channels, and finally (5) use deep customer knowledge to innovate. Based on business experience, it pays back significantly (EY 2015c).

*New business models.* New market/technologies/pro-market regulation are on the rise; and new business models are a requirement, not a nice to have. In 2014, largest gross margins came from energy efficiency, demand response, and smart metering (EY 2015a). By 2020, utilities expect 40% of revenues coming from new models (Ericsson 2016). 74% of them view this as an 'urgent' or 'growing concern (Engerati 2016). Conventional and renewables both have a role to play too, prompting an increasing number of corporate restructurings, spin-offs and JVs. Furthermore, competition in emerging markets will be around technology, leapfrogging fossil fuels (Fig. 23.5).

*Efficient operations.* Utilities are transforming;

- from grids with large centralized fossil power plants (PP) to a distributed microgrid environment
- from managing stable supply to fluctuating supply
- from demand driven to demand side integrated operations
- from analog coupling of load to digital coupling integrating supply and demand using digital communication.

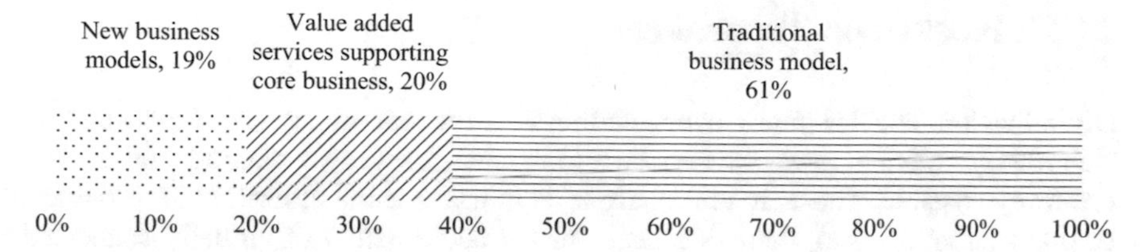


**Fig. 23.5** Expected composition of utilities' revenues (Ericsson 2016)

Both current and future operations require new technologies to come up with the best possible operations and efficiency.

### *23.2.3 The Role of Blockchain as a Disruptor and a Response to Disruption in Power and Utilities*

Hileman and Rauchs (2017) studied 132 blockchain use cases recently and revealed that 3% of the use cases belong to Power and Utilities sector. They also found out that 32% of Distributed Ledger Technology service providers target energy sector. When current and potential use cases are examined in the literature, it can be concluded that blockchain has a role in every aspect of responding to disruption mentioned in the previous sections. It can be incorporated to modernize the grid (Basden and Cottrell 2017; Horta et al. 2016) or it can be approached as a potential differentiator (Normandeau 2016) to develop new business models via its generic features also applicable for other industries (pwc 2015, 2017; Deloitte 2016; EY 2017b; Merz 2016; Aitzhan and Svetinovic 2016): Decentralized storage of transaction data, payments via cryptocurrencies and smart contracts can lead to a market without intermediaries namely data aggregators such as suppliers, clearing houses, brokers and energy trading platforms; enabling power generators or storage units (nodes) trade electricity autonomously. Even supplier switching process is subject to be an aspect of the past; for nodes will be able to buy electricity from whichever provider fits their contractual terms at any point of time, which can be named as "continuous switching".

Having explained the general context; the scope of this study is the distributed generation setting where domestic consumers can;

- produce electricity at their homes
- buy electricity from the grid or other consumers
- store the electricity at their homes and/or electric vehicles
- sell the surplus electricity to other consumers and/or to the grid
- buying and selling is performed in a peer to peer setting
- buying and selling includes smart contracts, where different prices can be set autonomously based on supply and demand or other factors.

## 23.3 Blockchain Technology

Digitalization enables more connectedness in the era of advanced information technology making new ways of working such as crowdsourcing and shared economy possible. Those, in return, are used to increase the utilization of resources. In this futuristic world, there is a need for the coordination of virtually connected people, machines and any kind of entity in a cost-effective and trustable way. To this end, blockchain is a promising data management technology that enables the distributed storage of data and the trustless transactions without the need for a central authority.

Blockchain is an emerging data management technology which allows self-organization of connected nodes as a network similar to the internet. The autonomous governance of the nodes is maintained using a distributed consensus mechanism. A blockchain constitutes a peer-to-peer network where there exists no central authority. This technology enables multiple parties to share the database secured by cryptographic signatures. The participants actually download a copy of the same ledger (database) which contains all of the transactions made in the database. It is a distributed system which is more transparent and free of various general costs created by the central management efforts. Instead of a central authority, the block chain uses a verification mechanism where nodes of the network independently check the state and integrity of the blockchain using the predefined consensus algorithm. This, in result, generates a safe and secure way for the nodes/participants to modify the database even if they don't know and trust each other. This also ensures that the power is distributed through the network (Fig. 23.6).

We can define five main components of the blockchain technology as follows (Hileman and Rauchs 2017):

- *Peer-to-peer network*: Network for peer discovery and data sharing in a peer-to-peer fashion
- *Consensus algorithm*: Algorithm used to reach an agreement on a value or transaction at the blockchain without the need to trust or rely on participants

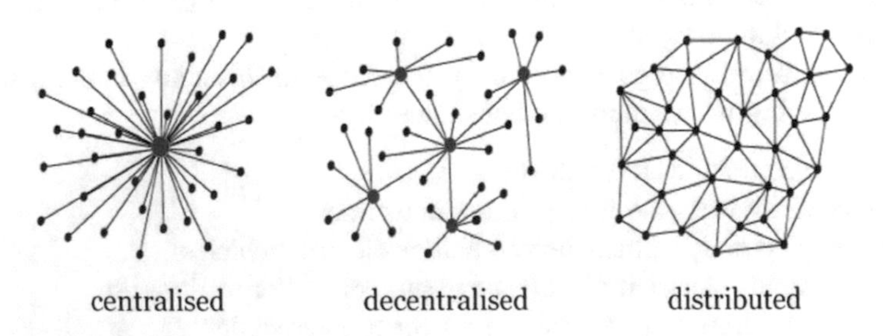


**Fig. 23.6** Illustration of various structures of networks (Hagström and Dahlquist 2017)

- *Cryptography*: techniques that ensure secure communication and transaction such as one-way hash functions, Merkle trees and public key infrastructure (private-public key pairs)
- *Ledger*: List of transactions bundled together in cryptographically linked 'blocks'
- *Validity rules*: Rules used to maintain the network (i.e., what transactions are considered valid, how the ledger gets updated, etc.)

These components may be specified differently for different choices of blockchain architecture and design.

Blockchain literally is represented by the transaction storage approach of the system which is named as blocks. A chain of transactions in the form of blocks and the components of a block in detail are illustrated in Fig. 23.7. When the verification of a transaction is completed it is stored as a block which is linked to the previous block in chronological order. The blockchain protocol which is the distributed consensus mechanism is a collection of rules specifying how the participants should verify the transactions for all of the nodes in the network. Once a block is added to the chain it cannot be changed, which is called as the immutability property of the blockchain technology. The chaining of the blocks uses cryptographic signatures (i.e. hashes) to ensure that the blockchain is reversible. A hash function is a mathematical process that takes input data of any size, then performs an operation on it, and returns output data of a fixed size.

The two main benefits of the blockchain technology are that it is trustless and distributed. Blockchain presents a trusted environment for its participants without any requirement of informal trust (e.g. handshake agreement) between its participants or traditional formal trust via central authority (e.g. courts). The way blockchain guarantees trust is by its pre-announced working rules and decision-making mechanisms thus it is confidential and transparent at the same time. The second benefit of blockchain technology is due to its distributed structure. This property enables a network to work without an intermediary or central authority. Instead, the power is shared among the distributed participants.

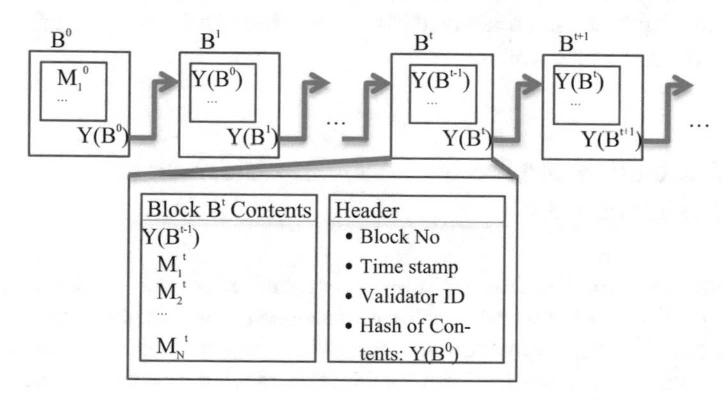


**Fig. 23.7** Symbolic representation of the data in a blockchain (Sikorski et al. 2017)

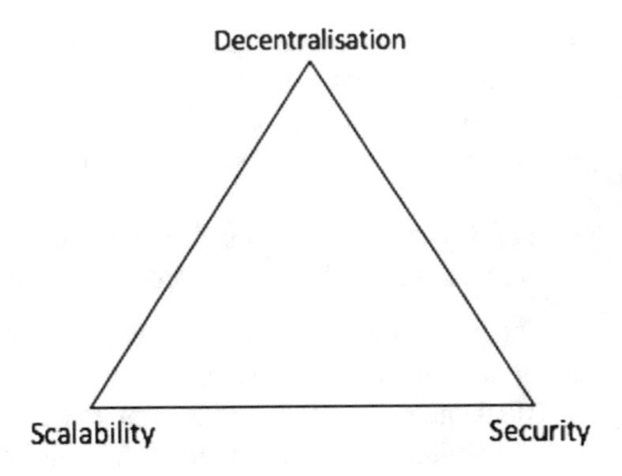


**Fig. 23.8** Trilemma triangle of blockchains

On the other side, blockchain has also limitations. The limitations of blockchain are represented by a trilemma illustrated in Fig. 23.8. Trilemma means that only two dimensions of the triangle can be achieved at a time (Hagström and Dahlquist 2017).

The three dimensions that make up the triangle are decentralization, scalability and security. The blockchains that are in use such as bitcoin and ethereum are decentralized and secure but they are not scalable. Scalability is commonly identified by the limits of throughput, together with latency. Compared to a traditional database, blockchain is relatively slow. The speed of transaction (i.e. latency) and the volume of transactions (i.e. throughput) in a blockchain depend on whether the blockchain is private or public which in return specifies the type of consensus algorithm used, block size limit and the time interval between each block, the number of verification nodes and the transaction type (small cryptocurrency transaction vs. large smart contract).

As discussed above blockchain performance depends on various design choices and factors. Moreover these factors are closely interlinked to each other. These factors are discussed extensively in the literature however there is not a consensus about the importance and interrelations of the factors. In Sect. 23.4, we propose a framework for evaluating the performance factors of a power and utilities blockchain to identify their interrelationships among them and also their effect on the performance of the blockchain.

## 23.4 Evaluating the Factors Affecting Blockchain Performance

In this section, we first identify the factors that affect the performance of a blockchain. Then, we evaluate the interrelationships among the factors using a methodology. For this evaluation we first collect expert opinion specified as linguistic variables. Then, we use fuzzy DEMATEL method to aggregate the opinions and draw results related to the performance factors. In the following subsections, first we will describe the Fuzzy Dematel methodology, then define the performance factors of a power and utilities blockchain and then using expert opinions we will present the obtained results.

### 23.4.1 Fuzzy DEMATEL Method

DEMATEL which stands for decision making trial and evaluation laboratory is a multi-criteria decision making method used to evaluate the interrelationships among factors of a system. Usually, one can identify various factors for any type of system. However, each factor may not be as important as the other because some factors behave as a cause to affect the status of another factor. Sometimes the relationship can be bilateral or sometimes some factors can be independent of any of the other factors. While designing systems effectively and/or trying to improve the system performance by changing the system factors, it will help to know about the interrelationships among the factors in order to obtain the intended results. Otherwise, the relationships among the factors may yield secondary effects as a result of changing one of the system factors.

DEMATEL was developed in Geneva Research Centre of the Battelle Memorial Institute. It uses expert opinion and draws results using the group knowledge of experts. While collecting expert opinions, experts are asked to evaluate the strength of the relationship of factors pairwise. Using linguistic evaluation helps human experts to express their knowledge because preferences of humans are often vague. To represent these linguistics evaluations, fuzzy logic is employed in Fuzzy DEMATEL method.

In fuzzy logic, a fuzzy set A in U is characterized by a membership function $\mu_A(x)$ which associates with each point in U a real number in interval [0,1], with the value of $\mu_A(x)$ at x representing "the grade of membership" of x in A (Zadeh 1965).

A formula for the membership function $\mu_A(x)$ of a triangular fuzzy number (TFN) $\tilde{x}$ which has a shape shown in Fig. 23.9, is given in Eq. 23.1, where a, b and c denotes real numbers (Ross 1995):

$$\mu_A(x) = (l,m,r) = \begin{matrix} \frac{x-a}{b-a} & ; & a \leq x \leq b \\ \frac{c-x}{c-b} & ; & b \leq x \leq c \\ 0 & ; & otherwise \end{matrix} \tag{23.1}$$

Algebraic operations for TFNs are given by (2)–(8) where all the fuzzy numbers are positive (here it is assumed that $a \geq 0, e \geq 0$) (Chen et al. 1992):

$$(a,b,c) + (d,e,f) \cong (a+d, b+e, c+f) \tag{23.2}$$

$$(a,b,c) - (d,e,f) \cong (a-f, b-e, c-d) \tag{23.3}$$

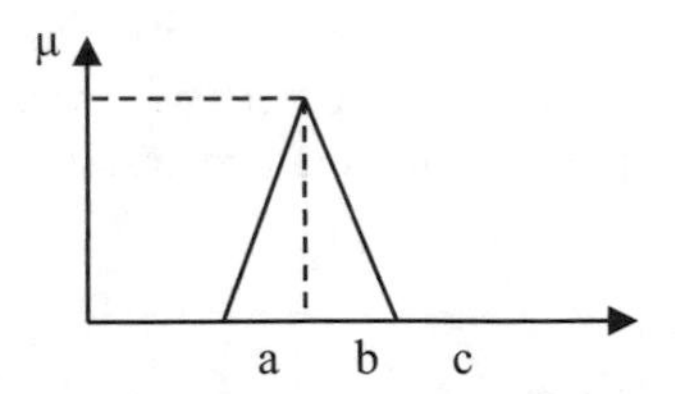


**Fig. 23.9** Membership function of a TFN

$$(a,b,c) \otimes (d,e,f) \cong (ad,be,cf) \tag{23.4}$$

$$(a,b,c) \div (d,e,f) \cong \left(\frac{a}{f},\frac{b}{e},\frac{c}{d}\right) \tag{23.5}$$

$$\lambda \otimes (a,b,c) \cong \begin{cases} (\lambda a, \lambda b, \lambda c), \\ (\lambda c, \lambda b, \lambda a), \end{cases} \quad if \begin{matrix} \lambda \geq 0 \\ \lambda \leq 0 \end{matrix}, \quad \forall \lambda \in \Re \tag{23.6}$$

$$\lambda \div (a,b,c) \cong \begin{cases} \left(\frac{\lambda}{c},\frac{\lambda}{b},\frac{\lambda}{a}\right), \\ \left(\frac{\lambda}{a},\frac{\lambda}{b},\frac{\lambda}{c}\right), \end{cases} \quad if \begin{matrix} \lambda \geq 0 \\ \lambda \leq 0 \end{matrix}, \forall \lambda \in \Re \tag{23.7}$$

$$(a,b,c)^{\lambda} \cong \begin{cases} \left(a^{\lambda},b^{\lambda},c^{\lambda}\right), \\ \left(\frac{1}{c^{\lambda}},\frac{1}{b^{\lambda}},\frac{1}{a^{\lambda}}\right), \end{cases} \quad if \begin{matrix} \lambda \geq 0 \\ \lambda \leq 0 \end{matrix}, \forall \lambda \in \Re \tag{23.8}$$

We employ the following steps of the fuzzy DEMATEL procedure presented by Ucal et al. (2012):

Step 1. Setting the goal and specification of the group of experts: Even though fuzzy DEMATEL is a method for evaluating the interrelationships among factors, one should initially set the goal of the study. Besides, group of experts are specified at the beginning of the study.

Step 2. Identifying the factors and the fuzzy linguistic scale: The factors which will be analyzed are identified by the analyst prior to the expert evaluations. The set of factors should include all possible factors which are related to the goal of the study. When identifying the factors, the analyst may also refer to the experts not to miss any of the factors which may be related. In this phase, the scale that will be used to judge the relationships among the factors by the experts are also specified. The scales used in fuzzy Dematel is linguistic such as {Very high, High, Low, Very low, No}. Each level of the scale is then represented by positive triangular fuzzy numbers $(l_{ij}, m_{ij}, r_{ij})$. An example of the scale is shown in Table 23.1.

Step 3. Aggregating the assessments of experts. Initially, the experts are asked to assess the strength of the pairwise relationship in terms of influences and directions between factors, $C = \{C_i | i = 1, 2, \cdots, n\}$. After collecting the fuzzy assessments of the experts, they are converted into fuzzy numbers. Finally those are defuzzified into crisp values and aggregated as a single value which represents the group opinion of the experts. We use the following CFCS (Converting Fuzzy data into Crisp Scores) method for the defuzzification and aggregation operations:

Let $\tilde{z}_{ij}^{k} = \left(z_{lij}^{k}, z_{mij}^{k}, z_{rij}^{k}\right)$ indicate the fuzzy assessment of evaluator $k(k = 1, 2, \cdots, p)$ about the degree to which the factor $i$ affects the factor $j$.

Normalize fuzzy assessments of evaluators $\tilde{z}_{ij}^{k} = \left(z_{lij}^{k}, z_{mij}^{k}, z_{rij}^{k}\right)$, the degree to which the factor $i$ affects the factor $j$, where $k(k = 1, 2, \cdots, p)$ is the evaluator k:

$$x_{lij}^{k} = \frac{(z_{lij}^{k} - \min z_{lij}^{k})}{\max z_{rij}^{k} - \min z_{lij}^{k}} \tag{23.9}$$

$$x_{mij}^{k} = \frac{(z_{mij}^{k} - \min z_{lij}^{k})}{\max z_{rij}^{k} - \min z_{lij}^{k}} \tag{23.10}$$

$$x_{rij}^{k} = \frac{(z_{rij}^{k} - \min z_{lij}^{k})}{\max z_{rij}^{k} - \min z_{lij}^{k}} \tag{23.11}$$

Compute left $\left(x_{lsij}^{k}\right)$ and right $\left(x_{rsij}^{k}\right)$ normalized values:

$$x_{lsij}^{k} = \frac{x_{mlj}^{k}}{(1 + x_{mij}^{k} - x_{lij}^{k})} \tag{23.12}$$

$$x_{rsij}^{k} = \frac{x_{rij}^{k}}{(1 + x_{rij}^{k} - x_{mij}^{k})} \tag{23.13}$$

Compute total normalized crisp value:

$$x_{ij}^{k} = \frac{\left[x_{lsij}^{k}\left(1 - x_{lsij}^{k}\right) + \left(x_{rsij}^{k}\right)^{2}\right]}{\left[1 - x_{lsij}^{k} + x_{rsij}^{k}\right]} \tag{23.14}$$

**Table 23.1** Fuzzy linguistic scale for fuzzy DEMATEL

| Linguistic terms | Triangular fuzzy numbers |
|---|---|
| Very high influence (VIH) | (0.75, 1.0, 1.0) |
| High influence (HI) | (0.5, 0.75, 1.0) |
| Low influence (LI) | (0.25, 0.5, 0.75) |
| Very low influence (VLI) | (0, 0.25, 0.5) |
| No influence (NI) | (0, 0, 0.25) |

Compute crisp values:

$$z_{ij}^{k} = \min z_{lij}^{k} + x_{ij}^{k}\left(\max z_{rij}^{k} - \min z_{lij}^{k}\right) \tag{23.15}$$

Integrate crisp values:

$$z_{ij} = \frac{1}{p}\left(z_{ij}^{1} + z_{ij}^{2} + \cdots + z_{ij}^{p}\right) \tag{23.16}$$

Step 4. Developing the structural model: The structural model divides the factors into two groups as cause group and effect group. First we use the initial direct-relation matrix $Z = \left[z_{ij}\right]_{nxn}$ to obtain the normalized direct-relation matrix $X = \left[x_{ij}\right]_{nxn}$ where $0 \le x_{ij} \le 1$, where $i,j = 1,2,\ldots,n$.

$$X = \frac{1}{\max\limits_{0 \le i \le 1} \sum_{j=1}^{n} z_{ij}} Z \tag{23.17}$$

Then, the total-relation matrix T is calculated by formula (23.18).

$$T = X(I - X)^{-1} \tag{23.18}$$

Finally, the following values are computed to generate the causal diagram by formulas (23.19–23.21).

$$T = t_{ij}, \qquad i,j = 1,2,\ldots,n \tag{23.19}$$

$$D = \sum_{j=1}^{n} t_{ij} \tag{23.20}$$

$$R = \sum_{i=1}^{n} t_{ij} \tag{23.21}$$

Finally a diagram is constructed to illustrate the cause-and-effect relations among the factors. To draw the causal diagram, "prominence" values which are calculated using adding up the D and R values and the "relation" values which are obtained by subtracting R from D values are represented at the horizontal and vertical axis respectively. Prominence presents the importance of the factors. Relation axis divides the factors into two groups as cause and effect factors. If the factor is a cause, it will take on positive relation values (i.e. D − R > 0), whereas factors that receives effect will have a negative relation value. Hence, causal diagrams can visualize the complicated causal relationships of factors into a visible structural model, providing valuable insight for problem-solving. Further, with the help of a causal diagram, we may make proper decisions by recognizing the difference between cause and effect factors.

### *23.4.2 Interrelationships of the Performance Factors of an Power and Utilities Blockchain*

We use fuzzy DEMATEL method to identify the interrelationships of the performance factors of an energy blockchain. Blockchain is a promising new technology which will enable collaboration among entities without a central authority. Thus, it is autonomous and has many advantages due to its independency of intermediaries, such as transparency and lower administrative costs. However blockchains also have their limitations which could have adverse effects on the development of the system. Understanding the factors which affect the performance of the blockchain and their interrelationships could play significant role to identify the critical components/factors of the system and designing the system accordingly.

Our aim is prioritizing the blockchain performance factors in this study. We collect cause-effect judgements of three experts on the factor relationships. The domain experts have of the following qualifications: the first expert is a senior manager in a well-known multinational consultancy company responsible of energy sector and emerging technologies, the second expert is an academician working on innovation and energy management and the third expert has a senior technical background on software development and is currently working at a technology company innovating on blockchain solutions. We have reviewed the blockchain literature to identify the factors that affects the performance of an energy blockchain. As a result, we specify four main dimensions: user satisfaction, productivity, collaboration and risk management. Each performance dimension is a result of various factors/characteristics of the blockchain. We propose the below framework for performance evaluation and specify the factors as follows:

| **User satisfaction** |
|---|
| **F1: Latency**: is the time between the transaction initiation and its approval and containment in a block |
| **F2: Service quality**: includes various quality dimension such as usability, increased monetary incentives and value-added services such as extended reports, solutions that can be serviced by possible intelligent extensions |
| **F3: Control and power for users**: is the level of empowerment of users which enable them to control all their information and transactions |
| **Productivity** |
| **F4: Cost minimization**: relates to the costs of relatively lower transaction and time plus the relatively high initial capital costs |

(continued)

(continued)

| |
|---|
| **F5: Throughput of the database**: is the number of transactions per second that the database can manage |
| **F6: Scalability and integration with internal and external systems**: is the capability of the system to be enlarged or downsized or integration capability with other systems |
| **F7: Sustainability**: specifies the needed resources and large energy consumption to maintain a blockchain system |
| **Collaboration** |
| **F8: Disintermediation**: specifies the level of self-governance characteristics of the system where parties are able to make an exchange without the oversight or intermediation of a third party, additionally the advantage of ecosystem simplification |
| **F9: Trust**: is the trust level that the blocksystem provides by transparency, confidentiality, privacy and traceability |
| **Risk Management** |
| **F10: Cyber security**: The cyber security level that the blockchain system provides by encryption, resilience, durability, reliability and immutability |
| **F11: Operational risk**: all operational risks due to blockchain being an emerging technology, resolving challenges such as transaction speed, the verification process, and data limits |
| **F12: Business risk**: All business risks for instance due to compliance with industry standards and regulations and fraud, immutability and the lack of qualified human resources |

We have used the scale presented at Table 23.1 for linguistic variables. Using a survey form, we collected the experts' judgements on the pairwise relationships of factors 1–12 in terms of the strength of the cause and effect. The judgments of the three experts are given in Table 23.2.

Then, we aggregated the expert judgements and employed the defuzzification method using Eqs. 23.8–23.16. Finally, the structural model is formulated using Eqs. 23.17–23.21. The total-relation matrix is given in Table 23.3.

Finally, the causal diagram is obtained for the performance factors of an energy blockchain as in Fig. 23.10. The prominence (D + R) and the relation (D − R) values of each factor are represented on the horizontal and vertical axis respectively.

The prominence value of a factor represents the importance of a factor. The higher the prominence value, the more important is a factor for the performance of the blockchain. The results show that factors 12, 11, 10, 9 and 8-namely business risk, operational risk, cyber security, trust and disintermediation—are relatively more important factors for the performance of an energy blockchain. The relation dimension (y-axis) is related to the causal relations. As the relation value increases, the factor becomes an influencer/cause factor. Whereas, as the relation value decreases the factor is an effect factor which is influenced by other factors. Cause group includes the factors 6, 5, 7, 1 and 3-namely scalability and integration, throughput of the database, sustainability, latency and control and power for users.

**Table 23.2** Expert judgements

| | F1 | F2 | F3 | F4 | F5 | F6 | F7 | F8 | F9 | F10 | F11 | F12 |
|---|---|---|---|---|---|---|---|---|---|---|---|---|
| **F1** | NI | HI | HI | VLI | VHI | VHI | LI | NI | HI | HI | HI | LI |
| **F2** | NI | NI | LI | HI | NI | LI | HI | NI | HI | LI | LI | HI |
| **F3** | VLI | HI | NI | HI | NI | LI | LI | HI | VHI | HI | HI | HI |
| **F4** | VLI | HI | LI | NI | NI | HI | HI | HI | NI | NI | LI | LI |
| **F5** | VHI | HI | LI | LI | NI | VHI | LI | LI | LI | LI | LI | LI |
| **F6** | HI | HI | LI | HI | HI | NI | HI | LI | HI | LI | HI | HI |
| **F7** | VLI | HI | HI | VLI | VLI | HI | NI | HI | HI | LI | VHI | VHI |
| **F8** | LI | LI | HI | HI | NI | HI | HI | NI | HI | HI | LI | HI |
| **F9** | NI | HI | HI | VLI | NI | VLI | HI | HI | NI | HI | HI | HI |
| **F10** | LI | HI | HI | LI | HI | HI | VHI | LI | VHI | NI | HI | HI |
| **F11** | HI | HI | HI | LI | LI | LI | HI | LI | HI | HI | NI | VHI |
| **F12** | VLI | HI | HI | LI | NI | HI | HI | HI | HI | HI | VHI | NI |

| | F1 | F2 | F3 | F4 | F5 | F6 | F7 | F8 | F9 | F10 | F11 | F12 |
|---|---|---|---|---|---|---|---|---|---|---|---|---|
| **F1** | NI | VLI | VLI | VHI | VHI | LI | VHI | LI | LI | LI | HI | VLI |
| **F2** | NI | NI | NI | NI | NI | NI | NI | NI | HI | NI | NI | NI |
| **F3** | NI | VHI | NI | LI | VLI | NI | HI | LI | VHI | NI | NI | NI |
| **F4** | NI | NI | NI | NI | NI | NI | NI | NI | NI | NI | HI | LI |
| **F5** | NI | LI | LI | VHI | NI | LI | VHI | LI | VLI | LI | HI | HI |
| **F6** | NI | VHI | LI | VLI | NI | NI | LI | NI | NI | HI | LI | LI |
| **F7** | VHI | NI | LI | VHI | NI | NI | NI | NI | NI | NI | HI | LI |
| **F8** | NI | HI | VHI | LI | LI | NI | VHI | NI | VHI | HI | LI | HI |
| **F9** | NI | NI | LI | LI | LI | NI | VHI | LI | NI | HI | NI | HI |
| **F10** | NI | NI | VLI | VLI | NI | HI | HI | NI | VHI | NI | VHI | VHI |
| **F11** | NI | HI | HI | HI | NI | HI | NI | HI | HI | HI | NI | VHI |
| **F12** | NI | LI | HI | LI | LI | HI | NI | NI | LI | NI | NI | NI |

| | F1 | F2 | F3 | F4 | F5 | F6 | F7 | F8 | F9 | F10 | F11 | F12 |
|---|---|---|---|---|---|---|---|---|---|---|---|---|
| **F1** | NI | HI | NI | VLI | HI | HI | LI | NI | NI | LI | VHI | VLI |
| **F2** | NI | NI | LI | LI | LI | HI | LI | VLI | LI | NI | VLI | VLI |
| **F3** | LI | LI | NI | LI | HI | NI | VLI | VHI | VHI | HI | VLI | NI |
| **F4** | HI | HI | VLI | NI | HI | HI | LI | NI | LI | LI | VLI | VLI |
| **F5** | NI | LI | NI | LI | NI | VHI | VLI | LI | VLI | VLI | HI | VLI |
| **F6** | NI | VHI | HI | HI | VLI | NI | HI | NI | VLI | NI | HI | HI |
| **F7** | VLI | VLI | NI | VHI | LI | NI | NI | NI | NI | NI | LI | LI |
| **F8** | HI | VLI | VHI | HI | HI | LI | HI | NI | HI | HI | HI | HI |
| **F9** | HI | VLI | LI | LI | HI | NI | LI | HI | NI | HI | LI | VLI |
| **F10** | VHI | NI | LI | LI | LI | LI | LI | HI | VHI | NI | HI | HI |
| **F11** | VLI | HI | LI | VLI | HI | HI | NI | HI | HI | LI | NI | HI |
| **F12** | LI | LI | HI | LI | LI | HI | NI | HI | LI | VLI | NI | NI |

**Table 23.3** The total-relation matrix

| | F1 | F2 | F3 | F4 | F5 | F6 | F7 | F8 | F9 | F10 | F11 | F12 |
|---|---|---|---|---|---|---|---|---|---|---|---|---|
| F1 | 0.16 | 0.34 | 0.29 | 0.32 | 0.30 | 0.33 | 0.34 | 0.23 | 0.32 | 0.29 | 0.37 | 0.32 |
| F2 | 0.09 | 0.15 | 0.18 | 0.20 | 0.13 | 0.18 | 0.20 | 0.13 | 0.24 | 0.15 | 0.18 | 0.20 |
| F3 | 0.18 | 0.33 | 0.22 | 0.31 | 0.21 | 0.23 | 0.30 | 0.28 | 0.37 | 0.26 | 0.28 | 0.28 |
| F4 | 0.15 | 0.24 | 0.20 | 0.17 | 0.15 | 0.22 | 0.22 | 0.16 | 0.20 | 0.16 | 0.23 | 0.23 |
| F5 | 0.19 | 0.33 | 0.28 | 0.33 | 0.17 | 0.33 | 0.32 | 0.25 | 0.30 | 0.26 | 0.33 | 0.32 |
| F6 | 0.17 | 0.35 | 0.30 | 0.31 | 0.21 | 0.21 | 0.31 | 0.21 | 0.29 | 0.24 | 0.31 | 0.33 |
| F7 | 0.19 | 0.25 | 0.25 | 0.29 | 0.18 | 0.22 | 0.19 | 0.19 | 0.25 | 0.19 | 0.29 | 0.29 |
| F8 | 0.24 | 0.37 | 0.40 | 0.38 | 0.27 | 0.32 | 0.40 | 0.24 | 0.42 | 0.35 | 0.37 | 0.41 |
| F9 | 0.18 | 0.28 | 0.30 | 0.29 | 0.22 | 0.23 | 0.33 | 0.27 | 0.25 | 0.29 | 0.29 | 0.32 |
| F10 | 0.24 | 0.32 | 0.34 | 0.33 | 0.26 | 0.34 | 0.37 | 0.28 | 0.42 | 0.24 | 0.39 | 0.40 |
| F11 | 0.21 | 0.38 | 0.36 | 0.34 | 0.25 | 0.34 | 0.32 | 0.31 | 0.40 | 0.32 | 0.28 | 0.41 |
| F12 | 0.17 | 0.31 | 0.31 | 0.29 | 0.20 | 0.30 | 0.26 | 0.24 | 0.31 | 0.23 | 0.27 | 0.23 |

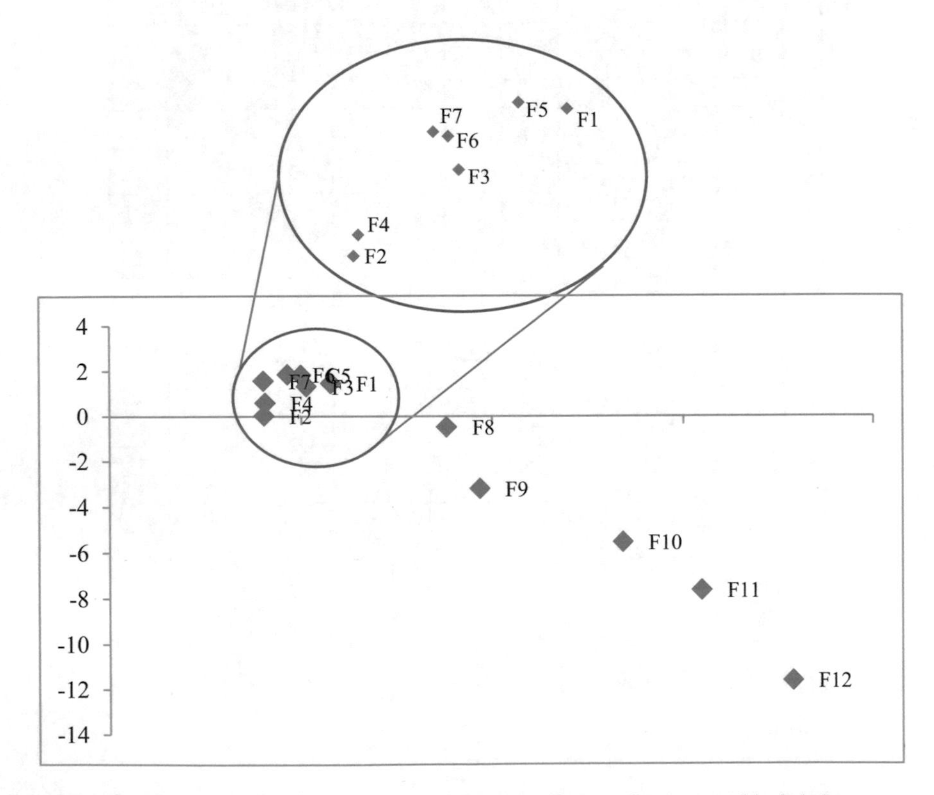


**Fig. 23.10** The causal diagram among the performance factors of an energy blockchain

## 23.5 Conclusion

Blockchain is an emerging whereas promising technology that is estimated to change many industries fundamentally. However, the limitations should be investigated very well to improve the system for the fulfillment of the crucial requirements of each industry. Because each industry, even each different segment of the industry may have various requirements. Thus, the expected performance should be analyzed well and the underlying factors for the performance should be evaluated well.

In this study, we first present the disruption in the power and utilities sector and the role blockchain can play in the transformation of the industry. Then we propose a framework for the evaluation of the performance factor of blockchain using fuzzy DEMATEL. The most important factors are found to be related to the risk dimension. Additionally, it is also concluded that productivity dimension is affecting the other dimensions of the performance for a power and utilities blockchain.

Future research can focus on simulating the performance evaluation using methods such as agent-based simulation. Besides, choosing the appropriate blockchain type/setting (i.e. public/private etc.) based on the performance requirements of a certain industry can be studied using multi-criteria decision making techniques.

## References


Aitzhan, N. Z., & Svetinovic, D. (2016). *Security and privacy in decentralized energy trading through multi-signatures, blockchain and anonymous messaging streams.* IEEE.

Basden, J., & Cottrell, M. (2017, March). How utilities are using blockchain to modernize the grid. Retrieved from hbr.org. https://hbr.org/2017/03/how-utilities-are-using-blockchain-to-modernize-the-grid.

Carmichael, J. (2016, March). *Arcade city is a blockchain-based ride-sharing uber killer.* Retrieved from www.inverse.com. https://www.inverse.com/article/13500-arcade-city-is-a-blockchain-based-ride-sharing-uber-killer.

Chen, S. J., Hwang, C. L., & Hwang, F. P. (1992). Fuzzy multiple attribute decision making: methods and applications. Lecture notes in economics and mathematical systems, Springer–Verlag, Berlin.

Christensen, C. M. (1997). *The innovator's dilemma: When new technologies cause great firms to fail.* Boston: Harvard Business School Press.

Deloitte. (2016). *Blockchain applications in energy trading.* Deloitte.

Engerati. (2016, August 25). Retrieved from http://www.engerati.com/article/role-utilities-leading-industry-transformation.

Ericsson. (2014). *Transforming industries: Energy and utilities.* Ericsson.

Ericsson. (2016). Shaping a new battleground in the IoT Era: Highlights from the 2016 IDC Pan–European Utilities Executive Summit. *IDC Energy Insights.* IDC.

European Commission. (2014). *Definition of a research and innovation policy leveraging cloud computing and IoT combination.* European Commission.

EY. (2015a). *Benchmarking European power and utility asset impairments: Testing times ahead.* EYGM Limited.
EY. (2015b). *Digital disruption of utilities: Will microgrids be utility killers or saviors?* EY.
EY. (2015c). *Customer plug in.* EY.
EY. (2016). *The upside of disruption: Megatrends shaping 2016 and beyond.*
EY. (2017a). *Future cities: Is the path to future growth through smart infrastructure?* London.
EY. (2017b). *Overview of blockchain for energy and commodity trading.* EY.
Financial Times. (2016). *Financial times.* Retrieved from Financial Times: https://www.ft.com/content/5b2dd030-1e93-11e6-b286-cddde55ca122.
Frost & Sullivan. (2017, November). *Visionary innovation (mega trends).* Retrieved from Visionary Innovation (Mega Trends): https://ww2.frost.com/research/visionary-innovation.
Greenough, J. (2015). *Energy companies are using the 'Internet of Things' to increase efficiency and save billions.* Retrieved from Business Insider: http://www.businessinsider.com/companies-utilities-save-with-iot-2015-5.
Hagström, L., & Dahlquist, O. (2017). Scaling blockchain for the energy sector. MSc. Thesis, Uppsala Universitet.
Hileman, G., & Rauchs, M. (2017). *Global blockchain benchmarking study.* Cambridge: Cambridge Centre for Alternative Finance.
Horta, J., Kofman, D., & Menga, D. (2016). *Novel paradigms for advanced distribution grid energy management.* Paris: Institut Mines-Télécom - membre de Télécom ParisTech.
Institute for Electric Innovation. (2015). *THOUGHT LEADERS SPEAK OUT: The evolving electric power industry.* Institute for Electric Innovation.
International Energy Agency. (2016). *Energy technology perspectives 2016.* International Energy Agency.
IPPR. (2014). *A new approach to electricity markets: How new, disruptive technologies change everything.* IPPR.
Korn Ferry|Hay Group. (2017, November). *The six global megatrends you must be prepared for.* Retrieved from www.haygroup.com: http://www.haygroup.com/en/campaigns/the-six-global-megatrends-you-must-be-prepared-for/#.
Laszka, A., Dubey, A., Walker, M., & Schmidt, D. (2017). *Providing privacy, safety, and security in IoT-Based transactive energy systems using distributed ledgers.* ACM.
Merz, M. (2016). Potential of blockchain technology in energy trading. In D. Burgwinkel, et al. (Eds.), *Blockchain technology introduction for business and IT managers.* Berlin: de Gruyter.
Metering and Smart Energy International. (2016a, May 30). *Residential IoT market to reach US $117.3bn in 2026.* Retrieved from Metering & Smart Energy International: https://www.metering.com/news/residential-iot-market-us117-3bn/.
Metering and Smart Energy International. (2016b, September). *RWE expands RE business with Belectric acquisition.* Retrieved from www.metering.com: https://www.metering.com/news/rwe-re-belectric/.
Metering and Smart Energy International. (2016c, May). *Residential IoT market to reach US $117.3bn in 2026.* Retrieved from Metering & Smart Energy International: https://www.metering.com/news/residential-iot-market-us117-3bn/.
Moore, M. (2015, December). *Npower launches smart home trial.* Retrieved from www.silicon.co.uk: http://www.silicon.co.uk/networks/m2m/npower-launches-smart-home-trial-181635.
Munsing, E., Mather, J., & Moura, S. (2017). *Blockchains for decentralized optimization of energy resources in microgrid networks.* eCal.
Normandeau, K. (2016, November). *Blockchain energy trading: What the future holds.* Retrieved from Microgrid Knowledge: https://microgridknowledge.com/blockchain-energy-trading/.
pwc. (2015). *Blockchain—An opportunity for energy producers and consumers?* pwc.
pwc. (2016). *Five megatrends and their implications for global defense & security.* pwc.
pwc. (2017). *Use cases for blokchain in energy & commodity trading.* pwc.
Robinson, D. (2015). *The scissors effect: How structural trends and government intervention are damaging major European electricity companies and affecting consumers.* The Oxford Institute for Energy Studies.

Ross, T. J. (1995). Fuzzy logic with engineering applications. McGraw-Hill, New York.
Sikorski, J. J., Haughton, J., & Kraft, M. (2017). Blockchain technology in the chemical industry: machine-to-machine electricity market.*Appl Energy,* 195, 234–246.
Singh, S. (2014, May). *The 10 social and tech trends that could shape the next decade*. Retrieved from www.forbes.com: https://www.forbes.com/sites/sarwantsingh/2014/05/12/the-top-10-mega-trends-of-the-decade/#175e7104a62c.
Trabish, H. K. (2014, May 26). *Utility dive*. Retrieved from Utility Dive: http://www.utilitydive.com/news/barclays-downgrades-entire-us-electric-utility-sector/266936/.
Ucal, I., Ugurlu, S., & Kahraman, C. (2012). Prioritization of Supply Chain Performance Measurement Factors by a Fuzzy Multicriteria Approach. In L. Benyoucef (Eds.), *Multi-Criteria and Game Theory Applications Manufacturing and Logistics* (pp. 161–180). Springer, London.
United Nations. (2015). *World urbanization prospects: The 2014 revision.* United Nations.
Utility Dive. (2016). *2016 state of the electric utility survey.* Utility Dive.
Utility Week. (2014). *Utility week.* Retrieved from The internet of things: http://utilityweek.co.uk/news/the-internet-of-things/1033912#.Wgi1l000xp9.
Utility Week. (2016, July 22). *Utility week.* Retrieved from Utility Week: http://utilityweek.co.uk/news/topic-non-traditional-business-models/1263432#.WbPe-E00xp9.
Vielmetter, G., & Sell, Y. (2014). *Leadership 2030: The six megatrends you need to understand to lead your company into the future.* AMACOM Div American Mgmt Assn.
Vodafone. (2016). *Vodafone IoT barometer 2016.* Vodafone.
Weller, C. (2016, July). *4 mega-trends that could change the world by 2030.* Retrieved from www.businessinsider.com: http://www.businessinsider.com/mega-trends-could-change-world-by-2030-2016-7/#1-individual-empowerment-1.
Zadeh, L. A. (1965). Fuzzy Sets, *Information and Control, 8,* 338–353.

# Chapter 24
# Energy Future: Innovation Based on Time, Synergy and Innovation Factors

**Eunika Mercier-Laurent and Gülgün Kayakutlu**

**Abstract** Computational intelligence has been widely used to analyse the complex problems in the energy field. Examples of using different methods in energy applications for economic, strategic and operational analysis in the energy field. Forecasting and Performance analysis examples are shown as a support for decision makers. This article is the conclusion of the book defining a new vision for the energy future based on innovation. A computational model is proposed to give a new dimension for the decision makers in the energy field. The novel mathematical model is defined to consider the energy future based on the innovation impacts complemented with the time, synergy and system approaches.

## 24.1 Introduction

This book edifies the complexity in the energy systems and introduces a variety of computational methods that are used in search for solutions. Energy complexity can be shown in different levels. That is why, the samples for collaborative and computational analysis are given in forecasting process, economic, strategic and operational decisions and performance calculations. The final section, future trends is reserved to cover novel approaches. The first chapter was the use of block chains which is new for the energy applications. This last chapter is focused on the wider use of intelligence based innovation.

Sustainable resources management, ecological production/distribution systems and eco-social-technological-ethical concerns in consumption are the issues of energy systems today. Complexity of the natural systems, the need for evolutionary

E. Mercier-Laurent (✉)
CReSTIC EA 3804 UFR Sciences Exactes et Naturelles,
Moulin de la Housse, BP 1039, 51687 Reims CEDEX 2, France
e-mail: eunika@innovation3d.fr

G. Kayakutlu
Istanbul Technical University Ardennes, Macka, Besiktas, Istanbul, Turkey
e-mail: kayakutlu@itu.edu.tr

© Springer International Publishing AG, part of Springer Nature 2018

C. Kahraman and G. Kayakutlu (eds.), *Energy Management—Collective and Computational Intelligence with Theory and Applications*, Studies in Systems, Decision and Control 149, https://doi.org/10.1007/978-3-319-75690-5_24

processes through improved reasoning and knowledge forge innovative solutions. Innovation in the energy industry was focused on the renewable energy resources and efficiency technologies. Planet concern, ecological and sustainable approaches in the energy supply chain has opened the path for new infrastructures and novel knowledge base.

The combined use of energy resources (a variety of renewables or renewable and fossil) and increasing type of energy services (utility, micro-grid, self-energy producing etc.) synergy between the energy systems urge to be a need (Kayakutlu and Mercier-Laurent 2017). Innovation investments in the energy field are multiplying to consider technological, ecological, social, economic and political aspects (Mercier-Laurent 2011).

This article aims to model the innovation based future of energy by considering all the innovation factors and impacts handled in time and considered with system and synergy approach. The proposed mathematical model is developed based on Mercier-Laurent's insight to eco-innovation (Mercier-Laurent 2011).

This paper is so organized that innovation concepts handled in the paper will be defined in next section followed by the evolution of innovation in the energy supply chain. The proposed mathematical model will be built in the fourth section and the last chapter will give the conclusions and recommendations for the future.

## 24.2 Innovation Concepts

Facing climatic change requires innovation in energy life cycle and education of a wise use of this resource in connection to others as water and air (United Nations 2017). Paradoxically modern technology requires energy to work. We produce a huge amount of data stored in data centers that we must cool. It is urgent to understand the impacts produced by human activity and try to reduce them with the aim to preserve our biosphere (Mercier-Laurent 2015). Such an approach requires considering production and use of energy not separately but as inter-influencing components of our biosphere and our wellbeing.

As a consequence, the general innovation process can be represented in a loop of ideas, evaluation of impacts, transformation, generation of values and integration of feedback at all stages as shown in Fig. 24.1.

The results include also sustainable success of all participants. The purpose of such innovation is not only quick generation of revenue, but also improving our quality of life while minimizing the impact on our biosphere.

While Corporate Social Responsibility consider three impacts: environmental, economic and social at least seven inter-related impacts (economic, technological, cultural, social, environmental, politic and cognitive) should be considered, Fig. 24.2.

Advance in technology offer a plethora of tools that may help us simulating and optimizing. New needs can also influence development of new technologies.

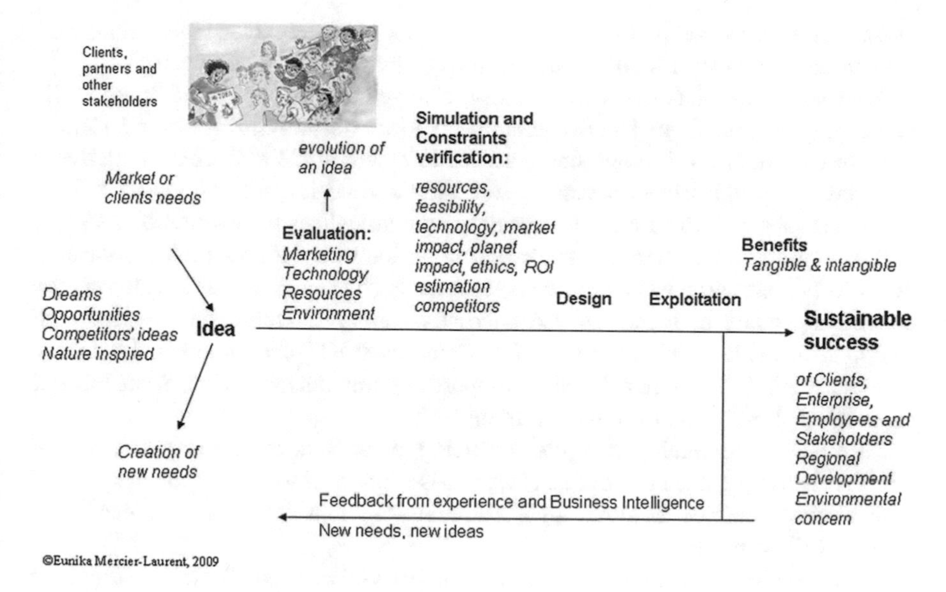


**Fig. 24.1** General innovation process

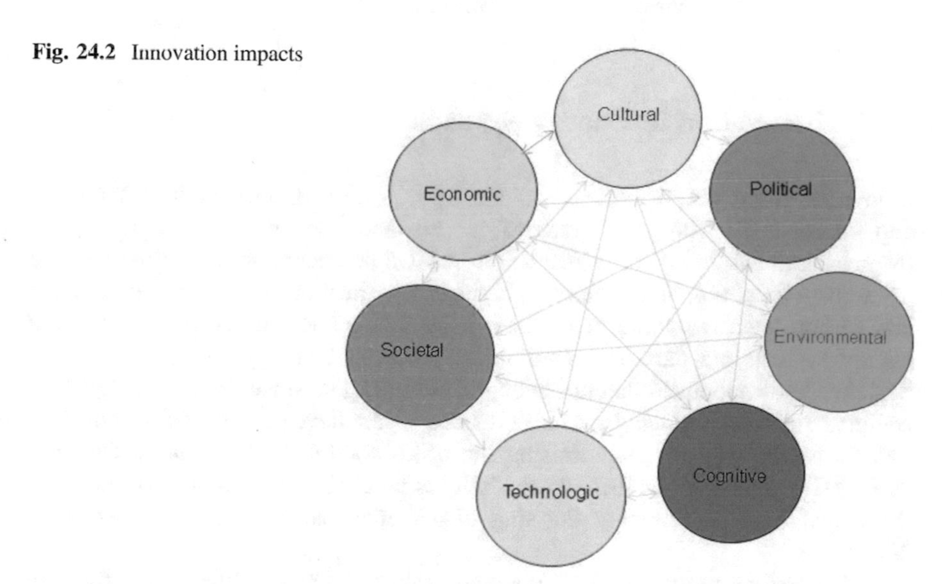


**Fig. 24.2** Innovation impacts

While energy providers and distributors focus mainly on economic impact—revenue—today trends and necessity of combining renewable energies introduce new difficulty in evaluating and sharing of produced values.

Policies related to energy are top-down, while the feedback from what happen on the bottom may be interesting to evolve them. Thanks to social networks we can know and reproduce experiments both on companies as on citizen levels in the new uses of energy. One of them is circular energy that is to be developed.

Easy access to solar energy (upowa.org) change the way people use it, especially in developing countries. UpOwa empowers local off-the-grid West and Central African communities through smart access to electricity. From users' experience they develop progressively access to education and health care.

We can observe how far the access to energy influences the evolution of society —from agriculture to start-up, empowering agriculture with technology, creating new professions and new uses. Connected objects (IOT) and drones are also energy consuming, many of them can use alternative energy at the condition that their designers are able to think "differently". Sometimes IOT and, automated vehicles, automated pilots and other decision support systems decide for us, what kill our capacity of thinking and problem solving.

New way of combining energies, practicing the circular energy is more complex but it leads to optimized use of this resource. Designers of toys combine mechanical with electric energies to move toys. Thinking "without frontiers" between fields gives excellent results.

Education has a great influence on how energy will be used. Today it is urgent to educate new attitudes—from waste of energy (wrong use of air conditioning systems—ex NY 17 °C inside while 40° outside).

## 24.3 Innovation in Energy Systems

Innovation improved in the energy industry by emission abatement technologies and continued with the renewable resource technology improvements (Mercier-Laurent 2009). As a result, the renewable energy sources have reduced investment costs and take more roles in the energy markets. Current trends towards less carbon intensive, more distributed, more interconnected and more intelligent energy systems are part of a continuous process of energy systems evolution. Sustainable development that has been the ultimate goal since the beginning of the century. Sustainability can be simply defined as the development that responds to today's needs without compromising the needs for future generations (Duxbury et al. 2017). It is obvious that sustainability is an evolutionary process growing to the potential of evolutionary thinking in resource management (Rammel et al. 2007).

Collaborative efforts for "a stable, safe, reliable" energy supply, ensuring the "energy balances in short and long term with a target to "promote social and environmental justice" presents a wealthier life for the citizens (Crandall et al. 2014). These are all common interest for the Globe and yet, economic priorities do not allow to realize them to the limits of human requests. Brasil has shown a dramatic experience of wrong priorities through the agricultural issues experienced as a result of the biomass investments. The focus on wind technology investments with the interest to construct the whole technology production in the country has been a sharp economical turn (Juárez et al. 2014). Malaysia, a developing country with a high speed and causing high increase in carbon emissions, is reconsidering the energy policies by integrating the power and gas markets; an innovative approach considering the social aspects of energy security (Shaikh et al. 2017). Hence, the energy innovation may be done at the industrial, regional, national or global level. Cross-country efforts benefit the feedbacks from the national policies. At that level trends are either in using a certain technology with transition through the cross border as in the case of geothermal energy investments; or, integrate the different industries like utility sector and tourism. These efforts lead for the discovery of knowledge insight of socio-political impacts in parallel with the technical and economic paradigms (Araújo 2014).

The recent innovative evolutions are inclined to consider energy systems as socio-technical systems (Johansen and Røyrvik 2014). When the impacts on humanity is taken into consideration time becomes the first important issue. Energy infrastructures are good examples to show the importance of time. There are utility investments of several decades which have impacts on the society and transportation. Paris sewerage system of 1370 years is a good example (Gaziulusoy and Brezet 2015). Sustainability of the investments made today need to be considered for the impacts of a lifetime ahead. Social, cultural and organisational levels of innovation are to be analysed in multi-level dynamics.

The transition of social behaviour in energy use is very much effected by the governing policies (Kuzemko et al. 2016). The politics changes as the energy systems evolve and the change of politics influence the society in terms of accepting and using the new systems.

Energy system integrations with the spread of the small grids have totally changed the utility business. Community energy systems concept is growing with the integration of micro-grids, hubs, virtual power plants, multi-carrier delivery and multi-faceted supply systems (Koirala et al. 2016). As these operational synergies drive for the research on local conditions for a city or region, there is a need for strategic synergy. After COP21 commitments country politics and organisational strategies have become “responsible, renewable and recyclable” (Peck and Parker 2015). Strategies are redesigned to be responsible to reduce energy consumption, with a goal of achieving 100% renewable energy use as soon as possible and recycle more than 50% of energy used.

Most recent research is on connected complexity of energy systems and innovation systems. This approach emphasizes combined study of three different pictures: first one is the concerns for responding to the demand analysed with behavioural economics; secondly, optimized operations to minimize the costs and maximize the benefits, and thirdly the evolutionary and institutional strategies considering infrastructures, network effects and learning (Grubb et al. 2017).

A mirror reflecting the future is expected to include

- Synergy of operational and strategic energy systems;
- Analysis should not only cover the current state but the history as far as the knowledge can be extracted and the forthcoming events;

- Culture for the next generations should be wiser to combine the green, the effective use and reutilization of different energy types;
- Policies should consider the social and cultural behaviours, human-technology must be one;
- Smart technologies will be evolved for the intelligent ones to include learning and adaptation to changes.

## 24.4 Computational Model Construction

Models on energy systems are basically designed for the energy markets and used by the grids or micro-grids. The new evolutions allow to combine and model the systems through a variety of single dimension, like environmental (Toba and Seck 2016), technological (Sgobbi et al. 2016), social (Miller et al. 2015), political (Burke and Stephens 2017) and cultural (Ruotsalainen et al. 2017).

Literature survey on innovation concepts and theories showed that there is also a need for a combined approach. The evolution of innovation in energy markets and systems has taken us to the categorisation that we should work on the energy future in three pillars: time, culture and synergy with the impact of seven innovation factors as given in Fig. 24.1. The seven factors are technological (x1), environmental (x2), social (x3), economic (x4), cultural (x5), political (x6) and cognitive (x7). They are all taken as independent factors as shown in Fig. 24.3. The seven innovation factors continuously are analysed within the impact of three generation in time and different in different levels of innovation.

On March 31, 2015, there was a general blackout in Turkey, in some cities the metro stopped for almost 8 h. The reasons are analysed to be equipment misfit in the transmission infrastructure of the compact national grid. Almost 30 million were effected. It is observed that the role of enabling systems, is likely to be more

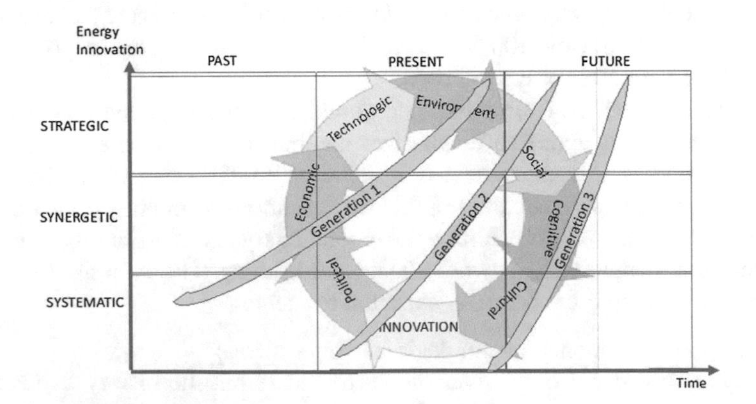


**Fig. 24.3** Innovation based future of energy

significant for systems-of-systems architecting and analysis (Adler and Dagli 2012). That will be the clue for our model to be concerned of different innovations systems (systematic, synergetic and strategic) in different time states.

Time is taken as the past, present and future because the knowledge of a system's present state does not imply any clue about the initial conditions or future states. In highly complex systems there is a need for defining the physical equations for the dynamics and the sensitivity of the system in order to find a solution (Kwapień and Drożdż 2012).

In our model t will stand for the time period (can be taken as a month, a year or a decade). The three different time periods will be taken as t − 1 for the past, t for present and t + 1 for the future. That means all the innovations, be systematic $\left(f_{t-1}^{s1}, f_t^{s1}, f_{t+1}^{s1}\right)$, synergetic $\left(f_{t-1}^{s2}, f_t^{s2}, f_{t+1}^{s2}\right)$, or strategic $\left(f_{t-1}^{s3}, f_t^{s3}, f_{t+1}^{s3}\right)$, will be defined in three time periods.

Defining $f_t^{si}$ (i = 1, 2, 3 and t = 1, .. n, n being the lifecycle of analysis) innovation functions as a combination of all the innovation factors cannot be defined linear as it will mean that technological or political impacts and the energy innovation will be increasing at the same rate, furthermore that will mean the elasticity = 1, which is not possible (Nagurney 1999). It could be defined as a quadratic function like

$$f_t^{si} = \beta_0 + \beta_1 X1_t + \beta_2 X2_t^2 + \beta_3 X3_t^2 + \varepsilon_t \tag{24.1}$$

But then, it would mean that the factors X2 and X3 which are the independent factors will have a homogeneous impact on energy innovation. As Lee and Chang has shown for energy intensity relation with the economic factors, it is not possible to have a total homogeneity (Lee and Chang 2008). Therefore a natural logarithmic relation is defined as in the energy intensity models.

$$\begin{aligned} lnf_t^{si} = {} & \beta_{0t}^{si} + \beta_{1t}^{si} lnX1_t + \beta_{2t}^{si} lnX2_t + \beta_{3t}^{si} lnX3_t + \beta_{4t}^{si} lnX4_t \\ & + \beta_{5t}^{si} lnX5_t + \beta_{6t}^{si} lnX6_t + \beta_{7t}^{si} lnX7_t + \varepsilon_t \end{aligned} \tag{24.2}$$

where, *si* is the innovation type for i = 1, 2, 3 indicating systematic, synergetic and strategic t is used as t = t − 1, t and t + 1 so that past, present and future states can be considered $\beta_{0t}^{si}$, $\beta_{jt}^{si}$ are constants for each type of innovation *si* at time *t* for j = 1, 2, 3, 4, 5, 6, 7 and $\varepsilon_t$ is an error term with minimal effect which is smaller than all the above constants.

Innovation type i is defined as a function of *XJ* with J (J = 1, 2, 3, 4, 5, 6, 7) being the factor number so that it can be expressed for the current generation as:

$$lnf_t^{si} = \beta_{0t}^{si} + \sum_{J=}^{7} \beta_{jt}^{si} lnXJ_t + \varepsilon_t \tag{24.3}$$

The impact of generations will be taken as a temporal evolution system of Euler-Lagrange as in (Navarro-González and Villacampa 2013); since, it is a strong belief that the knowledge of three generations are kept in balance in a family. States from A to B are defined as in (24.3) (Navarro-González and Villacampa 2013).

$$S_{AB} = \int_{A}^{B} \mathcal{L}(x1, x2..)dt \tag{24.4}$$

$$\partial S_{AB} = \frac{\partial \mathcal{L}}{\partial x} - \frac{\partial}{\partial t}\left(\frac{\partial \mathcal{L}}{\partial x}\right) = 0 \tag{24.5}$$

Since our objective is to maximize the innovation evolutions we should consider the Lagrange expression of the previous generation is stated for a function $g_t^{si}$ defined with the same factors:

$$\mathcal{L}_t^{si} = lnf_t^{si} + \sum_{j=1}^{7} \lambda_{jt}^{si} \frac{\delta g_t^{si}}{\delta xj} \tag{24.6}$$

where $0 \leq \lambda_{jt}^{si} \leq 1$ for all si (i = 1, 2, 3), for all j (1, 2, 3, 4, 5, 6, 7) and for all t (t = 1, .. n).

We can also include the previous to previous generation expressed by gg including the same factor impacts so (24.6) becomes:

$$\mathcal{L}_t^{si} = lnf_t^{si} + \sum_{j=1}^{7} \lambda_{jt}^{si} \frac{\delta g_t^{si}}{\delta xj} + \sum_{j=1}^{7} \gamma_t^{si} \frac{\delta^2 gg_t^{si}}{\delta xj^2} \tag{24.7}$$

where $0 \leq \gamma_t^{si} \leq 1$ and $0 \leq \lambda_{jt}^{si} \leq 1$ for all si (i = 1, 2, 3), for all j (1, 2, 3, 4, 5, 6, 7) and for all t (t = 1, 2, 3).

Now we can optimize the innovation either separately that makes three functions to be optimized in parallel or sum them up by multiplying with constants.

This has been an innovation function defined in a perfect world without constraints. This function is to be improved considering all the possible constraints based on the seven factors.

## 24.5 Conclusion

Energy is a perfect playground for innovation bridging renewable techniques, distributed energy resources and smart use of energy environmental impacts in our complex environment. Applied computational intelligence techniques have potential to improve decision making, planning, policy designing and operation of the energy systems.

This article proposes a computational model to maximize the innovation in energy systems defined by considering system, synergetic and strategic innovation in three time periods as past, present and future and taking into account the three generations. Each type of innovation is defined as a function of environmental, technological, economical, sociological, cultural, political and cognitive impacts. The proposed model should be considered as the initial step of creating computational intelligence for innovation in energy systems that we should consider in the future. This model will be generalized to allow the application to the other fields.

Perspectives for the nearest future include the study of constraints depending on all impacts so that all multiple objective functions can further be presented as second order differentiable functions.

This model demonstrates clearly the complexity of the innovation in energy systems. Even as proposed here we believe that it can be solved using the related reasoning methods. After identifying the suited version of the model, it should be converted to an agent based intelligent system so the model reasonable solutions.

## References


Adler, C. O., & Dagli, C. H. (2012). Enabling systems and the adaptability of complex systems-of-systems. *Procedia Computer Science, 12,* 31–36. https://doi.org/10.1016/j.procs.2012.09.025.

Araújo, K. (2014). The emerging field of energy transitions: Progress, challenges, and opportunities. *Energy Research & Social Science, 1,* 112–121. https://doi.org/10.1016/j.erss.2014.03.002.

Burke, M. J., & Stephens, J. C. (2017). Political power and renewable energy futures: A critical review. *Energy Research & Social Science*. https://doi.org/10.1016/j.erss.2017.10.018.

Crandall, K. et al. (2014). Turning a vision to reality: Boulder's utility of the future. *Distributed Generation and its Implications for the Utility Industry*, 435–452. https://doi.org/10.1016/b978-0-12-800240-7.00022-9.

Duxbury, N., Kangas, A., & De Beukelaer, C. (2017). Cultural policies for sustainable development: Four strategic paths. *International Journal of Cultural Policy, 23*(2), 214–230. https://doi.org/10.1080/10286632.2017.1280789.

Gaziulusoy, A. I., & Brezet, H. (2015). Design for system innovations and transitions: A conceptual framework integrating insights from sustainablity science and theories of system innovations and transitions. *Journal of Cleaner Production, 108,* 1–11. https://doi.org/10.1016/j.jclepro.2015.06.066.

Grubb, M., McDowall, W., & Drummond, P. (2017). On order and complexity in innovations systems: Conceptual frameworks for policy mixes in sustainability transitions. *Energy Research and Social Science*. https://doi.org/10.1016/j.erss.2017.09.016.

Johansen, J. P., & Røyrvik, J. (2014). Organizing synergies in integrated energy systems. *Energy Procedia, 58,* 24–29. https://doi.org/10.1016/j.egypro.2014.10.404.

Juárez, A. A., Araújo, A. M., Rohatgi, J. S., & de Oliveira Filho, O. D. Q. (2014). Development of the wind power in Brazil: Political, social and technical issues. *Renewable and Sustainable Energy Reviews, 39,* 828–834. https://doi.org/10.1016/j.rser.2014.07.086.

Kayakutlu, G. & Mercier-Laurent, E. (2017). 5-future of energy. *Intelligence in Energy*, 153–198. https://doi.org/10.1016/B978-1-78548-039-3.50005-5.

Koirala, B. P., Koliou, E., Friege, J., Hakvoort, R. A., & Herder, P. M. (2016). Energetic communities for community energy: A review of key issues and trends shaping integrated community energy systems. *Renewable and Sustainable Energy Reviews, 56,* 722–744. https://doi.org/10.1016/j.rser.2015.11.080.

Kuzemko, C., Lockwood, M., Mitchell, C., & Hoggett, R. (2016). Governing for sustainable energy system change: Politics, contexts and contingency. *Energy Research & Social Science, 12,* 96–105. https://doi.org/10.1016/j.erss.2015.12.022.

Kwapień, J., & Drożdż, S. (2012). Physical approach to complex systems. *Physics Reports, 515*(3–4), 115–226. https://doi.org/10.1016/j.physrep.2012.01.007.

Lee, C.-C., & Chang, C.-P. (2008). Energy consumption and economic growth in Asian economies: A more comprehensive analysis using panel data. *Resource and Energy Economics, 30*(1), 50–65. https://doi.org/10.1016/j.reseneeco.2007.03.003.

Mercier-Laurent, E. (2009). *Digital ecosystems for the knowledge economy, invited talk MEDES 09.* http://sigrappfr.acm.org/MEDES/09/keynotes.php.

Mercier-Laurent, E. (2011). *Innovation ecosystems*. Wiley. https://doi.org/10.1002/9781118603048.

Mercier-Laurent, E. (2015). *The innovation biosphere: Planet and brains in the digital era.* Wiley-ISTE. ISBN: 978-1-848-21556-6.

Miller, C. A., O'Leary, J., Graffy, E., Stechel, E. B., & Dirks, G. (2015). Narrative futures and the governance of energy transitions. *Futures, 70,* 65–74. https://doi.org/10.1016/j.futures.2014.12.001.

Nagurney, A. (1999). Network economics: A variational inequality approach. *Finance*. https://doi.org/10.1007/978-94-011-2178-1.

Navarro-González, F. J., & Villacampa, Y. (2013). Generation of representation models for complex systems using Lagrangian functions. *Advances in Engineering Software, 64,* 33–37. https://doi.org/10.1016/j.advengsoft.2013.04.015.

Peck, P., & Parker, T. (2015). The "sustainable energy concept"—Making sense of norms and co-evolution within a large research facility's energy strategy. *Journal of Cleaner Production.* https://doi.org/10.1016/j.jclepro.2015.09.121.

Rammel, C., Stagl, S., & Wilfing, H. (2007). Managing complex adaptive systems—A co-evolutionary perspective on natural resource management. *Ecological Economics, 63*(1), 9–21. https://doi.org/10.1016/j.ecolecon.2006.12.014.

Ruotsalainen, J., Karjalainen, J., Child, M., & Heinonen, S. (2017). Culture, values, lifestyles, and power in energy futures: A critical peer-to-peer vision for renewable energy. *Energy Research & Social Science, 34,* 231–239. https://doi.org/10.1016/j.erss.2017.08.001.

Sgobbi, A., Simões, S. G., Magagna, D., & Nijs, W. (2016). Assessing the impacts of technology improvements on the deployment of marine energy in Europe with an energy system perspective. *Renewable Energy, 89,* 515–525. https://doi.org/10.1016/j.renene.2015.11.076.

Shaikh, P. H., Nor, N. B. M., Sahito, A. A., Nallagownden, P., Elamvazuthi, I., & Shaikh, M. S. (2017). Building energy for sustainable development in Malaysia: A review. *Renewable and Sustainable Energy Reviews, 75,* 1392–1403. https://doi.org/10.1016/j.rser.2016.11.128.

Toba, A.-L., & Seck, M. (2016). Modeling social, economic, technical & environmental components in an energy system. *Procedia Computer Science, 95,* 400–407. https://doi.org/10.1016/j.procs.2016.09.353.

United Nations. (2017). Katowice announced as host venue of UN climate change conference COP 24 in 2018. Available at: http://newsroom.unfccc.int/unfccc-newsroom/katowice-announced-as-host-venue-of-un-climate-change-conference-cop-24-in-2018/. Accessed: December 10, 2017.

Printed in the United States
By Bookmasters